Dynamics of Infinite Dimensional Systems

NATO ASI Series

Advanced Science Institutes Series

A series presenting the results of activities sponsored by the NATO Science Committee, which aims at the dissemination of advanced scientific and technological knowledge, with a view to strengthening links between scientific communities.

The Series is published by an international board of publishers in conjunction with the NATO Scientific Affairs Division

A Life Sciences B Physics	Plenum Publishing Corporation London and New York
C Mathematical and Physical Sciences	D. Reidel Publishing Company Dordrecht, Boston, Lancaster and Tokyo
D Behavioural and Social Sciences E Applied Sciences	Martinus Nijhoff Publishers Boston, The Hague, Dordrecht and Lancaster
F Computer and Systems Sciences G Ecological Sciences H Cell Biology	Springer-Verlag Berlin Heidelberg New York London Paris Tokyo

Series F: Computer and Systems Sciences Vol. 37

Dynamics of Infinite Dimensional Systems

Edited by

Shui-Nee Chow

Department of Mathematics
Michigan State University
East Lansing, MI 48824-1027
U.S.A.

Jack K. Hale

Division of Applied Mathematics
Brown University
Providence, RI 02912
U.S.A.

Springer-Verlag
Berlin Heidelberg New York London Paris Tokyo
Published in cooperation with NATO Scientific Affairs Division

Proceedings of the NATO Advanced Study Institute on Dynamics of Infinite Dimensional Systems, held in Lisbon, Portugal, May 19–24, 1986

ISBN 3-540-18374-4 Springer-Verlag Berlin Heidelberg New York
ISBN 0-387-18374-4 Springer-Verlag New York Berlin Heidelberg

Library of Congress Cataloging in Publication Data. NATO Advanced Study Institute on Dynamics of Infinite Dimensional Systems (1986: Lisbon, Portugal) Dynamics of infinite dimensional systems. (NATO ASI series. Series F, computer and systems sciences; vol. 37) "Proceedings of the NATO Advanced Study Institute on Dynamics of Infinite Dimensional Systems held in Lisbon, Portugal, May 19–24, 1986"—T.p. verso. 1. Differential equations—Congresses. 2. Differential equations, Partial—Congresses. I. Hale, Jack K. II. Chow, Shui-Nee. III. Title. IV. Series. QA372.N38 1986 515.3'5 87-26383
ISBN 0-387-18374-4 (U.S.)

Printed in Germany

Printing: Druckhaus Beltz, Hemsbach; Bookbinding: J. Schäffer GmbH & Co. KG, Grünstadt
2145/3140-543210

PREFACE

The 1986 NATO Advanced Study Institute on Dynamics of Infinite Dimensional Systems was held at the Instituto Superior Tecnico, Lisbon, Portugal.

In recent years, there have been several research workers who have been considering partial differential equations and functional differential equations as dynamical systems on function spaces. Such approaches have led to the formulation of more theoretical problems that need to be investigated. In the applications, the theoretical ideas have contributed significantly to a better understanding of phenomena that have been experimentally and computationally observed.

The investigators of this development come with several different backgrounds - some from classical partial differential equations, some from classical ordinary differential equations and some interested in specific applications. Each group has special ideas and often these ideas have not been transmitted from one group to another.

The purpose of this NATO Workshop was to bring together research workers from these various areas. It provided a soundboard for the impact of the ideas of each respective discipline. We believe that goal was accomplished, but time will be a better judge.

We have included the list of participants at the workshop, with most of these giving a presentation. Although the proceedings do not include all of the presentations, it is a good representative sample.

We wish to express our gratitude to NATO, and to Dr. M. di Lullo of NATO, who unfortunately did not live to see the completion of this project.

We also greatly appreciate the additional financial support from the United States Air Force Office of Scientific Research, the United States Army Research Office, the Instituto Nacional de Investigaçâo Cientifica (Portugal) and the Junta Nacional de Investigaçâo Cientifica (Portugal).

We would also like to thank the Secretary of State for Higher Education, Professor Fernando Ferreira Leal, the Secretary of State for Scientific Research, Professor Arantes e Oliveira, and the President of the Executive Council and Scientific Council of the Instituto Superior Tecnico, Professor Diamantino Durão.

For hosting the conference, we wish to thank the Complexo Interdisciplinari of the Instituto Nacional de Investigaçâo Cientifica.

July 1987

Shui-Nee Chow, East Lansing
Jack K. Hale, Providence

Table of Contents

Semilinear Parabolic Systems Under Nonlinear Boundary Conditions

Herbert Amann
Mathematisches Institut
Universität Zürich
Rämistrasse 74
CH-8001 Zürich

We study problems of the form

$$\text{(A)} \qquad \dot{u} + \mathcal{A}u = F(u)\ , \quad \mathcal{B}u = G(u)\ , \quad u(0) = u_0\ ,$$

which can be considered as abstract counterparts to semilinear parabolic systems under nonlinear boundary conditions. Typical examples, to which our abstract theory applies, are of the form

$$\text{(P)} \qquad \begin{aligned} \partial_t u - \partial_j(a_{jk}\partial_k u) &= f(x,u,\partial u) && \text{in } \Omega \times (0,\infty)\ , \\ a_{jk}\nu^j\partial_k u &= g(x,u) && \text{on } \partial\Omega \times (0,\infty)\ , \\ u(\cdot,0) &= u_0 && \text{on } \Omega\ , \end{aligned}$$

where $\Omega \subset \mathbb{R}^n$ is a bounded smooth domain, $\nu = (\nu^1,\ldots,\nu^n)$ is the outer normal on $\partial\Omega$ and $u = (u^1,\ldots,u^N)$ is an N-vector valued function. We assume that (P) represents a parabolic system and f and g are smooth functions such that f satisfies some polynomial growth restriction with respect to ∂u. Instead of giving the precise assumptions we mention only that all as-

NATO ASI Series, Vol. F37
Dynamics of Infinite Dimensional Systems
Edited by S.-N. Chow, and J. K. Hale

sumptions are satisfied if (P) is of the very special form

$$\begin{aligned}
\partial_t u^1 - \alpha^{11}\Delta u^1 - \alpha^{12}\Delta u^2 &= f^1(u^1,u^2) \\
\partial_t u^2 - \alpha^{21}\Delta u^1 - \alpha^{22}\Delta u^2 &= f^2(u^2,u^2)
\end{aligned} \qquad \text{in } \Omega \times (0,\infty) ,$$

$$\begin{aligned}
\alpha^{11}\frac{\partial u^1}{\partial \nu} + \alpha^{12}\frac{\partial u^2}{\partial \nu} &= g^1(u^1,u^2) \\
\alpha^{21}\frac{\partial u^1}{\partial \nu} + \alpha^{22}\frac{\partial u^2}{\partial \nu} &= g^2(u^1,u^2)
\end{aligned} \qquad \text{on } \partial\Omega \times (0,\infty) ,$$

where f^1, f^2, g^1, g^2 are arbitrary smooth functions, provided $\alpha^{11} > 0$, $\alpha^{22} > 0$ and <u>either</u> $4\alpha^{11}\alpha^{22} > (\alpha^{12} + \alpha^{21})^2$ <u>or</u> $\alpha^{21} = 0$.

As for problem (A) we assume that W^1, W and ∂W^1 are Banach spaces such that $W^1 \hookrightarrow W$, where $\hookrightarrow$ means continuous injection. Moreover we assume (for simplicity) that

$$(\mathcal{A},\mathcal{B}) \in \mathrm{Isom}(W^1, W \times \partial W^1) .$$

Then we put $W^1_{\mathcal{A}} := \ker \mathcal{A}$ and $W^1_{\mathcal{B}} := \ker(\mathcal{B})$, we let $A := \mathcal{A} \mid W^1_{\mathcal{B}}$ and we assume that

> $-A$ is the infinitesimal generator of a strongly continuous analytic semigroup on W.

It follows that

$$R_1 := (\mathcal{B} \mid W^1_{\mathcal{A}})^{-1} \in \mathrm{Isom}(\, W^1, W^1_{\mathcal{A}}) .$$

Simple heuristic arguments lead then to the following "variation of constants formula" for problem (A):

$$(1) \qquad u(t) = e^{-tA}u_0 + \int_0^t e^{-(t-\tau)A}(F(u(\tau)) + A R_1 G(u(\tau)))d\tau , \quad 0 \leq t < \infty.$$

For a detailed description of the deduction of this formula we refer to [1]. Unfortunately (1) does not make any sense as it stands since $\mathrm{im}(R_1) = W_A^1$ and $\mathrm{dom}(A) = W_B^1$, and since it is easily verified that $W^1 = W_A^1 \oplus W_B^1$. However it can be shown that there exist appropriate "extrapolation spaces" ∂W^{-1} and W_B^{-1} and "extrapolated" operators R_0 and A_{-1} such that the following diagram commutes

$$\begin{array}{ccc}
\partial W^1 & \hookrightarrow & \partial W^{-1} \\
R_1 \downarrow & & \downarrow R_0 \\
W^1 = W_A^1 \oplus W_B^1 & \hookrightarrow & W \\
A \downarrow & & \downarrow A_{-1} \\
W & \hookrightarrow & W_B^{-1}
\end{array} \quad .$$

This shows that the composition $A_{-1}R_0$ is now a well defined bounded linear operator. In addition $-A_{-1}$ generates an analytic semigroup on W_B^{-1}. By interpolation it is possible to construct intermediate spaces and operators such that we have a commutative diagram

$$\begin{array}{ccccc}
\partial W^1 & \hookrightarrow & \partial W^{-1+2\beta} & \hookrightarrow & \partial W^{-1} \\
R_1 \downarrow & & R_\beta \downarrow & & \downarrow R_0 \\
W_A^1 \oplus W_B^1 & \hookrightarrow & W_B^\beta & \hookrightarrow & W \\
A \downarrow & & \downarrow A_{\beta-1} & & \downarrow A_{-1} \\
W & \hookrightarrow & W_B^{\beta-1} & \hookrightarrow & W_B^{-1}
\end{array} \quad .$$

Observe that this means, in particular, that $A_{\beta-1}R_\beta \in L(\partial W^{-1+2\beta}, W_B^{\beta-1})$. Moreover $-A_{\beta-1}$ generates an analytic semigroup on $W_B^{\beta-1}$. Now we replace formula (1) by the following *generalized variation-of-constants formula*

$$(2) \qquad u(t) = e^{-tA_{\beta-1}}u_0 + \int_0^t e^{-(t-\tau)A_{\beta-1}}(F(u(\tau)) + A_{\beta-1}R_\beta G(u(\tau))]d\tau, \quad 0 \leq t \leq \infty,$$

which is now well defined (in W_B^β).

In the concrete situation of problem (P) we can take

$$W^1 = W_p^2(\Omega,\mathbb{C}^N)\ , \quad W := L_p(\Omega,\mathbb{C}^N)\ , \quad \partial W^1 = W_p^{1-1/p}(\partial\Omega,\mathbb{C}^N)$$

where $p > n$ is arbitrary. Then one finds that one can choose $\beta = 1/2$ and that

$$W_B^{1/2} = W_p^1(\Omega,\mathbb{C}^N)\ .$$

Moreover it can be shown that u is a solution of (2) iff u is a weak solution (in the W_p^1-sense) of problem (P).

On the basis of our generalized variation-of-constants formula it is now easy to prove the following result (which we formulate for problem (P), for simplicity).

Theorem 1: Problem (P) *possesses for each* $u_0 \in W_p^1 := W_p^1(\Omega,\mathbb{C}^N)$ *a unique maximal weak solution* $u(\cdot,u_0) \in C([0,t^+(u_0)),W_p^1)$. *The function* $(t,u_0) \mapsto u(t,u_0)$ *defines a local semiflow on* W_p^1 *and bounded orbits are relatively compact. Moreover* $\lim_{t\to t^+(u_0)} \|u(t,u_0)\|_{1,p} = \infty$ *if* $t^+(u_0) < \infty$.

Of course, $\|\cdot\|_{s,q}$ denotes the norm in $W_q^s(\Omega,\mathbb{C}^N)$, $s \geq 0$, $1 \leq q \leq \infty$.

Being in possession of the generalized variation of constants formula it is now easy to prove results about the qualitative behaviour of the semiflow. For this one has only to notice that almost all results in the geometric theory of semilinear parabolic evolution equations under homogeneous linear boundary conditions are derived by means of the classical variation-of-constants formula (e.g. [4]).

In particular we can use the generalized variation-of-constants formula to obtain conditions guaranteeing the existence of global solutions. The following theorem gives a result of this type, where we suppose, for simplicity, that f is independent of ∂u.

Theorem 2: Suppose that $0 \leq s_0 \leq 1$, $1 \leq p_0 \leq \infty$, *and that* $s_0 = 0$ *if* $p_0 = 1$. *Moreover suppose that* f *and* g *satisfy growth restrictions of the form*

$$|f(x,\xi)| \leq c(1+|\xi|^{\lambda}) \ , \ |g(y,\xi)| \leq c(1+|\xi|^{\mu}) \ , \ x \in \bar{\Omega},\ y \in \partial\Omega, \xi \in \mathbb{C}^N ,$$

with

$$1 \leq \lambda < 1 + \frac{2p_0}{n-s_0p_0} \quad , \qquad 1 \leq \mu < 1 + \frac{p_0}{n-s_0p_0} \ ,$$

provided $n \geq s_0p_0$ *(and no growth restrictions if* $s_0p_0 > n$*). Moreover suppose that we know a priori that*

$$\|u(t,u_0)\|_{s_0,p_0} \leq c < \infty \ , \quad 0 \leq t < t^+(u_0)$$

for some $u_0 \in W_p^1$. *Then* $t^+(u_0) = \infty$, *that is,* $u(\cdot,u_0)$ *is a global solution. If, in addition, the spectrum of* $-A$ *is contained in the left half plane, then the orbit through* u_0 *is bounded in* W_p^1, *hence relatively compact in* W_p^1.

It should be observed that the growth restriction for f reduces to $1 \leq \lambda < (n+2)/(n-2)$ if an a priori bound in W_2^1, that is, in the "energy norm", is known. If we know only an L_{p_0}-a priori bound, then the growth restriction for f reduces to $1 \leq \lambda < 1 + 2p_0/n$. In this case recent results of Friedman and McLeod [3] imply that these bounds are optimal (except for the equality sign).

Detailed proofs for the above and more general and precise results are contained in [2].

References

[1] Amann, H: Semigroups and Nonlinear Evolution Equations. Linear Algebra & Appl. in press.

[2] Amann, H: Parabolic Evolution Equations and Nonlinear Boundary Conditions. to appear.

[3] Friedman, A. and McLeod, B: Blow-up of Positive Solutions of Semilinear Heat Equations. Ind. Univ. Math. J. 34(1985), 425-447.

[4] Henry, D: Geometric Theory of Semilinear Parabolic Equations. Lecture Notes in Math. #840, Springer-Verlag, Berlin, 1981.

THE SHADOWING LEMMA FOR ELLIPTIC PDE

Sigurd Angenent
Math. Dept. of the University of Leiden.

A. Introduction.

The purpose of this note is to show how certain ideas from dynamical systems theory can be generalised to nonlinear elliptic PDE.

The specific problem we shall discuss is the following: what can be said about the set of bounded solutions of the equation

$$(1) \qquad \Delta u + f(x,u) = 0.$$

which are defined on all of R^n. Here $f(x,u)$ is a given smooth nonlinearity, and $u : R^n \to R^m$ is the unknown (vectorvalued) variable.

If the space dimension n equals one then (1) actually is an ODE which may be rewritten as a first order system of equations

$$(2) \qquad u' = v, \quad v' = -f(x,u).$$

If, in addition, the nonlinearity is periodic in x, i.e. $f(x+1,u)=f(x,u)$, then one can construct the time-one map $F:R^{2m} \to R^{2m}$ of the system (2). This map is defined as follows. Given a point $(u(0),v(0))$ in R^{2m}, solve the ODE (2) with this point as initial data. Then $F(u(0),v(0))=(u(1),v(1))$.

The bounded solutions of (1) are now in one to one correspondence with the bounded orbits of the map F.

Roughly speaking, the shadowing lemma says this. Given a sequence of points p_k in R^{2m} ($k=0,\pm1,\pm2,\ldots$) such that p_{k+1} and $F(p_k)$ are close,

NATO ASI Series, Vol. F37
Dynamics of Infinite Dimensional Systems
Edited by S.-N. Chow, and J. K. Hale

uniformly in k, and if the p_k lie on a "hyperbolic invariant set for F", then there are points q_k close to the p_k such that $q_{k+1}=F(q_k)$. In other words, the "pseudo orbit" (p_k) is shadowed by the orbit (q_k) of the map F. we refer the reader to [GH, page 251] for a clear and precise statement.

In terms of the differential equation (1) with n=1 this means the following. Let the real line be covered by the intervals $I_k=[k,k+1]$, and suppose that we are given solutions $u_k:I_k\to R^m$ of the equation (1). Then under some condition "H" the following holds: if the solutions u_k and u_{k-1} do not differ to much at their common point of definition x=k (i.e. $u_k(k)-u_{k-1}(k)$ and $u_k'(k)-u_{k-1}'(k)$ are small, uniformly in k), there exists a solution u of (1) on R such that u is close to u_k on the interval I_k.

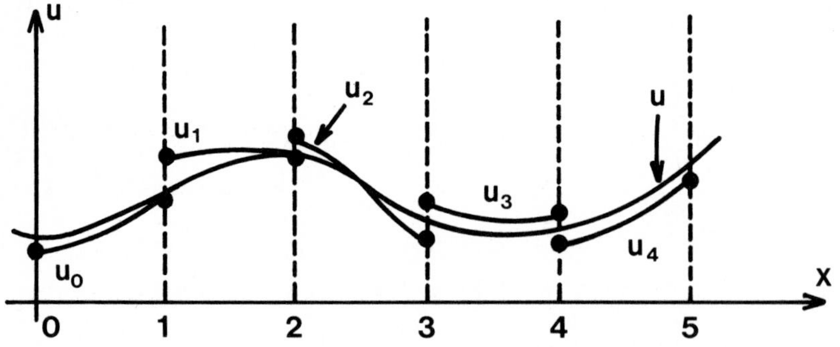

fig. 1 Shadowing local solutions.

The condition "H" involves the existence of exponential dichotomies. We refer the reader to [Pa] for a discussion from this point of view.

In the following we shall show how this last formulation can be generalised to arbitrary space dimensions.

B. The shadowing lemma.

We consider the equation (1) with arbitrary n,m>0. We assume that the nonlinearity f(x,u) has three continuous derivatives and we want the number

$$M_f = \sup \{ \| D_x^\alpha D_u^\beta f(x,u) \| \ : \ \|\alpha\|+\|\beta\| \leq 3, \ x \varepsilon R^n, u \varepsilon R^m \}$$

to be finite.

Now let $\{V_a\}$ be a locally finite covering of R^n by open sets (the index a runs through some countable index set) and suppose we are given a set of local solutions $u_a : V_a \to R^m$ of (1), one for each open set V_a. The statement we're after is that, if for any pair of overlapping sets V_a and V_b the difference $u_a - u_b$ restricted to the overlap $V_a \cap V_b = V_{ab}$ is small, then under some extra condition "H" there exists a solution u of (1) defined on all of R^n such that $u - u_a$ on V_a is small for each a.

Before we can make a precise statement we have to introduce some constants to quantify "small", and some concepts to formulate the condition "H".

First of all the covering has to be uniformly locally finite, i.e.

$$M_V = \max_b \{a : V_a \text{ and } V_b \text{ overlap}\}$$

should be finite (any open set in the covering intersects at most M_V other open sets in the covering).

Since the covering is locally finite it admits a partition of unity $\{f_a\}$. Thus

$$0 \leq f_a \ \varepsilon \ C^\infty(R^n)$$

$$\sum_a f_a(x) = 1 \text{ pointwise}$$

$$\text{supp}(f_a) \text{ is contained in } V_a.$$

It will be convenient to have the functions g_a defined by

$$g_a(x) = f_a(x) \ / \ \{\sum_b f_b(x)^2\}^{-\frac{1}{2}}$$

around. They satisfy $\sum_a g_a(x)^2 = 1$ pointwise.

We define the "radius of the partition" to be

$$R = \{\sup_{a,x,\alpha}(\|D^\alpha f_a(x)\|,\ \|D^\alpha g_a(x)\| \ :\ \|\alpha\|=3)\}^{-1/3}$$

and we assume it is positive. If the f_a are dimensionless then R has the dimension of a length, which is why we call it the "radius". It is a measure for the minimal size of the overlaps V_{ab}.

It follows from interpolation inequalities and $0\leq f_a, g_a\leq 1$ that

$$\sup_{a,x}(\|D^\beta f_a(x)\|,\ \|D^\beta g_a(x)\|) \leq C/R^{\|\beta\|} \qquad (\text{if } \|\beta\|\leq 3)$$

where the constant C depends on the space dimension n only.

For each open set V_a we consider the Banach spaces

$$X_a = C^{2,\lambda}(\overline{V}_a;\ R^m) \qquad Y_a = C^{0,\lambda}(\overline{V}_a;\ R^m)$$

$$X = C^{2,\lambda}(R^n;\ R^m) \qquad Y = C^{0,\lambda}(R^n;\ R^m)$$

where $0<\lambda<1$ is a constant. We define the spaces X_{ab} and Y_{ab} in a similar manner, and we shall write Y_a^c for the subspace of Y which contains the functions with support in V_a. Thus Y_a^c is also a subspace of Y_a.

Assume that the solutions u_a on V_a belong to X_a, and assume that

$$\sup_a \|u_a\|_{X_a} = M_u$$

is finite.

We shall say that the u_a form a <u>δ-pseudo solution</u> if

$$\sup_{a,b} \| u_a|_{V_{ab}} - u_b|_{V_{ab}}\|_{X_{ab}} \leq \delta$$

holds. If a function u in X satisfies

$$\sup_a \| u|_{V_a} - u_a\|_{X_a} \leq \varepsilon$$

then we shall say that the u_a are <u>ε-shadowed</u> by the function u.

In order to formulate the hypothesis "H" we linearise equation (1).

with each local solution we associate a linear operator A_a, given by

$$A_a = \Delta + q_a(x)$$

$$q_a(x) = f_u(x, u_a(x)).$$

Note that $q_a(x)$ is an m by m matrix valued function on $\overline{V}_a$, and that A_a is a bounded linear operator from X_a to Y_a.

We define a <u>local inverse</u> of A_a to be an operator $T_a : Y_a^c \to X_a$ such that

$$T_a\, A_a\, g_a = g_a$$

(3)

$$g_a\, A_a\, T_a = g_a$$

holds. Here g_a denotes the multiplication operator corresponding to the function g_a, and (3) should be read as an operator identity.

We can now formulate the hypothesis "H":

<u>"H". Each operator A_a has a local inverse T_a, and</u>

$$K = \sup_a \|T_a\|_{Y_a^c, X_a}$$

<u>is a finite number.</u>

The main result of this note is the following:

<u>Theorem. Let u_a be a set of local solutions of equation (1), with respect to some covering and partition of unity. Define the constants M_u, M_V, M_f and R as above and suppose that condition "H" is satisfied with constant K.</u>

<u>Then there is a constant R^*, depending on M_f, M_u, M_V and K only, such that if</u>

$$R \geq R^*$$

<u>then the following holds:</u>

for any $\varepsilon>0$ there is a $\delta>0$ such that if the u_a form a δ-pseudo solution then there exists a solution u of equation (1) which ε-shadows the u_a.

C. The proof.

Define the function v(x) by

$$v(x) = \Sigma f_a(x)u_a(x) \quad \text{(pointwise)}.$$

Then v belongs to X. The idea of the proof is to perturb v a little so that it becomes a solution of (1). This we do in three steps:

Step 1 Show that the operator A from X to Y given by

$$A = \Delta + f_u(x,v(x))$$

has a bounded inverse $T:Y\to X$ and estimate $\|T\|$.

Step 2 Define the nonlinear mapping F from X to Y to be

$$F(w) = \Delta w + f(x,w) \ .$$

Estimate F(v), F'(w) and F"(w) (for all w and our given v).

Step 3 Using the estimates of the first two steps apply a fixed point argument to the mapping $G(w)=w-TF(w)$, and verify that the fixed point of G is a solution of (1) which ε-shadows our given δ-pseudo solution.

Now let's work out the details.

From here on we shall use the letter C to denote any finite constant which depends on the values of M_u,M_V and M_f only, but may change from line to line in the text.

Step 1 Define $q(x)=f_u(x,v(x))$ and $q_a(x)=f_u(x,u_a(x))$ as before. It follows from

$$v-u_a = \sum_b f_b\,(u_b-u_a)$$

that we have the estimates

$$\|v|_{V_a} - u_a\|_{X_a} \leq C\delta$$

and
$$\|q|_{V_a} - q_a\|_{X_a} \leq C\delta.$$

Here we have used the C^3 norms the functions g_a to bound the norms of the corresponding multiplication operators g_a. Since we may assume, without loss of generality, that $R>1$ these C^3 norms can be estimated in terms of some constant C which depends on the space dimension only.

In order to construct an inverse of A from the local inverses T_a of A_a we define the operator

$$S = \sum_a g_a T_a g_a.$$

More precisely, for any w in Y we define the function

$$Sw = \sum_a S_a w$$

where
$$S_a w = g_a T_a(g_a w).$$

Since $S_a w$ has support in V_a the summation is locally finite and one has

$$\|Sw\|_X \leq M_V \sup_a \|S_a w\|_X .$$

So the operator S is well defined and bounded from Y to X. In fact we have the following estimate:

$$\|S\|_{Y,X} \leq C \sup_a \|g_a T_a g_a\|_{Y,X} \leq CK.$$

This operator S is not the exact inverse of A, but it is a good approximation. To see just how good it is we compare AS with 1_Y and SA with 1_X:

$$\begin{aligned} AS &= \sum A g_a T_a g_a \\ &= \sum \{[A,g_a]T_a g_a + g_a(A-A_a)T_a g_a + g_a^2\} \end{aligned}$$

$$= 1_Y + \Sigma\{[A,g_a] + g_a(q-q_a)\}T_a g_a$$

(where we've used $g_a A_a T_a = g_a$). Here [A,B] denotes the commutator of two operators A and B, i.e. AB−BA.

In the same way one finds

$$SA = 1_X + \Sigma\, g_a T_a\{[g_a,A] + (q-q_a)g_a\}.$$

Our estimates on $q-q_a$ imply

$$\| \Sigma\, g_a T_a(q-q_a)g_a \|_{X,X} \leq CK\,\delta\ .$$

Furthermore a short computation shows that

$$[A,g_a] = 2\nabla g_a \cdot \nabla + \Delta(g_a).$$

The estimates on the derivatives of the g_a then ensure

$$\| \Sigma\, g_a T_a[A,g_a] \|_{X,X} \leq CK/R.$$

So we see that if δ is small enough and R is large enough then

$$\| SA-1_X \|_{X,X} \leq CK(\delta+R^{-1}) \leq \tfrac{1}{2}$$

and similarly

$$\| AS-1_Y \|_{Y,Y} \leq CK(\delta+R^{-1}) \leq \tfrac{1}{2}\ .$$

It is here that we have to require R to be large. It should be clear from these last estimates that we want R to be larger than CK for some constant C. In view of our earlier assumption $R>1$ we get $R^*=\max(1,CK)$.

It follows that AS and SA are invertible, and that we have the estimates $\|(AS)^{-1}\|$, $\|(SA)^{-1}\|\leq 2$. Moreover A is invertible and if we define T to be the inverse of A then $T=(SA)^{-1}S=S(AS)^{-1}$, so that

$$\|T\|_{Y,X} \leq 2\, \|S\|_{Y,X} \leq CK.$$

Step 2 The nonlinear map $F:X\to Y$ is defined by $F(w)=\Delta w+f(x,w)$. Since f is C^3 the substitution operator $w\to f(x,w)$ is C^2 from X to Y, so that F is C^2.

The derivatives of F are given by

$$F'(w)\cdot\delta w = [\Delta + f_u(x,w(x))]\cdot\delta w$$

$$F''(w)\cdot(\delta w)^2 = f_{uu}(x,w(x))\cdot(\delta w)^2.$$

One easily obtains the estimates

$$\|F^{(k)}(w)\| \leq C\ (1+\|w\|_X) \qquad (k\leq 2)$$

for all w in X.

In order to estimate F(v) we note that on each V_a we have

$$\begin{aligned}\|\ F(v)\ \|_{Y_a} &= \|F(v)-F(u_a)\|_{Y_a}\\ &\leq \|F'(\Theta v+(1-\Theta)u_a)\| \cdot \|v-u_a\|_{X_a}\\ &\leq C\delta.\end{aligned}$$

Step 3 Consider the map $G:X\to X$ given by

$$G(w) = w - TF(w).$$

Fixed points of G are exactly the solutions of (1). In order to find a solution u close to v we define a sequence w_n:

$$w_{n+1} = G(w_n) \quad (n=0,1,2,\ldots),$$
$$w0 = v.$$

Since $G'(v) = 1_X - TF'(v) = 0$, G is a contraction near v. Therefore, if G(v) is close enough to v the sequence w_n will converge to a fixed point u of G.

To make this precise we observe that our estimates on T and $F''(w)$ imply that

$$\|G'(w)\|_{X,X} \leqslant CK\ \|w-v\|_X$$

so that $\|G'(w)\| \leqslant \frac{1}{2}$ if $\|w-v\| \leqslant (2CK)^{-1} \overset{\text{def}}{=} \rho$.

Furthermore we have

$$\|w_1-w_0\| = \|G(v)-v\| \leqslant CK\delta.$$

Using these estimates one can inductively show that if $\delta \leqslant CK^{-2}$ then

$$\|w_n-v\| \leqslant (1-2^{-n})\rho$$

$$\|w_n-w_{n-1}\| \leqslant 2^{-n}CK\delta \leqslant 2^{-n}\rho$$

holds for all $n=1,2,3,\ldots$.

Hence $u=\lim w_n$ exists and

$$\|u-v\|_X \leqslant CK\delta.$$

To conclude the proof we recall the estimate

$$\|v|_{V_a}-u_a\|_{X_a} \leqslant C\delta.$$

This implies for our solution u that

$$\|u|_{V_a}-u_a\|_{X_a} \leqslant C(1+K)\delta.$$

Therefore if δ is small enough the constructed solution u will ε-shadow the system of local solutions $\{u_a\}$.

D. An example.

In this last section we shall give an example of an equation to which our form of the Shadowing Lemma can be applied.

Consider the scalar equation

(10) $$\Delta u + f(x,u) = 0$$

where the nonlinearity depends periodically on the space variable x, i.e. $f(x+a,u)=f(x,u)$ for all integral vectors a (we shall call a vector integral if all it's coordinates are integers).

To emphasize the similarity with the ODE case we define the set B_m to be

$$B_m = \{ u \in L_\infty : u \text{ is a solution of (10) and } \|u\|_{L_\infty} \leq m \}.$$

If the nonlinearity is smooth (say f is C^3) then standard regularity theory for (10) gives us a priori L_∞ bounds on the first few derivatives of any u in B_m. Therefore the set B_m equipped with the topology of uniform convergence on compact subsets of R^n becomes a compact metrisable topological space. Furthermore, convergence in this topology implies $C^{2,\lambda}_{loc}$ convergence.

It follows from the periodicity of the nonlinearity that if $u \in B_m$ is a solution then the translates $u_a(x)=u(x+a)$, over integral vectors a, are also solutions. Thus we have a Z^n action on B_m, i.e. a homeomorphism $\tau_a : u \to u_a$ of B_m for each integral a, with the property that $\tau_{a+b}=\tau_a \tau_b$ holds for all a and b (of course one has to verify that the τ_a are homeomorphisms, but this is not so difficult).

This Z^n action generalises the "time-one map" one has in the ODE case, which we discussed in the introduction.

Now suppose that $f(x,0)=0$ and that equation (10) has a solution U(x) satisfying

(11) $$\lim_{x \to \infty} U(x) = 0.$$

Moreover assume that the operator A, given by

(12) $$A = \Delta + f_u(x,U(x))$$

from $C^{2,\lambda}$ to C^{λ} has a bounded inverse, say $T=A^{-1}$. This implies the invertibility of

(13) $$A_0 = \Delta + f_u(x,0)$$

and we shall write $T_0=(A_0)^{-1}$. The inverse T_0 can be obtained as a limit of translates of the operator T. Thus, given a w in C^{λ} we define $T^a w$ to be

$$T^a w = \tau_{-a} \cdot T \cdot \tau_a(w) \ .$$

Then, if we let a tend to infinity, $T^a w$ will converge in $C^{2,\lambda}_{loc}$ to $T_0 w$.

Ofcourse for any translate $\tau_a U = U_a$ the operator

$$A_a = \Delta + f_u(x,U_a(x))$$

also is invertible. In fact we have $(A_a)^{-1}=T^{-a}$.

In the topology of the space B_m the condition (11) is equivalent to

$$\lim_{a\to\infty} \tau_a U = 0.$$

Thus, in the set B_m, the U_a form a "net of points" which converges to the zero solution. In the ODE case this situation occurs whenever one has a transversal homoclinic orbit of the time-one map.

In this situation we can apply the Shadowing Lemma.

Let r>0 be an integer, and define

$$Q = (-3/4,+3/4)\times\cdots\times(-3/4,+3/4)$$
(an n dimensional cube)

$$V_a = r\cdot(a+Q) \quad \text{(for any a in } Z^n).$$

Thus V_a is a cube centered at ra with side 3r/2. Clearly the V_a form a uniformly locally finite covering of R^n (in fact one has $M_V=3^n$).

Next choose a function $0\leq\phi\varepsilon C_c^\infty(Q)$ such that

$$\Sigma_a \tau_a \phi = 1$$

holds pointwise, and define

$$f_a(x) = \phi(\frac{x}{r} - a)$$

Then the f_a form a partition of unity subordinate to the covering V_a, and a short computation shows that the "radius of the partition" is given by

$$R = Cr$$

for some constant $C>0$.

Finally we have to say to which solutions we want to apply the Shadowing Lemma. Here we can choose from a whole range of possibilities.

For each integral vector a we toss a coin and define j_a to be 0 or 1 depending on the outcome (say heads=1 and tails=0). Given such an array of zeros and ones we put

$$u_a(x) = \begin{cases} 0 & \text{if } j_a=0 \\ U(x-ra) & \text{if } j_a=1. \end{cases}$$

In other words, at the centre of each open cube V_a we either place a translate of our solution U or just the zero solution, depending on the value of j_a.

The Shadowing Lemma will allow us to find a solution of (10) which on each V_a is close to u_a. Let's check the hypotheses to see if this is really true.

We clearly have $C^{2,\lambda}$ bounds on the solutions u_a, and the covering is uniformly finite. If we assume that the third derivatives of the nonlinearity f(x,u) are bounded then the constants M_u,M_f and M_V are finite.

As a local inverse of $\Delta+f_u(x,u_a(x))$ we take T_0 if $j_a=0$ or $\tau_{-ra}\cdot T_1\cdot\tau_{ra}$ if $j_a=1$ (where τ_{ra} stands for the usual translation operator). So the hypothesis "H" is also satisfied and the constant K is given by

$$K = \max(\|T_0\|, \|T\|).$$

If we choose the integer r large enough then the radius of the partition will become so large that $R > R^*$ certainly holds, where R^* is the constant given by the Shadowing Lemma.

Thus, fixing some small $\varepsilon > 0$, there is a $\delta > 0$ such that, if the u_a form a δ-pseudo solution then there is a unique solution u of (10) which will ε-shadow the u_a's.

With a little effort one sees the condition (11) (i.e. U(x) vanishes at infinity) implies that for large enough r the u_a will form a δ-pseudo solution, so that we have our ε-shadowing solution u.

Observe that all estimates are independent of the particular array of zeros and ones our coin tossing produced. Therefore we have a solution u^j for every such array $\{j_a\} \varepsilon \{0,1\}^{Z^n}$. If the constant ε is small enough then all these solution will be different.

Apparently we have a mapping from the set of zero/one arrays, $\{0,1\}^{Z^n} = C$, into the set of solutions of (10), B_m (we have to choose m so large that B_m contains all these solutions). With the product topology the set C is homeomorphic to the Cantor set. In particular C is a compact metrisable space. We claim that the mapping which assigns the solution u^j to a zero/one array j is a homeomorphism from C into B_m. Indeed, since C is compact and since the mapping is one-to-one we only have to verify sequential continuity. Let j^n be a sequence of zero/one arrays converging to some j_∞ in C. The corresponding sequence of solutions u^{j_n} in B_m is precompact. Let u be any accumulation point of this sequence. Then one sees that u is a solution of (10) which ε-shadows the system of local solutions corresponding to the limit array j_∞. Hence $u = u^{j_\infty}$, and it follows that the sequence u^{j_n} converges in B_m to u^{j_∞}.

Thus the set B_m contains a Cantor set $C = \{0,1\}^{Z^n}$ and the translations τ_{ra} (a in Z^n) act as "multishifts" on this Cantor set.

This whole construction was based on the existence of the solution U and the fact that we could invert the operator $\Delta + f_u(x, U(x))$. We shall conclude this note by briefly indicating an example of a nonlinearity which admits such a solution. In order to keep the length of this note within reasonable bounds proofs will be presented on some other occasion (they are available on request).

We start with the equation

(14) $$\Delta u + f(u) = 0$$

where $$f(u) = \surd(\kappa + x^2) - \surd(\kappa + 1)$$

and $\kappa > 0$ is small. It is known that under these conditions (14) has a unique (upto translations) positive radial solution which vanishes at infinity. We call this solution U (see [BLP] and [PS]).

The set of translates $U_y = \tau_y U$ of U (y need not be integral) gives us a smooth n dimensional manifold in B_m. It's closure is homeomorphic to an n-sphere since $U(x) \to 0$ as $\|x\| \to \infty$. In the ODE case we would speak of a nontransversal homoclinic orbit.

Next we perturb the equation to

(15) $$\Delta u + f(u) + \alpha p(x) u = 0$$

where α is a small number and $p \varepsilon C^\infty$ satisfies $p(x+a) = p(x)$ for all integral a.

The solutions of (15) which decay at infinity are exactly the critical points of the functional

$$W(u) = \int_{R^n} [\tfrac{1}{2}(\nabla u)^2 - F(u) - \tfrac{1}{2}\alpha p(x) u^2]\, dx$$

with $F'(u) = f(u)$, $F(0) = 0$.

For $\alpha = 0$ the U_a form an n dimensional manifold of critical points of W (in some suitable function space on which W is defined). A bifurcation analysis along the lines suggested by Weinstein ([W, section 2]) will show the following. Suppose the function

$$M(x) = \int_{R^n} p(y)U(x+y)^2\, dy$$

(defined on R^n) has a nondegenerate critical point, say x_0. Then for small $\alpha \neq 0$ equation (15) will have a solution V close to the solution U_{x_0} of (14) which is radially symmetric about x_0.

This solution will be nondegenerate in the sense that $\Delta + f_u(x,V(x))$ will have a bounded inverse from C^λ to $C^{2,\lambda}$.

In the ODE case the solution V corresponds to a transversal homoclinic orbit of the time one map. The gradient of M(x) replaces the Melnikov function familiar from the ODE case.

Clearly the solution V can be used for the construction we sketched in the first part of this section.

References.

[BLP] BERESTYCKI H., LIONS P. & L.A.PELETIER, An ODE approach to the existence of positive solutions for semilinear problems in R^n, Indiana Math. Journ. 30 141-157 (1981)

[GH] GUCKENHEIMER J. & P.HOLMES, Nonlinear Oscillations, Dynamical Systems and Bifurcations of Vectorfields, Appl.Math.Sci. vol. 42 Springer-Verlag, New York 1983.

[Pa] PALMER K.J., Exponential Dichotomies and Transversal Homoclinic Points, JDE 55 225-256 (1984)

[PS] PELETIER L.A. & J.SERRIN, Uniqueness of positive solutions of semilinear equations in R^n, Arch.Rat.Mech.Anal. 81 181-197(1983)

[W] WEINSTEIN A., Bifurcations and Hamilton's principle, Math Zeitschrift 159 235-248 (1978).

Current address (1986/1987): Dept. of Math. California Institute of Technology, Pasadena, California.

COAGULATION-FRAGMENTATION DYNAMICS

J. M. Ball and J. Carr
Department of Mathematics
Heriot-Watt University
Riccarton
Edinburgh EH14 4AS
Scotland, U.K.

1. INTRODUCTION

The dynamics of cluster growth has attracted considerable interest in many apparently unrelated areas of pure and applied science. Examples include polymer science, colloidal and aerosol physics, atmospheric science, astrophysics and the kinetics of phase transformations in binary alloys [5,6,8,10,12,13]. The common link in all these examples is that they can be considered as a system of a large number of clusters of particles that can coagulate to form larger clusters or fragment to form smaller ones.

If $c_r(t) \geqslant 0$, $r = 1,2,\ldots,$ denotes the expected number of r-particle clusters per unit volume at time t, then the discrete coagulation-fragmentation equations are

$$\dot{c}_r = \frac{1}{2}\sum_{s=1}^{r-1}[a_{r-s,s}c_{r-s}c_s - b_{r-s,s}c_r] - \sum_{s=1}^{\infty}[a_{r,s}c_rc_s - b_{r,s}c_{r+s}] \qquad (1.1)$$

for $r = 1,2,\ldots,$ where the first sum is absent if $r = 1$. The coagulation rates $a_{r,s}$ and fragmentation rates $b_{r,s}$ are non-negative constants with $a_{r,s} = a_{s,r}$ and $b_{r,s} = b_{s,r}$. This model neglects (among other things) the geometric location of clusters and considers only binary collisions of clusters. For derivations of this and similar equations see [10].

The dependence of the rate coefficients $a_{r,s}, b_{r,s}$ on r and s depends on the particular application. In this paper we shall concentrate on the Becker-Döring equations in which $a_{r,s} = b_{r,s} = 0$ if both r and s are greater than 1. In this case we can write the equations in the form

NATO ASI Series, Vol. F37
Dynamics of Infinite Dimensional Systems
Edited by S.-N. Chow, and J. K. Hale

$$\dot{c}_r = J_{r-1} - J_r, \quad r \geqslant 2,$$
$$\dot{c}_1 = -J_1 - \sum_{r=1}^{\infty} J_r \tag{1.2}$$

where $J_r = a_r c_1 c_r - b_{r+1} c_{r+1}$. To see that (1.2) is a special case of (1.1) take $a_r = a_{r,1}, b_{r+1} = b_{r,1}$ for $r \geqslant 2$ and $2a_1 = a_{1,1}$, $2b_2 = b_{1,1}$. For ease of notation, from now on all summations will be over the positive integers unless stated otherwise.

In sections 2-3 we discuss equation (1.2). The asymptotic behaviour of solutions is especially interesting, both mathematically and for applications. For example, in the binary alloy problem the essence of the phase transition is the formation of larger and larger clusters as t increases. Mathematically this can be identified with a weak but not strong convergence as $t \to \infty$. We only outline the main ideas involved in this investigation; full details appear in [3]. The general equation (1.1) is more difficult to analyse since solutions may have singularities not present in (1.2). In section 4 we briefly discuss some of the difficulties and state a new result on density conservation.

2. BASIC IDEAS

We first review some facts concerning convergence in a space of sequences. Let $X = \{c = (c_r): \Sigma\, r|c_r| < \infty\}$ and let $\{c^j\}$ be a sequence of elements in X. We say that c^j converges strongly to $c \in X$ (symbolically $c^j \to c$) if $\Sigma\, r|c_r^j - c_r| \to 0$ as $j \to \infty$. It is also useful to have another notion of convergence in X. We say that c^j converges weak * to $c \in X$ (symbolically $c^j \xrightarrow{*} c$) if (i) $\sup\{\Sigma\, r|c_r^j|: j = 1,2,\ldots\} < \infty$ and (ii) $c_r^j \to c_r$ as $j \to \infty$ for each $r = 1,2,\ldots$. Thus weak * convergence is in a sense pointwise convergence. The justification of the terminology comes from functional analysis (cf. [9], p374). Clearly strong convergence implies weak * convergence. However, the converse is false in general; for example take $c^j = (j^{-1}\delta_{rj})$ where $\delta_{rj} = 1$ if $r = j$ and 0 otherwise. Then c^j converges weak * to the zero sequence but it does not converge strongly. We can express the weak * convergence as convergence in a metric space. For $\rho > 0$ let $B_\rho = \{(y_r) \in X: \Sigma\, r|y_r| \leqslant \rho\}$. Then (B_ρ, d) is a metric space where $d(y,z) = \Sigma\, |y_r - z_r|$. Clearly a sequence $\{y^j\} \subset B_\rho$ converges weak * to $y \in X$ if and only if $y \in B_\rho$ and $d(y^j,y) \to 0$ as $j \to \infty$. Weak * convergence is useful because B_ρ is compact; equivalently, any bounded sequence in X has a weak * convergent subsequence.

In order to arrive quickly at the most interesting questions concerning (1.2) we give a rapid review of its properties. The density is given by $\Sigma\, rc_r(t)$ and since matter is neither created or destroyed in an interaction it is a conserved quantity. Thus we look for equilibrium solutions $c^\rho = (c_r{}^\rho)$ with $\rho = \Sigma\, rc_r{}^\rho$. From (1.2) we must have $J_r(c^\rho) = 0$ for all r so that

$$c_r{}^\rho = Q_r(c_r{}^\rho)^r \tag{2.1}$$

where $Q_1 = 1$, $Q_{r+1} = Q_r a_r/b_{r+1}$, $r \geq 1$. It remains to identify $c_1{}^\rho$. To do this let

$$F(z) = \Sigma\, rQ_r z^r$$

In the binary alloy problem the above series has finite radius of convergence z_s and $F(z_s) = \rho_s < \infty$. In this paper we shall describe our results for this case; for other cases see [3]. Since F is an increasing function of z, the equation $F(z) = \rho$ has a unique solution $z = c_1{}^\rho$ if $0 \leq \rho \leq \rho_s$ and no solution if $\rho > \rho_s$. Thus if $0 \leq \rho \leq \rho_s$ there is a unique equilibrium c^ρ with density ρ, while if $\rho > \rho_s$ there is no equilibrium with density ρ. Let

$$V(c) = \Sigma\, c_r[\ln(c_r/Q_r) - 1]. \tag{2.2}$$

The 'free-energy' function V is a Lyapunov function for (1.2), that is it is non-increasing along solutions. Also, for $0 \leq \rho \leq \rho_s$, the equilibrium c^ρ is the unique minimizer of V on the set $X^\rho = \{c = (c_r)\colon c_r \geq 0$ for all r, $\Sigma\, rc_r = \rho\}$.

Suppose that the initial data for (1.2) has density ρ_0. If $\rho_0 \leq \rho_s$ the above results suggest that the corresponding solution $c(t) \to c^\rho$ strongly in X as $t \to \infty$ and this is indeed the case. If $\rho_0 > \rho_s$ the asymptotic behaviour is not so clear since there is no equilibrium with density ρ_0. Since V is non-increasing along solutions it is natural to consider the behaviour of minimizing sequences of V on X^{ρ_0}. The basic result here is that if $\rho_0 > \rho_s$ and c^j is a minimizing sequence of V on X^{ρ_0} then c^j converges weak * to c^{ρ_s} in X but not strongly. The main result on asymptotic behaviour says that the solution $c(t)$ of (1.2) with density ρ_0 is minimizing for V on X^{ρ_0} as $t \to \infty$, so that, for $0 \leq \rho_0 \leq \rho_s$, $c(t) \to c^{\rho_0}$ strongly in X and, for $\rho_0 > \rho_s$, $c(t) \xrightarrow{*} c^{\rho_0}$ in X. Note that for the case $\rho_0 > \rho_s$ we have that

$$\rho_0 = \Sigma\, rc_r(t) > \Sigma\, r \lim_{t\to\infty} c_r(t) = \Sigma\, rc_r^{\rho_s} = \rho_s.$$

The excess density $\rho_0 - \rho_s$ corresponds to the formulation of larger and larger clusters as t increases, i.e. condensation.

To obtain results on the asymptotic behaviour of a solution $c(t)$ we have to exploit the Lyapunov function V. To do this we apply the invariance principle for evolution equations endowed with a Lyapunov function (cf. [7] for a survey). To apply this method we need to find a metric with respect to which V is continuous, the positive orbit $\{c(t)\colon t \geq 0\}$ is relatively compact and solutions depend continuously on initial data. It might seem natural to try and use the metric induced by strong convergence on X, that is $d(y,z) = \Sigma\, r|y_r - z_r|$. However, in the case $\rho_0 = \Sigma\, rc_r(0) > \rho_s$ the positive orbit cannot be relatively compact with this metric since there is no equilibrium with density ρ_0. Moreover, since the only obvious global estimate is density conservation we have to use the metric induced by weak * convergence on bounded subsets of X to achieve relative compactness of positive orbits. Unfortunately, V defined by (2.2) is not continuous in this metric. Fortunately, however, because density is conserved,

$$V_z(c) = V(c) - \ln z\, \Sigma\, rc_r$$

is a Lyapunov function for each z, and for exactly one value of z, namely $z = z_s$, V_z is sequentially weak * continuous. Thus we can apply the invariance principle to prove that $c(t) \xrightarrow{*} c^{\rho}$ as $t \to \infty$ for some ρ, $0 \leq \rho \leq \min(\rho_0,\rho_s)$ where ρ_0 is the density of the initial data. We then prove the result described above by using a maximum principle for (1.2) in the case $\rho < \rho_s$. At this stage of the proof in [3] we made certain hypotheses on the initial data; a more refined argument shows that these hypotheses are not needed [4].

3. EXISTENCE AND DENSITY CONSERVATION

We prove existence of solutions to (1.2) by taking a limit of solutions of the finite-dimensional system

$$\begin{aligned} \dot{c}_r &= J_{r-1} - J_r, \qquad 2 \leqslant r \leqslant n-1 \\ \dot{c}_1 &= -J_1 - \sum_{r=1}^{n-1} J_r, \quad \dot{c}_n = J_{n-1}. \end{aligned} \tag{3.1}$$

Solutions of (3.1) satisfy $\sum_{r=1}^{n} rc_r(t) = \sum_{r=1}^{n} rc_r(0)$ so that $c_r = O(r^{-1})$ for all n. Hence if $a_r, b_r = o(r)$ then for each r, $\dot{c}_r$ is bounded. Thus by applying the Arzela-Ascoli Theorem and passing to the limit in the equations we get a simple global existence proof. In fact since fragmentation can be thought of as a dissipative mechanism we do not need any hypotheses on b_r and by working harder we need only assume $a_r = O(r)$ to get global existence. If $r^{-1}a_r \to \infty$ as $r \to \infty$, there is in general no solution of (1.2) even on a short time interval.

We remarked earlier that formally the density $\Sigma\, rc_r(t)$ is a constant of the motion. This is always true for (1.2); it is not true in general for (1.1) (cf. section 4). To prove it for (1.2) we consider partial sums. Now

$$\sum_{r=1}^{n} r[c_r(t) - c_r(0)] = -\int_0^t \left[nJ_n(c(s)) + \sum_{r=n}^{\infty} J_r(c(s))\right] ds. \tag{3.2}$$

For a solution of (1.2) we require that $\Sigma\, J_r(c)$ converges so that

$$\int_0^t \sum_{r=n}^{\infty} J_r(c(s))ds \to 0 \quad \text{as} \quad n \to \infty.$$

Also, from (1.2)

$$n\int_0^t J_n(c(s))ds = n \sum_{r=n+1}^{\infty} (c_r(t) - c_r(0)) \to 0 \quad \text{as} \quad n \to \infty$$

since $n \sum_{r=n+1}^{\infty} c_r \leqslant \sum_{r=n+1}^{\infty} rc_r$. Thus letting $n \to \infty$ in (3.2) proves that the density is conserved.

4. THE GENERAL DISCRETE COAGULATION-FRAGMENTATION EQUATIONS

We first discuss equations (1.1) when both coagulation and fragmentation are included. In this case it is usual to assume the detailed balance condition. This demands that (i) an equilibrium solution $\bar{c} = (\bar{c}_r)$ with $\bar{c}_r > 0$ exists and (ii) at equilibrium the net rate of conversion of r and s clusters to $r + s$ clusters is zero, so that $b_{r,s}\bar{c}_{r+s} = a_{r,s}\bar{c}_r\bar{c}_s$. This places the following restriction on $a_{r,s}$, $b_{r,s}$;

$$a_{r,s}Q_rQ_s = b_{r,s}Q_{r+s} \tag{4.1}$$

for some Q_r. Assuming (4.1), it follows that equilibria of (1.1) have the form $c_r = Q_r(c_1)^r$ and a formal calculation shows that $V(c) = \Sigma\, c_r[\ln(c_r/Q_r) - 1]$ is a Lyapunov function for (1.1). This is the same as for the Becker-Döring equation (cf. equations (2.1) and (2.2)) and so we expect to get similar results. The analysis for (1.1) however is even more complicated than that needed for (1.2). Most of the analysis has been completed [4] but there are still some technicalities to be finalised. To reveal some of these difficulties we look at some special cases of (1.2). In particular, we show that the density $\Sigma\, rc_r(t)$, which formally is a constant of the motion, need not in fact be conserved.

(a) Let $a_{r,s} = 0$, $b_{r,s} = 1$ for all r and s so that

$$\dot{c}_r = \sum_{s=1}^{\infty} c_{r+s} - \frac{1}{2}(r-1)c_r. \tag{4.2}$$

A solution of (4.2) is

$$c_r(t) = (e^{-t/2})^{r-1}\left[c_r(0) + \sum_{n=r+1}^{\infty} c_n(0)(2(1-e^{-t/2}) + (1-e^{-t/2})^2(n-r-1))\right] \tag{4.3}$$

and it is easy to check that for this solution the density is a conserved quantity [2]. However, for any $\lambda > 0$, (4.2) has a solution

$$c_r = e^{\lambda t}r^{-3}x_r, \tag{4.4}$$

where x_r is defined by $x_1 = 1$, $x_{r+1} = (1 + \alpha_r)x_r$, and

$$\alpha_r = \frac{6\lambda r^2 + (6\lambda - 2)r + 2\lambda - 1}{r^3(2 + r + 2\lambda)}.$$

Since $\alpha_r = O(r^{-2})$, x_r is bounded and $\Sigma\, rc_r(t) = e^{\lambda t}\Sigma\, rc_r(0)$. The special solutions (4.4) also show that for any initial data, solutions of (4.2) are not unique. Clearly, the solutions given by (4.4) are unphysical. In this case it is easy to pick out the correct unique solution by placing extra requirements on the definition of a solution (cf. [1] for the continuous case of (4.2)). However, in more complicated situations it is useful to know conditions on the fragmentation coefficients which prohibit non-uniqueness.

(b) Let $b_{r,s} = 0$ for all r and s so that we are only considering coagulation processes. In this case the density conservation can break down at a finite time t_c, a phenomenon known as gelation [11]. The gel point t_c is characterised as the first time for which $\sum_{r,s} ra_{r,s}c_rc_s$ diverges and is interpreted as the formation of a super-particle (gel phase). In particular, this phenomenon occurs when $a_{r,s} = (rs)^{\alpha}$, $\alpha > 1/2$. For $t > t_c$ it may be necessary to modify the equations to account for interactions of the gel phase with finite clusters.

For applications to phase transitions, one set of conditions suggested by O. Penrose on the coagulation and fragmention rates is that $a_{r,s} = O(r^{1/3} + s^{1/3})$ and that $b_{r,s} = a_{r,s}Q_{r+s}^{-1}Q_rQ_s$, where $Q_r \sim z_s^{-r}\exp(-\alpha r^{1/3})$ with α, z_s positive constants. Note that in this case $b_{r,s} \sim r^{1/3}$ for r large and s bounded while for r and s large with $r - s$ small, $b_{r,s}$ is small. The physical motivation here is that surface area considerations show that it is unlikely that a large cluster of size $r + s$ will split into two large clusters of size r and s (and hence increase the surface energy by a large amount). It turns out that under these conditions we can show that density is conserved. More generally we have:

<u>Theorem</u>

Suppose that for some $n_0 \geqslant 1$ and $k > 0$ we have that

(i) $a_{r,s} \leqslant k(r + s)$ for all $r,s \geqslant n_0$,

(ii) $\frac{n}{r}\sum_{j=n}^{r-n} b_{r-j,j} \leqslant k$ for all r,n with $r \geqslant 2n \geqslant 2n_0$,

(iii) $\frac{1}{r} \sum_{j=n_0}^{m} jb_{r-j,j} \leq k$ for all r and n with $r \geq n + n_0$

where $m = \min(n, r - n)$.

Then if c is a solution of (1.1) on $[0,T)$ with $\rho_0 = \Sigma\, rc_r(0) < \infty$, $\Sigma\, rc_r(t) = \rho_0$ for all $t \in [0,T)$.

The proof of the above result is given in [4].

REFERENCES

1. Aizenman, M., Bak, T.A.: Convergence to equilibrium in a system of reacting polymers. Commun. Math. Phys. **65**, 203-230 (1979).

2. Bak, T.A., Bak, K.:Acta Chem. Scand. 1997, **13**, (1959).

3. Ball, J.M., Carr, J., Penrose, O.: The Becker-Döring cluster equations: basic properties and asymptotic behaviour of solutions. Commun. Math. Phys. **104**, 657-692 (1986).

4. Ball, J.M., Carr, J., Penrose, O.: In preparation.

5. Binder, K., Heermann, D.W.: Growth of domains and scaling in the late stages of phase separation and diffusion-controlled ordering phenomena. Preprint.

6. Cohen, R.J., Benedek, G.B.: Equilibrium and kinetic theory of polymerization and the sol-gel transition. J. Phys. Chem. **86**, 3696-3714 (1982).

7. Dafermos, C.M.: Contraction Semigroups and Trend to Equilibrium in Continuum Mechanics, in Lecture Notes in Mathematics 503 pp 295-306, Springer-Verlag (1976).

8. Drake, R.: In: Topics in current aerosol research. International reviews in aerosol physics and chemistry, Vol. 2, Hidy, G.M., Brock J.R.(eds.). Oxford: Pergamon Press 1972.

9. Dunford, N., Schwartz, J.T.: Linear operators, Part I. New York: Interscience 1958.

10. Friedlander, S.K.: Smoke, Dust and Haze. Wiley (1977).

11. Hendricks, E.M., Ernst, M.H., Ziff, R.M.: Coagulation equations with gelation. J. Stat. Phys. **31**, 519-563 (1983).

12. Penrose, O., Buhagiar, A.: Kinetics of nucleation in a lattice gas model: Microscopic theory and simulation compared. J. Stat. Phys. **30**, 219-241 (1983).

13. Pruppacher, H.R., Klett, J.D.: Microphysics of clouds and precipitation, Reidel (1978).

FUNCTIONAL DIFFERENTIAL EQUATIONS AND JENSEN'S INEQUALITY

L.C. Becker
Department of Mathematics
Christian Brothers College
Memphis, Tennessee 38104

T.A. Burton
S. Zhang*
Department of Mathematics
Southern Illinois University
Carbondale, Illinois 62901
(* on leave from Anhui University, PRC)

1. Introduction

We consider a system of functional differential equations

$$x'(t) = F(t,x_t), \tag{1}$$

where x_t is the function defined by $x_t(s)=x(t+s)$ for $-h\leq s\leq 0$ $(h>0)$. Let $(C, \|\cdot\|)$ denote the Banach space of continuous functions $\psi:[-h,0]\to R^n$ with the norm $\|\psi\| = \sup_{-h\leq s\leq 0} |\psi(s)|$, where $|\cdot|$ is any convenient norm in R^n. For $H>0$, let C_H designate the open ball in C for which $\|\psi\|<H$. It is assumed that $F:[0,\infty) \times C_H \to R^n$ is continuous and takes bounded sets into bounded sets.

Under the above conditions, it is known that for each $(t_0,\psi)\in[0,\infty) \times C_H$ there is a solution $x(t_0,\psi)$ satisfying (1) on an interval $[t_0,t_0+\alpha)$, for some $\alpha>0$, with $x_{t_0}(t_0,\psi)=\psi$. The value of such a solution at t is denoted by $x(t,t_0,\psi)$. Furthermore, if there is an $H_1<H$ such that $|x(t,t_0,\psi)|\leq H_1$ for all $t\geq t_0$ for which $x(t_0,\psi)$ is defined, then $\alpha=\infty$. The details supporting these statements are found in [2; pp. 187-191] or [3; pp. 36-45].

In the classical theory of Liapunov's direct method, the task is to find a scalar functional $V:[0,\infty) \times C_H \to [0,\infty)$ whose derivative along solutions of (1), defined by

$$V'_{(1)}(t,\psi) = \limsup_{\delta \to 0^+} [V(t+\delta,x_{t+\delta}(t,\psi)) - V(t,\psi)]/\delta \quad , \tag{2}$$

satisfies relations which will yield information about solutions of (1) without actually finding these solutions. A typical problem may be formulated as follows, but first a definition.

NATO ASI Series, Vol. F37
Dynamics of Infinite Dimensional Systems
Edited by S.-N. Chow, and J.K. Hale

DEF.1. A wedge is a continuous, strictly increasing function $W_i:[0,\infty)\to[0,\infty)$ such that $W_i(0)=0$ and $W_i(r)\to\infty$ as $r\to\infty$.

PROBLEM. Suppose we have a Liapunov functional $V:[0,\infty) \times C_H \to [0,\infty)$ such that

(i) $W_1(|\psi(0)|) \le V(t,\psi)$, with $V(t,0)=0$,

and

(ii) $V'_{(1)}(t,\psi) \le -[1/(t+1)]W_2(|\psi(0)|)$.

Standard theory (cf. [2], [3], [5]) implies that the zero solution is stable. But can we also conclude that the zero solution is asymptotically stable?

Difficulties that arise in trying to prove asymptotic stability may be described as follows. Let $x(t)=x(t,t_0,\psi)$ be a solution of (1) on $[t_0,\infty)$ with $|x(t)|<H$. A consequence of (ii) is

$$V(t,x_t) \le V(t_0,\psi) - \int_{t_0}^{t} [1/(s+1)]W_2(|x(s)|)ds \; . \tag{3}$$

If $x(t) \not\to 0$ as $t \to \infty$, then there is an $\varepsilon>0$ and a strictly increasing sequence $\{t_n\} \to \infty$ with $|x(t_n)|\ge\varepsilon$. The usual argument uses this sequence in conjunction with an inequality similar to (3) and ends with $V(t,x_t)$ being driven to $-\infty$, which contradicts (i). However, there are two difficulties preventing us from using (3) in such a manner. First, $F(t,x_t)$ may be unbounded for bounded x_t so that the integral in (3) remains small because $|x(t)|$ quickly returns to values near zero. But even if $F(t,x_t)$ is bounded for bounded x_t so that $|x(t)|\ge\varepsilon/2$ on the intervals $[t_n,t_n+\beta]$ for some positive constant β independent of n, the integral in (3) may still be small if $t_{n+1}-t_n$ is large.

What is needed is a technique to separate the integral in (3) into two parts, one involving $1/(t+1)$ and the other $W_2(|x(t)|)$, while still maintaining the direction of the inequality in (3). A Schwarz inequality reverses the direction of such an inequality. However, an application of Jensen's inequality seems to be just what is needed.

2. Stability

Proofs of the following facts about convex functions are

found in Natanson [4].

DEF.2. A function $G:[a,b]\to(-\infty,\infty)$ is said to be convex downward if $G([t_1+t_2]/2) \le [G(t_1)+G(t_2)]/2$ for any $t_1,t_2 \in [a,b]$.

LEMMA. If $f:[a,b]\to(-\infty,\infty)$ is increasing, then $F(t)=\int_a^t f(u)du$ is convex downward.

COR. If $W(r)$ is a wedge, then $W_1(r)=\int_0^r W(s)ds$ is a wedge; and on $[0,1]$, $W_1(r)\le W(r)$. Hence, if $V'_{(1)}(t,\psi)\le -W(|\psi(0)|)$, we may assume without loss of generality that W is convex downward.

THEOREM (Jensen's Inequality). Let $W:[0,\infty)\to[0,\infty)$ be convex downward. If $|x(t)|$ and $\eta(t)\ge 0$ are continuous on $[0,\infty)$ and $\int_a^b \eta(s)ds > 0$, then

$$\int_a^b \eta(s)W(|x(s)|)ds \ge \int_a^b \eta(s)ds\ W\left[\int_a^b \eta(s)|x(s)|ds \Big/ \int_a^b \eta(s)ds\right].$$

APPLICATION. Suppose $V'_{(1)}(t,x_t)\le -\eta(t)W(|x(t)|)$. Let I_j denote the interval $[(j-1)h,\ jh]$ and $V(t)=V(t,x_t)$. Then for $t\ge nh$, we have

$$V(t) \le V(0) - \sum_{j=1}^{n} \int_{I_j} \eta(s)W(|x(s)|)ds$$

$$\le V(0) - \sum_{j=1}^{n} \int_{I_j} \eta(s)ds\ \ W\left[\int_{I_j} \eta(s)|x(s)|ds \Big/ \int_{I_j} \eta(s)ds\right],$$

assuming that $\int_{t-h}^{t} \eta(s)ds > 0$ for $t\ge h$. If, in addition, $\eta(t)$ is nonincreasing and if there is an $M>0$ such that $\int_{t-h}^{t} \eta(s)ds \le M\eta(t)$ for $t\ge h$, then the above inequality could be extended even further to

$$V(t) \le V(0) - \sum_{j=1}^{n} \int_{I_j} \eta(s)ds\ W\left[(1/M) \int_{I_j} |x(s)|ds\right],$$

yielding the separation we had sought. An example of such a function is $\eta(t)=1/(t+1)$ mentioned earlier. This motivates the following definition.

DEF.3. A continuous function $\eta:[0,\infty)\to[0,\infty)$ is said to be a J-function if η is non-increasing, $\eta \notin L^1[0,\infty)$, and if for each

$h>0$ there exists an $M>0$ with $\int_{t-h}^{t} \eta(s)ds \le M\eta(t)$ for $h \le t < \infty$.

THEOREM 1. Let both $V:[0,\infty) \times C_H \to [0,\infty)$ and $\eta:[0,\infty)\to[0,\infty)$ be continuous with

(i) $W_1(|\psi(0)|) \le V(t,\psi)$, where $V(t,0) = 0$;

(ii) $V'_{(1)}(t,x_t) \le -\eta(t)[W_2(|x(t)|)+W_3(|F(t,x_t)|)]$;

(iii) W_3 convex downward;

(iv) η, a J-function; and

(v) $x=0$ uniformly stable (U.S.).

Then $x=0$ is asymptotically stable (A.S.).

Condition (ii) might seem a bit extravagant, but it is not. The purpose of the following example is to illustrate how such inequalities can be obtained from the standard relations.

EXAMPLE A. Consider the scalar equation

(A1) $$x'(t) = b(t)x(t-h) - c(t)x(t)$$

with $b,c:[0,\infty)\to(-\infty,\infty)$ continuous and b bounded. Assume there is a number $a>1$ such that

(A2) $$c(t) > a|b(t+h)|$$

and that

(A3) $$c(t) \ge \eta(t)$$

for some J-function $\eta \le 1$. Then $x=0$ is A.S. (We remark that $c(t)$ may be unbounded, and it may approach 0 along a sequence tending to ∞.)

PROOF. Let $\bar{a} = (a+1)/2$ and use the functional

$$V(t,x_t) = |x(t)| + \bar{a}\int_{t-h}^{t} |b(u+h)||x(u)|du$$

so that

$$V'(t,x_t) \le |b(t)x(t-h)|-c(t)|x(t)|+\bar{a}(|b(t+h)||x(t)| - |b(t)||x(t-h)|) \le -\delta c(t)|x(t)| ,$$

where $\delta=(a-1)/2a>0$. At this stage, this does not resemble condition (ii); however, if we define the functional

$$H(t,x_t,x_{t-h}) = V(t,x_t) + V(t-h,x_{t-h}) ,$$

then

$$\begin{aligned} H'(t,x_t,x_{t-h}) &\leq -\delta c(t)|x(t)| - \delta c(t-h)|x(t-h)| \\ &\leq -(\delta/2)c(t)|x(t)| - (\delta/2)c(t)|x(t)| - \delta|b(t)x(t-h)| \\ &\leq -(\delta/2)\eta(t)(|x(t)|+|x'(t)|) , \end{aligned}$$

which is of the form required by the theorem. Since the function b is bounded, it is evident that x=0 is U.S.

DEF.4. A measurable function $\eta:[0,\infty)\to[0,\infty)$ is said to be uniformly integrally positive with parameter h (UIP(h)) if there is a $\delta>0$ such that $\int_{t-h}^{t} \eta(s)ds \geq \delta$ for $t\geq h$.

THEOREM 2. Let $V:[0,\infty) \times C_H\to[0,\infty)$ and $\eta_1,\eta_2:[0,\infty)\to[0,\infty)$, where $\eta_1 \notin L^1[0,\infty)$ and η_2 is UIP(h). If

(i) x=0 is U.S. and

(ii) $$V'_{(1)}(t,x_t) \leq -\eta_1(t)\left\{W_1\left[\int_{t-h}^{t}|F(s,x_s)|ds\right] + W_2\left[\int_{t-h}^{t}\eta_2(s)W_3(|x(s)|)ds\right]\right\},$$

then x=0 is A.S.

EXAMPLE B. Consider the scalar equation

(B1) $$x'(t) = b(t)x(t-h) ,$$

where the function $-b:[0,\infty)\to(0,\infty)$ is continuous and UIP(h). Suppose there are positive constants α, Γ_0, and K such that

(B2) $$\alpha-|b(t+h)| \overset{\text{def}}{=} \eta_1(t) \geq 0 ,$$

(B3) $$|b(t+h)|\left[-2 + \alpha h + \int_{t-h}^{t}|b(u+h)|du\right] \leq -\Gamma_0 ,$$

(B4) $$\int_{t-h}^{t}|b(s+h)|ds \leq K$$

and

(B5) $$\bar{\eta}(t) \overset{\text{def}}{=} \min[\eta_1(t), \eta_1(t-h)] \notin L^1[0,\infty).$$

Then x=0 is U.S. and A.S.

PROOF. First, we rewrite (B1) as

(B1)' $$x' = b(t+h)x - (d/dt) \int_{t-h}^{t} b(u+h)x(u)\,du$$

and define the functional

$$V(t,x_t) = (x + \int_{t-h}^{t} b(u+h)x(u)\,du)^2 + \alpha\int_{-h}^{0} \int_{t+s}^{t} |b(u+h)|x^2(u)\,du\,ds,$$

which, upon differentiation, yields

$$V'(t,x_t) \leq |b(t+h)|[-2 + \alpha h + \int_{t-h}^{t} |b(u+h)|du]x^2(t) - (\alpha - |b(t+h)|) \int_{t-h}^{t} |b(u+h)|x^2(u)\,du.$$

From (B2) and (B3), we see that

(B6) $$V'(t,x_t) \leq -\eta_1(t) \int_{t-h}^{t} |b(u+h)|x^2(u)\,du.$$

Now apply Jensen's inequality to the right-hand side of (B6) to obtain

$$V'(t,x_t) \leq -[\eta_1(t)/\int_{t-h}^{t} |b(s+h)|ds][\int_{t-h}^{t} |b(s+h)||x(s)|ds]^2 ,$$

which by (B4) implies

(B7) $$V'(t-h,x_{t-h}) \leq -[\eta_1(t-h)/K][\int_{t-h}^{t} |b(s)x(s-h)|ds]^2 .$$

Finally, define the functional

$$\Omega(t,x_t,x_{t-h}) = V(t,x_t) + KV(t-h,x_{t-h})$$

and use (B5), (B6), and (B7) to show that the conditions of Theorem 2 are met. The proof is complete when uniform stability is established with an argument that involves the functionals $V(t,x_t)$ and $H(t,x_t) = x^2(t) + \int_{t-h}^{t} |b(u+h)|x^2(u)\,du$.

The use of the same functionals, with some minor adjustments, plus a couple of applications of Jensen's inequality can be used to establish the validity of the next example.

EXAMPLE C. Consider the scalar equation

$$x'(t) = -a(t)x(t) + b(t)x(t-h) \tag{C1}$$

where $a,b:[-h,\infty)\to R$ are continuous and λ, defined by $\lambda(t) = \max[|a(t)|,|b(t+h)|]$, is UIP(h). Assume there is an $\alpha>0$ such that

$$\begin{aligned} 2[b(t+h) - a(t)] &+ \alpha h\lambda(t) \\ &+ |b(t+h) - a(t)|\int_{t-h}^{t} |b(u+h)|du \overset{\text{def}}{=} \Gamma(t) \le 0; \end{aligned} \tag{C2}$$

$$\alpha - |b(t+h) - a(t)| \overset{\text{def}}{=} \eta_1(t) \ge 0; \tag{C3}$$

$$\bar{\eta}(t) \overset{\text{def}}{=} \min[\eta_1(t),\eta_1(t-h)] \notin L^1[0,\infty); \tag{C4}$$

and

$$0 < \int_{t-h}^{t} |b(s+h)|ds \le K, \quad 0 < \int_{t-h}^{t} |a(s)|ds \le K \tag{C5}$$

for some $K>0$. Then U.S. implies A.S. If, in addition, there is a positive constant Γ_0 such that $\Gamma(t) \le -\Gamma_0$ and if $-2a(t) + |b(t)| + |b(t+h)|$ is bounded above, then $x=0$ is U.S.

EXAMPLE OF EXAMPLE C. Let

$$x'(t) = -a(t)x(t) + b(t)x(t-h)$$

with $a(t) = a_0 + a_1\cos(2\pi t/h)$ and $b(t) = -b_0 + a_1\cos(2\pi t/h)$, where $b_0 \ge a_0 > 0$ and $a_1>0$. Then $b_0 + a_1 < 1/h$ implies A.S.

The point of this contrived example is that we can let $a_1\to\infty$ as $h\to 0$ thereby showing that $a(t)$ and $b(t)$ change their signs while, at the same time, $-a(t) + b(t+h) = -(a_0 + b_0) < 0$.

The foregoing is just one type of problem that can be attacked with the aid of Jensen's inequality. A comprehensive treatment of both finite and infinite delay problems will appear in [1].

3. References

1. Becker, L.C., Burton, T.A., and Zhang, S., Functional differential equations and Jensen's inequality, preprint.

2. Burton, T.A., Stability and Periodic Solutions of Ordinary and Functional Differential Equations, Academic Press, Orlando, Florida, 1985.

3. Hale, J.K., Theory of Functional Differential Equations, Springer-Verlag, New York, 1977.

4. Natanson, I.P., Theory of Functions of a Real Variable, Vol. II, Ungar, New York, 1960.

5. Yoshizawa, T., Stability Theory by Liapunov's Second Method, Math. Soc. Japan, Tokyo, 1966.

METHOD OF UPPER AND LOWER SOLUTIONS FOR NONLINEAR INTEGRAL EQUATIONS AND AN APPLICATION TO AN INFECTIOUS DISEASE MODEL.

A. Cañada
Departamento de Análisis Matemático
Universidad de Granada, 18071,
Granada, Spain.

1. Introduction.

Let us consider the integral equation

$$x(t) = \int_{t-\tau}^{t} f(s,x(s))\,ds, \quad t \in R \tag{1}$$

where τ is a positive constant and $f : R\times R \to R$, $(t,x) \to f(t,x)$, is continuous and ω-periodic ($\omega > 0$) in the variable t. Equation (1) can be interpreted as a model for the spread of some infectious diseases with periodic contact rate (see Cooke and Kaplan [1]). In this case, x(t) means the proportion of infectives in the population at time t; f(t,x(t)) means the proportion of new infectives per unit time ($f(t,0) = 0$, $\forall t \in R$) and τ is the lenght of time an individual remains infectious. Cooke and Kaplan considered functions f which generalize functions of the type $f(t,x) = a(t)x(1-x)$, with $x \in [0,1]$ and a(t) is the effective contact rate at time t. If the constant τ satisfies

$$\tau > 1/a(t), \quad \forall t \in R \tag{2}$$

(where the function a(t) is supposed to be continuous, ω-periodic and positive), Cooke and Kaplan, by using the compression of the cone theorem (Krasnoselskii [2]), proved the existence of positive periodic solutions of (1) (that is, ω-periodic functions $x : R \to R^+$, which are continuous and satisfy (1) for all $t \in R$).

In [3], Leggett and Williams improved the compression of the cone theorem applying their result to (1) and obtaining the existence of positive and ω-periodic solutions, even if the effective contact rate $\check{a}(t) = \dfrac{\delta f(t,0)}{\delta x}$ is small (even zero) during some time (See also Nussbaum [4] where a more general class of integral equations have been studied by means of global

NATO ASI Series, Vol. F37
Dynamics of Infinite Dimensional Systems
Edited by S.-N. Chow, and J. K. Hale

bifurcation theorems).

In this work we use the method of upper and lower solutions (see [5]) to study the existence of ω-periodic and positive solutions of (1), obtaining, in some cases, better results than those of [1], [3]. In particular, we deduce from (2) that if $m = \inf_{t \in R} a(t)$, (2) is equivalent to

$$m > 1/\tau \qquad (3)$$

It is said in [1] that it would be interesting to obtain a theorem where the quantity m could be replaced by the average value of a(t). Assuming that $\tau \geq \omega$, this is possible and easy to get with our approach.

2. Main Results.

Let $C = \{ x : R \to R \mid x$ is continuous and ω-periodic $\}$ with $\|x\| = \sup \{ |x(t)| : t \in [0, \omega] \}$.

A function $x_0 \in C$ is said to be a lower solution of (1) if

$$x_0(t) \leq \int_{t-\tau}^{t} f(s, x_0(s))\, ds, \quad \forall t \in R \qquad (4)$$

Similarly, a function $x^0 \in C$ is said to be an upper solution of (1) if

$$x^0(t) \geq \int_{t-\tau}^{t} f(s, x^0(s))\, ds, \quad \forall t \in R \qquad (5)$$

It is known that the existence of a lower and an upper solution of (1) verifying $x_0(t) \leq x^0(t), \forall t \in R$, is not sufficient to prove the existence of a solution x(t) of (1) satisfying $x_0(t) \leq x(t) \leq x^0(t)$, $\forall t \in R$. So that additional assumptions are necessary to prove the existence of a solution of (1). Usually, one requires some monotone property of the function f in the variable x (see Theorem 1), or if f does not satisfy this monotone property, one needs a more restrictive notion of upper and lower solution (see Theorem 2).

<u>Theorem 1</u>. Let us suppose that there is $\varepsilon_0 > 0$ verifying:

i) $f(t,0) = 0$, $\forall t \in R$ and f is nondecreasing respect to $x \in [0, \varepsilon_0]$ for fixed $t \in R$.

ii) There exists the function $\frac{\delta f(t,x)}{\delta x}$ and it is continuous in $R \times [0, \varepsilon_0]$. Moreover, the function $g : R \times (0, \varepsilon_0] \to R$ defined by

$$g(t, \varepsilon) = \frac{1}{\varepsilon} \int_0^{\varepsilon} \left[\int_{t-\tau}^{t} \frac{\delta f(s,u)}{\delta x} \, ds \right] du$$

satisfies $g(t, \varepsilon) \geqq 1$, $\forall (t, \varepsilon) \in R \times (0, \varepsilon_0]$, ε sufficiently small.

iii) There exists $\beta \in (0, \varepsilon_0]$ such that

$$\int_{t-\tau}^{t} f(s, \beta) \, ds \leqq \beta \qquad \forall\, t \in R .$$

Then, equation (1) has a ω-periodic and positive solution.

Proof. Clearly $x^0(t) \equiv \beta$ is an upper solution of (1). Let $\alpha \in (0, \varepsilon_0]$ be satisfying $\alpha < \beta$ and α sufficiently small. Then, we have

$$\int_{t-\tau}^{t} f(s, \alpha) \, ds = \int_{t-\tau}^{t} (f(s, \alpha) - f(s,0)) \, ds =$$

$$\int_{t-\tau}^{t} \left[\int_0^{\alpha} \frac{\delta f(s,u)}{\delta x} \, du \right] ds = \int_0^{\alpha} \left[\int_{t-\tau}^{t} \frac{\delta f(s,u)}{\delta x} \, ds \right] du \geqq \alpha ,$$

from ii). Therefore $x_0(t) \equiv \alpha$ is a lower solution of (1). Now, the sequences of ω-periodic functions $\{x_n\}$, $\{x^n\}$, defined by

$$x_n(t) = \int_{t-\tau}^{t} f(s, x_{n-1}(s)) \, ds, \ \forall t \in R, \ n \geqq 1$$

$$x^n(t) = \int_{t-\tau}^{t} f(s, x^{n-1}(s)) \, ds, \ \forall t \in R, \ n \geqq 1$$

satisfy

$$\alpha \leqq x_n(t) \leqq x_{n+1}(t) \leqq \beta \quad , \ \forall\, t \in R, \quad n \geqq 1$$

$$\beta \geqq x^n(t) \geqq x^{n+1}(t) \geqq \alpha \quad , \ \forall t \in R, \quad n \geqq 1.$$

Moreover $x_n(t) \leqq x^n(t)$, $\forall\, t \in R$, $n \geqq 1$ and the sequences $\{x_n\}$, $\{x^n\}$ are uniformly bounded and equicontinuous in $[0, \omega]$. So that $\{x_n\}$ converges uniformly from below to a solution x of (1), $\alpha \leqq x(t) \leqq \beta$, $\forall t \in R$; while $\{x^n\}$ converges uniformly from above to a solution y of (1), $\alpha \leqq y(t) \leqq \beta$, $\forall\, t \in R$ ($x(t) \leqq y(t)$, $\forall t \in R$).

Remarks:

1.- A sufficient condition for ii) is

ii') $\displaystyle\int_{t-\tau}^{t} \frac{\delta f(s,0)}{\delta x}\, ds > 1 \ , \ \forall t \in R.$

Trivially, if $m = \inf_{t \in R} \frac{\delta f(t,0)}{\delta x} > 0$ and $m > 1/\tau$, ii') is satisfied (such as it is supposed in [1]); but we allow $\frac{\delta f(t,0)}{\delta x}$ to be zero on some subintervals of R.

2.- In this remark we are going to see that, in some cases, it is possible to replace m by the average value of $\frac{\delta f(t,0)}{\delta x}$. In fact, if $\frac{\delta f(t,0)}{\delta x} \geqq 0$, $\forall t \in R$ and $\tau \geqq \omega$, we have

$$\int_{t-\tau}^{t} \frac{\delta f(s,0)}{\delta x}\, ds = \int_{t+\omega-\tau}^{t+\omega} \frac{\delta f(s,0)}{\delta x}\, ds \geqq \int_{t}^{t+\omega} \frac{\delta f(s,0)}{\delta x}\, ds =$$

$\displaystyle\int_{0}^{\omega} \frac{\delta f(s,0)}{\delta x}\, ds$. So that, if $\tau \geqq \omega$, $\displaystyle\int_{0}^{\omega} \frac{\delta f(s,0)}{\delta x}\, ds > 1$ is sufficient for ii').

If $\tau < \omega$, the problem is more difficult and we must use the general theory about the minimum of a function just to get that the function $g : R \rightarrow R$, defined by

$$g(t) = \int_{t-\tau}^{t} \frac{\delta f(s,0)}{\delta x}\, ds$$

has a minimum value greater than one (It will involve, in general, the second derivatives of the function f).

3.- Cooke and Kaplan [1] do not suppose that f is nondecreasing respect to $x \in [0,\varepsilon_0]$ for fixed $t \in R$, but in a subset of the form $R\mathrm{x}[0,\varepsilon_1]$ where ε_1 is, in general, less than ε_0. This is our unique disadvantage, and we are not able to obtain better results from the method of upper and lower solutions (Think in the case $\tau = \omega = 1$, $f = f(x)$. Then, the solutions of (1) are the solutions of the scalar equation

$$c = f(c) \qquad (6)$$

where c is a constant, and we know that if f is not nondecreasing, we may not obtain the solutions of (6) from a lower and an upper solution).

4.- A sufficient condition for iii) is $f(t,x) \leq \beta/\tau$, $\forall (t,x) \in R\mathrm{x}(0,\beta]$. This condition was assumed by Leggett and Williams [3].

5.- It must be remarked that the method used here provides an iteration scheme where the difference between the upper and lower iterates (x_n, x^n) is a good error estimate respect to the solutions x,y, of (1).

If we delete the hypothesis that f is nondecreasing respect to $x \in [0, \varepsilon_0]$ for fixed $t \in R$, we need a more restrictive notion of upper and lower solution. Taking into account that the operator $T : C \to C$, defined by

$$Tx(t) = \int_{t-\tau}^{t} f(s,x(s))\,ds$$

is completely continuous, we have the following theorem :

<u>Theorem 2</u>. Assume that there exist two functions $x_0, x^0 \in C$, with $x_0(t) \leqq x^0(t)$, $\forall t \in R$, such that

$$x_0(t) \leqq \int_{t-\tau}^{t} f(s,x(s))\,ds \leqq x^0(t), \quad \forall t \in R \qquad (7)$$

whenever $x \in C$ satisfying $x_0(t) \leqq x(t) \leqq x^0(t)$, $\forall t \in R$. Then, (1) has a solution x such that, $x_0(t) \leqq x(t) \leqq x^0(t)$, $\forall t \in R$.

<u>Remark</u>. If there are two constants $\beta > \alpha > 0$ such that

$$\alpha/\tau \leqq \inf_{\substack{t \in R \\ x \in [\alpha,\beta]}} f(t,x) \leqq \sup_{\substack{t \in R \\ x \in [\alpha,\beta]}} f(t,x) \leqq \beta/\tau ,$$

equation (1) has a positive and ω-periodic solution.

3. An Example.

Let $f(t,x) = a(t)g(x)$ be a function, defined in RxR, where $a : R \to R$ is continuous, 1-periodic ($\omega = 1$), nonnegative and satisfying : $\int_0^1 a(s)\,ds = 2$, $\sup_{t \in R} a(t) = \sup_{t \in [0,1]} a(t) = M > 5$. Also, $g : R \to R$ is a C^1 nondecreasing function such that $g(0) = 0$, $g'(0) > 1/2$.

Then, if $2 \geqq \tau \geqq 1$ and if there exists $\beta > 0$ satisfying $g(\beta) \leqq \beta/4$, equation (1) has a positive and 1-periodic solution.

In fact, i) of Theorem 1 is trivially verified from the hypotheses on the functions a and g. On the other hand, $\int_0^\omega \frac{\delta f(s,0)}{\delta x}\,ds = \int_0^\omega a(s)g'(0)\,ds > 1$, and therefore ii) of Theorem 1 is also satisfied.

Lastly,
$$\int_{t-\tau}^{t} f(s,\beta)\,ds = \int_{t-\tau}^{t} a(s)\,g(\beta)\,ds = g(\beta)\int_{t-\tau}^{t} a(s)\,ds$$
$$\leq g(\beta)\int_{t-2}^{t} a(s)\,ds = 4g(\beta) \leq \beta$$
, that is hypothesis iii) of Theorem 1.

Remark. This example can not be studied from the results by Cooke and Kaplan [1], because the function a may be zero on some subintervals of R. Also, the results of Legget and Williams [3], do not apply, because the condition H3 in [3] implies $a(t)g(x) \leq R'/\tau$, $\forall (t,x) \in R\times[0,R']$, i.e., $Mg(R') \leq R'$, for some positive R', and this is not necessarily satisfied by our function g.

ACknowledgment. This work was done while the author visited the Department of Mathematics of the University of Texas at Arlington (Texas, USA), with a grant of the "Comité Hispano-Norteamericano para la cooperación científica y tecnológica". The author thanks R. Kannan for his suggestion to apply the method of upper and lower solutions to this kind of problems.

REFERENCES

1.- Cooke, K.L. and Kaplan, J. A periodicity threshold theorem for epidemics and population growth. Math. Biosciences, 31, (1.976), 105-120.

2.- Krasnosel'skii, M.A. Positive solutions of operator equations. Nordhoff, (1.964).

3.- Leggett, R.W. and Williams, L.R. A fixed point theorem with application to an infectious disease model. J. Math. Anal. Appl. 76, (1.980), 91-97.

4.- Nussbaum, R. A periodicity threshold theorem for some nonlinear integral equations. Siam J. Math. Anal. 9, (1.978), 356-376.

5.- Pachpatte, B.G. Method of upper and lower solutions for nonlinear integral equations of Urysohn type. J. Integral Equations, (1.984), 119-125.

Competition of Azimuthal Modes and Quasi-Periodic Flows in the Couette-Taylor Problem

P. CHOSSAT, Y. DEMAY, G. IOOSS
U.A. n°168 du C.N.R.S.
Département de Mathématiques
Parc Valrose - F.06034 NICE CEDEX

1. Introduction

The classical Couette Taylor experiment consists in looking at the flow of a viscous fluid between two coaxial cylinders which rotate at constant (but independant) speeds. The speed of rotation of the inner cylinder is responsible for the destabilization mechanism of the basic laminar flow (called the Couette flow). The great variety of possible regimes before turbulence [1] and the relative simplicity of the model (related to its symmetries), make this problem a very interesting one for the study of transition to turbulence, and also for testing techniques of computation of flows and their stability. After the classical work of G.I. Taylor in the twenties [17] and the introduction of bifurcation techniques by Kirchgäßner and Sorger in the sixties [12], progress has been done recently by introducing bifurcation with symmetry and Centre manifold techniques. Using group theoretic methods, Golubitsky and Stewart have classified the possible time-periodic flows which occur from the interaction of an axisymmetric disturbance of the Couette flow with a non-axisymmetric one [9].

In the present paper we want to show how Centre Manifold technique associated to group theory allow to analytically reach complicated regimes (quasi-periodic on 2-tori or 3-tori) and to compute their stability diagrams. The situation which has been chosen is an interaction between two non-axisymmetric disturbances of the Couette flow. The numerical study of linear stability of Couette flow shows the existence of such interactions [4]. This corresponds to critical values of the rotation speeds of the cylinders, at which the linear part of the equations for a perturbation of the Couette flow possess two pairs of purely imaginary eigenvalues : $\pm i\eta_1$, $\pm i\eta_2$, $\eta_1 \neq \eta_2$. An important remark here is that eigenvalues have to be double, due to the reflectionnal symmetry through a plane perpendicular to the axis of cylinders (as will be explained in the next section). The Centre Manifold in this case is therefore 8-dimensional. Here the use of group invariance of the model is crucial for understanding what can happen. A complete analysis of this problem was done in [4]. We shall not repeat all these calculations, but rather emphasize the techniques. Finally we show bifurcation and stability diagrams which were obtained by computing numerically the relevant coefficients in a typical case, with a recapitulative table of all solutions occuring in this situation.

The next section contains recalls on the basic results (linear stability of Couette flow,...) and on the methods, while in section 3 we will concentrate on solutions of a particular type (spirals and interpenetrating spirals). The interest of these spiraling flows will appear in the numerical example of section 4.

NATO ASI Series, Vol. F37
Dynamics of Infinite Dimensional Systems
Edited by S.-N. Chow, and J. K. Hale

2. Classical results and recalls on Centre Manifolds reduction

2.1 – The mathematical model

The fluid flow is governed by Navier-Stokes equations. Angular velocities of the inner, resp. outer cylinders are Ω_1 and Ω_2, and their radii are R_1, resp. R_2. We assume cylinders of infinite length, but with given periodicity of lenght h. (We shall come back in 2.3 on this hypothesis). Scales are chosen as follows : unit of length= R_1, unit of time = Ω_1^{-1}, unit of pressure = $\rho R_1^2 \Omega_1^2$ (ρ = density of fluid). Then, in cylindrical coordinates (r,θ,z), the classical Couette solution is

$$(1) \qquad V^o = (0,v^o,0) \;,\; v^o = \left(\frac{a^2\omega-1}{a^2-1}\right) r \left[1 + \left(\frac{a^2(1-\omega)}{a^2\omega-1}\right)\frac{1}{r^2}\right],$$

where $a = R_2/R_1$ and $\omega = \Omega_2/\Omega_1$. The pressure $p^o(r)$ has a simple expression too.

Let $(\mathcal{U},q)$ be a perturbation of Couette flow, then it satisfies the following system :

$$(2) \qquad \begin{cases} \dfrac{\partial \mathcal{U}}{\partial t} + (\mathcal{U}.\nabla)V^o + (V^o.\nabla)\mathcal{U} + (\mathcal{U}.\nabla)\mathcal{U} + \nabla q = \dfrac{1}{\mathcal{R}}\Delta\mathcal{U} \\ \nabla.\mathcal{U} = 0 \\ \mathcal{U}_{|r=1,a} = 0, \end{cases}$$

where $\mathcal{R} = \dfrac{\Omega_1 R_1}{\nu}$ is the Reynolds number (ν is the viscosity).

2.2 – Symmetries

The solution (1), and therefore the equations (2), are invariant under translations along the axis Oz of the cylinders, reflections through the plane $z = 0$, and rotations about Oz. More precisely : we define the following representations of these three isometries in the affine space $\mathcal{R}^3$.

$$(3) \qquad \begin{cases} [\tau_s \mathcal{U}](r,\theta,z) = \mathcal{U}(r,\theta,z+s) \\ [S\,\mathcal{U}](r,\theta,z) = (u_r(r,\theta,-z), u_\theta(r,\theta,-z), -u_z(r,\theta,-z) \\ [\mathcal{R}_\varphi \mathcal{U}](r,\theta,z) = \mathcal{U}(r,\theta+\varphi,z) \end{cases}$$

and corresponding formulas for the scalar field q. Because we assume periodicity along Oz, τ_s is equivalent to a representation of the rotation group SO(2), and hence τ_s and S span a representation of O(2) (group of symmetries of the circle). Also, $\mathcal{R}_\varphi$ is a representation of

SO(2). Therefore eqs (2) commute with the resulting representation of O(2) × SO(2).

2.3 – Classical linear analysis and mode interactions

We look at solutions of eqs (2) without nonlinear term, of the following form (Fourier expansion in θ and z justified from z-periodicity assumption) :

$$(4) \qquad \begin{cases} \mathcal{U} = \hat{\mathcal{U}}(r)\, e^{i(\alpha z+m\theta)+\sigma t} \\ q = \hat{q}(r)\, e^{i(\alpha z+m\theta)+\sigma t} \end{cases}$$

where α is an integer multiple of $\frac{2\pi}{h}$: $\alpha = n_0 \frac{2\pi}{h}$, $n_0 \in \mathbb{N}$.

We then obtain an eigenvalue problem for σ, which reduces to a system of O.D.E.'s with respect to the variable r. When a and ω are fixed, we know that Re $\sigma < 0$ for $\mathcal{R}$ small enough, which corresponds to stable Couette flow. The critical value of $\mathcal{R}$ is computed by solving first the equation Re $\sigma = 0$, which gives

$$\mathcal{R} = \mathcal{R}_0(\alpha,m,a,\omega).$$

Then, the critical value is

$$(5) \qquad \mathcal{R}_c(a,\omega,h) = \inf \mathcal{R}_0(n_0 \tfrac{2\pi}{h},m,a,\omega).$$

Classicaly $\mathcal{R}_c$ is obtained with α not restricted to multiplies of $\frac{2\pi}{h}$ [7]. When ω is negative enough, $\mathcal{R}_c$ corresponds to purely imaginary critical eigenvalues $\pm\sigma_0$, with some optimal wave number $m_c \neq 0$ (azimuthal wave number) and α_c (axial wave number). When ω becomes more negative, m_c increases and α_c varies, either with small discontinuities when jumping from m_c to m_c+1 (classical case [7]), or in the set $\{n_0\, 2\pi/h,\ n_0 \in \mathbb{N}^*\}$ if one imposes the period length h along Oz. In the latter case, values of a, ω and h can be found such that the same $\alpha_c = n_0\, 2\pi/h$ corresponds to identical $\mathcal{R}_c$ with different azimuthal wave numbers :

$$\mathcal{R}_c(a,\omega,h) = \mathcal{R}_0(n_0\, 2\omega/h,\, m_1,a,\omega) = \mathcal{R}_0(n_0\, 2\pi/h,\, m_2,a,\omega).$$

It is numerically relevant to take $m_2 = m_1+1$, although the analysis was done in a more general case in [4]. This situation corresponds to the interaction of the two azimuthal modes m_1 and m_2 with the same axial wave number α_c. We could also consider interactions with different α_c's, but we shall restrict our attention here to the case which would appear to be the most common.

The following table I shows an example of critical values

$a=\frac{R_2}{R_1}$	$\omega_c=\frac{\Omega_2}{\Omega_1}$	$\mathcal{R}_c$	α_c	m_1	m_2	η_1	η_2
1.33	-1.156	600.2	13	2	3	-.7483	-.9048

Table I - The numerical method is described in [6].

Remark : The assumption of periodicity of length h along axis Oz might appear very artificial. It is however experimentally relevant outside a small neighborhood of the ends of cylinders. One should take enough length h, in order to avoid a too large number of interactions between excited axial modes.

2.4 - The Centre Manifold reduction for eqs. (2)

We now assume the existence, for $\mathcal{R} = \mathcal{R}_c$ and $\omega = \omega_c$ (the other parameters being fixed) of two pairs of eigenvalues σ of the form $\pm i\eta_1$, $\pm i\eta_2$, with eigenvectors

$$(6) \qquad \begin{cases} \text{for } i\eta_1 : \zeta_0 = \hat{U}(r)\, e^{i(m_1\theta+\alpha z)} \,, \zeta_1 = S\hat{U}(r)\, e^{i(m_1\theta-\alpha z)} \\ \text{for } i\eta_2 : \zeta_2 = \hat{V}(r)\, e^{i(m_2\theta+\alpha z)} \,, \zeta_3 = S\hat{V}(r)\, e^{i(m_2\theta-\alpha z)} \end{cases}$$

(we note $\alpha_c = \alpha$ from now on).

We set

$$(7) \qquad \mu = 1/\mathcal{R}_c - 1/\mathcal{R} \ , \ \ \nu = \omega - \omega_c \,.$$

In an appropriate functional frame [11], eqs (2) take the form

$$(8) \qquad \frac{dU}{dt} = L_0 U + \mu L_1 U + \nu L_2 U + M(U,U)$$

where U belongs to some "good" Hilbert space containing the boundary conditions and the divergence-free condition (in order to eliminate the pressure q by projection, see [11]). In (8), the operators are defined as follows :

$$(9) \qquad \begin{cases} L_0 U = R_c^{-1} \Delta U -(U.\nabla)V^o{}_c-(V^o{}_c \nabla)U - \nabla q \\ L_1 U = -\Delta U - \nabla q \\ L_2 U = -(U.\nabla)V^1-(V^1.\nabla)U-\nabla q \\ M(U,V) = -{}^1/{}_2[(U.\nabla)V + (V.\nabla)U]-\nabla q \end{cases}$$

where the ∇q's are here for the left part to be divergence free and with a zero normal component at the boundary. $V^o{}_c$ is the Couette solution at $\omega = \omega_c$ and

$$V^1 = (0, v^1, 0) \ , \ v^1 = \frac{a^2}{a^2-1}\left(r - \frac{1}{r}\right).$$

At $\mu = \nu = 0$, L_0 has two pairs of double eigenvalues $\pm i\eta_1$, $\pm i\eta_2$, the other eigenvalues having a negative real part. Then we can apply the Center Manifold Theorem to (8) [13]. For (μ,ν) close to $(0,0)$, there exists a Center Manifold, localy invariant and attracting in a neighborhood of $U = 0$, the Taylor expansion of which is obtained in the following way : we set

$$\text{(10)} \quad \begin{cases} U = X + Y \\ X = \sum_{j=0}^{3} (x_j \zeta_j + \bar{x}_j \bar{\zeta}_j) \\ Y = \Phi(\mu,\nu,X) = \sum_{p,q,\underline{r}} \mu^p \nu^q \Phi_{pq\underline{r}} x_0^{r_0} \bar{x}_0^{s_0} x_1^{r_1} \bar{x}_1^{s_1} x_2^{r_2} \bar{x}_2^{s_2} x_3^{r_3} \bar{x}_3^{s_3} \end{cases}$$

where $\underline{r} = (r_0, s_0, \ldots, r_3, s_3)$ and
$\Phi_{pq0} = 0$, $\Phi_{00\underline{r}} = 0$ if $|\underline{r}| = 1$.

Then the dynamical behavior is asymptoticaly identical, in a neighborhood of 0, to the one of the equation

$$\text{(11)} \qquad \frac{dX}{dt} = F(\mu,\nu,X)$$

obtained by projection of the vector field on the Centre Manifold $Y = \Phi(\mu,\nu,X)$. The Taylor expansion of (11) can be computed near 0 by an identification rule (see [6]).

In addition, the Center Manifold can be constructed to be invariant under the group actions defined in §2.2 [15], so that $F(\mu,\nu, .)$ commutes with these group representations restricted to the space $V = \text{span}\{\zeta_j\}$.

3. Bifurcation with symmetry and applications to eq.(2-11)

3.1 – Symmetry breaking bifurcations

Let Γ be the group of transformations acting in the space V and generated by (2.3) (rep. of $O(2) \times SO(2)$). Let Σ be a subgroup of Γ and $V^\Sigma = \{X \in V \,/\, \gamma X = X, \ \forall \gamma \in \Sigma\}$ (fixed point subspace associated to the isotropy subgroup Σ). Because $F(\mu,\nu,.)$ commutes with $\gamma \in \Gamma$, it is easy to check that V^Σ is stable by eq. (2.11). We can therefore restrict (2.11) to the lower dimensional spaces V^Σ, for every isotropy subgroup Σ, and then find solutions with these isotropies. A group-theoretic approach of this property was first developped by D.Sattinger [16] and L.Michel [14].

Also, remark that looking for the normal form of a vector field whose singularity is a couple of semi-simple imaginary eigenvalues, is the same as looking for polynomial vector fields which commute with an SO(2) action. In fact, using the Lyapunov-Schmidt method, we can add to the initial symmetry group of the problem the group SO(2) [8]. This idea was used for the Couette-Taylor problem in [9].

In our case, the SO(2) symmetry is already present, due to the geometry of the problem. If the eigenvalues $i\eta_1$ and $i\eta_2$ are not rationnaly dependent : $\eta_1/\eta_2 \notin \mathbb{Q}$, the normal form is a polynomial vector field which commutes with a torus action $SO(2) \times SO(2)$ (see [5]). The case with initial O(2) symmetry was studied in [5] (our present situation is different from the one studied in [5], because we assume identical axial modes and different azimuthal modes, while in the frame of [5] it would be exactly the inverse !).

The foregoing remarks about bifurcation of solutions with some isotropy subgroup symmetry, lead to the idea of "spontaneous symmetry breaking". Applying these ideas to our 8-dimensional problem leads to very complex diagrams of bifurcation and stability. In section 4 we give a list of solutions which have been found in [4]. Stability conditions were also derived and the method is explained in §3.4, but are complicated to express (we refer the interested reader to [4]).

3.2 - Spirals and Interpenetrating Spirals

Among all solutions listed in table II, spirals and IPS are of special interest because flows with these characteristics are often observed in experiments.

Lemma 1. The space $x_1 = x_2 = 0$ is a 4-dim fixed point subspace V^Σ of V, with Σ the subgroup of Γ spanned by the transformation $(\theta + \frac{2\pi}{m_1+m_2}, z + \frac{2\pi m_2}{\alpha(m_1+m_2)})$. The subspaces of V^Σ

$$V^{\Sigma_0} = \{x_0 = 0\} \cap V^\Sigma \text{ and } V^{\Sigma_3} = \{x_3 = 0\} \cap V^\Sigma ,$$

are 2-dim fixed point subspaces, with Σ_0 and Σ_3 the subgroups resp. spanned by transformations

$(\theta,z) \longrightarrow (\theta+\varphi, z-\frac{m_1}{\alpha}\varphi)$ and $(\theta,z) \longrightarrow (\theta+\varphi, z+\frac{m_2}{\alpha}\varphi)$, $\varphi \in S^1$.

To check lemma 1 is an exercise for the reader, by using the relations (2.6) in order to write down the fixed-point (i.e. invariance) conditions for the action of Σ, Σ_0 and Σ_3.

We can now reduce eqs (2.11) to V^Σ, and even to V^{Σ_0} or V^{Σ_3}.

Lemma 2. We set $x_j = r_j e^{i\psi_j}$ $(j = 0,3)$ and, for X in V^Σ,
$F(\mu,\nu,X) = f_0(\mu,\nu,X)\zeta_0 + f_3(\mu,\nu,X)\zeta_3 + \text{c.c.}$ Then :
$f_j(\mu,\nu,X) = r_j e^{i\psi_j} \tilde{f}_j(\mu,\nu,r_0,r_3)$,
where $\tilde{f}_j$ is an even function of r_0 and r_3.

Proof : We set $\Phi_0 = m_1\varphi + \alpha s$, $\Phi_3 = m_2\varphi - \alpha s$, $(\varphi,s) \in S^1 \times \mathbb{R}/h\mathbb{Z}$.
Since F commutes with the Γ-action in V, we easily get from this the following relations, where μ and ν arguments are not written and $X = (x_0,x_3)$ is set :

(1) $$f_j(e^{i\Phi_0} x_0 \,,\, e^{i\Phi_3} x_3) = e^{i\Phi_j} f_j(x_0,x_3) \,,\ j = 0,3 .$$

Replacing successively Φ_j by $-\psi_j$ ($j = 0,3$), Φ_0 by π and Φ_3 by π in relations (1), we easily get the result. ∎

We can now write the equations restricted to V^{Σ}:

(2) $$\begin{cases} \dfrac{dx_0}{dt} = x_0\, \tilde{f}_0(\mu,\nu,r_0,r_3) \\ \dfrac{dx_3}{dt} = x_3\, \tilde{f}_3(\mu,\nu,r_0,r_3) \end{cases} \quad \text{(and complex conjugate).}$$

Setting $x_3 = 0$ or $x_0 = 0$ in (2) leads to equations restricted to spaces V^{Σ_0} or V^{Σ_3} respectively. There are two "obvious" types of solutions bifurcating from $X = 0$ in V^{Σ_0} and V^{Σ_3}:

1) <u>m_1-spirals</u>

The resolution of the equation

$$\frac{dx_0}{dt} = x_0\, \tilde{f}_0(\mu,\nu,r_0,0)$$

is very standard, writing for ex. the equations for amplitude and phase. We obtain the bifurcated solutions $r_0 = r_0(\mu,\nu)$ (independent of t) and $\psi_0 = \Omega_0(\mu,\nu)t + \varphi_0$ (φ_0 arbitrary phase), where $\Omega_0(0,0) = \eta_0$. These solutions are therefore rotating waves. We can precise their structure by using (2.6):

(3) $$\begin{cases} X = r_0[\hat{U}(r)\, e^{i(\alpha z + m_1\theta + \Omega_0 t + \varphi_0)} + \text{c.c.}] \\ U = X + \Phi(\mu,\nu,X) = U(r, \alpha z + m_1\theta + \Omega_0 t + \varphi_0) \end{cases}$$ thanks to the property of propagation of symmetries).

It readily follows from (3) that the corresponding fluid flow has a spiral structure, with m_1 spires in intervals of lenght $2\pi/\alpha$.

2) <u>m_2-spirals</u>

Exactly as m_1-spirals, but solving equation

$$\frac{dx_3}{dt} = x_3\, \tilde{f}_3(\mu,\nu,0,r_3).$$

We get a spiral flow

(4) $$\begin{cases} X = r_3[SV(r)\, e^{i(-\alpha z + m_2\theta + \Omega_3 + \varphi_3)} + \text{c.c.}] \\ U = U(r, -\alpha z + m_2\theta + \Omega_3 t + \varphi_3) \end{cases}$$

where $\Omega_3 \sim \eta_2$ and φ_3 is the (arbitrary) phase at $t=0$.

If we now set $r_0 r_3 \neq 0$, then we get a branch of solutions which (under following conditions) bifurcates from the primary spiral flows :

3) Interpenetrating spirals (I.P.S.)

We remark that equations for amplitudes r_0 and r_3 do not depend on phases ψ_0, ψ_3 in (2). We therefore get solutions by solving the system

$$(5) \qquad \begin{cases} \operatorname{Re} \tilde{f}_0(\mu,\nu,r_0,r_3) = 0 \\ \operatorname{Re} \tilde{f}_3(\mu,\nu,r_0,r_3) = 0 . \end{cases}$$

Under a (generic) non-degeneracy condition for (5) (see [4]), we get a branch of solutions $r_0(\mu,\nu)$, $r_3(\mu,\nu)$ of (5), with time dependent phases

$$(6) \qquad \begin{cases} \psi_0 = \Omega_0(\mu,\nu)t + \varphi_0 \\ \psi_3 = \Sigma_3(\mu,\nu)t + \varphi_3 \end{cases}$$

where $\Omega_0(0,0) = \eta_1$ and $\Omega_3(0,0) = \eta_2$. These solutions are quasiperiodic with two independent frequencies $\Omega_0(\mu,\nu)$ and $\Omega_3(\mu,\nu)$, and bifurcate from m_1-spirals and m_2 spirals. The structure of the flow on the Centre manifold is given by :

$$(7) \quad X = r_0[\hat{U}(r)\, e^{i(\alpha z+m_1\theta+\Omega_0 t+\varphi_0)} + c.c.] + r_3[S\hat{V}(r)\, e^{i(-\alpha z+m_2\theta+\Omega_3 t+\varphi_3)} + c.c.]$$

It is therefore a (pure) superposition of spiral modes, which propagate in opposite directions if $\eta_1 \,.\, \eta_2 > 0$. This kind of structure (superposition of rotating waves) was described in [2] and [3]. Moreover it propagates to the fluid flow itself :

Lemma 3. The I.P.S. have the form :

$$U(r,\theta,z,t) = \tilde{U}(r,\alpha z + m_1\theta + \Omega_0 t + \varphi_0) - \alpha z + m_2\theta + \Omega_3 t + \varphi_3),$$

where $\varphi_0, \varphi_3 \in \mathbb{R}/2\pi\mathbb{Z}$ and $\tilde{U}$ is 2π-periodic in φ_0 and φ_3.

This lemma is proved in [4] (Appendix).

3.3 – Stability of Spirals and Interpenetrating Spirals

1) *Spirals.* Spirals are rotating waves, i.e. can be written (in space V) : $X_0(t) = R_{\omega t}\tilde{X}_0$ where X_0 is independent of time and (R_φ) is the rotation group representation in V. E.g. for m_1-spirals, we take $\omega = \frac{\Omega_0}{m_1}$ and $X_0 = x_0\,\zeta_0 + \overline{x}\,\zeta_0$. Without loss of generality, we can look at perturbation of $X_0(t)$ of the form $R_{\omega t}\,Y(t)$. Then the stability of $X_0(t)$ is given by the asymptotic behaviour of $Y(t)$, the equation for which is :

$$(8) \qquad \frac{dY}{dt} + \omega J Y = F(\mu,\nu,\tilde{X}_0+Y) - F(\mu,\nu,\tilde{X}_0),$$

where $J = dR_\varphi/d\varphi|_{\varphi=0}$. The asymptotic behavior of Y is then given by the eigenvalues of $D_X F(\mu,\nu,\tilde{X}_0)$ (autonomous operator). We can interpret this rule as changing the frame of reference, from the inertial to a rotating one, so that the rotating wave actually look stationnary. Details of calculation are in [4].

2) *I.P.S.* Here again we can reduce the problem to an eigenvalue problem for an autonomous operator. For this we first define in V a representation $S^{(0)}{}_{\Phi_3} . S^{(3)}{}_{\Phi_3}$ of $SO(2) \times SO(2)$, which is <u>equivalent</u> to $R_\varphi.\tau_s$, by setting, as in proof of lemma 2,

$$\Phi_0 = m_1\varphi + \alpha s \quad , \quad \Phi_3 = m_2\varphi - \alpha s.$$

Then the action of $S^{(0)}{}_{\Phi_0} . S^{(3)}{}_{\Phi_3}$ is diagonal in V^Σ, and the I.P.S. can be written :

$$X_0(t) = S^{(0)}{}_{\Phi_0 t} . S^{(3)}{}_{\Phi_3 t}\, \tilde{N}_0$$

where $\tilde{X}$ is independent of t. We now look for perturbations of $X_0(t)$ of the form $S^{(0)}{}_{\Omega_0 t} . S^{(0)}{}_{\Omega_3 t}\, Y(t)$, and because of the invariance properties of eq.(2.11), the equation for Y is :

(9)
$$\frac{dY}{dt} + (\Omega_0 J_0 + \Omega_3 J_3)Y = F(\mu,\nu,\tilde{X}_0 + Y) - F(\mu,\nu,\tilde{X}_0) ,$$

where $J_k = dS_\Phi{}^{(k)}|_{\Phi=0}$. Hence the asymptotic behavior of I.P.S. is given by eigenvalues of the (autonomous) operator $D_X F(\mu,\nu,\tilde{X}_0) - \Omega_0 J_0 - \Omega_3 J_3$. Remark that I.P.S. appear, at fixed (μ,ν), as a (φ_0,φ_3)-family of solutions (a torus of solutions). The consequence is that 0 is always an eigenvalue of the foregoing operator, with eigenvectors $J_0 \tilde{X}_0$ and $J_3 \tilde{X}_3$. The stability of I.P.S. is therefore of <u>orbital</u> type [11]. Details of calculation are in [4].

4. A numerical example

Stability conditions for solutions of (2.11) are expressed as relations between coefficients in the Taylor expansion of this equation, such that eigenvalues or Floquet exponents of some linearized operators have a negative real part (see §3.4). The computation of these coefficients is not a simple tool, and increases very fast with the order in series expansion. A numerical code has been written for such computations in the Couette-Taylor problem (see [6]) and applied to the present situation of mode interactions. Here we show diagrams corresponding to the critical parameters listed in Table I. Figure 1 is the bifurcation picture in the (μ,ν)-plane. All symbols are explained in table II. Figure 2 is the bifurcation and stability diagram for someone who moves parameters around the origin in the (μ,ν)-plane. The main interest of this picture is to show the sequence of stable flows : m_2-Spirals —> I.P.S. —> QP_3^*. QP_3^* are solutions on a 3-torus, with possibly two locked frequencies, and are stable if supercritical (this has not been computed).

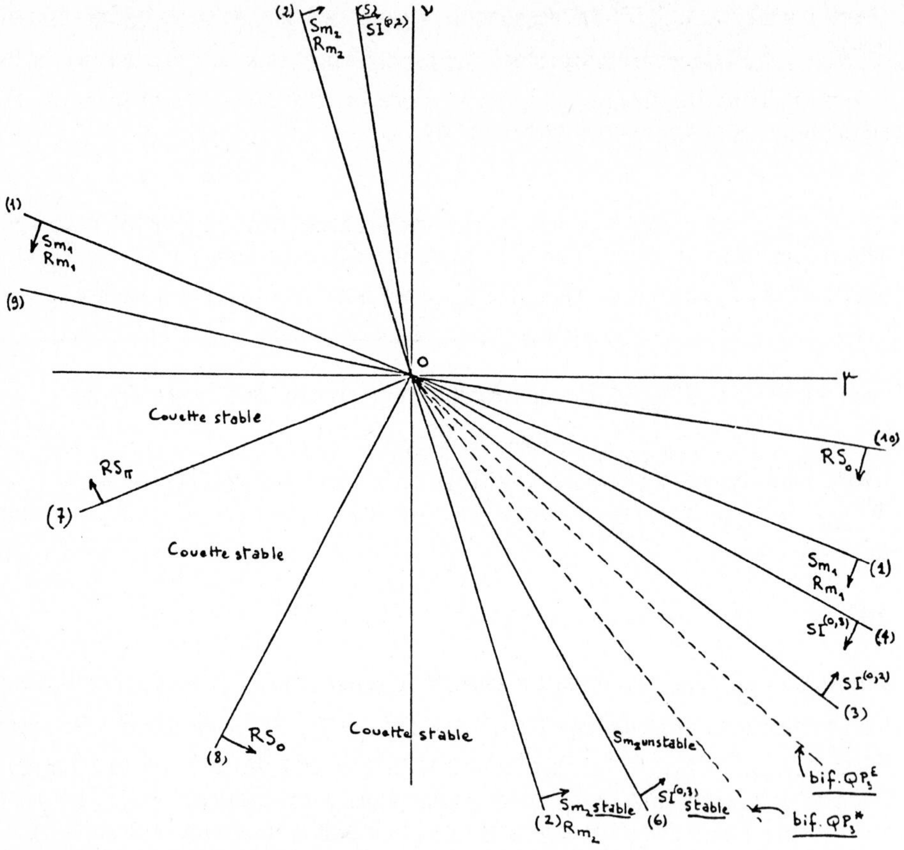

Figure 1. Bifurcation picture in (μ,ν)-plane
* Arrows indicate direction of bifurcation. * Symbols are explained in table II.
* Diagram of bifurcation shown in figure 2.

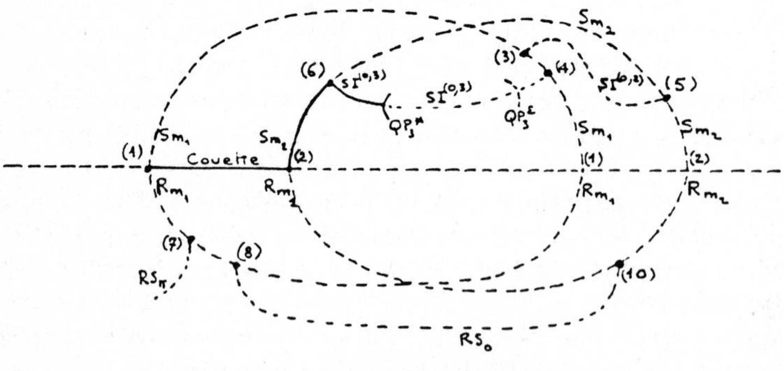

Figure 2. Bifurcation diagram
* Obtained by turning around O in parameters plane. * Stable branches in full lines.

Symbol	Name	Time dependance	Structure
S_{m_j}	m_j-Spirals (j=1,2)	periodic	rotating wave in θ and z (m_j-waves)
R_{m_j}	m_j-Ribbons (j=1,2)	periodic	rotating wave in θ (m_j-waves) and flat cells (invariant under S)
$SI^{(0\ 3)}$	I.P.S.	2 frequencies	superposed Sm_1 and Sm_2 spirals, opposite spires
$SI^{(0\ 2)}$	I.P.S.	idem	interacting Sm_1 and Sm_2 spirals (not a pure superposition as $SI^{(0\ 3)}$)
RS_0	Superposed ribbons	idem	Interacting R_{m_1} and R_{m_2}, flat cells
RS_{π}	Superposed ribbons	idem	Interacting R_{m_1} and R_{m_2}, no flat cells
$QP_3{}^*$		3 frequencies(or 2 if phase locking)	Frequencies $\Omega_0 \sim \eta_1$, $\Omega_3 \sim \eta_2$ and $\tilde{\eta}_3 \sim \frac{2m_2}{m_1+m_2}\eta_1 - \frac{2m_1}{m_1+m_2}$
$QP_3{}^{\varepsilon}$		3 frequencies (or 2 if phase locking)	Frequencies $\Omega_0 \sim \eta_1$, $\Omega_3 \sim \eta_2$ and $\eta = O(\lvert x_0 x_3 \rvert)$

Table II – Solutions appearing in figure 3 and 4, see [4] for detailed description.

References

[1] **C.D. Andereck, S.S. Liu, H.L. Swinney,** *Flow regimes in a circular Couette system with independently rotating cylinders,* J. F. M. (1986) 164, 155-183
[2] **P. Chossat,** *Bifurcation d'ondes rotatives superposées,* C.R. Acad. Sci. Paris (1985) 300, I, 7, 209-213
[3] **P. Chossat,** *Interaction d'ondes rotatives dans le problème de Couette-Taylor,* C.R. Acad. Sci. Paris (1985) 300, I, 8, 251-255

[4] **P. Chossat, Y. Demay, G. Iooss,** *Interaction de modes azimutaux dans le problème de Couette-Taylor,* to appear (1986)

[5] **P. Chossat, M. Golubitsky, B. Keyfitz,** *Hopf-Hopf interaction with O(2)-symmetry,* to appear (1986)

[6] **Y. Demay, G. Iooss,** *Calcul des solutions bifurquées pour le problème de Couette-Taylor avec les deux cylindres en rotation,* J. Méca. Théorique et Appliquée, vol. spécial "Bifurcations et comportements chaotiques" (1985)

[7] **R.C. Di-Prima, R.C. Grannick,** *A non-linear investigation of the stability of flow between counter-rotating cylinders,* IUTAM Symp. on Instability and Continuous Systems (1971), Springer

[8] **M. Golubitsky, I. Stewart,** *Hopf bifurcation in the presence of symmetry,* Arch. Rat. Mech. Anal. (1985) 87, 2, 107-165

[9] **M. Golubitsky, I. Stewart,** *Symmetry and stability in Taylor-Couette Flow,* SIAM J. Math. Anal. (1986) 17,2, 249-288

[10] **J. Guckenheimer, P. Holmes,** *Nonlinear Oscillations, Dynamical Systems and Bifurcations of Vector Fields,* (1983) Applied Math. Sci. Series, Springer

[11] **G. Iooss,** *Bifurcation and Transition to Turbulence in Hydrodynamics,* (1983) CIME Session on Bif. Theory and Applications, Lec. Notes in Math. 1057, Springer

[12] **K. Kirchgässner, P. Sorger,** *Branching Analysis for the Taylor Problem,* (1969), Quart. J. Mech. Appl. Math. 22, 183

[13] **J. Marsden, M. Mc Cracken,** *The Hopf bifurcation and its applications,* (1976), Aplied Math. Sci. series, 19, Springer

[14] **L. Michel,** *Symmetry defects and broken symmetry - Hidden Symmetry,* (1980) Review of Mod. Phys. 52, 3, 617-651

[15] **D. Ruelle,** *Bifurcations in the presence of a symmetry group,* (1983) Arch. Rat. Mech. Anal. 51, 136-152

[16] **D. Sattinger,** *Group Theoretic Methods in Bifurcation Theory,* (1979), Lect. Notes in Math. 762, Springer

[17] **G.I. Taylor,** *Stability of a viscous liquid contained between two rotating cylinders,* (1923), Phil. Trans. Roy. Soc. London A, 223 289-343.

ON BIFURCATION FOR VARIATIONAL PROBLEMS

Shui-Nee Chow*
Department of Mathematics
Michigan State University
Wells Hall
East Lansing, MI 48824

and

Reiner Lauterbach**
Institut für Mathematik
Universität Augsburg
Memminger Str. 6
D-89 Augsburg
West Germany

1. INTRODUCTION.

In this lecture we would like to present a bifurcation theorem for problems having a variational structure. It generalizes earlier results by Rabinowitz [5], Böhme [1] and Marino [4]. If one combines our result with a recent result of Maddocks [3] one can discuss global stability assignments for a given branch of solutions. Before we present our theorem in section II, we want to describe this application to stability analysis. Further applications to problems in mathematical physics may be found in Maddocks' paper [3].

Let E be a Hilbert space and

$$f\colon E \times \mathbb{R} \to E \tag{1.1}$$

be a C^1-map such that

$$F\colon E \times \mathbb{R} \to E \tag{1.2}$$

is a potential for f, i.e.,

$$f(x,\lambda) = \nabla_x F(x,\lambda). \tag{1.3}$$

A zero of f is generally called a critical point of F. We want to study critical points of F and the (linearized) stability of these points. Let us assume that we know a curve

$$S = \{(x(s),\ \lambda(s)),\ s \in \mathbb{R} \tag{1.4}$$

of critical points of F. Let us further assume that

(A) we know the (linearized) stability of one point

$$(x_0,\lambda_0) = (x(s_0),\ \lambda(s_0)) \in S,$$

*Research partially supported by National Science Foundation Grant DMS 8401719

**Supported by Deutsche Forschungsgemeinschaft LA 525/1-1.

NATO ASI Series, Vol. F37
Dynamics of Infinite Dimensional Systems
Edited by S.-N. Chow, and J.K. Hale

i.e., the number of eigenvalues of the bounded linear operator $A_o = D_x f(x_o,\lambda_o)$ in the right half-plane. (We assume that the part of the spectrum $\sigma(A_o)$ of A_o with real part $\geq -\varepsilon$, for some positive ε, consists of finitely many eigenvalues, each with finite multiplicity);

(B) we know the geometry of S and furthermore, that for an interval Λ with

$$s_o \in \Lambda \subset \mathbb{R},$$

there are no bifurcation points along $S|_\Lambda$.

The question which we want to consider is the following: Can we obtain stability information for points $(x,\lambda) \in S|_\Lambda$? We can do this if we

(1) could relate the eigenvalue perturbation near a fold with the orientation of the fold,

(2) had a relation between nonzero crossing of eigenvalues through zero, and bifurcation of critical points.

The first question was addressed in the paper by Maddocks. He proves the following result: in a special coordinate system the change of stability is connected to the shape of the fold. If the space E is the abscissa and $-F_\lambda$ is the ordinate, the following conclusions hold: if the diagram looks like

(a)

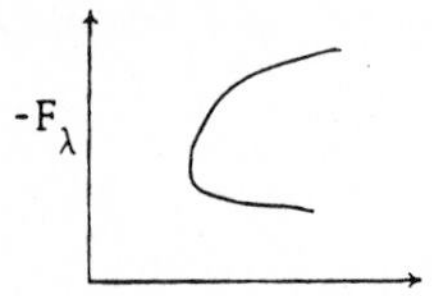

then the lower branch is always unstable,

(b)

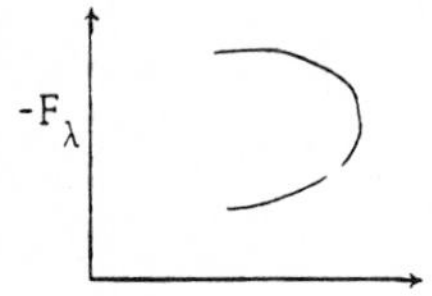

then the upper branch is always unstable,

(c)

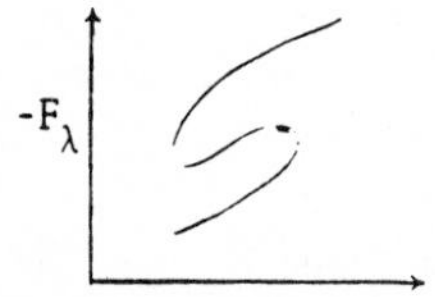

if the lowest branch is stable, then the most upper branch is also stable.

Using these pictures one can follow the stability assignment along the curve $S|_\Lambda$ provided eigenvalues can cross the imaginary axis only at fold points. We would like to conclude this from (B). Degree theory gives immediately that (x_0,λ_0) is a bifurcation point if there is an odd number of eigenvalues of $A(s) = D_x f(x(s), \lambda(s))$ passing through zero. In special cases Rabinowitz [5], Böhme [1], and Marino [4] have shown that also the change of sign on an even, nonzero number of eigenvalues leads to bifurcation. Using the variational structure of the problem, we want to show that this is true in general.

2. BIFURCATION THEOREM.

For the sake of simplicity let us assume $x(s) = 0$ for all $s \in \Lambda$. Let $A(\lambda) = D_x f(0,\lambda)$ and assume

$$\dim \ker A(0) = n > 0. \tag{2.1}$$

Furthermore, let 0 be an isolated point in $\sigma(A(0))$ and assume that none of the eigenvalues of $A(\lambda)$ is zero for $\lambda \neq 0$ and small. Denote the collection of all eigenvalues of $A(\lambda)$ which converge to zero as $\lambda \to 0$ by

$$\mathrm{eig}_0 A(\lambda). \tag{2.2}$$

Let $r(\lambda)$ be the number of elements in $\mathrm{eig}_0 A(\lambda)$ which have positive real part. Assume that

$$r^+ = \lim_{\substack{\lambda\to 0\\ \lambda>0}} r(\lambda)$$

and

$$r^- = \lim_{\substack{\lambda\to 0\\ \lambda<0}} r(\lambda)$$

both exist. Then we have the following result:

Theorem (Chow & Lauterbach [2]).

Let $F: E \times \mathbb{R} \to \mathbb{R}$ be C^2 and $f(x,\lambda) = \nabla_x F(x,\lambda)$. Suppose that $f(0,\lambda) = 0$ for all $\lambda \in \mathbb{R}$ and 0 is an eigenvalue of $A(0) = D_x f(0,0)$.

If r^+, r^- exist and

$$r^+ - r^- = 0,$$

then $(0,0)$ is a bifurcation point for the equation

$$f(x,\lambda) = 0.$$

Remark 1: The difference between our result and earlier results is the generality in the dependence on λ.

Remark 2: Some modification allows us to apply this theorem to problems in elasticity having convex energy functionals.

BIBLIOGRAPHY.

[1] Böhme, R.: Die Lösung der Verzweigungsgleichung für nicht-Lineare Eigenwertprobleme, Math. Z. 27(1972), 105-126.

[2] Chow, S.-N. & Lauterbach, R.: A Bifurcation Theorem for Critical Points of Variational Problems, Preprint 1985.

[3] Maddocks, J.: Preprint 1985.

[4] Marino, A.: La biforcazione nel caso variationale, Conf. Sem. Mat. Bari 1973.

[5] Rabinowitz, P.H.: A bifurcation theorem for potential operators, J. Funct. Anal. 25(1977), 412-424.

Nilpotent Normal Form in Dimension 4

Richard Cushman
Jan A. Sanders

Department of Mathematics and Computer Science
Free University
Amsterdam

ABSTRACT

The normal form for a vectorfield in R^4 with irreducible nilpotent linear part is given and we prove that it is minimal. A counting argument suffices to show that we have indeed all the terms in the normal form. The normal form is not unique.

1. Introduction

In this paper we want to give the normal form of a formal power series vectorfield in $\mathbb{R}^4$ with respect to an irreducible nilpotent linear part. For the definition of the nilpotent normal form we refer to (Cu1986a), where one can also find the normal forms for the irreducible 2- and 3-dimensional problems and the reduced (2×2) 4-dimensional. An alternate treatment, more in the spirit of Lyapunov-Schmidt reduction theory on inner product spaces, has recently been given in (El1986a). The normal form we give here fits in both frameworks. We shall concentrate mainly on the independence proof, since the sufficiency can easily be checked by writing down the table function and differentiating.

Let us quickly summarize the theory. Let N be the linear part of a vectorfield at some equilibrium point. Extend N to the Lie algebra $sl_2(\mathbb{R})$ consisting of the triple $\{N, M, H\}$. This is always possible by the Jacobson-Morozov theorem. Moreover, M and H can be easily computed once N is in Jordan normal form.

A vectorfield is said to be in **normal form** if it sits in a complement to $im\ ad_N$. It follows from the representation theory of $sl_2(\mathbb{R})$ that $ker\ ad_M$ is a *natural* complement. Given a nilpotent N, which we bring in the normal form

$$N = \begin{pmatrix} 0 & 0 & 0 & 0 \\ 1 & 0 & 0 & 0 \\ 0 & 2 & 0 & 0 \\ 0 & 0 & 3 & 0 \end{pmatrix},$$

NATO ASI Series, Vol. F37
Dynamics of Infinite Dimensional Systems
Edited by S.-N. Chow, and J. K. Hale

we define the **normal form space** to be $ker\ ad_M$ where M is defined by

$$M=\begin{pmatrix}0&3&0&0\\0&0&2&0\\0&0&0&1\\0&0&0&0\end{pmatrix}.$$

The strange normalization for N and M is chosen for to make the step to invariant theory more natural. It does have the advantage over the usual Jordan normal form that one can read the dimension of the block from its extremal elements.

In order to give the minimal normal form, we define:

$$p_1=x_4, \tag{1.1a}$$

$$p_2=x_2x_4-x_3^2, \tag{1.1b}$$

$$p_3=x_1x_4^2+2x_3^3-3x_2x_3x_4, \tag{1.1c}$$

$$p_4=6x_1x_2x_3x_4-x_1^2x_4^2-4x_1x_3^3+3x_2^2x_3^2-4x_2^3x_4. \tag{1.1d}$$

We observe that $p_i \in ker\ L_M$ for $i=1,\ldots,4$ and we find that following relation holds:

$$p_3^2+4p_2^3+p_1^2p_4=0.$$

Let

$$v_1=\begin{pmatrix}x_1\\x_2\\x_3\\x_4\end{pmatrix},v_2=\begin{pmatrix}3x_1x_2x_3-2x_2^3-x_1^2x_4\\2x_1x_3^2-x_2^2x_3-x_1x_2x_4\\x_1x_3x_4-2x_2^2x_4+x_2x_3^2\\x_1x_4^2+2x_3^3-3x_2x_3x_4\end{pmatrix}$$

and

$$v_3=\begin{pmatrix}3(x_1x_3-x_2^2)(x_1x_4-x_2x_3)\\3x_1^2x_3^2-2x_1x_3^3-2x_2^3x_4+x_1^2x_4^2\\3(x_1x_4-x_2x_3)(x_2x_4-x_3^2)\\6(x_2x_4-x_3^2)^2\end{pmatrix},v_4=\begin{pmatrix}3(x_1x_3-x_2^2)\\x_1x_4-x_2x_3\\x_2x_4-x_3^2\\0\end{pmatrix},v_5=\begin{pmatrix}1\\0\\0\\0\end{pmatrix}.$$

Notice that v_3 starts with the square of an element in $ker\ L_M$. The normal form is now given by

$$\begin{aligned}\dot{x}=&Nx+F_1(p_1,p_2,p_4)v_1+F_2(p_1,p_2,p_4)v_2+F_3(p_2,p_4)v_3\\&+F_4(p_1,p_2,p_4)Mv_1+F_5(p_1,p_2,p_4)Mv_2+F_6(p_1,p_2,p_4)v_4\\&+F_7(p_1,p_2,p_4)M^2v_1+F_8(p_1;p_2,p_4)Mv_4+F_9(p_1,p_2,p_4)v_5.\end{aligned} \tag{1.2}$$

Notice that F_3 is not a function of p_1. For the sake of completeness we give the table function here:

$$T(t,u)=\frac{u^3+t(1+u^2+u^4)+ut^2(1+u^2)+t^3(1+u^2)}{(1-u^3t)(1-u^2t^2)(1-t^4)}+\frac{ut^4}{(1-u^2t^2)(1-t^4)}.$$

The reader can check that indeed

$$\frac{\partial}{\partial u}uT(t,u)\,|_{u=1}=\frac{4}{(1-t)^4}$$

as required.

2. Algebraic independence

In this section we shall consider $ker\ L_M$, and give its Cohen-Macaulay decomposition.

Consider the mapping $\Phi:\mathbb{R}^4\rightarrow\mathbb{R}^4:x\rightarrow p=(p_1,\ldots,p_4)$ where the $p_i,i=1,\ldots,4$ are given by (1.1). Since

$$p_3^2+4p_2^3+p_1^2p_4=0 \tag{2.1}$$

the image of Φ is contained in the variety V defined by (2.1). In fact, Φ is surjective.

Proposition 1

Let $\mathcal{R}$ be the ring of polynomial functions on the variety V Then

$$ker\ L_M=\Phi^*\mathcal{R} \tag{2.2}$$

and $\mathcal{R}$ has the Cohen-Macaulay decomposition

$$\mathcal{R}=\mathbb{R}[p_1,p_2,p_4](1\oplus p_3). \tag{2.3}$$

Proof

We refer to (St1979a) for the definition of Cohen-Macaulay rings and further references to the literature. Since $\mathcal{R}=\mathbb{R}[p_1,\ldots,p_4]/I$, where I is the ideal generated by $p_3^2+4p_2^3+p_1^2p_4$, it is clear that any element element $F\in\mathcal{R}$ may be written

$$F(p_1,\ldots,p_4)=F_1(p_1,p_2,p_4)+p_3F_2(p_1,p_2,p_4). \tag{2.4}$$

To prove the proposition we have to show that F_1 and F_2 are unique. In other words, if $F_1+p_3F_2=0$, we must show that $F_1=F_2=0$. We shall prove this using induction on the sum of the degrees in p_4 of F_1 and F_2. Suppose that the total p_4-degree is zero, then differentiating

$$0=\Phi^*(F_1(p_1,p_2,p_4)+p_3F_2(p_1,p_2,p_4)) \tag{2.5}$$

with respect to x_1 gives

$$0=\frac{\partial p_3}{\partial x_1}F_2=p_1^2F_2$$

Thus $F_2=0$ and therefore $F_1=0$. In general, differentiating (2.5) with respect to x_1 gives

$$0=-2p_3\frac{\partial F_1}{\partial p_4}+p_1^2F_2-2p_3^2\frac{\partial F_2}{\partial p_4} \tag{2.6}$$

$$=-2p_3\frac{\partial F_1}{\partial p_4}+p_1^2F_2+2(4p_2^3+p_1^2p_4)\frac{\partial F_2}{\partial p_4}$$

$$=F_1^*+p_3F_2^*$$

Since the total degree of F_1^* and F_2^* is less than that of F_1 and F_2, by the induction hypothesis,

$$\frac{\partial F_1}{\partial p_4}=0$$

and thus by (2.6)

$$p_1^2F_2+2(4p_2^3+p_1^2p_4)\frac{\partial F_2}{\partial p_4}=0. \tag{2.7}$$

Substituting

$$F_2=\sum_{k=0}^{N}F_{2k}(p_1,p_2)p_4^k$$

into (2.7) and equating the coefficients of powers of p_4 to zero gives $F_{2N}=0$ and hence $F_2=0$ by recursion. This again implies $F_1=0$. □

3. The minimal normal form

We are now in a position to prove that the normal form (1.2) is minimal. The method used to find this normal form is not the subject of this paper. One can either use the method described in (Cu1986a) or first compute the Molien function of a tensor representation and then use the symbolic calculus. The last one is the more systematic way of doing things. Here we shall only be concerned with the proof that the proposed normal form is minimal, in the sense that it

(1) sits in *ker* ad_M;

(2) there are no relations inside the normal form.

Proposition 2

The decomposition of the *ker* L_M-module *ker* ad_M into

$$\mathbb{R}[p_1,p_2,p_4](v_1\oplus v_2\oplus Mv_1\oplus Mv_2\oplus v_4\oplus M^2v_1\oplus Mv_4\oplus v_5)\oplus\mathbb{R}[p_2,p_4]v_3$$

is a direct sum decomposition. This decomposition is not necessarily unique.

Proof

It is mechanical to check that the expression (1.2) is indeed in $ker\ ad_M$.

To prove minimality put

$$0=F_1(p_1,p_2,p_4)v_1+F_2(p_1,p_2,p_4)v_2+F_3(p_2,p_4)v_3 \tag{3.1}$$
$$+F_4(p_1,p_2,p_4)Mv_1+F_5(p_1,p_2,p_4)Mv_2+F_6(p_1,p_2,p_4)v_4$$
$$+F_7(p_1,p_2,p_4)M^2v_1+F_8(p_1,p_2,p_4)Mv_4+F_9(p_1,p_2,p_4)v_5.$$

The fourth component of (3.1) is

$$F_1(p_1,p_2,p_4)p_1+F_2(p_1,p_2,p_4)p_3+6F_3(p_2,p_4)p_2^2=0 \tag{3.2}$$

Using proposition 1 we immediately see that $F_2=0$. Thus from (3.2) we conclude that

$$F_1(p_1,p_2,p_4)p_1+6F_3(p_2,p_4)p_2^2=0. \tag{3.3}$$

(3.3) cannot hold for nonzero F_1 and F_3 for then p_1,p_2 and p_4 would be algebraically dependent or it would be identically zero. But the last possibility is ruled out by the fact that the first term of the equation has a factor p_1. It follows that $F_1=F_3=0$.

The third component of (3.1) is

$$0=F_4(p_1,p_2,p_4)p_1+F_5(p_1,p_2,p_4)p_3+F_6(p_1,p_2,p_4)p_2. \tag{3.4}$$

Again by proposition 1 we may immediately conclude that $F_5=0$. Thus (3.4) leads to

$$0=F_4(p_1,p_2,p_4)p_1+F_6(p_1,p_2,p_4)p_2. \tag{3.5}$$

Putting

$$F_4=p_2G_1\ ,\ F_6=-p_1G_1 \tag{3.6}$$

for any polynomial $G_1(p_1,p_2,p_4)$ solves (3.5). Using (3.6) and $F_i=0$ for $i=1,2,3,5$, the second component of (3.1) reads

$$0=2p_1F_7(p_1,p_2,p_4)+2p_2F_8(p_1,p_2,p_4) \tag{3.7}$$
$$+2x_3p_2G_1(p_1,p_2,p_4)-(x_1x_4-x_2x_3)p_1G_1(p_1,p_2,p_4)$$

Since

$$2x_3p_2-(x_1x_4-x_2x_3)p_1=2x_3(x_2x_4-x_3^2)-(x_1x_4-x_2x_3)x_4=-p_3,$$

(3.7) is equivalent to

$$0=2p_1F_7(p_1,p_2,p_4)+2p_2F_8(p_1,p_2,p_4)-p_3G_1(p_1,p_2,p_4)$$

Again by proposition 1 it follows that $G_1=0$. Therefore by (3.6) $F_4=F_6=0$. Again we can solve

$$0=2p_1F_7(p_1,p_2,p_4)+2p_2F_8(p_1,p_2,p_4)$$

by putting

$$F_7 = p_2 G_2 \ , \ F_8 = -p_1 G_2 \tag{3.8}$$

where G_2 is a polynomial in p_1,p_2 and p_4. Using (3.8) and $F_i=0$ for $i=1, \ldots, 6$, the first component of (3.1) reads

$$F_9(p_1,p_2,p_4) = 3(x_1x_4 - x_2x_3)p_1G_2(p_1,p_2,p_4) - 6x_3p_2G_2(p_1,p_2,p_4)$$
$$= 3p_3G_2(p_1,p_2,p_4).$$

Thus $G_2=F_9=0$ and we have shown that $F_i=0$ for $i=1, \ldots, 9$. This completes the proof of minimality. □

Acknowledgement

The authors wish to acknowledge the (partial) support by DARPA during their respective stays in East Lansing at MSU in 1985 and 1986. In this period the research for this paper took place.

References

Cu1986a. Cushman,R. and Sanders,J.A., "Nilpotent normal forms and representation theory of sl(2,R)," in *Multi-parameter Bifurcation Theory*, ed. M. Golubitsky and J. Guckenheimer, American Mathematical Society, Contemporary Mathematics 56, Providence (1986).

El1986a. Elphick,C., Brachet,M.E., Coullet,P., and Iooss,G., "A simple global characterization for normal forms of singular vector fields," Preprint (1986).

St1979a. Stanley,R.P., "Invariants of finite groups and their applications to combinatorics," *Bulletin of the AMS* **1**(3), pp. 475-511 (1979).

Perturbed Dual Semigroups and Delay Equations

Odo Diekmann

Centre for Mathematics and Computer Science
P.O. Box 4079, 1009 AB Amsterdam, The Netherlands
and
Institute of Theoretical Biology
University of Leiden
Groenhovenstraat 5
2311 BT Leiden, The Netherlands

The theory of dual semigroups on non-reflexive Banach spaces can be used to define a natural generalization of the notion of a bounded perturbation of the generator and a new version of the variation - of - constants formula. This approach was developed in joint work with Ph. Clément, M. Gyllenberg, H.J.A.M. Heijmans and H.R. Thieme, motivated by some applications to physiologically structured population growth models. In this paper it is shown that delay differential equations fit very well into exactly the same functional analytic framework.

1. Introduction.

The variation - of - constants formula is a very convenient starting point for the derivation of many results in the local stability and bifurcation theory of ordinary and partial differential equations. This statement is equally valid for retarded functional differential equations, but here the variation - of - constants formula shows a peculiar feature. Indeed, in the book of HALE [6] we find that the formula involves the so-called fundamental matrix solution X which is, by definition, the solution corresponding to the special *discontinuous* initial condition $X(t) = 0$ for $t < 0$ and $X(0) = I$ (the identity matrix) and which, therefore, does not "live" in the state space C. As a consequence one has to interpret the convolution integral which figures in the variation - of - constants formula as a family (parametrized by the independent variable of the functions in C) of integrals in Euclidean space. Thus the formula becomes symbolic rather than functional analytic (it does not fit into the standard semigroup framework).

The difficulty resolves to some extent if one embeds C into the product space $M_p = \mathbb{R}^n \times L_p$ (see, for instance, KAPPEL and SCHAPPACHER [8]). The fundamental matrix solution is well-defined in the M_p context and so is the convolution integral which maps C into C (here we identify C with its embedding into M_p). Actually the fact that the variation - of - constants formula makes sense in M_p is the main motivation for introducing this space (see DELFOUR and MANITIUS [2] and the references given there).

Another class of equations for which the variation - of - constants formula is not directly available comes up in population biology. If one considers age-dependent population growth models with possibly nonlinear birth terms the stability and bifurcation theory is troubled by annoying technical difficulties, as one may notice by reading the work of GURTIN and MACCAMY [5], PRÜß [11] and WEBB [13]. There is, actually, much more similarity with the case of delay equations than appears from these works. The solution which has the (Dirac) measure concentrated at age zero as initial condition is well-defined and plays exactly the same role as the fundamental solution for delay equations. Again the trouble is that this specially important initial condition does not exist in the conventional state space $L_1(\mathbb{R}_+)$. An identical picture emerges : even though the integrand in the convolution term of the variation - of - constants formula does not live in the state space X but in some larger space the convolution integral itself defines an element of X.

This observation triggered Ph. Clément, M. Gyllenberg, H.J.A.M. Heijmans, H.R. Thieme and myself to investigate the phenomenon from a functional analytic point of view. Motivated by the duality between the Kolmogorov forward and backward formulations of linear age dependent population dynamics we studied the existing theory of dual semigroups on (non-reflexive) Banach spaces. We found that within this framework one can give a very natural generalization of the notion of a bounded perturbation of the generator and that this leads to a new version of the variation - of - constants formula. In sections 2 and 3 I will give a summary of the preprint [1]. In section 4 I show that delay equations fit into the framework. It will appear that one can "construct" the space M_∞ starting from C and the very simple semigroup generated by $\dot{x} = 0$ considered as a delay equation, that it is

NATO ASI Series, Vol. F37
Dynamics of Infinite Dimensional Systems
Edited by S.-N. Chow, and J. K. Hale

natural to embed C into M_∞ , and that one has to interpret the convolution integral in a weak $\star$ sense. Thus we arrive at a functional analytic underpinning of the standard variation - of - constants formula for delay equations.

There is yet another way in which the basic difficulty manifests itself in both delay equations and age dependent population growth equations. The domain of the semigroup generator involves much more than just a smoothness condition. In fact for delay equations all information about the particular equation is contained in the domain, whereas the action of the generator does not depend on the equation at all. This aspect of the problem will be explained as well. The domain of the weak $\star$ generator on M_∞ is independent of the particular equation and involves a smoothness condition only. When taking the restriction to C we have to restrict the range of the generator and this causes a shift of information from action to domain.

The problems motivating [1] are identical to some of the problems motivating recent work by DESCH and SCHAPPACHER (see [3] and the references given there), but the solution we propose is different (although there are some common characteristics).

2. Dual semigroups.

Let $\{T(t)\}$ be a strongly continuous semigroup of bounded linear operators on a Banach space X and let A denote its generator. The adjoint operators $T^*(t)$ form a semigroup on the dual space X^*. The semigroup $\{T^*(t)\}$ is weak $\star$ continuous. But if we equip X^* with the usual norm topology $\{T^*(t)\}$ need not be strongly continuous (unless X is reflexive). The operator A^* , the adjoint of A , is the weak $\star$ generator of $\{T^*(t)\}$ but need not be densely defined.

In their classic treatise HILLE and PHILLIPS [7] showed that the dialogue of a space and a semigroup demands a duality theory which is made to measure. We need a special star, called sun and represented by the symbol $\odot$. Let

$$X^{\odot} = \left\{x^* \in X^* \,|\, \lim_{t\downarrow 0}\|T^*(t)x^* - x^*\| = 0\right\}. \tag{2.1}$$

Then $X^{\odot}$ is the maximal invariant subspace on which $\{T^*(t)\}$ is strongly continuous, $X^{\odot}$ is norm closed and $X^{\odot} = \overline{D(A^*)}$. Let $\{T^{\odot}(t)\}$ denote the strongly continuous semigroup on $X^{\odot}$ which is obtained by restriction of $\{T^*(t)\}$ and let $A^{\odot}$ denote its generator. Then $A^{\odot}$ is the part of A^* in $X^{\odot}$, i.e. the largest restriction of A^* with both domain and range in $A^{\odot}$.

Repeating the same procedure we obtain a weak $\star$ continuous semigroup $\{T^{\odot *}(t)\}$ on $X^{\odot *}$, the dual space of $X^{\odot}$, with weak $\star$ generator $A^{\odot *}$. Let

$$X^{\odot\odot} = \left\{x^{\odot *} \in X^{\odot *} \,|\, \lim_{t\downarrow 0}\|T^{\odot *}(t)x^{\odot *} - x^{\odot *}\| = 0\right\}. \tag{2.2}$$

A straightforward reinterpretation of the duality pairing between elements of X and $X^{\odot}$ yields a natural embedding of X into $X^{\odot *}$ and henceforth we identify X and this embedding into $X^{\odot *}$. Then X becomes a subspace of $X^{\odot\odot}$.

DEFINITION : *X is called $\odot$-reflexive with respect to A iff $X = X^{\odot\odot}$.*

A slightly more careful formulation is obtained if we first equip X with the equivalent norm

$$\| x \|' = \sup\{|<x,x^{\odot}>| : x^{\odot} \in X^{\odot} , \| x^{\odot} \| \leq 1\} .$$

But if $\{T(t)\}$ is a contraction semigroup, which it is in our application to delay equations, the two norms are actually the same.

It is known that X is $\odot$-reflexive with respect to A iff $(\lambda I - A)^{-1}$ is $X^{\odot}$-weakly compact. Moreover, X is $\odot$-reflexive with respect to A iff $X^{\odot}$ is $\odot$-reflexive with respect to $A^{\odot}$.

3. Perturbation theory.

Let $\{T_0(t)\}$ be a strongly continuous semigroup on X generated by A_0 and assume that X is $\odot$-reflexive with respect to A_0. We want to perturb the generator A_0 by a linear operator B, where B is bounded as an operator from X into $X^{\odot *}$. To this end we consider the variation - of - constants equation

$$T(t)x = T_0(t)x + \int_0^t T_0^{\odot *}(t-\tau)\, B\, T(\tau)x d\tau \tag{3.1}$$

Here the integral has to be understood in the weak $\star$ sense, i.e.

$$<\int_0^t T_0^{\odot\star}(t-\tau)\, B\, T(\tau)x d\tau \,,\, x^{\odot}> \,:=\, \int_0^t <BT(\tau)x \,,\, T_0^{\odot}(t-\tau)x^{\odot}>d\tau$$

for arbitrary $x^{\odot}\in X^{\odot}$. So in principle the integral takes values in $X^{\odot\star}$ but one can show that in fact it takes values in the closed subspace $X^{\odot\odot} = X$. Within this setting the standard contraction arguments apply and one infers that (3.1) admits a unique solution $\{T(t)\}$. By duality and restriction we obtain semigroups $\{T^{\star}(t)\}$, $\{T^{\odot}(t)\}$ and $\{T^{\odot\star}(t)\}$ on, respectively $X^{\star}$, $X^{\odot}$ and $X^{\odot\star}$, since it can be shown that the spaces of strong continuity do not depend on B ! Similarly the domains of the weak $\star$ generators on the "big" spaces $X^{\star}$ and $X^{\odot\star}$ are independent of B.

THEOREM. *The operator* $Ax = A_0^{\odot\star}x + Bx$ *with* $D(A) = \{x\in D(A_0^{\odot\star}) : A_0^{\odot\star}x + Bx\in X\}$ *is the generator of a strongly continuous semigroup* $\{T(t)\}$ *on* X *and the variation - of - constants formula (3.1) holds.*

Next assume that the operator B has finite dimensional range. So let there be given $r_1^{\odot\star}, \ldots, r_n^{\odot\star}\in X^{\odot\star}$ and $r_1^{\star}, \ldots, r_n^{\star}\in X^{\star}$ such that

$$Bx = \sum_{i=1}^{n} <r_i^{\star} \,,\, x>r_i^{\odot\star} \,. \tag{3.2}$$

Let Q denote a $n \times n$-matrix-valued function with entries

$$q_{ij}(t) = <r_i^{\star} \,,\, \int_0^t T_0^{\odot\star}(\tau)r_j^{\odot\star}d\tau > \tag{3.3}$$

A simple estimate shows that Q is Lipschitz continuous. As a consequence we have a representation of the form

$$Q(t) = \int_0^t K(\tau)d\tau \tag{3.4}$$

where the entries k_{ij} of K belong to L_{∞}. By a roundabout way we thus gave a meaning to $<r_i^{\star},T_0^{\odot\star}(t)r_j^{\odot\star}>$, though as a function of t and not pointwise in t !

Define the n-vector $y(t)$ by

$$y_i(t) = <r_i^{\star} \,,\, T(t)x> \tag{3.5}$$

where $T(t)x$ is the solution of (3.1). Equation (3.1) and a little technical calculation (to avoid undefined expressions) imply that y satisfies the *renewal equation*

$$y = h + K \star y \tag{3.6}$$

where the n-vector valued forcing function h is given by

$$h_i(t) = <r_i^{\star} \,,\, T_0(t)x> \tag{3.7}$$

and $K \star y$ denotes the convolution product of K and y. Conversely, given any solution y of (3.6) with h of the form (3.7) we can recover $T(t)x$ from

$$T(t)x = T_0(t)x + \sum_{j=1}^{n}\int_0^t T_0^{\odot\star}(t-\tau)r_j^{\odot\star}y_j(\tau)d\tau \tag{3.8}$$

It appears that solving (3.1) is reduced to solving (3.6).

Let $B^{\star} : X^{\odot}\rightarrow X^{\star}$ denote the (restriction of the) adjoint of B. The semigroup $\{T^{\odot}(t)\}$ satisfies the "adjoint" variation - of - constants equation

$$T^{\odot}(t)x^{\odot} = T_0^{\odot}(t)x^{\odot} + \int_0^t T_0^{\star}(t - \tau)B^{\star}T^{\odot}(\tau)x^{\odot}d\tau \,. \tag{3.9}$$

So if

$$B^{\star}x^{\odot} = \sum_{i=1}^{n} r_i^{\star}<x^{\odot} \,,\, r_i^{\odot\star}> \tag{3.10}$$

then the n-vector valued function z defined by

$$z_i(t) = \langle T^{\odot}(t)x^{\odot}, r_i^{\odot *}\rangle \tag{3.11}$$

satisfies the "adjoint" renewal equation

$$z = g + K^T * z \tag{3.12}$$

where K^T denotes the transpose of K (if the entries are complex we have to take complex conjugates as well) and the forcing function g is defined by

$$g_i(t) = \langle T_0^{\odot}(t)x^{\odot}, r_i^{\odot *}\rangle \tag{3.13}$$

Again one can recover the full semigroup from a knowledge of z only.

The symmetry of the present framework can be expressed in a diagram :

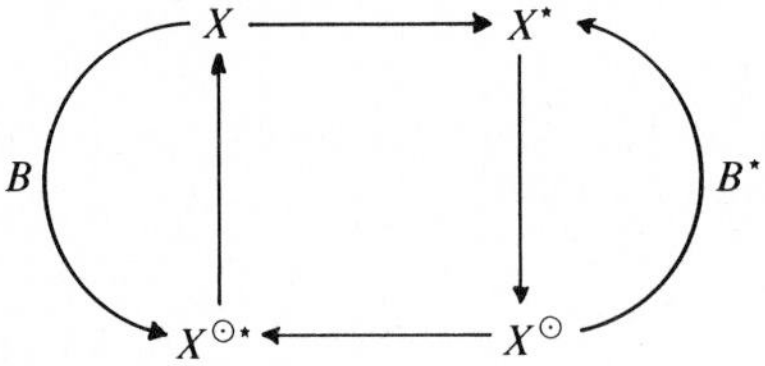

where horizontal arrows indicate going over to the dual space (taking adjoints) and vertical arrows indicate taking restrictions to the maximal space of strong continuity. When X is not $\odot$-reflexive with respect to A_0 this symmetry is disturbed. Nevertheless similar results hold. A canonical embedding of $X^{\odot\odot}$ into X^{**} seems to play a leading part, but it is not yet precisely clear how the most elegant and efficient argumentation proceeds, so we refrain from further discussion here.

4. Retarded functional differential equations.

Let ζ be a given $n \times n$-real-matrix-valued function of bounded variation such that $\zeta(\theta) = 0$ for $\theta \leqslant 0$ and $\zeta(\theta) = \zeta(1)$ for $\theta \geqslant 1$. Here and in the following we assume that all bounded variation functions are normalized such that they are right continuous on $(0,1)$, zero on $(-\infty,0]$ and constant on $[1,\infty)$. We consider the linear retarded functional differential equation

$$\dot{x}(t) = \int_0^1 d\zeta(\tau)x(t-\tau) \tag{4.1}$$

with initial condition

$$x(\theta) = \phi(\theta) \quad , \quad -1 \leqslant \theta \leqslant 0 \, , \tag{4.2}$$

where $\phi \in X = C\left([-1,0] \, ; \mathbb{R}^n\right)$. The semigroup $\{T_\zeta(t)\}$ on X is defined by

$$(T_\zeta(t)\phi)(\theta) = x_t(\theta \, ; \phi) \tag{4.3}$$

where $x(t; \phi)$ denotes the unique solution of (4.1) - (4.2) and, as usual, $x_t(\theta, \phi) = x(t + \theta; \phi)$. The action of $T_\zeta(t)$ is built from two constituents : translation and a rule for extension. An easy prototype is obtained by making the rule for extension as simple as possible, that is by taking $\zeta \equiv 0$. So let us take as the unperturbed problem the equation $\dot{x} = 0$ considered as a delay equation. The semigroup

$$(T_0(t)\phi)(\theta) = \begin{cases} \phi(\theta+t) & , \quad \theta+t \leqslant 0 \\ \phi(0) & , \quad \theta+t \geqslant 0 \end{cases} \tag{4.4}$$

on X is generated by

$$A_0\phi = \phi' \quad , \quad D(A_0) = \{\phi \in C^1 : \phi'(0) = 0\} \tag{4.5}$$

Let X^* be represented by NBV ([0,1] ; $\mathbb{R}^n$) with the pairing

$$\langle f, \phi\rangle = \int_0^1 df(\tau)\phi(-\tau)$$

Then

$$(T_0^*(t)f)(\sigma) = f(\sigma+t)\ ,\ \sigma>0\ , \tag{4.6}$$

and

$$A_0^*f = f'\ ,\ D(A_0^*) = \{f : f\in AC\ ,\ f'\in NBV\}\ , \tag{4.7}$$

where AC abbreviates "absolutely continuous". Hence

$$X^{\odot} = \overline{D(A_0^*)} = AC = \mathbb{R}^n \oplus AC_0 = \{f : f(t) = c + \int_0^t g(\tau)d\tau\ ,\ g\in L_1,\ \text{supp}\ g\subset[0,1]\} \tag{4.8}$$

It is sometimes convenient to work with the couple (c, g) to represent f. This amounts to representing $X^{\odot}$ by $\mathbb{R}^n \times L_1$, where $L_1 = \{g\in L_1(\mathbb{R}_+) : \text{supp} g\subset[0,1]\}$, with norm $\|(c, g)\| = |c| + \|g\|_{L_1}$. In these coordinates we have

$$T_0^{\odot}(t)(c,g) = \left[c + \int_0^t g(\tau)d\tau\ ,\ g(t+\cdot)\right] \tag{4.9}$$

$$A_0^{\odot}(c,g) = (g(0),g')\ ,\ D(A_0^{\odot}) = \{(c,g) : g\in AC\} \tag{4.10}$$

Next we take the representation $X^{\odot *} = M_\infty = \mathbb{R}^n \times L_\infty([-1,0]\ ;\ \mathbb{R}^n)$ with norm $\|(\alpha,\phi)\| = \sup\{|\alpha|, \|\phi\|_{L_\infty}\}$ and pairing

$$<(c,g)\ ,\ (\alpha,\phi)> = c\alpha + \int_0^1 g(\tau)\phi(-\tau)d\tau\ .$$

It follows that $T_0^{\odot *}(t)$ is the shift of the α-extended ϕ :

$$T_0^{\odot *}(t)(\alpha,\phi) = (\alpha,\phi_t)\ \text{ where }\ \phi_t(\theta) = \begin{cases}\phi(t+\theta) & ,\ t+\theta\leq 0\\ \alpha & ,\ t+\theta>0\end{cases} \tag{4.11}$$

and

$$A_0^{\odot *}(\alpha,\phi) = (0,\phi')\ ,\ D(A_0^{\odot *}) = \{(\alpha,\phi) : \phi\in \text{Lip}(\alpha)\} \tag{4.12}$$

where $\text{Lip}(\alpha)$ denotes the class of elements of L_∞ which contain a Lipschitz continuous function which assumes the value α at $\theta = 0$. Finally,

$$X^{\odot\odot} = \overline{D(A_0^{\odot *})} = \{(\alpha,\phi) : \phi\in C(\alpha)\} \tag{4.13}$$

where $C(\alpha)$ denotes the class of elements of L_∞ which contain a continuous function which assumes the value α at $\theta = 0$. The embedding of X into $X^{\odot *}$ is described by $j\phi = (\phi(0)\ ,\ \tilde{\phi})$, where $\tilde{\phi}$ denotes the L_∞ equivalence class to which ϕ belongs. Clearly $j(X) = X^{\odot\odot}$. We conclude that X is $\odot$-reflexive with respect to A_0, a fact which can also be deduced from the compactness of $(\lambda I - A_0)^{-1}$.

So far we have used the semigroup $\{T_0(t)\}$ to construct a dual space $X^{\odot *} = M_\infty$ in which $X = C$ lies embedded. Next we are going to perturb the generator by changing the rule for the extension of the function. The space C is too small to describe this perturbation but the space M_∞ is large enough.

Define $B : C \to M_\infty$ by

$$B\phi = (<\zeta\ ,\ \phi>\ ,\ 0) \tag{4.14}$$

Clearly the results of section 3 apply. In particular the theorem implies that for given ζ we have a semigroup $\{T(t)\}$ generated by the operator A with domain

$$D(A) = \{\phi\in\text{Lip} : \phi'\in C(<\zeta,\phi>)\} = \{\phi\in C^1 : \phi'(0) = <\zeta,\phi>\}$$

and action $A\phi = \phi'$. It is well-known that the same operator A generates the semigroup $\{T_\zeta(t)\}$ introduced at the beginning of section 4 and consequently the semigroups are really the same. We prefer to give another more direct proof of this fact which does not require any knowledge about the generator of $\{T_\zeta(t)\}$.

The element r_i^* is the i-th row of ζ and $r_i^{\odot *} = (e_i,0)$, where e_i denotes the i-th unit column vector in $\mathbb{R}^n$. It is convenient to combine these into matrices $r^* = \zeta$ and $r^{\odot *} = (I,0)$. Then

$$T_0^{\odot *}(t)r^{\odot *} = (I\ ,\ H(t+\cdot)I)\ , \tag{4.15}$$

where H denotes the Heaviside function. Substituting this into (3.8) we find

$$(T(t)\phi)(\theta) = (T_0(t)\phi)(\theta) + \int_0^{\max\{0,t+\theta\}} y(\tau)d\tau \tag{4.16}$$

So if we define for $t \geqslant 0$

$$x(t;\phi) = (T(t)\phi)(0) \tag{4.17}$$

then $x(t;\phi) = \phi(0) + \int_0^t y(\tau)d\tau$ from which it follows that $y(t) = \dot{x}(t;\phi)$ and, moreover, (4.16) then implies that for $t+\theta \geqslant 0$

$$(T(t)\phi)(\theta) = \phi(0) + \int_0^{t+\theta} y(\tau)d\tau = x(t+\theta;\phi)\,. \tag{4.18}$$

It remains to calculate the kernel K and the forcing function h in the renewal equation (3.6) for y. From (4.15) we deduce that

$$\int_0^t T_0^{\odot *}(\tau) r^{\odot *} d\tau = (t + \cdot)_+ \, I \tag{4.19}$$

where $(\theta)_+ := \max\{0,\theta\}$. Hence

$$M(t) = <\zeta\,,\, (t + \cdot)_+ I> = \int_0^t d\zeta(\tau)(t - \tau) = \int_0^t \zeta(\sigma)d\sigma \tag{4.20}$$

from which it follows that

$$K(t) = \zeta(t)\,. \tag{4.21}$$

Finally

$$h(t) = <r^\star\,,\, T_0(t)\phi> = \int_t^1 d\zeta(\tau)\phi(t - \tau) + \zeta(t)\phi(0)\,. \tag{4.22}$$

Now observe that one may start from (4.1) - (4.2) and manipulate as follows :

$$\begin{aligned}\dot{x}(t) &= \int_0^t d\zeta(\tau)x(t - \tau) + \int_t^1 d\zeta(\tau)\phi(t - \tau)\\ &= \int_0^t d\zeta(\tau)\, \dot{x}\,(t - \tau)d\tau + \zeta(t)\phi(0) + \int_t^1 d\zeta(\tau)\phi(t - \tau)\end{aligned}$$

to obtain exactly the same renewal equation for $\dot{x} = y$. Once more we attain to the conclusion that our semigroup is the solution semigroup corresponding to the problem (4.1) - (4.2).

In conclusion of this section I show how the present approach yields rather directly a suitable interpretation of the dual semigroup. The action of $r^{\odot *}$ corresponds to taking the limit from above in zero. So (3.13) implies that for $t>0$

$$g(t) = (T_0^{\odot}(t)f)\,(0+) = f(t+) = f(t) \tag{4.23}$$

or, in other words, the forcing function in the renewal equation and the state in our dynamical framework are one and the same thing ! According to (3.9) and (3.11) we have

$$(T^{\odot}(t)f)(\sigma) = f(t + \sigma) + \int_0^t \zeta^T(t - \tau + \sigma)z(\tau)d\tau\,. \tag{4.24}$$

On the other hand we may start from the renewal equation $z = f + \zeta^T \star z$ and define a semigroup $\{S(t)\}$ by requiring that

$$z_t = S(t)f + \zeta^T \star z_t\,, \tag{4.25}$$

that is , $S(t)f$ is the new forcing function in the renewal equation for the translated function z_t. A straightforward computation shows that $T^{\odot}(t) = S(t)$. Note, finally, that the renewal equation $z = f + \zeta^T \star z$ is obtained from the delay equation $\dot{z}(t) = <\zeta^T\,,\, z_t>$ by integrating the renewal

equation for $\dot{z}$.
Further comments on duality for delay equations may be found in [2,4] and the references given there.

5. Concluding remarks.
The present note concentrates on linear equations. It should be clear, however, that one can deal in a similar spirit with Lipschitz continuous operators from X into $X^{\odot *}$ and that results on linearized stability, the center manifold etc. can be proved in the standard manner. Nonlinear retarded functional differential equations are *semilinear* in the sense of section 3 !

A minor but curious point is that the duality framework shows so easily that even differential-difference equations such as $\dot{x}(t) = x(t - \frac{1}{2})$ admit well-defined solutions when the initial function is given as an element of M_∞ only.

Equations with infinite delay don't require the full machinery of the non-$\odot$-reflexive case simply because X is again invariant under the perturbed semigroup. As yet I have made no attempt to eleborate the details. Recently NAITO [10] and MURAKAMI [9] have employed the second dual of X and the weak $\star$ integral to define the variation - of - constants formula for equations with infinite delay.

My unfamiliarity with neutral equations keeps me from investigating whether or not the present framework has anything to offer for those and I welcome any aficionado who is willing to do so.

In the recent paper [12] and in work in progress Verduyn Lunel" studies the renewal equations (3.6) and (3.12) using the Laplace transform and complex function theory as his main tools. As corollaries he obtains strong results concerning the existence or non-existence of solutions which vanish identically after a given finite time and concerning the characterization of $\overline{\mathfrak{M}}$, the closure of the linear span of all eigenfunctions.

REFERENCES

[1] PH. CLÉMENT, O. DIEKMANN, M. GYLLENBERG, H.J.A.M. HEIJMANS and H.R. THIEME, Perturbation theory for dual semigroups. I. The $\odot$-reflexive case. Preprint.

[2] M.C. DELFOUR and A. MANITIUS, The structural operator F and its role in the theory of retarded systems. J. Math. Anal. Appl. I. **73** (1980) 466-490, II. **74**, (1980) 359-381.

[3] W. DESCH and W. SCHAPPACHER, Spectral properties of finite-dimensional perturbed linear semigroups. J. Diff. Equ. **59** (1985) 80-102.

[4] O. DIEKMANN, A duality principle for delay equations, Proceedings of Equadiff 5, M. GREGUŠ (ed.), Teubner **47** (1982) 84-86.

[5] M.E. GURTIN and R.C. MACCAMY, Nonlinear age-dependent population dynamics, Arch. Rat. Mech. Anal. **54** (1974) 281-300.

[6] J.K. HALE, Theory of Functional Differential Equations, Springer, 1977.

[7] E. HILLE and R.S. PHILLIPS, Functional Analysis and Semi-groups, Amer. Math. Soc., Providence, 1957.

[8] F. KAPPEL and W. SCHAPPACHER, Non-linear functional differential equations and abstract integral equations, Proc. Roy. Soc. Edinburgh, **84 A** (1979) 71-91.

[9] S. MURAKAMI, Linear periodic functional differential equations with infinite delay, to appear.

[10] T. NAITO, A modified form of the variation - of - constants formula for equations with infinite delay, Tôhoku Math. J. **36** (1984) 33-40.

[11] J. PRÜß, Stability analysis for equilibria in age-specific population dynamics, Nonl. Anal., Th., Meth. Appl. **7** (1983) 1291-1313.

[12] S.M. VERDUYN LUNEL, A sharp version of Henry's theorem on small solutions, J. Diff. Equ. **62** (1986) 266-274.

[13] G.F. WEBB, Theory of Nonlinear Age-Dependent Population Dynamics, Marcel Dekker, 1985.

ON OPERATORS WHICH LEAVE INVARIANT A HALF-SPACE

José M.Ferreira
Centro de Física da Matéria
Condensada
1699 Lisboa-Codex
Portugal

The purpose of this talk is to give a geometric characterization of those linear continuous operators $A:D(A)\subset E\to E$ ($D(A)$ dense in E), where E is a real locally convex topological vector space, whose adjoint $A^*:D(A^*)\subset E^*\to E^*$ (E^* the dual space of E) has a nonnegative real eigenvalue.

Some problems in control theory and retarded functional differential equations, require the converse of the property above. In fact, for a discrete control system $x_{k+1}=Ax_k+Bu_k$ on a Banach space, with A and B linear bounded operators and controls u_k constrained to a convex set Ω, the local controllability of the system requires that the adjoint operator A^* of A, have no real nonnegative eigenvalue (cf. [5]). For a linear retarded functional differential equation $\dot{x}(t)=\int_{-r}^{0} d\eta(\theta)\, x(t+\theta)$, where $r>0$ and $\eta(\theta)$ is a function of bounded variation, if we denote by A the infinitesimal generator of the strongly continuous semigroup on $C[-r,0]$ associated to the equation, one can state that all its unbounded solutions are oscillatory if and only if the adjoint operator A^* of A has no real nonnegative eigenvalue (see [1]).

In the following, by $<.,.>$ we will denote the usual dual product in $E\times E^*$. A closed half-space S in E is always

NATO ASI Series, Vol. F37
Dynamics of Infinite Dimensional Systems
Edited by S.-N. Chow, and J. K. Hale

given by $S=\{x \in E : <x,e> \geq 0\}$, where $e \in E^*$ is uniquely determined up to nonnegative scalar multiples. A set Ω E is said to be invariant for A if $A(\Omega$ $D(A))$ Ω.

Then we can state the following theorem .

Theorem 1 - If A leaves invariant a closed half-space $S = \{x \in E : <x,e> \geq 0\}$, $e \in D(A^*)$, $e \neq 0$, then A^* has a real nonnegative eigenvalue having e as corresponding eigenvector. If A^* has a real nonnegative eigenvalue then A leaves invariant a closed half-space.

Proof. If $A^*e=0$ then $\lambda=0$ is obviously an eigenvalue of A^* and e its corresponding eigenvector.

Assume now $A^*e \neq 0$ and consider the closed half-space of E given by $R=\{x \in E : <x,A^*e> \geq 0\}$. Since $<x,A^*e>=<Ax,e>$ for every $x \in D(A)$, we have that $x \in R$ $D(A)$ if and only if $x \in D(A)$ and $Ax \in S$. If A leaves invariant the closed half-space S we conclude that S D(A) R D(A). As $\overline{D(A)}=E$, then we have S R.

In such circunstances, every x in R which do not be in S, is necessarily in the supportting hyperplane of R. Therefore S and R must have the same supportting hyperplane and consequently e and A^*e are proportional vectors. That is, $A^*e=\lambda e$ for some $\lambda>0$. This proves the first part of the theorem.

Conversely, assume that $\lambda \geq 0$ is a real eigenvalue of A^* and take an eigenvector $e \in D(A^*)$, $e \neq 0$, associated to λ. Then $S = \{x \in E : <x,e> \geq 0\}$ is an invariant closed half-space for A, since $<Ax,e> = <x,A^*e> = \lambda<x,e>$ for every $x \in D(A)$. This completes the proof.

Through the preceding arguments one sees that the eigenvalue λ is positive if and only if $A^*e\neq 0$ and this is equivalent to have the open half-space $S_o=\{x \in E : \langle x,e\rangle>0\}$ invariant for A. Hence the following corollary holds.

<u>**Corollary**</u> - If A leaves invariant the open half-space $S_o = \{x \in E : \langle x,e\rangle>0\}$, $e\neq 0$, $e \in D(A^*)$, then A^* has a positive real eigenvalue having e as corresponding eigenvector. If A^* has a positive real eigenvalue then A leaves invariant an open half-space.

Remarks : **1.** Note that when A has an invariant open half-space S_o, then $S=\bar{S}_o$ is a closed half-space invariant for A and this operator maps the supporting hyperplane H, of S, into itself.

2. When D(A)=E, in [**3**, Corollary 2.6, p.267] more restrictive conditions on A and E are required, in order to conclude that A^* has a nonnegative real eigenvalue.

3. When E is a Banach space and A a bounded linear operator on E, the if part of the corollary can follow by [**2**, Theorem 3.3].

It is well known that an eigenvalue λ of A^* is also an eigenvalue of A, if E is a Hilbert space and $A:E\to E$ is a normal linear bounded operator. The same holds in finite dimension and for a general linear bounded operator on a Banach space, some compactness properties must be verified.

If E is a Banach space and $A:E\to E$ is a linear

bounded compact operator on E, a nonzero eigenvalue of A^* is also an eigenvalue of A and conversely. Therefore A has a positive eigenvalue if and only if A leaves invariant an open half-space S_o. However it can happen that the corresponding eigenvector be in the supporting hyperplane H of S_o. This fact depends only upon the behavior of A on that hyperplane. In the following theorem, a type of cross condition, as those introduced in [5], is given, in order that a similar situation do not occur.

Theorem 2 - Let $A:E \to E$ be a linear bounded compact operator on the Banach space E. Then A has a real positive eigenvalue if and only if A leaves invariant an open half-space. Moreover, if H is the supporting hyperplane of an open half-space $S_o = \{x \in E : \langle x,e\rangle > 0\}$ invariant for A and

(a) for every $x \in H$, $x \neq 0$, there exists $f \in E^*$ such that $\langle x,f\rangle = 0$ and $\langle Ax,f\rangle \neq 0$

then: (i) λ is a simple eigenvalue of A;

(ii) e is the unique (up to scalar multiples) eigenvector of A^*;

(iii) the unique (up to scalar multiples) eigenvector of A is in S_o.

Proof. Assumption (a) implies that A cannot have any eigenvector in the hyperplane H. So, an eigenvector u of A associated with λ, can be assumed to lie in S_o.

Denote by U the subspace generated by u. Then U and H are supplementary topological subspaces of E and for any other eigenvector v of A associated to λ, we have $v=w+\mu u$ for some $\mu\neq 0$ and $w\in H$, uniquely determined. Therefore $Av=Aw+\mu Au =$ $= \lambda w+\lambda\mu u$ and as $Au=\lambda u$, we conclude that A has an eigenvector in H, which is contradictory. Thus w=0 and u is the unique (up to scalar multiples) eigenvector of A corresponding to λ. Hence (i), (ii) and (iii) follow.

Remarks: **4**. When E is a Hilbert space and $A:E\to E$ is normal, under the corresponding assumption (a), (i)-(iii) hold for the nonnegative eigenvalue given by Theorem 1. The same holds in the finite dimensional case.

5. In [**3**, Theorem 2.7, p.67] a different compactness assumption is used in order to conclude that A has a nonnegative real eigenvalue, when A has an invariant cone.

6. For A in the circunstances of Theorem 2, if A has an invariant proper cone and its spectral radius $\rho(A)$ is positive, then $\rho(A)$ is an eigenvalue of A, by the Krein-Rutman theorem ([2,3,4,7]). Therefore A has an invariant open half-space. The converse of this, is not true, in general. A counter-example can easily be obtained in finite dimension.

References

1. **J.M.Ferreira** and **I.Györi**, Oscillatory behavior in linear retarded functional differential equations, J.Math.Anal. Appl. (in press).
2. **M.G.Krein** and **M.A.Rutman**, Linear operators leaving invariant a cone, Amer.Math.Soc.Transl.Ser.I 10(1962)
3. **H.H.Schaefer**, Topological vector spaces, Springer-Verlag 1971.

4. **H.H.Schaefer**, Banach lattices and positive operators, Springer-Verlag 1974.

5. H.**Schneider** and **M.Vidyasagar**, Cross positive matrices, SIAM J.Numer.Anal. 7(1970), 508-519.

6. **N.K.Son**, Controllability of linear discrete-time systems with constrained controls in Banach spaces, Control and Cybernetics 10(1981), 5-16.

7. **J.Schröder**, Operator inequalities, Ac.Press 1980.

(Reasearch partially supported by the L.N.E.T.I., Department of Electronics)

GLOBAL HOPF BIFURCATION IN REACTION DIFFUSION SYSTEMS WITH SYMMETRY

by

Bernold Fiedler*
Inst. of Applied Mathematics
Im Neuenheimer Feld 294
D - 6900 Heidelberg
W-Germany

Abstract

For differential equations with a compact symmetry group Γ some results on global bifurcation of time periodic orbits are presented. In particular it is investigated how the spatial and temporal action of the group Γ on a time periodic orbit may vary along global bifurcation branches, e.g. at period doubling bifurcations. The results are obtained geometrically by generic but equivariant approximation, rather than by topological techniques. As an application periodic solutions in reaction diffusion systems with symmetry D_n or $O(3)$ are discussed.

* This work was supported by the Deutsche Forschungsgemeinschaft, SFB 123.

NATO ASI Series, Vol. F37
Dynamics of Infinite Dimensional Systems
Edited by S.-N. Chow, and J.K. Hale

§ 1. Introduction

Suppose we know the stationary solutions of some nonlinear dynamical system, but we are interested in time periodic solutions. We may obtain them through local Hopf bifurcation, in one real parameter λ, under the usual conditions on linearizations at stationary solutions [8,10,18,29]. But we find only small amplitude oscillations around equilibrium, this way. Global Hopf bifurcation [2,14,25,27,28] tries to follow branches or continua of periodic solutions, far away from the stationary solutions where they originate. In other words, global Hopf bifurcation investigates how such branches might bifurcate or terminate. It is one possible approach to analyze the local secondary bifurcations of periodic orbits and, in addition, the rules of their interplay in a global Hopf bifurcation diagram [14,27,28] . The former is an analysis of certain purely singularities, whereas the latter may be accomplished by a suitably defined index of periodic orbits. If the dynamical system is equivariant, i.e. the (semi) flow commutes with the action of some group Γ, then periodic solutions may show some symmetry. It is our main objective here to study how such symmetries may change along global branches of periodic solutions. The case of no symmetry, i.e. of $\Gamma=\{id\}$, was studied e.g. in [3,4,5,9,14,27,28]. For a general background on equivariant problems see [18,35,36,42].

To be specific we will consider the following concrete reaction diffusion system

(1.1.a) $$\begin{pmatrix}1 & 0\\ 0 & \lambda\end{pmatrix}u_t = \Delta u + f(u) , \quad x\in\Omega\subseteq\mathbb{R}^N$$

(1.1.b) $$\partial_\nu u = 0 \qquad x\in\partial\Omega ,$$

with $u=(u_1,u_2)$, smooth $\partial\Omega$, $\lambda>0$, and Brusselator kinetics f [30]. Let $\Gamma\subseteq O(N)$ be the compact Lie group of orthogonal matrices γ which leave the bounded domain Ω invariant. Then Γ acts on functions $u(x)$ by

(1.2) $(\gamma u)(x) := u(\gamma^{-1}x)$,

and this action commutes with the (local) semiflow φ_t of (1.1)

(1.3) $\varphi_t(\gamma u_0) = \gamma\ \varphi_t(u_0)$.

As usual, [23], this fits into the framework of analytic semigroups

(1.4) $d_t u = A(\lambda)u + f(\lambda,u)$

on some (Sobolev type) Hilbert space $u \in X$, $\lambda \in \Lambda = \mathbb{R}$, with sufficiently smooth $A(\cdot)$, $f(\cdot,\cdot)$. We require A to have fixed domain $\mathcal{D}(A)$, independently of λ, and A^{-1} to be compact. See [14,(1.2)] for more details on the technical setting. Abstractly, we require (1.2) to be equivariant with respect to the orthogonal action of some compact Lie group Γ on X, i.e.

(1.5)
$$A(\lambda)\gamma u = \gamma A(\lambda)u, \qquad u \in \mathcal{D}(A)$$
$$f(\lambda,\gamma u) = \gamma f(\lambda,u) ,$$

for all $\gamma \in \Gamma$.

Given a periodic solution u(t) of (1.2) with minimal period $p > 0$, we describe its symmetry in the spirit of [19] as follows. Let

$$H := \{h \in \Gamma :\ hu(t) = u(t+\Theta(h,t)\cdot p),\ \text{for some } t \text{ and some } \Theta(h,t)\}.$$

Then $\Theta(h,t)$ is independent of t and exists for all t, once it exists for some t. We write

(1.6.a) $H = \{h \in \Gamma : hu(t) = u(t+\Theta(h)\cdot p)\}$

Note that

(1.6.b) $\Theta : H \to S^1 := \mathbb{R}/\mathbb{Z}$

is a group homomorphism, all by equivariance (1.3) of the

semiflow of φ_t of (1.4). Let

(1.6.c) $K := \ker \Theta = \{h \in \Gamma : hu(t) = u(t)\}$

and note that

$$H/K \cong \operatorname{im} \Theta \cong \begin{cases} \mathbb{Z}_n , & \text{or} \\ S^1 \end{cases}$$

has to be a cyclic factor in the original group Γ. We call the triple

(1.7) (H,K,Θ)

the <u>symmetry of the periodic solution u(t)</u>; it may depend on the (time-) orbit $u(t)$, of course. Given a subgroup K_0 of Γ (we denote this by $K_0 \leq \Gamma$) let

(1.8) $X^{K_0} := \{u \in X : K_0 u = u\}$

denote the space of K_0-fixed vectors; given $u_0 \in X$ let

(1.9) $\Gamma_{u_0} := \{\gamma \in \Gamma : \gamma u_0 = u_0\}$

denote the isotropy of u_0. If $u(t)$ has symmetry (H,K,Θ), e.g., then

(1.10) $u(t) \in X^K$ and $\Gamma_{u(t)} = K$

for all t. In general

(1.11) $u \in X^{K_0} \Longleftrightarrow \Gamma_u \geq K_0$.

It is particularly useful that each subspace X^{K_0} is invariant under the whole semiflow φ_t, again by equivariance (1.3).

In our main result, theorem 2.1 below, we fix an arbitrary cyclic factor $H_0/K_0 \cong \mathbb{Z}_n$ of Γ. A priori, this factor need not correspond to any periodic solution at all. Next we choose a certain subset

$$d \subsetneq \mathbb{Z} / n \mathbb{Z} .$$

Examples for admissible sets d will be given in § 2, see

table 2.1. These sets reflect that Θ,K may necessarily vary along global branches. Then we define a <u>global Hopf index</u>

$$\mathcal{H}^{d}_{H_0,K_0} .$$

This integer valued index is computed easily from some rudimentary information on the linearization at stationary solutions. Essentially $\mathcal{H}^{d}_{H_0,K_0} \neq 0$ implies global Hopf bifurcation of periodic solutions $(\lambda,u(t))$ with "symmetry" at least (H_0,K,Θ) , $K_0 \leq K \leq H_0$, and

$$\Theta \in d \quad (\mathrm{mod}\ n).$$

Here we view Θ as an integer (mod n) which represents the homomorphism from $H_0/K_0 \cong \mathbb{Z}_n = \{0,\frac{1}{n},\ldots,\frac{n-1}{n}\}$ to $S^1=\mathbb{R}/\mathbb{Z}$. We postpone further details to §2.

Suffice it to emphasize that we have some control on what Θ is doing, globally. This sharpens recent results of Alexander and Auchmuty [1] who consider a ring of n linearly coupled ODE oscillators with dihedral symmetry $\Gamma=D_n$. They already obtain a global result in this particular setting, but can only keep $\Theta\neq 0$. See §3 for further discussion.

This paper is organized as follows. In §2 we present our main result, theorem 2.1, in its technically precise form, but sacrificing generality for simplicity to some extent. As an example we discuss the Brusselator with diffusion (1.1) in §3. We choose domains Ω with symmetry group D_n (the symmetry of a regular n-gon in $\mathbb{R}^2$) and $O(3)$ (the symmetry of a ball in $\mathbb{R}^3$). Following the basic idea of Mallet-Paret, Yorke [27,28], the principal tool in the proof of theorem 2.1 is a global bifurcation diagram for generic, but equivariant, approximations of the nonlinearity f. This is the place where local singularities and global indices amalgamate. We sketch some aspects of this in §4. The discussion in §5 highlights some limitations and extensions of theorem 2.1 and, finally, leads to some open questions.

Acknowledgement. Many people have influenced this paper in various ways. Let me just mention my debts to J. Alexander, S. v. Gils, M. Golubitsky, W. Jäger, R. Lauterbach, J. Mallet-Paret, M. Medveď, C. Pospiech, A. Vanderbauwhede, J. Yorke, and to the organizers of this meeting.

§2. Main result

In this section we state our main result on global equivariant Hopf bifurcation for an analytic semigroup setting

$$(2.1)\qquad d_t u = A(\lambda)u + f(\lambda,u) \quad , \quad (\lambda,u)\in\Lambda\times X,$$

equivariant with respect to the compact Lie group Γ. We define a global Hopf index $\mathcal{H}^d_{H_0,K_0}$. Nonzero index implies global bifurcation (theorem 2.1). Alternatives (2.8.a-c) fix our use of the word "global", here, and definition 2.2 explains what we mean by "virtual period" and "virtual symmetry". In proposition 2.3 we indicate how these "virtual" objects come in naturally through a generic approximation process.

Fix two subgroups $H_0 \geq K_0$ of Γ such that K_0 is normal in H_0 and $H_0/K_0 \cong \mathbb{Z}_n$ is a finite cyclic group. We write $\mathbb{Z}_n = \{0, \frac{1}{n},\ldots,\frac{n-1}{n}\}$ additively (mod 1). We are interested in global Hopf bifurcation in $\Lambda\times X^{K_0}$ of periodic orbits u(t) with symmetry at least (H_0,K,Θ). This means that the symmetry $(\hat{H},\hat{K},\hat{\Theta})$ of u(t) as defined in (1.6.a-c), (1.7) should satisfy

$$(2.2)\qquad \hat{H} \geq H_0 \quad\text{and}\quad \hat{\Theta}|_{H_0} = \Theta .$$

Note that $K = \hat{K} \cap H_0$ and $K_0 \leq K \leq H_0$, cf. (1.6.c), (1.10), (1.11). If periodic orbits $(\lambda,u(t))$ with symmetry at least (H_0,K,Θ) bifurcate from a stationary solution (λ_0, U_0), then $U_0 \in X^{H_0}$.

We will need some conditions on the linearization of (2.1) at stationary solutions $(\lambda_0,U_0)\in\Lambda\times X^{H_0}$. The linearization is the unbounded operator on X

$$Lu := A(\lambda_0)u + D_u f(\lambda_0,U_0)u$$

with domain $\mathcal{D}(A)$; this operator has compact resolvent just as $A(\lambda_0)$ itself. Note that L restricts to a sectorial operator L^{K_0} on X^{K_0}. We call (λ_0,U_0) a <u>center</u> if L^{K_0} has some purely imaginary nonzero eigenvalues. We assume

(2.3.a) there exists a smooth " trivial branch " $(\lambda, U(\lambda)) \in \Lambda\times X^{H_0}$ of stationary solutions such that the eigenvalues of the linearization $L^{K_0}(\lambda)$ at $(\lambda,U(\lambda))$ are always nonzero;

(2.3.b) the set of centers on the trivial branch is bounded

(2.3.c) the linearization $L^{K_0}(\lambda)$ at $(\lambda,U(\lambda))$ depends analytically on λ near centers.

The restrictive aspect of assumption (2.3.a) is discussed in §5, (2.3.c) is motivated at the end of §4, but (2.3.b,c) are not particularly aggravating.

Defining the index $H^d_{H_0,K_0}$ requires some preparation. First we note that $H_0/K_0 \cong \mathbb{Z}_n$ acts canonically on X^{K_0}, and $L^{K_0}(\lambda)$ commutes with this $\mathbb{Z}_n$-action. The inequivalent irreducible types ρ_r of real representations of $\mathbb{Z}_n$ are given, in complex notation, by

(2.4) $$\rho_r(h)z = e^{2\pi irh}z\ , \quad h\in\mathbb{Z}_n,\ z\in\mathbb{C},\ r\in\mathbb{Z},\ 0\le r\le n/2\ ,$$

cf. [38]. Accordingly, X^{K_0} decomposes into representation spaces

$$X^{K_0} = \bigoplus_{0\le r\le n/2} X_r^{K_0}\ ,$$

and $L(\lambda)$ again restricts to a sectorial operator $L_r^{K_0}(\lambda)$ on each $X_r^{K_0}$. Denote:

(2.5.a) $E^r(\lambda)$: the number of eigenvalues of $L_r^{K_0}(\lambda)$ with positive real part, counting algebraic multiplicity

(2.5.b) $\chi^r := \lim_{\lambda\to\infty} \frac{1}{2}(E^r(\lambda) - E^r(-\lambda))$,
the net number of pairs of eigenvalues of $L_r^{K_0}(\lambda)$ crossing the imaginary axis from left to right as λ sweeps through $\mathbb{R}$ ("net crossing number").

By assumptions (2.3.a,b), χ^r is a well-defined integer.

Previewing §4 we note here that a generic Hopf bifurcation from eigenvalues in $X_r^{K_0}$ leads to small amplitude periodic orbits with $\Theta=\pm r$ (theorem 4.1); secondary bifurcations may lead to a Θ-doubling or -halving (corollary 4.3). This motivates the following algebraic structure. Let $n=2^\nu\cdot n'$ with $\nu,n'\in\mathbb{Z}$, n' odd. The numbers ±2 act, by multiplication, as an automorphism on $\mathbb{Z}/n'\mathbb{Z}$. We denote orbits under this action of $\{\pm2\}$ on $\mathbb{Z}/n'\mathbb{Z}$ by $d\subseteq\mathbb{Z}/n'\mathbb{Z}$, and the set of all orbits by $D(n')$. Because $\mathbb{Z}/n\mathbb{Z}$ is isomorphic to $(\mathbb{Z}/n'\mathbb{Z})\times(\mathbb{Z}/2^\nu\mathbb{Z})$ we may extend this definition: $D(n)$ consists of all subsets d of $\mathbb{Z}/n\mathbb{Z}$ of the form

$$d = \{\Theta\in\mathbb{Z}/n\mathbb{Z} : \Theta\in d' \pmod{n'}\}$$

for some $d'\in D(n')$. For some examples see table 2.1.

n'	$\lvert D(n')\rvert$	$d\in D(n')$
1	1	{0}
3	2	{0}, {1,2}={±1}
5	2	{0}, {±1,±2}
7	2	{0}, {±1,±2,±3}
9	3	{0}, {±1,±2,±4},{±3}
15	4	{0}, {±1,±2,±4,±7}, {±3,±6}, {±5}
31	4	{0}, {±1,±2,±4,±8,±15}, {±3,±6,±7,±12,±14}, {±5,±9,±10,±11,±13},
127	10	{0}, {±1,±2, ...,±64} etc.

Table 2.1: Orbits under multiplication by ±2 in $\mathbb{Z}/n'\mathbb{Z}$.

For any cyclic factor $H_0/K_0 \cong \mathbb{Z}_n$ of Γ, and any $d\in D(n)$ we now define the <u>global equivariant Hopf index</u> of the branch $(\lambda,U(\lambda))\in\Lambda\times X^{K_0}$ by

(2.6) $$\mathcal{H}^{d}_{H_0,K_0} := (-1)^{E^0(\lambda)} \sum_{r \in d (\mathrm{mod}\ n)} \chi^r$$

where $E^0(\lambda)$, χ^r depend on H_0, K_0 and are defined in (2.5.a,b). Note that $E^0(\lambda)$ is independent of λ by assumption (2.3.a)

With these preparations in mind we now state our main result. The terms "virtual symmetry" and "virtual period" will be explained in definition 2.2.

<u>2.1 Theorem:</u>

If the analytic semigroup (2.1) satisfies assumptions (2.3.a-c) and if

(2.7) $$\mathcal{H}^{d}_{H_0,K_0} \neq 0 ,$$

then there exists a <u>global</u> continuum $\mathcal{C} \subseteq \Lambda \times X^{K_0}$ of periodic solutions, and centers in X^{H_0}, with <u>virtual symmetry</u> at least

$$(H_0, K, \Theta) , \qquad \Theta \in d \ (\mathrm{mod}\ n) .$$

$\mathcal{C}$ contains both (uncountably many) periodic solutions and (at least one) center on the trivial branch $(\lambda, U(\lambda))$.

"Global" means that

(2.8.a) $\mathcal{C}$ is unbounded, or

(2.8.b) $\mathcal{C}$ contains periodic solutions with arbitrarily large <u>virtual periods</u>, or

(2.8.c) $\mathcal{C}$ contains a stationary solution outside of the trivial branch.

<u>2.2 Definition</u>

Let $u(t)$ be a center of (2.1), or a periodic solution with minimal period $p>0$. We call $q>0$ a <u>virtual period</u> of $u(\cdot)$ and $(\hat{H}, \hat{K}, \hat{\Theta})$ a <u>virtual symmetry</u> of $u(\cdot)$, if there exists a solution $v(t)$ of the variational equation

(2.1)' $\quad d_t v(t) = A(\lambda)v(t) + D_u f(\lambda,u(t))v(t)$

such that the pair $(u(t),v(t)) \in X \times X$ has minimal period q and symmetry $(\hat{H},\hat{K},\hat{\Theta})$, with the obvious diagonal action of Γ on $X \times X$.

Note that the minimal period resp. the symmetry of u(t) is always a virtual period resp. a virtual symmetry, taking $v \equiv 0$. But for fixed u(t) there may be finitely many multiples of the minimal period resp. restrictions of the symmetry, occuring as virtual periods resp. virtual symmetries. However, we expect that usually the word "virtual" can be dropped for most orbits on C. Virtual periods first appear in [28], see [4,5,9,15] for further discussion of the non-equivariant case $\Gamma = \{0\}$.

Unlike minimal period and symmetry, the notions of virtual period and virtual symmetry are closed under limits. This is significant because it opens the door for generic approximations.

2.3 Proposition

Let $u_n(t)$ be solutions of

$(2.1)_n \quad d_t u_n = A_n(\lambda_n)u_n + f_n(\lambda_n,u_n)$

with virtual period q_n and virtual symmetry $(\hat{H},\hat{K},\hat{\Theta})$. Assume that $A_n, f_n, \lambda_n, u_n(\cdot), q_n$ converge in suitable norms to A, f, λ, $u(\cdot), q$ such that u(t) solves (2.1).

Then q is a virtual period of $u(\cdot)$ and $(\hat{H},\hat{K},\hat{\Theta})$ is a virtual symmetry.

§3. Example

In this section we apply theorem 2.1 to reaction diffusion systems

$$(1.1.a) \qquad \begin{pmatrix} 1 & 0 \\ 0 & \lambda \end{pmatrix} u_t = \Delta u + f(u) , \qquad x \in \Omega \quad \mathbb{R}^N, \lambda \in \mathbb{R}^+$$

$$(1.1.b) \qquad \partial_\nu u = 0 \qquad , \qquad x \in \partial\Omega$$

with domains Ω of symmetry $\Gamma = D_n$ (e.g. a "smoothed" regular n-gon in $\mathbb{R}^2$) or $\Gamma = O(3)$ ($\Omega = B_R(0)$, a ball of radius R in $\mathbb{R}^3$). See [7,19,21,24,35,36,39,40,42,43] for a detailed analysis of local stationary and Hopf bifurcations with such symmetries. For a kinetics we choose the Brusselator [30], for simplicity: admittedly this is not the most realistic model for the Belousov-Zhabotinskii reaction, cf. [17,30,41] . In §5 we will mention some additional applications.

Because the parameter λ in (1.1) just multiplies a time derivative, assumptions (2.3.a,c) are automatically satisfied for any stationary solution U of (1.1) with only nonzero eigenvalues at $\lambda=1$, say. Any such solution yields a "trivial branch" (λ, U). However, we need $\lambda > 0$ instead of $\lambda \in \mathbb{R}$ - just a minor discrepancy with §2 if we map $\mathbb{R}^+$ onto $\mathbb{R}$. Due to the Neumann boundary condition, we may even take U independent of $x \in \Omega$ to be a zero of f

$$(3.1) \qquad f(U) = 0 ;$$

in particular this implies $U \in X^\Gamma$ (cf. (1.2),(1.8)). The linearization becomes

$$L(\lambda) = \begin{pmatrix} 1 & 0 \\ 0 & 1/\lambda \end{pmatrix} \cdot (\Delta + B) , \text{ where}$$

$$B = (b_{ij}) = f'(U)$$

is a constant 2×2 matrix. A short calculation shows that $\mu \in \text{spec}\,(L(\lambda))$ iff

$$\lambda\mu^2 + ((\mu_\Delta - b_{11})\lambda + (\mu_\Delta - b_{22}))\mu + p_B(\mu_\Delta) = 0$$

for some $\mu_\Delta \in \mathrm{spec}\,(-\Delta)$; here $p_B(\mu_\Delta)=\mu_\Delta^2-\mathrm{tr}B\cdot\mu_\Delta+\det B$ stands for the characteristic polynomial of B, and μ_Δ is nonnegative. Thus (λ,U) is a center iff

$$(3.2.a) \qquad \lambda = -\frac{\mu_\Delta - b_{22}}{\mu_\Delta - b_{11}} > 0 \ , \text{ and}$$

$$(3.2.b) \qquad p_B(\mu_\Delta) > 0 \ ,$$

for some $\mu_\Delta \in \mathrm{spec}\,(-\Delta)$. For our particular Brusselator kinetics

$$f(u_1,u_2) = (1-(b+1)u_1+u_1^2u_2,\ bu_1-u_1^2u_2)$$

we obtain $U=(1,b)$, $b_{11}=b-1$, $b_{12}=1$, $b_{21}=-b$, $b_{22}=-1$, and positivity conditions (3.2.a), (3.2.b) are satisfied, respectively, iff

$$(3.3.a) \qquad \mu_\Delta + 1 < b$$

$$(3.3.b) \qquad b < \mu_\Delta + 2 + 1/\mu_\Delta$$

At each such center the purely imaginary pair $\mu,\bar{\mu}$ crosses the imaginary axis from left to right, as λ increases, because

$$\frac{d}{d\lambda}\,\mathrm{Re}\ \mu(\lambda) = \frac{\mu_\Delta - b_{22}}{2\lambda^2} = \frac{\mu_\Delta + 1}{2\lambda^2} > 0 \ .$$

Summarizing the above discussion, we observe that assumptions (2.3.a-c) all hold and, moreover,

$$(2.7) \qquad H^d_{H_0,K_0} \neq 0$$

provided that

(3.4.a) $-\Delta$ has an eigenvalue μ_Δ on Ω such that (3.3.a,b) hold, and there exists some $r \in d$ (mod . n) such that the representation r of H_0/K_0 occurs in the eigenspace

$$\mathrm{Eig}\,(-\Delta,\mu_\Delta) \cap X^{K_0} \ , \text{ and}$$

(3.4.b) each $\mu'_\Delta \in \mathrm{spec}\,(-\Delta)$ satisfies

$$\mu'_\Delta + 2 + 1/\mu'_\Delta \neq b \ .$$

Given $b>1$, any cyclic factor H_0/K_0 and any $d\in D(n')$ these conditions certainly hold, if we rescale Ω homothetically to become suitably large.

But what, then, are the conclusions? Take $\Omega\subseteq\mathbb{R}^2$ with symmetry D_n, first. Choose $H_0 = \mathbb{Z}_n$, $K_0 = \{id\}$, $d\in D(n)\setminus\{0\}$. We obtain a global continuum C with virtual symmetry (H_0,K,Θ) (not just "at least" (H_0,K,Θ)), $\Theta\in d \pmod n$. Indeed, suppose we have a virtual symmetry $(\hat{H},\hat{K};\hat{\Theta})$ with $\hat{H}\geq H_0$, $\hat{\Theta}|_{H_0} = \Theta$. Then $\hat{H}=H_0=\mathbb{Z}_n$, or else $\hat{H}=D_n$ which leads to a contradiction as follows. The commutator subgroup $\hat{H}'$ of $\hat{H}$ (generated by all $g^{-1}h^{-1}h$ with $g,h\in\hat{H}$) is $\mathbb{Z}_n$ or $\mathbb{Z}_{n/2}$, $\hat{K}\geq\hat{H}'$ because $\hat{H}/\hat{K}$ is abelian, hence $\hat{K}=\mathbb{Z}_n$ or D_n because $\hat{H}/\hat{K}$ is cyclic, and thus $K=\hat{K}\cap H_0=H_0$, in contradiction to $\Theta\neq 0$. Thus we obtain a global branch of periodic solutions with (virtual) symmetry

$$(\mathbb{Z}_n,K,\Theta), \qquad \Theta\in d \pmod n.$$

These solutions could be interpreted as discrete analogues of rotating waves.

Alexander and Auchmuty [1] consider a discrete ring of diffusively coupled Brusselator CSTRs. In our language, they obtain global continua of periodic solutions with $\Theta\neq 0$ (using the word "global" in a slightly different sense). Complementarily, we find several global continua corresponding to different d, if $D(n')\setminus\{0\}$ has more than just one element. By table 2.1 this is the case, e.g., if 9, 15, 31 or 127 divides n.

Next we consider briefly $\Omega=B_R(0)\subseteq\mathbb{R}^3$, $\Gamma=O(3)$, with the help of [19,§§14,15]. Pick $H_0=\mathbb{Z}_n^c \oplus T$, $K_0=D_2$, $H_0/K_0\cong\mathbb{Z}_6$ where T denotes the tetrahedral subgroup of $SO(3)$, $\mathbb{Z}_2^c = \{\pm id\}$ is the center of $O(3)$, and $D_2\leq T$ is generated by rotations over π around the three axes joining opposite midedges of the tetrahedron. Our notation follows [19,24]. By arguments analogous to the above discussion we obtain branches with virtual symmetry

$$(\mathbb{Z}_2^c \oplus T,K,\Theta), \qquad \Theta=\pm 1 \pmod 3),$$

$D_2 \leq K \leq \mathbb{Z}_2^c \oplus D_2$.

As a side remark we note that we can even obtain generic global bifurcations of such periodic solutions, with dimension ≥ 4 of the associated isotropy subspace (cf. [19]). This occurs for eigenvalues μ_Δ which belong to irreducible representations of dimension $2\ell+1$, provided that $\ell \geq 8$, ℓ even, or $\ell \geq 11$, ℓ odd.

§4. Generic bifurcations for $\mathbb{Z}_n$

Skipping proofs, this section contains some main ideas for the proof of theorem 2.1. The heart consists of a local analysis of all generic $\mathbb{Z}_n$-equivariant Hopf and secondary bifurcations plus a global orbit index Φ. For the case $\Gamma = \{id\}$ see [27,28] for ODEs, and [14] for analytic semigroups.

To obtain an equivariant degree theory for just stationary solutions, e.g. zeros of

$$f(\lambda,\cdot) : X \to X , \quad \dim X<\infty ,$$

we may consider the collection of $f^{K_0} := f|_{X^{K_0}}$ for (isotropy) subgroups K_0 of Γ, and their respective Brouwer degrees $\deg^{K_0}$. We may construct Brouwer degree and prove homotopy invariance by generic approximation of f^{K_0}, ignoring the equivariance with respect to the action of the normalizer $N(K_0)$ on X^{K_0} in this approximation. In some cases (e.g. for X a ball) the $\deg^{K_0}$ already determine the homotopy classes of f, see [12, theorem 8.4.1].

Similarly (but without a similar justification) we approximate f in

$$(4.1) \qquad d_t u = A(\lambda)u + f(\lambda,u) , \quad u \in X, \ \dim X=\infty,$$

by generic but only H_0-equivariant f_n on X^{K_0}. Thus <u>generic</u> means, below, that something holds for f in a countable intersection of open dense sets of H_0-equivariant maps $\Lambda\times X^{K_0} \to X^{K_0}$ in a suitable (uniform) C^k-topology, cf. [14, (1.2)]. In this framework it is sufficient to prove a generic global theorem for $\mathbb{Z}_n$-equivariance. But, ignoring all symmetry above H_0, we can only expect (virtual) symmetries at least (H_0,K,Θ) for our global continuum. Analogously, for stationary solutions, $\deg^{K_0}$ yields only global continua with isotropy at least K_0, cf. (1.11), essentially by the global theorem of Rabinowitz [32] (see [7,31]).

To simplify notation a bit let us write X instead X^{K_0}, $G = \mathbb{Z}_n = \{0,\frac{1}{n},\ldots,\frac{n-1}{n}\}$ for H_0 and $\{0\}$ for K_0, when formulating our $\mathbb{Z}_n$-generic results.

We call (λ_0,U_0) a generic H-center, $H\le H_0\le\mathbb{Z}_n$, if $G_{U_0}=H$ and the linearization L at (λ_0,U_0) has only one complex pair of purely imaginary eigenvalues, which we require to be simple and to cross the imaginary axis transversely along the stationary branch. Let r denote the representation of H on their eigenspace.

4.1 Theorem:

For generic f any center is a generic H-center, for some $H\le\mathbb{Z}_n$.

At a generic H-center a unique local branch of periodic solutions bifurcates (Hopf-bifurcation). These periodic solutions have symmetry (H,K,Θ) with

$$(4.2) \qquad \Theta \equiv \pm r \qquad (\mathrm{mod}\,|H|).$$

For example, consider the case $\Gamma = D_n$, $H = H_0 = \mathbb{Z}_n$ again. Γ-generically, pairs of double imaginary eigenvalues occur [21]. But H-generically, i.e. if we destroy the reflection symmetry from D_n, such double eigenvalues split, become simple and yield solutions rotating in opposite directions with $\Theta = \pm r$.

We begin to describe generic secondary bifurcations of periodic solutions $(\lambda,(u(t))$ with symmetry (H,K,Θ), $H \leq G = \mathbb{Z}_n$. Following [28] we call $(\lambda,U(t))$ a <u>type 0</u> orbit if the only Floquet-multiplier which is a root of unity is the simple trivial one. Note that tori may bifurcate from a type 0 orbit, but not periodic solutions which are our main concern here. A first periodic bifurcation possibility is the <u>turn</u>: two periodic solutions $(\lambda, u(t))$, $(\lambda,\tilde{u}(t))$ with the same symmetry approach each other and annihilate as λ increases (or decreases); this was called "type 1" for $H = \{0\}$ in [28].

To describe other possibilities, let Π be a Poincaré-map of the primary orbit $(\lambda,u(t))$ with Poincaré section $S \perp \langle u_t(0)\rangle$. Choose $h \in H$, generating H, such that $\Theta(h)$ generates $\Theta(H)$, i.e. $\Theta(h) \equiv \Theta \cdot h \equiv 1/m \pmod 1$ with $m := |H/K|$. This defines a local "Poincaré"-map $\tilde{\Pi}_0 : S \to hS$, and we put $\Pi_0 := h^{-1}\tilde{\Pi}_0$,

$$\Pi_0 : S \to S .$$

This map Π_0 has two advantages: it is a generic, but only K-equivariant, iteration on S for generic f; and it contains all information on Π via

$$\Pi = h^m \, \Pi_0^m \tag{4.3}$$

which holds due to H-equivariance of the semiflow. Note that $u(0) \in S$ is a fixed point of Π, Π_0 and $h^m \in K$.

We call a pitchfork bifurcation of Π_0, or Π_0^2, a <u>flip</u> resp. <u>flop</u> resp. <u>flip-flop</u> according to table 4.1. It lists the simple eigenvalue ± 1 of $D_x\Pi_0$ where this bifurcation occurs, together with the trivial (+) or nontrivial (-) one-dimensional real representation of K on that eigenspace. The case +1, + is a turn.

name	eigenvalue	representation
flip	-1	+
flop	+1	-
flip-flop	-1	-

Table 4.1: Generic bifurcations of Π_0.

4.2 Theorem:

Let f be generic. Then any periodic solution is one out of the list in table 4.2.

Table 4.2 uses the following terminology. "Pitchfork" and "doubling" refers to Π,Π^2, whereas "flip", "flop" and "flip-flop" belong to Π_0,Π_0^2. Quantities p,H,K,Θ,m,h belong to the primary orbit $(\lambda,u(t))$; $\tilde{p},\tilde{H},\tilde{K},\tilde{\Theta}$ refer to the secondary, bifurcating branch. We write $|K| = 2^{\kappa}\cdot k'$, k' odd, and $\tilde{p}/p$ for the limiting ratio of minimal periods at he bifurcation point. Turn and flip pitchfork are also discussed in [26], assuming n=2.

m	geometry	name	$\tilde{p}/p$	$\lvert\tilde{H}\rvert$	$\lvert\tilde{K}\rvert$	$\tilde{\Theta}$	$\tilde{m}$	$\tilde{h}$
-		type 0	-	-	-	-	-	-
-		turn	1	$\lvert H\rvert$	$\lvert K\rvert$	Θ	m	h
-	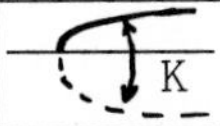	flop doubling	2	$\lvert H\rvert$	$\lvert K\rvert/2$	$\Theta/2$	2m	h
m odd	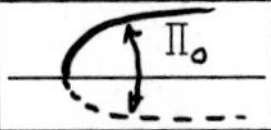	flip doubling	2	$\lvert H\rvert$	$\lvert K\rvert$	$\tilde{\Theta}(h)=\frac{1}{2m}+\frac{1}{2}$	m	(k'm+2)h
m even	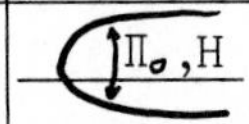	flip pitchfork	1	$\lvert H\rvert/2$	$\lvert K\rvert$	$\tilde{\Theta}(2h)=2\cdot\frac{1}{m}$	m/2	2h
m odd	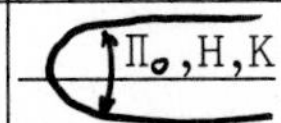	flip-flop pitchfork	1	$\lvert H\rvert/2$	$\lvert K\rvert/2$	$\tilde{\Theta}(2h)=2\cdot\frac{1}{m}$	m	(k'm+1)h,or (3k'm+1)h
m even	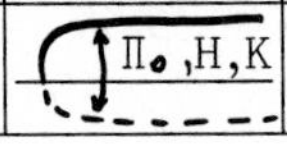	flip-flop doubling	2	$\lvert H\rvert$	$\lvert K\rvert/2$	$\tilde{\Theta}(h)=\frac{1}{2m}+\frac{1}{2}$	2m	(k'm+1)h

Table 4.2: Generic periodic orbits

4.3 Corollary:

Consider bifurcations from table 4.2 with $H = G = \mathbb{Z}_n$. Then

(4.4) $\Theta \in \{\tilde{\Theta}, 2\tilde{\Theta}\} \pmod{n'}$.

This was the local story. The global aspect is encoded in an equivariant orbit index Φ defined for periodic orbits $(\lambda, u(t))$ with minimal period p and symmetry (H_0, K, Θ). We consider Floquet exponents η; i.e. eigenvalues η of

(4.5) $d_t v - A(\lambda)v - D_u f(\lambda, u(t))v = -\eta v(t)$

on the spaces E_b^a , $a, b \in \{\pm 1\}$, of $v \in C^1(\mathbb{R}, X)$ satisfying

(4.6.a) $y(t+p) = ay(t)$

(4.6.b) $hy(t) = by(t+\Theta(h)\cdot p)$.

Denote

(4.7) σ_b^a : the number of positive eigenvalues η of (4.5) on the space E_b^a , counting algebraic multiplicity.

Then the equivariant orbit index Φ is defined as

(4.8) $$\Phi = \begin{cases} (-1)^{\sigma_+^+} & , \text{ if all } \sigma_b^- \text{ are even} \\ 0 & \text{otherwise.} \end{cases}$$

The index Φ has some charming properties. It is homotopy invariant (just like Brouwer degree). It ignores flip and flip-flop pitchforks, where H would break. Near generic H - centers (λ_0, U_0) it is given by

(4.9) $$\Phi = (-1)^{E^0(\lambda_0)} \chi_r^{loc}(\lambda_0)\cdot \text{sign}\,(\lambda-\lambda_0)$$

where $\chi_r^{loc}(\lambda_0) = \pm 1$ depending on the crossing direction of the purely imaginary pair of eigenvalues at (λ_0, U_0). These properties are reminiscent of [14,28].

Piecing these ingredients together one finds a global theorem for generic f_n, similar to theorem 2.1 but dropping the word "virtual". Passing to the nongeneric limit $f_n \to f$, "virtual" gets introduced by proposition 2.3. At last, assumption (2.8.c) becomes effective: it avoids that C consists only of centers, cf. [14,§4].

§5. Comments

We discuss two kinds of extensions of theorem 2.1, here. For simplicity of presentation in the preceding sections we fixed a particular technical frame which we are leaving behind now. Then we probe into the bounds of the scope of our result, including open questions.

In assumption (2.3.a) we require existence of a smooth "trivial branch" with nondegenerate linearization, and theorem 2.1 allows a global continuum C to terminate at another stationary branch. We may in fact consider several nondegenerate branches adding their respective contributions, defined as in (2.6), to a global Hopf index $H^d_{H_0,K_0}$. Only now, the sign convention in (2.6) becomes effective: if $H\neq 0$ then C may not just connect stationary branches which contribute to H.

However, the nondegeneracy assumption (2.3.a) for stationary branches is not really innocent. The group action of Γ may actually force degeneracies for certain isotropies Γ_U of a stationary U, at least if $N(\Gamma_U)/\Gamma_U$ is a continuous group for the normalizer $N(\Gamma_U)$ of Γ_U (e.g. $N(\mathbb{Z}_n) = O(2)$ in $O(2)$). In our examples we have chosen U with $\Gamma_U = \Gamma$, for this reason and also because homogeneous U were the easiest to compute. Still, we could have considered $\Gamma = O(3)$, $\Gamma_U \geq \mathbb{Z}_2^c \oplus T$ and avoided the Γ-orbit complication.

Restricting our attention to $H_0/K_0 = \mathbb{Z}_n$, finite, excludes an important class of examples: <u>rotating waves</u>, defined by

$H_0/K_0 \cong \text{im}\,\Theta \cong S^1$. Denoting the infinitesimal generator of this S^1 action on X^{K_0} by R, rotating waves satisfy

$$\alpha R u = A(\lambda)u + f(\lambda,u)$$

for some real $\alpha = (\Theta p)^{-1}$, $\Theta \in \mathbb{Z} \setminus \{0\}$, $p>0$, resp. $\alpha=0$ for stationary solutions (standing waves). In this setting there is again a global result, analogous to theorem 2.1, but in terms of (λ,u) and α instead of (λ,u) and minimal period p. Note that α is bounded, a priori! The set $D(\infty)\setminus\{0\}$ $(n'=\infty)$ is the set of all

$$d = \{\pm 2^{\kappa} d' : \kappa \geq 0\} , \quad d' \text{ odd},$$

this time; just like the Šarkovskii sequence.

Many other examples are amenable to our present analysis, moreover, they come up naturally in a symmetric arrangement. Catalysis in spherical pellets originally motivated our Lewis-number type scaling. It exhibits rotating hot spots, which can be treated along the lines of [13]. From fluid dynamics we mention Taylor-Couette flow as a prominent example [20,33]; this example motivates one of our open questions below. A general source of ODE-examples are graphs of coupled oscillators: identical oscillators sitting on the vertices are coupled along the edges of a graph with symmetry group Γ. The action is permutation of appropriate vertices and edges. Coupling may be by linear diffusion [1,37,41], by nonlinear (e.g. electric) coupling [11] or, leaving the ODE-setting, by time convolution as in neural nets (cf. [15, example 4.2]).

As some limitations, which are intrinsic to global Hopf bifurcation results, we would like to mention stability, homoclinic orbits and chaotic motions. For a stable periodic solution $\Phi = +1$, cf. (4.8), but the converse need not be true. Thus theorem 2.1 may predict hosts of periodic solutions, most of which are born unstable and remain unstable for all λ. Only numerics can help here. Perturbing symmetries above H slightly, we may in fact read generic lists like table 4.2 as a guide to numerical pathfollowing schemes. Another limitation comes from alternative (2.8.b) for global continua C which allows (minimal, virtual) periods to blow up. This phenomenon is intimately related to homoclinic orbits, sequences of period doublings and, e.g., flow plugs; for more discussion see [2,16, 22]. Thus global Hopf bifurcation may at best hint at, but cannot penetrate into, regions with complicated dynamics.

We finish with three open questions, returning to equivariant Hopf bifurcation. A less implicit but theoretically more satisfying definition of H is by sums of (equivariant) center indices of generic approximations, analogously to [16]. This would allow to drop nondegeneracy conditions like (2.3.a), but it may yield conflicting definitions of H, typically if a multiple eigenvalue zero occurs. This obstruction is due to the fact that H is not necessarily homotopy invariant as the (one-parameter) nonlinearity $f(\lambda,u)$ is varied - a two-parameter problem like [16] but with equivariance. The first question is:

(5.1) What are the obstructions to homotopy invariance of $H^{d}_{H_0,K_0}$?

Recall that the group Γ itself may already force degeneracies, because stationary solutions may occur in manifolds even for fixed λ. Related to (5.1), this leads to:

(5.2) What are suitable equivariant center indices for manifolds of centers?

Here "suitable" refers to the above construction of H, of course.

In Taylor-Couette flow, e.g. one can see "rotating waves" with an additional modulating frequency called "modulated wavy vortices". These correspond to invariant tori foliated both by periodic group orbits and by quasi-periodic solution orbits, differing from the group orbits, cf. [33,34] . Usually tori are quite fragile and may break in many ways, cf. [6]. But plain rotating waves behave more like stationary solutions from a global point of view. "Thus" foliated tori may behave more like just periodic solutions, and we ask:

(5.3) What can be said about periodically foliated tori, globally?

Making any of these question precise is part of the question.

References:

1. J.C. Alexander, J.F.G. Auchmuty: Global bifurcations of phase-locked oscillators, to appear in Arch. Rat. Mech. Anal.
2. J.C. Alexander, J.A. Yorke: Global bifurcations of periodic orbits, Amer. J. Math. 100 (1978), 263-292.
3. K.T. Alligood, J. Mallet-Paret, J.A. Yorke: Families of periodic orbits: local continuability does not imply global continuability, J. Diff. Geom. 16 (1981), 483-492.
4. K.T. Alligood, J. Mallet-Paret, J.A. Yorke: An index for the continuation of relatively isolated sets of periodic orbits, in [44], 1-21.
5. K.T. Alligood, J.A. Yorke: Families of periodic orbits: virtual periods and global continuability, J. Diff. Eq. 55 (1984), 59-71.
6. D.G. Aronson, M.A. Chory, G.R. Hall, R.P. McGehee: Bifurcations from an invariant circle for two-parameter families, Comm. Math. Phys. 83 (1982), 303-354.
7. G. Cerami: Symmetry breaking for a class of semilinear elliptic problems, Nonlin . Analysis TMA 10 (1986), 1-14.
8. S.-N. Chow, J.K. Hale: Methods of Bifurcation Theory, Springer-Verlag, New York 1982.
9. S.-N. Chow, J. Mallet-Paret, J.A. Yorke: A periodic orbit index which is a bifurcation invariant, in [44], 109-131.
10. M.G. Crandall, P.H. Rabinowitz: The Hopf bifurcation theorem in infinite dimensions, Arch. Rat. Mech. Anal. 67 (1977), 53-72.
11. M.F. Crowley, R.J. Field: Electrically coupled Belousov-Zhabotinsky oscillators: a potential chaos generator, in "Nonlinear Phenomena in Chemical Dynamics", C. Vidal, A. Pacault (eds.), Springer-Verlag, Berlin 1981.
12. T. tom Dieck: Transformation Groups and Representation Theory, Lect. Notes in Math. 766, Springer-Verlag, Berlin 1981.

13. B. Fiedler: Global Hopf bifurcation in porous catalysts, in "Equadiff 82", H.W. Knobloch, K. Schmitt (eds.), Lect. Notes in Math. 1017, Springer-Verlag, Berlin 1983, 177-183.

14. B. Fiedler: An index for global Hopf bifurcation in parabolic systems, J. reine u. angew. Math. 359 (1985), 1-36.

15. B. Fiedler: Global Hopf bifurcation for Volterra integral equations, to appear in SIAM J. Math. Analysis.

16. B. Fiedler: Global Hopf bifurcation of two-parameter flows, to appear in Arch. Rat. Mech. Analysis.

17. R.J. Field, R.M. Noyes: Oscillations in chemical systems IV. Limit cycle behavior in a model of a real chemical system, J. Chem. Phys. 60 (1974), 1877-1884.

18. M. Golubitsky, D.G. Schaeffer: Singularities and Groups in Bifurcation Theory I, Springer-Verlag, New York 1985.

19. M. Golubitsky, I. Stewart: Hopf bifurcation in the presence of symmetry, Arch. Rat. Mech. Analysis 87 (1985), 107-165.

20. M. Golubitsky, I. Stewart: Symmetry and stability in Taylor-Couette flow, preprint 1984.

21. M. Golubitsky, I. Stewart: Hopf bifurcation with dihedral group symmetry: coupled nonlinear oscillators, preprint 1985.

22. J. Harrison, J.A. Yorke: Flows on S^3 and $\mathbb{R}^3$ without periodic orbits, in [44], 401-407.

23. D. Henry: Geometric Theory of Semilinear Parabolic Equations, Springer-Verlag, New York 1981.

24. E. Ihrig, M. Golubitsky: Pattern selection with O(3) symmetry, Physica 13D (1984), 1-33.

25. J. Ize: Bifurcation Theory for Fredholm Operators, AMS memoir 174, Providence 1976.

26. A. Klič: Bifurcations in the systems with involutory symmetry, preprint 1985.

27. J. Mallet-Paret, J.A. Yorke: Two types of bifurcation points: sources and sinks of families of periodic orbits, in "Nonlinear Dynamics", R.H.G. Helleman (ed.), Ann. NY Acad. Sc. 357, New York 1980, 300-304.

28. J. Mallet-Paret, J.A. Yorke: Snakes: oriented families of periodic orbits, their sources, sinks, and continuation, J. Diff. Eq. 43 (1982), 419-350.

29. J.E. Marsden, M. McCracken: The Hopf Bifurcation and its Applications, Springer-Verlag, New York 1976.

30. G. Nicolis, I. Prigogine: Self-Organization in Nonequilibrium Systems, John Wiley & Sons, New York 1977.

31. C. Pospiech: Globale Verzweigungen mit Symmetriebrechung für ein Dirichlet-Problem, Diplomarbeit, Heidelberg 1984.

32. P.H. Rabinowitz: Some global results for nonlinear eigenvalue problems, J. Fct. Anal. 7 (1971), 487-513.

33. D. Rand: Dynamics and Symmetry. Predictions for modulated waves in rotating fluids, Arch. Rat. Mech. Analysis 79 (1982), 1-37.

34. M. Renardy: Bifurcation from rotating waves, Arch. Rat. Mech. Analysis 79 (1982), 49-84.

35. D.H. Sattinger: Group Theoretic Methods in Bifurcation Theory, Lect. Notes Math. 762, Springer-Verlag, Heidelberg 1979.

36. D.H. Sattinger: Branching in the Presence of Symmetry, Reg. Conf. Ser. Appl. Math. 40, SIAM, Philadelphia 1983.

37. I. Schreiber, M. Marek: Strange attractors in coupled reaction-diffusion cells, Physica 5D (1982), 258-272.

38. J.-P. Serre: Linear Representations of Finite Groups, Springer-Verlag, New York 1977.

39. J.A. Smoller, A.G. Wasserman: Symmetry-breaking for positive solutions of semilinear elliptic equations, to appear in Arch. Rat. Mech. Analysis.

40. J.A. Smoller, A.G. Wasserman: Symmetry-breaking for solutions of semilinear elliptic equations with general boundary conditions, preprint 1985.

41. J.J. Tyson: Oscillations, bistability, and echo waves in models of the Belousov-Zhabotinskii reaction, in "Bifurcation Theory and Applications in Scientific Disciplines", O. Gurel, O.E. Rössler (eds.), Ann. NY Acad. Sc. 316, New York 1979, 279-295.

42. A. Vanderbauwhede: Local Bifurcation and Symmetry, Res. Notes Math. 75, Pitman, London 1982.

43. A. Vanderbauwhede: Symmetry-breaking at positive solutions, preprint 1986.

44. J. Palis Jr. (ed.): Geometric Dynamics, Lect. Notes Math. 1007, Springer-Verlag, New York 1983.

LONGTIME BEHAVIOR FOR A CLASS OF ABSTRACT INTEGRODIFFERENTIAL EQUATIONS

W. E. Fitzgibbon
Department of Mathematics
University of Houston
Houston, Texas 77004/USA

We are concerned with semilinear integrodifferential equations of the form,

$$\dot{x}(\psi)(t) = AX(\psi)(t) + \int_{-\infty}^{t} g(t,\ s\ , x(\psi)(s))ds \qquad (1.\ a)$$

$$x(\psi)(\theta) = \psi(\theta) \qquad \theta\in(-\infty,\ 0] \qquad (1.\ b)$$

$$\psi \in C_\alpha \qquad \text{For some } \alpha\in(0,\ 1) \qquad (1.\ c)$$

Here A is required to be the infinitesional generator of an analytic semi-group $\{T(t)\mid t \geq 0\}$ acting on a Banach space $\underline{\overline{X}}$. We further stipulate that $0 \in \rho(A)$. For $\alpha\in(0,1)$, A^α denotes the fractional power of the operator A and $\underline{\overline{X}}_\alpha$ represents the interpolation space defined by the α power of A, i.e.

$$\underline{\overline{X}}_\alpha = \{x \mid x \in D(A^\alpha)\}$$

with

$$\|x\|_\alpha = \|A^\alpha x\|.$$

The space C_α is the space of bounded uniformly continuous functions from $(-\infty, 0]$ to $\underline{\overline{X}}_\alpha$ endowed with the supremum norm,

$$\|\psi\|_{C_\alpha} = \sup\ \{\|\psi(\theta)\|_\alpha \mid \theta\in(-\infty,\ 0]\}.$$

The nonlinearity $g(\ ,\ ,)$ is a mapping of $R^+ \times R \times \underline{\overline{X}}_\alpha$ to $\underline{\overline{X}}$.

An integrodifferential equation of the form (1 a-c) is frequently said to have infinite delay. There is an extensive recent literature concerning equations with infinite delay; a sample might include [1], [2], [3], [7], [8], [9], [12], [14], [19]. Closely related are papers on abstract functional equations with finite delay; c.f. [4], [5], [6], [7], [11], [17], [18], [19], [22], [25], [26], [29]. The contribution of the present work is that we are able to obtain asymptotic convergence results for equations having initial data in spaces of bounded, uniformly continuous functions; a more elaborate and detailed development of the results will appear in [7] and [8]. A simple

NATO ASI Series, Vol. F37
Dynamics of Infinite Dimensional Systems
Edited by S.-N. Chow, and J.K. Hale

integral comparison principle adopted from Redlinger [22] underpins our work.

In order to describe the comparison principle we need to introduce a scalar integral operator. Let $h_1(\ )$ and $h_2(\ )$ be nonnegative scalar functions such that $\int_0^c h_1(s)ds$, $\int_{-\infty}^c h_2(s)ds < \infty$ for all c; let $p(\ )$ be a continuous scalar function on $[0, \infty)$. If $y(\ )$ is a continuous non-negative function on $(-\infty, T)$, $(0 < T \le \infty)$, then the integral operator S is given by:

$$(S_y)(t) = p(t) + \int_0^t h_1(t-s) \int_{-\infty}^s h_2(s-r)y(r)drds$$

We have the following lemma:

Lemma 2. Let S be defined as above and act on continuous nonnegative functions y and $z(\ )$ for $t\in(-\infty, T)$, $(0 < T \le \infty)$.

If

$$y(t) - (S_y)(t) < z(t) - (S_z)(t) \qquad \text{for } t > 0$$

and

$$y(t) < z(t) \qquad \text{for } -\infty < t < T$$

Proof. If we set $t_0 = \inf\{t : y(t) = z(t)\}$, we may observe that $z(t_0) = y(t_0) < y(t_0)+(S_y)(t_0)-(S_z)(t_0) < z(t_0)$ and reach a contradiction.

In order to obtain a global existence result (1.a–c) we place restrictions upon our nonlinear kernel $g(,\ ,)$:

(G.1) The function $g(,\ ,) : R^+ \times R \times \overline{X}_\alpha \longrightarrow \overline{X}$ is continuous, $g(t,s,0) = 0$ for all $s \le t$ and then exists an $L > 0$ and $\gamma > 0$ so that

$$\|g(t,s,x)-g(t,s,y)\| \le e^{-\gamma(t-s)}L\|x-y\|_\alpha \tag{3}$$

and

$$\|g(t,s,x)-g(t',s,x)\| \le |t-t'|L\|x\|_\alpha \tag{4}$$

We state the global result.

Theorem 5. Let A and $g(,\ ,)$ satisfy the foregoing conditions. If $\psi \in C_\alpha$ and $\psi(0) \in A^\beta$ for some $\beta > \alpha$ then there exists a unique function $x(\psi)(\ ):R\longrightarrow\overline{X}$ such that

$$x(\psi)(t) = T(t)\psi(0) + \int_0^t T(t-s) \int_{-\infty}^s g(s,r,x(\psi)(r))drds. \tag{6}$$

Moreover, $x(\psi)(t)$ is continuously differentiable for $t > 0$ and satisfies (1.a–c).

Equation (6) is an integrated or variation of parameters form of the solution to (1.a-c). A local result is obtained via the expected Banach Fixed Point Theorem argument. The local result is extended by establishing bounds for $\|x(\psi)(t)\|_\alpha$. This is effected by using the comparison principle in a manner similar to that which will be used in obtaining covergence results. The regularity follows from the regularity theory for the abstract in homogeneous Cauchy problem

$$\dot{u}(t) = Au(t) + f(t)$$

which may be found in [20].

To obtain convergence results, we place a stronger condition on the nonlinearity $g(, ,)$ - - effectively this places a upper bound on the Lipschitz constant. We also need to use the fact that the existence of positive constants K_α and $\omega > 0$ so that

$$\|A^\alpha T(t)\| \le K_\alpha t^{-\alpha} e^{-\omega t}.$$

The stronger condition on $g(, ,)$ follows:

(G.2) In addition to (G.1) assume that

$$K_\alpha L\Gamma(1-\alpha)\gamma^{-1}\omega^{\alpha-1} < 1$$

Here $\Gamma(\ )$ denotes the Eulerian Gamma Function. Our asymptotic convergence result follows.

Theorem 7. Suppose that $g(, ,)$ satisfies (G.2) and A is as before. If $\psi, \psi \in C_\alpha$ there exists a $\delta > 0$ and $D > \|\phi-\psi\|_{C_\alpha}$ so that

$$\|x(\phi)(t) - x(\psi)(t)\|_\alpha < De^{-\delta t}$$

Proof. Applying A^α to the difference $x(\phi)(t) - x(\psi)(t)$ we obtain the inequality

$$\|x(\phi)(t) - x(\psi)(t)\|_\alpha \le \|A^\alpha T(t)\phi(o) - T(t)\psi(o)\|$$
$$+ \left\|\int_0^t A^\alpha\ T(t-s) \int_{-\infty}^s (g(s,r,x(\phi)(r) - g(s,r,x(\psi)(r)))drds\right\|.$$

Using the decay property of the semigroup and $g(, ,)$ we have

$$\|x(\phi)(t) - x(\psi)(t)\|_\alpha \le K_\alpha\ t^{-\alpha}\ e^{-\omega t}\ \|x(\psi)(o) - x(\psi)(o)\|_\alpha$$
$$+ \int_0^t K_\alpha(t-s)^{-\alpha}\ e^{-\omega(t-s)} \int_{-\infty}^s L\ e^{-\gamma(s-r)}\ \|x(\phi)(r) - x(\psi)(r)\|_\alpha drds$$

If $t > T_0 > 0$ then there exists $K > \|\phi-\psi\|_{C_\alpha}$ such that,

$$\|x(\phi)(t) - x(\psi)(t)\|_\alpha \le K\ e^{-\omega t}$$
$$+ K_\alpha\ L \int_0^t (t-s)^{-\alpha}\ e^{-\omega(t-s)} \int_{-\infty}^s e^{-\gamma(s-r)}\|x(\phi)(r) - x(\psi)(r)\|_\alpha drds.$$

We select a comparison function

$$z(t) = De^{-\delta t} \qquad t \in (-\infty, \infty)$$

and observe that if $0 < \delta < \min(\omega, \gamma)$ and $D > 0$ are chosen so that $D > K + K_\alpha L\Gamma(1-\alpha)(\gamma-\delta)^{-1}(\omega-\delta)^{\alpha-1} D$, then we have

$$K e^{-\delta t} + e^{-\delta t}K_\alpha LD\int_0^t (t-s)^{-\alpha} e^{-(\omega-\delta)(t-s)} \int_{-\infty}^s e^{-(\gamma-\delta)} drds$$

$$\le K e^{-\delta t} + e^{-\delta t} K_\alpha L\Gamma(1-\alpha)(\gamma-\delta)^{-1}(\omega-\delta)^{-\alpha-1} D \le D e^{-\delta t} = z(t)$$

We let $y(t) = \|x(\phi)(t) - x(\psi)(t)\|$ and compare the previous inequalities to see that

$$y(t) - K_\alpha L\int_0^t (t-s)^{-\alpha} e^{-\omega(t-s)} \int_{-\infty}^s e^{-\gamma(s-r)} y(r) drds$$

$$< z(t) - K_\alpha L\int_0^t (t-s)^{-\alpha} e^{-\omega(t-s)} \int_{-\infty}^s e^{-\gamma(s-r)} z(r) drds.$$

Because $y(t) < z(t)$ for $t < 0$ we can use Lemma 2 to conclude that

$$\|x(\phi)(t) - x(\psi)(t)\|_\alpha < De^{-\delta t}.$$

We remark that Condition G.2 is what those familiar with abstract convergence results would expect. The foregoing development may be applied to partial integrodifferential differential equations. For example, we may consider the following:

$$\partial u/\partial t = \Delta u + \int_{-\infty}^t g(t-s)f(\nabla u)ds$$

$$u(x,t) = 0 \qquad (x,t) t \partial\Omega \times \mathbb{R}$$

$$u(x,\theta) = \psi(x,\theta) \qquad \theta \in (-\infty, 0]$$

Where Ω is a bounded subset of R^3 with piecewise smooth boundary. The function $f(\ ) : R^3 \to R$ is required to be uniformly Lipschitz continuous and the kernel satisfies a decay condition of the form $|g(s)| \le C e^{-\gamma s}$.

We let $\underline{X} = L^2(\Omega)$ and

$$Au = \Delta u$$

with

$$D(A) = H^2(\Omega) \cap H_0^1(\Omega)$$

If $\alpha > 3/4$ we define $g(\ ,\ ,) : R^+ \times R \times \underline{X}_\alpha \to \underline{X}$ by setting $g(t,s,u) = g(t-s)f(\nabla u)$.

Theorem 5 will guarantee the existence solutions to the equation and Theorem 7 provides criteria for convergence of solutions. These involve the first eigenvalue of the Laplacian, the Lipschitz constant and the decay rate of $g(\ )$. Because $\underline{X}_\alpha \subset L^\infty(\Omega)$ convergence of solutions in $\underline{X}_\alpha$ insures uniform convergence. Analysis of this example appears in a more general context in [7] and the reader is referred thereto for details.

The results described here readily extend to the case of t independent $A(t)$, [7]. A similar approach is applied to abstract functional differential equations in finding memory spaces, [8]. We conclude by remarking that this simple integral comparison technique should prove helpful for a variety of semilinear problems.

REFERENCES

[1] Brewer, D.W., "A nonlinear semigroup for a functional differential equation," Trans. Amer. Math. Soc. 236(1978), 173-191.

[2] ——, "The asymptotic stability of a nonlinear functional differential equation of infinite delay," Hou. J. Math. 6(1980), 321-330.

[3] Coleman, B.D. and V.J. Mizel, "On the stability of solutions of functional differential equation," Arch. Rat. Mech. Anal. 30 (1968), 173-196.

[4] Dyson and R. Villella-Bressan, "Functional differential equations and nonlinear evolution operators," Proc. Royal. Soci., Edinburgh 75A (1975-1976).

[5] —— and ——, "Nonlinear functional differential equations in L^1 spaces," Nonlinear Analysis TMA 1 (1977), 383-395.

[6] —— and ——, "Semigroups of translations associated with functional and functional differential equations," Proc. Royal Soc., Edinburgh 82A (1979), 171-188.

[7] Fitzgibbon, W.E., "Asymptotic, behavior of solutions to a class of integrodifferential equations," J. Math. Anal. Appl. (to appear).

[8] ——, "Convergence theorems for semilinear Volterra Equations with infinite delay," J. Integral Equations 8 (1985), 264-272.

[9] ——, "Nonlinear Volterra equations with infinite delay," Monat. für Math. 84(1977), 275-288.

[10] ——, "Semilinear functional differential equations in Banach space," J. Differential Equations 29 (1979), 1-14.

[11] ——, "Stability for abstract nonlinear Volterra equations involving finite delay," J. Math. Anal. 60 (1977), 429-434.

[12] Friedman, A., Partial Differential Equations, Holt, Rhinehart and Winston, New York, 1969.

[13] Hale, J., Functional Differential Equations, Vol. 3, Appl, Math. Series.

[14] ——, "Functional differential equations with infinite delays," J. Math. Anal. 48 (1974), 276-283.

[15] Henry, D., Geometric Theory of Semilinear Parabolic Equations, Lecture Notes in Math 840, Springer-Verlag, New York, 1981.

[16] Kappel, F. and W. Schappacher, "Some contributions to the theory of infinite delay equations," J. Differential Equations 36(1980), 71-91.

[17] Kartsatos, A.G. and M.E. Parrott, "A simplified approach to existence and stablility of a functional evolution equation in a general Banach space," Infinite Dimensional Systems, (ed. F. Kappel and W. Schappacher), Lecture Notes in Mathematics, 1076, Springer-Verlag, Berlin, 1984, 115-121.

[18] ——, and ——, "Convergence of Kato approximations for nonlinear evolution equations involving functional perturbations," J. Differential Equations 37 (1983), 358-377.

[19] Kunish, K., "A semigroup approach to partial differential equations with delay," Abstract Cauchy Problems and Functional Differential Equations (F. Kappel and W. Schappacher Ed.) Pittman Press, Research Notes in Mathematics 48, London, 53-70.

[20] Pazy, A., Semigroups of Linear Operators and Applications to Partial Differential Equations, Vol. 44, Appl. Math. Series, Springer-Verlag, New York, 1983.

[21] Plant, A., "Stability of functional differential equations using weighted norms," Report 70, Fluid Mechanics Research Institute, University of Essex, 1976.

[22] Rankin, S.M., "Existence and asymptotic behavior of a functional differential equation in Banach space," J. Math. Anal. Appl. 88 (1982), 531-542.

[23] Redlinger, R., "On the asymptotic behavior of a semilinear functional equation in Banach space," J. Math. Anal. Appl., (to appear).[24]

[24] Travis, C.C. and G.F. Webb, "Existence and stability for partial functional differential equations," Trans. Amer. Math. Soc. 200 (1974) 531-542.

[25] —— and ——, "Partial differential equations with deviating arguments in the time variable, J. Math. Anal. Appl. 56 (1976), 397-409.

[26] —— and ——, "Existence stability and compactness in a α-norm for partial functional differential equations," Trans. Amer. Math. Soc. 240 (1978), 129-143.

[27] Villella-Bressan, R., "Flow invariant sets for functional differential equations," Abstract Cauchy Problems and Functional Differential Equations (F. Kappel and W. Schuppacher, Ed.), Pittman Press, Research Notes in Mathematics 48, London 213-229.

[28] Walter, W.A., Differential and Integral Inequalities, Springer-Verlag, New York, 1970.

[29] Webb, G.F., "Asymptotic stability for abstract nonlinear functional differential equations," Proc. Amer. Math. Soc. 54(1976), 225-230.

DESCRIBING THE FLOW ON THE ATTRACTOR OF ONE DIMENSIONAL REACTION DIFFUSION EQUATIONS BY SYSTEMS OF ODE

G. Fusco
Dipartimento di Metodi Matematici
Università di Roma "La Sapienza"
Via A. Scarpa, 10 - 00161 - Roma

INTRODUCTION

Many systems of reaction diffusion equations, delay equations, damped wave equations define nonlinear semigroups $T(t):X \to X$, $t \geq 0$, (X a function space) which admit a compact attractor i.e. a compact connected invariant set $A \subset X$ which is uniformly asymptotically stable and attracts all points in X [1]. In this situation understanding the dynamic of the semigroup T(t) may be considered equivalent to the description of the flow on A. Since from abstract theorems on dissipative semigroups [2],[3] it follows that the attractor is a finite dimensional set, once this point of view has been adopted, it is natural to ask if the infinite dimensional semigroup T(t) can, in some sense, be considered equivalent to the finite dimensional dynamical system generated by a suitable system of ODE. The most direct way one can try for constructing such a system of ODE is to show that the attractor is contained in a finite dimensional invariant manifold M. When this is the case one obtains the sought finite dimensional dynamical system simply by restricting T(t) to M. In general this construction is only formal because M cannot be computed explicitly and therefore the dynamical system describing the flow on M is only approximatly known and the problem of structural stability arises. Therefore instead of asking for a finite dimensional dynamical system which coincides with T(t) on A it may be more natural to require only topological equivalence in the sense of the following definition [4].

DEFINITION 1. *Let* X,Y *be Banach spaces and* $S_t:X \to X, T_t:Y \to Y$ $t \geq 0$ *two semiflows which possess global attractors* A^S, A^T, *then*

NATO ASI Series, Vol. F37
Dynamics of Infinite Dimensional Systems
Edited by S.-N. Chow, and J.K. Hale

S, T *are said to be equivalent if there is a homeomorphism* $h: A^S \to A^T$ *which preserves orbits and their orientation.*

In the following we present a specific example of infinite dimensional system which allows for an explicit construction a system of ODE which is equivalent to the given system according to definition 1 . The problem we whish to consider is the following scalar parabolic equation

$$(1) \qquad \begin{cases} u_t = (au_x)_x + f(x,u) , & x \in (0,1) \\ -\rho u + (1-\rho) au_x = 0 , & x = 0,\ t \geq 0 \\ \sigma u + (1-\sigma) au_x = 0 , & x = 1,\ t \geq 0, \end{cases}$$

where $f: [0,1] \times \mathbb{R} \to \mathbb{R}$ is C^2, $a: [0,1] \to \mathbb{R}$ is a positive C^1 function and $\rho, \sigma \in [0,1]$. We assume there exist a continuous function $g: \mathbb{R} \to \mathbb{R}$ and a number $\xi > 0$ such that

$$g(s)s , \qquad |s| > \xi$$

$$|f(x,s)| > |g(s)| , \qquad |s| > \xi$$

Then it can be shown [5] that 1) generates a semiflow $T(t)$ in the fractional power space X^α, $(\alpha \geq 1/2)$ associated with the operator $Au = -(au_x)_x$ with the boundary conditions, and the set $A = \{\phi | \phi \in X^\alpha$; *the solution through* ϕ *is defined in* $(-\infty,\infty)$ *and bounded*$\}$ is compact connected, finite dimensional and it is a global attractor for the semiflow. How a system of ODE describing the flow of 1) on A can be constructed?

It is reasonable to believe that such an ODE can be obtained by a suitable chosen continuous deformation a_ν, $\nu \in [0,1]$ of the conductivity function $a = a_1$ into a singular function a_o of the type

$$2) \qquad \begin{cases} a_o(x_1) = 0 & i = 0,1,\dots,n \\ a_o(x) = \infty & x \neq x_i,\ i = 0,1,\dots,n \end{cases}$$

where $0 = x_o < x_1 < \dots < x_n = 1$ is a partition of the interval $[0,1]$ which should depend on a, f, σ, ρ. To substantiate this conjecture consider the family $a_\nu = \frac{1}{\nu} a$. Then, for $\nu << 1$, u is near its average $z = \int_0^1 u$ and it can be shown [6] that $1)_\nu$ [1) with $a = a_\nu$] is equivalent to the equation

$$\dot{z} = \int_0^1 f(x,z)dx.$$

Since the attractor of this equation has a very simple structure, it must be expected that the structure of A_ν becomes simpler and simpler when the conductivity function increases in the whole interval $[0,1]$. On the other hand it can be shown [7] that, if $a_\nu(x) \to 0$ in some interval $[\bar{x}-\varepsilon, \bar{x}+\varepsilon] \subset [0,1]$ as $\nu \to 0$, then the number of equilibria of $1)_\nu$ grows undoundedly and therefore the structure of the attractor undergoes drastic changings. Thus if we want "to transform" 1) into a system of ODE by changing the function a without affecting the topological structure of A we must consider a family a_ν such that as $\nu \to 0$, $a_\nu(x) \to \infty$ almost everywhere and $a_\nu(x) \to 0$ in a suitable chosen set of zero measure. The problem of constructing a family a_ν which transforms 1) into an ODE can be divided in two parts: (i) *without changing the topological structure of* A *transform the given function* a *into a function* $a_{\bar{\nu}}$ *which can be considered a perturbation of a function* a_o *of type* 2), (ii) *show that, if* $a_{\bar{\nu}}$ *is near* a_o *in a certain sense, then the flow on* $A_{\bar{\nu}}$ *is topologically equivalent to the flow on the attractor* A_o *of a system of* ODE *which is naturally associated to* a_o.

About problem (i) we only note that it is a global problem which in view of the fact that 1) defines a Morse Smale system [8] reduces to the problem of transforming a into $a_{\bar{\nu}}$ without bifurcation of equilibria. Here we concentrate on problem (ii).

2. DEFINITION OF a_ν FOR SMALL ν AND THE MAIN RESULT

Let $0 = x_o < x_1, \dots, < x_n = 1$ be a partition of the interval $[0,1]$ and let ℓ_i, $0 \le i \le n$; α_i, $0 \le i \le n$ be two sets of n+1 positive constants and ℓ_i', $0 \le i \le n$; α_i', $0 \le i \le n$ two sets of functions of $\nu \in (0,1]$ such that $\ell_1' > \ell_i$, $\alpha_i' \ge \alpha_i$. Then, if e_i, $1 \le i \le n$ are other n positive constants, for ν sufficiently small we can consider a C^1 function $a_\nu:[0,1] \to \mathbb{R}$ which satisfies

$$a_\nu(x) \ge \frac{e_i}{\nu} \quad , \quad x_{i-1} + \nu\ell_{i-1}' \le x \le x_i - \nu\ell_i' \; ,$$

$$a_\nu(x) \ge \nu\alpha_i \quad , \quad x_i - \nu\ell_i' \le x \le x_i + \nu\ell_i' \; ,$$

$$a_\nu(x) \le \nu\alpha_i' \quad , \quad x_i - \nu\ell_i \le x \le x_i + \nu\ell_i \; .$$

Moreover we can assume that ℓ_i', α_i', a_ν satisfy the following hypothesis

$$\text{H)} \begin{cases} \ell' - \ell = 0(\nu^q) \quad , \quad \textit{for some} \quad 0 < q < 1 \; , \\[2ex] \dfrac{\alpha_i' - \alpha_i}{\ell_i' - \ell_i} \to 0 \quad , \quad \text{as} \quad \nu \to 0 \; , \\[2ex] \displaystyle\int_0^1 \frac{a_{\nu x}^2}{a_\nu^4}\, dx = 0(\nu^{-(3+q+\varepsilon)}) , \; \textit{for some given } \varepsilon \ge 0. \end{cases}$$

The integral condition on a_ν in this hypothesis puts a restriction on the derivative of a_ν in the intervals like $[x_i+\nu\ell_i$, $x_i+\nu\ell_i']$ and says that in this interval the derivative of a_ν is, in some sense, the smallest possible compatible with the fact that $a_\nu(x_i+\nu\ell_i) \le \nu\alpha_i$, $a_\nu(x_i+\nu\ell_i') \ge e_i/\nu$. The number ε must be chosen so small so that certain estimates needed for the proof of the following theorem 1 hold. It is not difficult to see that, given an integer N there exist functions a_ν which satisfy. H with $\varepsilon < 1/N$.

It turns out that for small ν the function a_ν just defined can be considered a perturbation of the singular function a_o in the sense outlined in the introduction, we have in fact

THEOREM 1. *Generically in* $(\rho,\sigma,\alpha_i,\ell_i,f)$ *there is a* $\bar{\nu} > 0$ *such that for* $\nu \le \bar{\nu}$ *the semiflow defined by* $1)_\nu$ *is equivalent, in the sense of definition* 1, *to the semiflow generated by the following system of* ODE

$$3) \quad \dot{z}_i=\mu_i(z_{i+1}-z_i)-\tilde{\mu}_{i-1}(z_i-z_{i-1})+f_i(z_i), \quad 1 \le i \le n$$

where $z_o = z_{n+1} = 0$, *and*

$$\mu_i = \frac{\alpha_i}{2\ell_i L_i}, \quad 1 \le i \le n-1$$

$$\mu_n = \frac{\alpha_n}{\ell_n L_n} \quad \frac{\sigma}{\sigma+(1-\sigma)\frac{a_n}{\ell_n}},$$

$$\tilde{\mu}_o = \frac{\alpha_o}{\ell_o L_1} \quad \frac{\rho}{\rho+(1-\rho)\frac{a_o}{\ell_o}}$$

$$\tilde{\mu}_i = \frac{\alpha_i}{2\ell_i L_{i+1}}, \quad 1 \le i \le n-1$$

$$L_i = x_i - x_{i-1},$$

$$f_i(s) = \frac{1}{L_i}\int_{x_{i-1}}^{x_i} f(x,s)dx.$$

PROOF. A complete proof of this theorem will appear elsewhere. Here we only indicate the main points:

a) Prepare equation $1)_\nu$ for application of center manifold theory. The natural idea is to write $u = z+v$ where $z=\sum_i \chi_i z_i$ with $z_i = L_i^{-1}\int_{x_{i-1}}^{x_i} u dx$, χ_i the characteristic function of the interval $[x_{i-1},x_i)$ and derive from $1)_\nu$ a system of two equations for z,v. This idea cannot be applied in this simple form because z is a step function and therefore, in general $z \notin X^\alpha$ and consequently $v \notin X^\alpha$. To overcome this difficulty we introduce an extraterm and write $u = z+\zeta(z)+v$ where $\zeta(z):[0,1] \to \mathbb{R}$ is a function such that $(z+\zeta(z)) \in X^\alpha$. We choose $\zeta(z)$ to be the solution of the problem

$$4)\quad \begin{cases} c_i = (a_\nu \zeta_x)_x + f(x,z_i)\ , & x \in (x_i, x_{i-1})\ , \\ z_i+(\zeta(z))(x_i^-) = z_{i+1}+(\zeta(z))(x_i^+)\ , & 1 \le i \le n-1 \\ (\zeta_x(z))(x_i^-) = (\zeta_x(z))(x_i^+) & ,\ 1 \le i \le n-1 \\ \displaystyle\int_{x_{i-1}}^{x_i} \zeta dx = 0. \end{cases}$$

Problem 4) plus the boundary conditions

$$5)\quad \begin{cases} -\rho(z_1+\zeta) + (1-\rho)a_\nu \zeta_x = 0\ , & x = 0, \\ \sigma(z_n+\zeta) + (1-\sigma)a_\nu \zeta_x = 0\ , & x = 1, \end{cases}$$

can be considered as a linear system of 3n equations in 3n unknowns, namely the 2n numbers $(\zeta(z))(x_i^+), (\zeta_x(z))(x_i^+)$, $0 \le i \le n-1$ and the n constants c_i. A direct computation shows that for small ν the determinant of this system is nonzero and therefore that equations 4),5) uniquely determine $\zeta(z)$. By making the change of variable $u = z+\zeta(z)+v$, by averaging equation $1)_\nu$ in the interval (x_{i-1}, x_i) and by using the expressions of $c_i, \zeta(z)$ given by 4),5) one obtains

$$6)\quad \dot{z}_i = \mu_i(z_{i+1}-z_i)+\tilde{\mu}_{i-1}(z_i-z_{i-1})+f_i(z_i)+\gamma_i(z) +$$

$$L_i^{-1}\{a_\nu v_x)(x_i)-(a_\nu v_x)(x_{i-1})+\int_{x_{i-1}}^{x_i} [f(x,z_i+\zeta+v)-f(x,z_i)]dx,$$

where $\gamma_1(z) = 0(\nu^q)$. By subtracting equations 6) from $1)_\nu$ one gets the equation

$$7)\ v_t-(a_\nu v_x)_x+\sum_{i=1}^{n}\chi_i\frac{(a_\nu v_x)(x_i)-(a_\nu v_x)(x_{i-1})}{L_i}=f(x,z+\zeta+v)-f(x,z)-$$

$$\sum_{i=1}^{n}\frac{\chi_i}{L_i}\int_{x_{i-1}}^{x_i}[f(x,z_i+\zeta+v)-f(x,z_i)]\,dx-\frac{\partial\zeta}{\partial z}G(v,z),$$

where $G(v,z)$ is the vector of the right hand sides of 6). Equations 6),7) define a semiflow on $\mathbb{R}^n\times Y^\alpha$, $(\alpha>3/4)$, where Y^α is the fractional power space of order α associated to the operator $B:D(B)\to Y$ defined by

$$B\phi=-(a\phi_x)_x-\sum_{i=1}^{n}\chi_i\frac{(a\phi_x)(x_i)-(a\phi_x)(x_{i-1})}{L_i}$$

plus boundary conditions; $Y\subset L^2(0,1)$ is the subspace $Y=$

$$=\{\phi\,|\,\phi\in L^2(0,1),\ \int_{x_{i-1}}^{x_i}\phi=0,\ 1\le i\le n\}.$$

b) Obtain estimates for the size of the attractor. Use these estimates for modifying equations 6),7) outside a region containing A_ν so that center manifold theory can be applied to the modified equations for obtaining an exponentially stable C^1 locally invariant manifold $S_\nu=\{(v,z)\,|\,v=\eta(z)\}$ ($\eta:\mathbb{R}^n\to Y^\alpha$ a C^1 function) containing A_ν.

c) Show that the vector field $1)_\nu$ restricted to S_ν approaches the vector field 3) in the C^1 topology as $\nu\to 0$.

d) Show that generically 3) defines a Morse-Smale system. This part of the proof is contained in [9].

3. STABLE NON CONSTANT EQUILIBRIA FOR 1) WITH NEWMANN BOUNDARY CONDITIONS

It is known that when the conductivity function a has a deep well around some point $x_1\in(0,1)$, then 1) with Newmann boundary conditions has a stable non constant equilibrium. The equivalence between $1)_\nu$ and 3) stated in theorem 1 yields an easy proof of this fact. For $n = 2$ and $\rho = \sigma = 0$, assuming for simplicity $x_1 = \frac{1}{2}$ and $f = u(1-u^2)$, theorem 1 implies that for ν small the semiflow defined by $1)_\nu$ is equivalent to the semiflow defined in $\mathbb{R}^2$ by the system

$$8)\qquad \begin{cases} \dot{z}_1 = \mu(z_2-z_1) + z_1(1-z_1^2) \\ \dot{z}_2 = -\mu(z_2-z_1) + z_2(1-z_2^2). \end{cases}$$

where $\mu = a_1/\ell_1$. By Keeping in mind the definition of the function a_ν we see that a deep well in the graph of a_ν corresponds to a small value of μ. The simplicity of 8) allows for a complete description of the set of equilibria for each $\mu > 0$. The equilibria

$$9)\qquad z_1 = z_2 = \begin{matrix} 1 \\ 0 \\ -1 \end{matrix} ,$$

exist for all $\mu > 0$. The zero equilibrium is unstable and the other two equilibria are stable. For $\mu \geq \frac{1}{2}$ there is no other equilibrium. At $\mu = \frac{1}{2}$ two new equilibria bifurcate from the zero solution

$$10)\qquad z_1 = \pm(1-2\mu)^{1/2} , \quad z_2 = -z_1.$$

These equilibria are unstable for $\frac{1}{2} > \mu \geq \frac{1}{3}$ and stable for $\mu < \frac{1}{3}$. At $\mu = \frac{1}{3}$ a secundary bifurcation take place and each of the equilibria 10) undergoes a pitchfork bifurcation and four new unstable equilibria appear which, for $\varepsilon = \frac{1}{3} - \mu$ small; are ot the form

$$\begin{pmatrix} z_1 \\ z_2 \end{pmatrix} = \pm(\tfrac{1}{3} + 2\varepsilon)^{1/2}\begin{pmatrix} 1 \\ -1 \end{pmatrix} \pm K\varepsilon^{1/2}\begin{pmatrix} 1 \\ 1 \end{pmatrix},$$

for same number K.

REFERENCES

[1] J.K. Hale: Asymptotic Behavior in Infinite Dimensions, LCDS Report 85.

[2] J. Mallet-Paret: Negatively invariant sets of compact maps and an extension of a theorem of Cartwright, J. Differential Equations 22 (1976) pp. 331-348.

[3] R. Mañé: On the dimension of the compact invariant sets of certain non-linear maps, Lect. Notes in Math., Vol. 898 (1980) pp. 230-242.

[4] W.M. Oliva: Stability of Morse Smale maps -RT-MAP-8301 IME, Un. of São Paulo.

[5] D. Henry: *Geometric theory of Semilinear Parabolic Equations*, Lect. Notes in Math. 240, Springer-Verlag 1981.

[6] J.K. Hale: Large Diffusivity and Asymptotic Behavior in Parabolic Systems, LCDS Report 85-1, January 1985.

[7] G. Fusco, J.K. Hale: Stable Equilibria in a Scalar Parabolic Equation with Variable Diffusion, SIAM Journal of Mathematical Analysis, November 1985.

[8] D. Henry: Some infinite-dimensional Morse-Smale systems defined by parabolic partial differential equations, Journal of Differential Equations.

[9] G. Fusco, W.M. Oliva: Jacobi Matrices and Transversality, to appear.

ASYMPTOTIC BEHAVIOR OF GRADIENT DISSIPATIVE SYSTEMS

Jack K. Hale
Lefschetz Center for Dynamical Systems
Brown University,
Providence RI 02912

In this paper, we summarize recent results on dissipative gradient systems and give applications to some hyperbolic equations.

Let X be a Banach space, $T(t): X \to X$ be a C^r-semigroup, $r \geq 1$; that is, $T(t)$ is a C^0-semigroup and $T(t)x$ has continuous derivatives up through order r in x. Let E be the set of equilibrium points of $T(t)$; that is, $x \in E$ if and only if $T(t)x = x$ for $t \geq 0$. The following definition is in Hale [1985] and is a weakened version of a definition given by Babin and Vishik [1983]. It is a special case of Conley [1972].

<u>Definition 1</u>. The C^r-semigroup $T(t)$ is said to be a <u>gradient system</u> if

(i) Each bounded orbit is precompact.

(ii) There exists a Liapunov function for $T(t)$; that is, there is a continuous function $V : X \to R$ with the property that

(ii_1) $V(x)$ is bounded below,

(ii_2) $V(x) \to \infty$ as $|x| \to \infty$

(ii_3) $V(T(t)x)$ is nonincreasing in t for each x in X,

(ii_4) If x is such that $T(t)x$ is defined for $t \in R$ and $V(T(t))x = x$ for $t \in R$, then $x \in E$.

An $x \in E$ is <u>hyperbolic</u> if the spectrum $\sigma(DT(t)x)$ does not intersect the unit circle with center zero in the complex plane. If $x \in E$ is hyperbolic, we let $W^u(x), W^s(x)$ denote the stable and unstable sets of x.

<u>*Theorem 2*</u>. *If $T(t)$ is a gradient system, then*

(i) The ω-limit set $\omega(x)$ for each x belongs to E,

(ii) if the negative orbit $\gamma^-(x)$ through x exists and is precompact, then the α-limit set $\alpha(x)$ of x belongs to E

(iii) if $V(T(t)x) < V(x)$ for $t > 0$ and $x \notin E$,

$T(t)$ is one-to-one and $DT(t)x$ is an isomorphism for each $x \in E$, $x_0 \in E$ with

NATO ASI Series, Vol. F37
Dynamics of Infinite Dimensional Systems
Edited by S.-N. Chow, and J. K. Hale

$\dim W^u(x_0) < \infty$, then $W^u(x_0)$, $W^s(x_0)$ are embedded submanifolds of X.

For a proof of (i) and (ii), see Hale [1985]. For a proof of (iii), see Henry [1981].

We need a few more definitions, A set $A \subset X$ is <u>invariant</u> under $T(t)$ if $T(t)A = A$ for $t \geqslant 0$. The set A is said to be a <u>compact attractor</u> if A is compact, invariant and $\omega(B) \subset A$ for each bounded set $B \subset X$ where

$$\omega(B) = \bigcap_{\tau \geqslant 0} \mathrm{Cl} \bigcup_{t \geqslant \tau} T(t)B.$$

The semigroup $T(t)$ is said to be <u>asymptotically smooth</u> if, for any closed bounded set $B \subset X$, $B \neq \phi$, $T(t)B \subset B$, $t \geqslant 0$, there is a compact set $J \subset B$ such that $\mathrm{dist}(T(t)B, J) \to 0$ as $j \to \infty$. A special case of an asymptotically smooth map is an α - contraction. The semigroup $T(t)$ is an <u>α - contraction</u> if there is a continuous function $k(t) \to 0$ as $t \to \infty$ such that $\alpha(T(t)B) \leqslant k(t)\alpha(B)$ for any bounded set $B \subset X$ where

$$\alpha(B) = \inf\{d > 0: B \text{ has a finite cover of diameter } < d\}.$$

Examples of α – contracting semigroups are

(i) $T(t)$ compact for $t > 0$

(ii) $T(t) = S(t) + U(t)$ where $U(t)$ is compact for $t \geqslant 0$ and $S(t)$ is linear with $|S(t)| \leqslant k\exp(-\beta t), t \geqslant 0$, for some constants $k, \beta > 0$.

(iii) There is a constant $g \in [0,1)$ and a compact pseudo metric ρ on X such that

$$|T(1)x - T(1)y| \leqslant g|x - y| + \rho(x,y)$$

for all $x,y \in X$ (see Lopez and Ceron [1984]).

<u>Theorem 3</u>. If $T(t)$, $t \geqslant 0$, is a gradient system for which $V(T(t)x) < V(x)$ for $t > 0$ if $x \notin E$, asymptotically smooth with orbits of bounded sets being bounded and E is bounded, then there is a compact connected attractor A for $T(t)$ and A $W^u(E) = \{y \in X: T(-t)y$ is defined for $t \geqslant 0$ and $T(-t)y \to E$ as $n \to \infty\}$. If each element of E is hyperbolic and $T(t)$ is one-to-one on A and $DT(t)(y)$, $y \in X$, is an isomorphism, then E is a finite set and

$$A = \bigcup_{x \in E} W^u(x).$$

The proof may be obtained from Hale [1985] and Theorem 1.

As a first application, consider the boundary value problem

$$u_{tt} + 2\alpha u_t - \Delta u = f(u) \quad \text{in} \quad \Omega \tag{1}$$

$$u = 0 \text{ on } \partial\Omega \tag{2}$$

where $\alpha > 0$, f is a C^2 - function and there are constants γ, C such that

$$|f''(u)| \leq C(|u|^{\gamma} + 1), \quad u \in R, \tag{3}$$

$$\overline{\lim}_{|u| \to \infty} f(u)/u \leq 0. \tag{4}$$

In the following, we use the notation L^2, H^1, etc to denote $L^2(\Omega)$, $H^1(\Omega)$, etc. and assume that $\Omega \subset R^3$.

Theorem 4. Suppose $0 < \gamma < 1$.

(i) there is a compact connected attractor A in $X = H_0^1 \times L^2$ and the representation in Theorem 3 holds.

(ii) $A \subset Y = (H^2 \cap H_0^1)$ and A is a compact attractor in Y.

This result is a consequence of Babin and Vishik [1983],[1984], Hale [1985], Haraux [1985]. Let us indicate the proof. If we write (1) as a system; letting $\omega = (u,v)$, then the equation on X is equivalent to

$$\omega_t = C\omega + f^e(u) \tag{5}$$

$$C = \begin{bmatrix} 0 & I \\ \Delta & -2\alpha I \end{bmatrix}, \quad f^e(u)(x) = \begin{bmatrix} 0 \\ f(u(x)) \end{bmatrix}$$

For $(\varphi, \psi) \in X$, let

$$V(\varphi,\psi) = \int_\Omega \left(\frac{1}{2}|\text{grad } \varphi|^2 + |\psi|^2\right)dx$$

This function is to play the role of the Liapunov function in the definition of the gradient system. One shows that $dV(\omega(t))/dt = -\int_\Omega u_t^2 dx$ along a smooth dense set of solutions. This allows one to obtain a priori bounds on the solutions showing that they are globally defined and the corresponding semigroup is well defined. Also, the inequality shows that orbits of bounded sets are bounded. This implies that $(ii_1), (ii_2), (ii_3)$ of Def. 1 are satisfied. To show (i) of Def.1, one uses the variation of constants formula

$$\omega(t) = e^{Ct}\omega_0 + \int_0^t e^{C(t-s)} f^e(u(s)) ds.$$

Since $\alpha > 0$, $\|e^{Ct}\| \leq k\exp(-\beta t)$, $t \geq 0$ for some positive constants k, β.

Since $0 < \gamma < 1$, f^e is a compact map and the integral term is a compact operator. Therefore, $T(t)$ is an α - contraction. This proves (i) of Def.1 and the above formula for $dV(\omega(t)/dt$ then gives $(ii)_4$ of Def.1. Thus, the system is a gradient system and $T(t)$ is an α - contraction. One can also show that $T(t)$ is a group and $DT(t)(y), y \in X$, is an isomorphism. Finally condition (4) implies E is bounded. Theorem 3 now implies the result (i) of Theorem 4.

The proof of the second part proceeds in a similar way using the fact that $A \subset Y$ (see Haraux [1985]) and orbits of bounded sets in Y are bounded (see Babin and Vishik [1983])

The case $\gamma = 1$ in (3) is much more difficult and only partial results are known. This is due to the fact that f is not a compact map from $H_0^1 \to L^2$.

<u>*Theorem 5*</u>. *Suppose* $\gamma = 1$ *and* $X = H_0^1 \times L^2$, $Y = (H^2 \cap H_0^1) \times H_0^1$. *Then*

(i) $T(t)$ is bounded dissipative in X

(ii) $\omega(\varphi, \psi) \subset E$ for any $(\varphi, \psi) \in X$

(iii) there is a compact attractor A in Y.

We say $T(t)$ is <u>bounded dissipative</u> if there is a bounded set $B \subset X$ such that, for any bounded set U in X, there is a t_0 such that $T(t)U \subset B$ for $t \geq t_0$. We say A is an <u>(X,Y) attractor</u> if, for any bounded set B in Y, the ω-limit set of B in X is in A. Part (i) of Theorem 5 has been proved by Lopes and Ceron [1984], Babin and Vishik [1985], and Haraux [1985], part (ii) by Ball [1973], part (iii) by Hale and Raugel [1986]. The following result is due to Lopes and Ceron [1984].

<u>*Theorem 6*</u> *Suppose* $0 < \gamma < 1$ *and the term* $2\alpha u_t$ *in (1) is replaced by* $h(u_t)$ *where there are positive constants* $\alpha > 0$, $\beta > 0$ *such that*

$$0 < \alpha \leq h'(v) \leq \beta \quad \text{for} \quad v \in R.$$

Then the conclusion of part (i) of Theorem 4 is true.

The proof of Theorem 6 will again be an application of Theorem 3 and the ideas in the proof are similar. The bounded dissipativeness is proved by using an appropriate modification of the Liapunov function used before. The novelty in the result for nonlinear damping is in the way that one proves the appropriate compactifying property of the semigroup. They show the semigroup is an α - contraction but do this by showing that the map $T(1)$ satisfies

$$|T(1)x - T(1)y| \leq g|x - y| + \rho(x,y) \tag{6}$$

for all $x,y \in X$ where g is a constant, $0 \leq g < 1$ and ρ is a compact pseudo-metric.

As another application, let us mention the beam equation

$$u_{tt} + 2\alpha u_t + a u_{xxxx} = (\beta + k\int_0^1 u_x^2(\xi,t)d\xi)u_{xx}, \quad 0 < x < 1, \tag{7}$$

with the clamped conditions

$$u = u_x = 0 \quad \text{at } x = 0,\ 1. \tag{8}$$

Let $X = (H^2 \cap H_0^1)xL^2$

<u>Theorem 7</u>. *There is a compact connected attractor for (6),(7) in X and the corresponding semigroup is an α - contraction.*

This result is due to Lopes and Ceron [1984] and generalizes considerably the previous results of Ball [1973a,b,c]. Their result also holds for the nonlinear damping $h(v)$ as in Theorem 6. This result is surprising because at the abstract level, there appear the same difficulties as in the case of equation (1) with $\gamma = 1$; that is, the nonlinear term is not a compact map. However, using the special form of the nonlinearity, one show that $T(1)$ satisfies (5).

<u>References</u>

A.V. Babin and M.I. Vishik [1983], Regular attractors of semigroups of evolutionary equations. J. Math. Pures et Appl. **62**, 441-491.

A.V. Babin and M.I. Vishik [1985], Attracteurs maximaux dans les equations aux derivees partielles. College de France, 1984, Pittman, 1985.

J.M. Ball [1983a] Saddle point analysis for an ordinary differential equation in a Banach space and an application to dynamic buckling of a beam. Nonlinear Elasticity (Ed. Dickey), Academic Press.

J.M. Ball [1973b], Stability theory for an extensible beam. J. Differential Equations **14**, 399-418.

J.M. Ball [1973c], Initial boundary value problems in an extensible beam. J. Math. Anal. Appl. **41**, 61-90.

C. Conley [1972], The gradient structure of a flow: I, RC 3932, Math. Sci. IBM, Yorktown Heights.

J.K. Hale [1985], Asymptotic behavior and dynamics in infinite dimensions. p.1 - 42 of Nonlinear Differential Equations (Eds. Hale and Martinez-Amores) Res. Notes in Math. Vol. 132, Pittman.

J.K. Hale and G Raugel [1986], Upper semicontinuity of the attractor for a singularly perturbed hyperbolic equation. J. Differential Equations. Submitted.

A. Haraux [1985], Two remarks on dissipative hyperbolic problems. College de France, 1984. Pittman, 1985.

D. Henry [1981], Geometric Theory of Semilinear Parabolic Equations. Lect. Notes in Math. Vol 840, Springer.

O. Lopes and S.S. Ceron [1984], Existence of forced periodic solutions of dissipative semilinear hyperbolic equations and systems. Annali di Mat. Pura Appl. Submitted.

GENERIC PROPERTIES OF EQUILIBRIUM SOLUTIONS BY PERTURBATION OF THE BOUNDARY

Daniel B. Henry

Instituto de Matemática e Estatística,
Universidade de São Paulo,
São Paulo (S.P.), Brasil.

1. Introduction. We study equilibrium solutions of

(1) $$u_t = \Delta u + f(x,u,\nabla u) \text{ in } \Omega \subset R^n$$

(which may be a system, $u=\mathrm{col}\,(u_1,\ldots,u_p)$), with boundary conditions.

(2) $$u = 0 \quad\text{or}\quad \partial u/\partial N = g(x,u) \text{ on } \partial\Omega.$$

and of the corresponding damped wave equation with "$u_{tt}+r(x)u_t$" in place of "u_t". The equilibrium problem

(3) $\Delta u + f(x,u,\nabla u) = 0$ in Ω, with boundary conditions,

are the same in each case, but the eigenvalue problem for the linearization

(4) $$\Delta v + \sum_{j=1}^{n} b_j(x)\,\partial v/\partial x_j + (c(x)-g(x,\lambda))v=0 \text{ in } \Omega$$

has $g(x,\lambda)=\lambda$ in the parabolic case, $g(x,\lambda)=\lambda^2+r(x)\lambda$ for the wave equation. Here

$$(b_j,c)(x) = \left(\frac{\partial f}{\partial \beta_j},\frac{\partial f}{\partial \gamma}\right)(x,u(x),\nabla u(x))$$

where u solves (3) for $f\,(x,\gamma,\beta_1,\ldots,\beta_n)$. If $v(x)$ is a nontri-

NATO ASI Series, Vol. F37
Dynamics of Infinite Dimensional Systems
Edited by S.-N. Chow, and J. K. Hale

vial solution of (4) for some $\lambda \in C$, then $e^{\lambda t}v(x)$ is a nontrivial solution of the linearization of (1), or of the corresponding wave equation, about the equilibrium.

Under various hypotheses about f and g, we can prove that -for most choices of the bounded smooth region Ω - all equilibrium solutions u are simple and (with more restrictive hypotheses), all equilibria are hyperbolic, i.e. (4) has no non-trivial solutions when $\mathrm{Re}\,\lambda = 0$.

The results are far from complete, but seem sufficient to demonstrate that generecity with respect to perturbation of the boundary -holding f and g fixed- is a very strong condition, worthy of further investigation. In many problems, it is also reasonable to require genericity with respect to perturbations of f and g; some studies of this kind are due to Uhlenbeck [5], Saut and Temam [4],Foias and Temam [1]. Some problems -such as the Navier-Stokes equation- are quite rigid, and it seems the only infinite-dimensional class of perturbations naturally allowed is perturbation of the boundary. In any case, we concentrate on perturbing the boundary, though the Navier-Stokes problem is still out of reach.

Results will merely be sketched, with no attempt at proof; details of the argument will be published in [2].

2. Differential calculus of boundary perturbations.

Given a bounded open set $\Omega_o \subset R^n$, consider the collection of all regions C^k-diffeomorphic to Ω_o ($k \geq 1$). We introduce a topology by defining (a sub-basis of the) neighborhoods of a given Ω as

$$\{h(\Omega) \mid \text{ h is in a small } C^k(\Omega,R^n)\text{-neighborhood of the inclusion } i_\Omega : \Omega \subset R^n \} .$$

When $\|h-i_\Omega\|_{C^k}$ is small, h is a C^k imbedding of Ω in R^n, a C^k diffeomorphism to its image $h(\Omega)$. Micheletti [3] shows this topology is metrizable, and the set of regions C^k-diffeomorphic to a given bounded C^k region may be considered a separable complete metric space. We say a function F defined on this space (with values in a Banach space) is of class C^r or C^∞ or analytic if $h \to F(h(\Omega))$ is C^r or C^∞ or analytic as a map of Banach spaces (h near i_Ω in $C^k(\Omega, R^n)$). Thus, for example, a simple eigenvalue of the Laplacian, for the Dirichlet or Neummann, problem in a bounded C^2 region $\Omega \subset R^n$, is an analytic function of Ω (in the space of regions C^2-diffeomorphic to Ω).

In this sense, we may express problems of perturbation of the boundary (or, of the domain of definition) of a boundary-value problem as problems of differential calculus in Banach spaces. Specifically consider a non-linear formal differential operator F_Ω .

$$F_\Omega(u)(x) = f(x, Lu(x)) \text{ for } x \in \Omega,$$

where L is a constant coefficient linear differential operator of order m, say

$$Lu(x) = (u(x);\ \frac{\partial u}{\partial x_j}(x)\ \ (1 \leqslant j \leqslant n);\ \frac{\partial^2 u}{\partial x_j \partial x_k}(1 \leqslant j,k \leqslant n);\text{etc.})$$

and $f(x,\lambda)$ is a given smooth function. We may consider F_Ω as a map from $C^m(\Omega)$ to $C^o(\Omega)$, or from $W_p^m(\Omega)$ to $L_p(\Omega)$ (under appropriate hypotheses). Then if $h:\Omega \to h(\Omega) \subset R^n$ is a C^m imbedding, it induces isomorphisms $h^*: C^k(h(\Omega)) \to C^k(\Omega)$ [or $W_p^k(h(\Omega)) \to W_p^k(\Omega)$] for $0 \leqslant k \leqslant m$, by

$$h^*\varphi = \varphi o h \quad \text{(the pull-back of } \varphi \text{ by h)}$$

and instead of $F_{h(\Omega)}: C^m(h(\Omega)) \to C^o(h(\Omega))$ we study

$h^*F_{h(\Omega)}h^{*-1}: C^m(\Omega) \to C^o(\Omega)$ acting in spaces independent of h. The degree of differentiability of $(h,u) \to h^*F_{h(\Omega)}h^{*-1}(u)$, in appropriate function spaces, follows from the chain rule and if $h(x,t) = x+t\overset{\circ}{h}(x)+o(t)$, $u(x,t)=u(x)+t\dot{u}(x)+o(t)$, as $t \to o$, we have

$$\frac{\partial}{\partial t}(h^*F_{h(\Omega)}h^{*-1}(u))\Big|_{t=o} = F'_\Omega(u)\dot{u} + \overset{\circ}{h}\circ\nabla(F_\Omega(u)) - F'_\Omega(u)\overset{\circ}{h}\circ\nabla u$$

where $F'_\Omega(u)v(x) = \frac{\partial f}{\partial\lambda}(x,Lu(x))Lv(x)$ for $x \in \Omega$, $\overset{\circ}{h}\circ\nabla = \sum_{j=1}^{n}\overset{\circ}{h}_j\partial/\partial x_j$.

Note that, when F_Ω is linear, the contribution to the derivative from variation of Ω is simply the commutator of F_Ω and $\overset{\circ}{h}\circ\nabla$.

3. The transversality theorem.

Our "generic" results are obtained by applying the transversality (or transversal density) theorem. In the usual formulation, for a sufficiently smooth map (C^k) of separable Banach manifolds $f: X\times Y \to Z$, with $\partial f/\partial x\,(x,y): T_xX \times T_yY \to TZ_{f(x,y)}$. Fredholm or semi-Fredholm with index strictly less than k, if $\xi \in Z$ is a regular value of f, then it is also a regular value of $x \to f(x,y)$ for "most" fixed $y \in Y$, the exceptional set being small in the sense of Baire category. The hypothesis says whenever $f(x,y)=\xi$, the derivative $Df(x,y):(\dot{x},\dot{y}) \to \frac{\partial f}{\partial x}\dot{x}+\frac{\partial f}{\partial y}\dot{y}$ is surjective. Thus the range $\mathcal{R}(\partial f/\partial y)$ must make up any deficit in $\mathcal{R}(\partial f/\partial x\,(x,y))$.

However, many cases arise where $\partial f/\partial x$ has index $-\infty$, so $\mathcal{R}(\partial f/\partial x)$ has infinite codimension and this hypothesis is difficult to verify. In fact, it is sufficient that the quotient space

$$\mathcal{R}(Df(x,y))\,/\mathcal{R}(\frac{\partial f}{\partial x}(x,y))$$

has sufficiently high dimension at each point of $f^{-1}(\xi)$ -in practice, we show it is infinite- dimensional (by contradic tion). This extension of the usual transversality theorem is crucial for most of the results below.

4. Generic simplicity of equilibria (scalar case).

For the scalar Dirichlet problem

$$\Delta u + f(x,u,\nabla u) = 0 \text{ in } \Omega \qquad u = 0 \text{ in } \partial\Omega ,$$

given $f: R^n x R x R^n \to R$ of class C^2, for most bounded C^2 regions Ω (in the sense of Baire category), all solutions u are simple; that is, the linerization

$$\Delta v + \sum_{j=1}^{n} \frac{\partial f}{\partial \beta_j}(x,u(x),\nabla u(x)) \frac{\partial v}{\partial x_j} + \frac{\partial f}{\partial \gamma}(xu,\nabla u)v = 0 \text{ in } \Omega, \quad v = 0 \text{ on } \partial\Omega$$

has only the trivial solution $v \equiv 0$.

When $f(x,0,0) \equiv 0$, this may be proved with the usual transversality theorem (and was proved by the author and by Saut and Temam [4], who inadvertantly omitted this hypothesis). In the general case, one must show the over determined problem.

$$\Delta u + f(x,u,\nabla u) = 0 \text{ in } \Omega, \; u = 0 \; \underline{\text{and}} \; \frac{\partial u}{\partial N} = 0 \text{ on } \partial\Omega$$

has no solution, if $f(x,0,0) \not\equiv 0$ on $\partial\Omega$. It is easy to construct counter-examples; but one may prove that (for fixed f), there are no solutions for most choices of Ω, excluding only a closed set, of infinite codimension. This suffices to prove the result claimed. Note $u \to \Delta u : H_0^2(\Omega) \to L_2(\Omega)$ has index $-\infty$, all harmonic polynomials (for example) being orthogonal to the image. This is a typical example of a problem with Fredholm index $-\infty$.

We have the same conclusion if the boundary condition requires u=0 on certain components of $\partial\Omega$, while (for example) $\frac{\partial u}{\partial N} = g(x,u)$ on other components. It is sufficient to perturb only the "Dirichlet" components of $\partial\Omega$ to obtain simple solutions.

If there are no "Dirichlet" components, the problem is more complicated. We mention only one example, motivated by applications to population genetics. Suppose $h:R \to R$ has only simple zeros, and $S:R^n \to R$ is C^3 with

$$\{x \mid S(x)=0.\ \partial S/\partial x_i(x)=0,\ \partial^2 S/\partial x_j \partial x_k(x)=0,\ \partial^3 S/\partial x_m \partial x_p \partial x_q\ (x)=0\}$$

for all i,j,k,m,p,q

empty, or of dimension <n-1.

Then for most bounded connected C^2 regions $\Omega \subset R^n$, all solutions u of

$$\Delta u + S(x)h(u)=0 \text{ in } \Omega, \quad \partial u/\partial N = 0 \text{ on } \partial\Omega$$

are simple. This problem typically has "trivial" solutions $u \equiv$ constant.

5. Generic simplicity of equilibria for a system.

Let $p \geq 2$ and suppose $f:R^n \times R^p \to R^p$ is smooth and consider the problem

$$\Delta u + f(x,u)=0 \quad \text{in } \Omega,\ u=0 \text{ on } \partial\Omega$$

where $u=\text{col}\ (u_1,\ldots,u_p)$. Under various hypotheses about f (besides smoothness), it may be proved most solutions u are simple.

Specifically suppose f and f' are at least C^1, where

$$f'(x,u) = [\,\partial f_i(x,u)/\partial u_j\,]_{i,j=1}^{p} \in R^{p\times p}$$

and that $x \to f(x,0)$, $f'(x,0)$ are at least C^2. Define linear spaces $\mathcal{F} \subset C(R^n,R^p)$ and $\mathcal{M} \subset C(R^n,R^{p\times p})$ as follows:

$$\mathcal{F} = \bigcup_{j \geq 0} \mathcal{F}_j \quad , \quad \mathcal{M} = \bigcup_{j \geq 0} \mathcal{M}_j$$

$$\mathcal{F}_o = \text{span}\{f(x,0)\}, \quad \mathcal{M}_o = \text{span}\{I, f'(x,0)\}$$

$$\mathcal{M}_{j+1} = \text{span}\left\{ \mathcal{M}_j;\ \partial A/\partial x_k \text{ for } 1 \leq k \leq n,\ A \in \mathcal{M}_j \cap C^1;\ f'(x,0)A(x)+A(x)f'(x,0) \text{ for } A \in \mathcal{M}_j \cap C^2 \right\}$$

$$\mathcal{F}_{j+1} = \text{span}\left\{\mathcal{F}_j;\ \partial a/\partial x_k \text{ for } 1 \leq k \leq n,\ a \in \mathcal{F}_j \cap C^1;\ A(x)f(x,0) \text{ for } A \in \mathcal{M}_j \cap C^1;\ f'(x,0)a(x) \text{ for } a \in \mathcal{F}_j \cap C^2 \right\}.$$

For example, $\mathcal{M}$ contains (at least) all polynomials in $f'(x,0)$, and $\mathcal{F}$ contains $\{(\text{polynomial in } f'(x,0)).f(x,0)\}$.

(1) If $\{a(x) \mid a \in \mathcal{F}\} = R^p$ for a dense set of $x \in R^n$, then for most bounded C^2 regions $\Omega \subset R^n$, all solutions are simple.

(2) If $f(x,0) \equiv 0$ then $\mathcal{F} = \{0\}$; but suppose for a dense set of $x \in R^n$, that

$$\alpha,\beta \in R^p,\quad \alpha \circ A(x)\beta = 0 \text{ for all } A \in \mathcal{M} \Rightarrow |\alpha||\beta| = 0.$$

Then for most bounded C^2 $\Omega \subset R^n$, all solutions are simple.

(3) If $f(x,u) = f(u)$ is independent of x and $f(0)=0$, then $\mathcal{F} = \{0\}$, $\mathcal{M} = \{\sum_{j=0}^{p-1} c_j f'(0)^j \mid c_j \in R\}$ and hypothesis (2)

rarely holds. (Perhaps the only exception being when $p=2$ and $f'(0)$ has complex eigenvalues).

In this case, assume f is C^4, $f'(0)$ has only simple eigenvalues and (writing $\frac{1}{k!} f^{(k)}(0)\beta^k$ for the k-order homogeneous polynomial) in the Taylor expansion for $f(\beta)$) assume

$$\alpha,\beta \in R^p, \quad \alpha\circ f'(0)^j\beta = 0, \quad \alpha^\circ f'(0)^j(f''(0)\beta^2)=0 \text{ and } \circ f'(0)^j(f'''(0)\beta^3)=0$$

for all $j \geqslant 0$, imply $|\alpha||\beta| = 0$.

Then for most choices of bounded C^2 $\Omega \subset R^n$, all solutions are simple.

For example, in case $f(x,u)=f(u)$ is independent of x but $f(0)\neq 0$, hypothesis (1) is verified when $f'(0)^j f(0)$ $(0\leqslant j\leqslant p-1)$ are linearly independent; and this holds for most choices of $f(0)$ provided $f'(0)$ has only simple eigenvalues. If $f(0)=0$ and $f'(0)$ has two distinct real eigenvalues, for example, then hypothesis (2) fails: there exist $\alpha\neq 0$ and $\beta\neq 0$ in R^p such that

$${}^T f'(0)\alpha=\lambda\alpha, \quad f'(0)\beta = \mu\beta(\lambda\neq\mu) \quad \text{and then}$$

$$\alpha\circ f'(0)^j\beta = 0 \text{ for all } j \geqslant 0 \text{ but } |\alpha||\beta|\neq 0.$$

Consider the system with $p=2$

$$\begin{cases} \Delta u_1 + \lambda u_1 + g_1(u^2)+k_1(u^3) + 0(|u|^4) = 0 \\ \Delta u_2 + \mu u_2 + g_2(u^2)+k_2(u^3) + 0(|u|^4) = 0 \end{cases}$$

with $\lambda\neq\mu$, where the g_j [or k_j] are homogeneous quadratic [or cubic] polynomials. If $e_1 = \text{col}(1,0)$, $e_2=\text{col}(0,1)$

$$|g_1(e_2^2)| + |k_1(e_2^3)| \neq 0$$

$$\text{and} \quad |g_2(e_1^2)| + |k_2(e_1^3)| \neq 0$$

Then hypothesis (3) holds and solutions (u_1, u_2) of this Dirichlet problem are simple in most bounded C^2 regions.

If $p \leqslant 3$, the conditions of hypothesis (3) hold for most choices of $f''(0), f'''(0)$, given $f'(0)$; but if $f'(0)$ has four distinct real eigenvalues (for example), the conditions hold on an open, but not dense, set of $f''(0), f'''(0)$.

6. Generic hyperbolicity of equilibria.

In some cases, hyperbolicity follows from simplicity. For example, suppose the (scalar) problem

$$\Delta v + (c(x) - g(x,\lambda))v = 0 \text{ in } \Omega \text{ , } v = 0 \text{ or } \partial v/\partial N = \gamma(x)v \text{ on } \partial\Omega$$

has a non-trivial solution $v(x)$ for some λ with $\mathrm{Re}\,\lambda = 0$. We assume $c(x)$ and $\gamma(x)$ are real-valued, and then multiplication by $\bar{v}$ and integration by parts yields $\int_\Omega (\mathrm{Im}\,g)|v|^2 = 0$. In the parabolic case $g(x,\lambda) = \lambda$ we conclude $\lambda = 0$; for the damped wave equation $g(x,\lambda) = \lambda^2 + r(x)\lambda$ we conclude either $\lambda = 0$ or $\int_\Omega r(x)|v|^2 = 0$, so if $r(x) > 0$ somewhere in each component Ω, we again conclude that $\lambda = 0$, and simplicity implies hyperbolicity.

Consider now some problems which are not (formally) self-adjoint.

For the linear problem

$$\Delta v + \sum_{j=1}^{n} b_j(x)\, \partial v/\partial x_j + (c(x) - g(x,\lambda))v = 0 \text{ in } \Omega, \ v = 0 \text{ on } \partial\Omega$$

with b_j, c, m, r real-valued and C^2, $g(x,\lambda) = m(x)\lambda^2 + r(x)\lambda$, if either $m \equiv 0$ and $r(x) \neq 0$ on a dense set (the parabolic case) or $m(x) \neq 0$ and $r(x) \neq 0$ on a dense set (the damped wave equation), then for most choices of Ω, the above problem has no nontrivial solutions with $\mathrm{Re}\lambda = 0$. This assumes the b_j, c, m, r are given independent of Ω.

The more interesting case is when the equation is obtained by linearization about some equilibrium of a nonlinear problem.

Consider the equilibrium problem

$$\Delta u + \sum_{j=1}^{n} b_j(x,u)\,\partial u/\partial x_j + c(x,u) = 0 \quad \text{in} \quad \Omega, \; u=0 \text{ on } \partial\Omega,$$

where the $b_j(x,u), c(x,u)$ are given C^3 functions, so we assume linear dependence on ∇u in $f(x,u,\nabla u)$ - this greatly simplifies the problem, making if sometimes solvable. Specifically if $C(x,0)\neq 0$ on a dense set and $g(x,\lambda)=m(x)\lambda^2+r(x)\lambda$ as before with $r(x)\neq 0$ on a dense set, then for most bounded C^2 regions $\Omega \subset R^n$, all solutions u of the above non-linear Dirichlet problem are hyperbolic, so the linearization of

$$m(x)u_{tt}+r(x)u_t = \Delta u + b(x,u)\circ\nabla u + c(x,u) \text{ in } \Omega, \; u=0 \text{ on } \partial\Omega$$

about any equilibrium has no non-trivial solutions of the form $e^{\lambda t}v(x)$. If $c(x,0)\equiv 0$, the zero solution will be hyperbolic for most Ω, but I have no information about the non-zero equilibria (unless we also perturb c). There is also no information about the case when $f(x,u,\nabla u)$ has non-linear dependence on ∇u. Clearly, much remains to be done.

References

[1] C. FOIAS and R. TEMAM, Structure of the set of stationary solutions of the Navier-Stokes equations, Comm. Pure Appl. Math. 30 (1977), p. 149-164.

[2] D. HENRY, Perturbation of the boundary for boundary value problems of partial differential equations.

[3] A. M. MICHELETTI, Metrica per famiglie di domini limitati e proprieta generiche degli autovalori, Annali della Scuola Norm. Sup. Pisa, v. 26 (1972), p. 683-694.

[4] J. C. SAUT and R. TEMAM, Generic properties of nonlinear boundary value problems, Comm. in PDE, 4 (1979), p. 293-319.

[5] K. UHLENBECK, Generic properties of eigenfunctions, Amer. J. Math. 98 (1976), p. 1059-1078.

Complex Analytical Methods in RFDE Theory

S. M. Verduyn Lunel
Centre for Mathematics and Computer Science
P.O. Box 4079, 1009 AB Amsterdam, The Netherlands

Laplace transformation of a linear autonomous retarded functional differential equation (RFDE) with finite delay yields an analytic equation that has to be solved over a ring of meromorphic functions. This paper presents the technics to study the equation and gives an overview of the recently obtained results related to completeness and convergence of the generalized eigenfunctions of the infinitesimal generator associated to a linear autonomous RFDE.

1980 Mathematics Subject Classification : 47D05, 34K05
Key Words & Phrases : C_0-semigroups, completeness, exponential series, generalized eigenfunctions, retarded functional differential equations, small solutions, Volterra convolution equation.

1. Introduction

By applying the Laplace transform to a certain Volterra convolution equation complex function theory can be made to bear on a linear autonomous retarded functional differential equation (RFDE) with finite delay.

In combination with the functional analytical information obtained in [3, 7], theorems such as the Paley-Wiener theorem and the Ahlfors-Heins theorem turn out to be quite useful in the study of RFDEs.

Although the matrix equation obtained by Laplace transformation has to be solved over a ring of meromorphic functions, it turns out that information about the exponential growth of the elements involved in the equation is sufficient to prove existence of solutions with certain properties.

This paper will give an overview, detailed results can be found in [8, 9, 10].
The organization of the paper is as follows. In section 2 we consider the linear autonomous RFDE, give definitions and motivate the basic questions before we start with the Volterra convolution equation in section 3. In section 4 we consider the Laplace transformed version of the Volterra convolution equation and present convergence results for the exponential series expansion of the solution.

2. The linear autonomous RFDE

Consider the linear autonomous RFDE with finite delay given by

$$\begin{aligned} \dot{x}(t) &= L x_t , \\ x_0 &= \phi, \\ \phi &\in C = C[-h,0], \end{aligned} \tag{2.1}$$

where $t \geqslant 0$, $x_t = x(t+\cdot)$ on [-h, 0] and L is a continuous mapping from C into $\mathbb{R}^n$, and hence is given by

$$L\phi = \int_0^h d\zeta(\theta)\phi(-\theta), \tag{2.2}$$

NATO ASI Series, Vol. F37
Dynamics of Infinite Dimensional Systems
Edited by S.-N. Chow, and J. K. Hale

where ζ is a matrix valued function on $\mathbb{R}^n$ that is left continuous, of bounded variation and constant on the interval [h, ∞).

Translation along the solution induces a C_0-semigroup

$$T(t)\phi = x(t + \cdot\,; \phi) = : x_t\,, \tag{2.3}$$

with infinitesimal generator

$$A\phi = \dot{\phi},$$

defined on

$$\mathcal{D}(A) = \{\phi \in C : \dot{\phi} \in C \text{ and } \dot{\phi}(0) = L\phi\}.$$

From the explicit formula for the resolvent [7]

$$R(\lambda, A) = (\lambda I - A)^{-1}$$

one can deduce that the resolvent is a compact operator. Consequently the spaces

$$\mathcal{N}((\lambda I - A)^k), \tag{2.4}$$

$\lambda \in \sigma(A)$, are finite dimensional, become stable for finite $k = m_\lambda$, the order of the eigenvalue λ, and the spectral decomposition

$$C = \mathcal{N}((\lambda I - A)^{m_\lambda}) \oplus \mathcal{R}((\lambda I - A)^{m_\lambda}) \tag{2.5}$$

holds.

The spectral projection corresponding to the spectral decomposition is given by

$$P_\lambda \phi = \int_{\Gamma_\lambda} R(z, A)\phi\, dz \tag{2.6}$$

where Γ_λ is a contour enclosing λ but no other point of the discrete set $\sigma(A)$.
Note that, in general, the spectral decomposition is not orthogonal.

Let $\mathcal{M}_C$ denote the linear subspace generated by

$$\mathcal{N}((\lambda I - A)^{m_\lambda}),\ \lambda \in \sigma(A).$$

The linear subspace $\mathcal{M}_C$ will be called the generalized eigenspace of A.

Define the ascent of a semigroup $\{T(t)\}$ by the value

$$\inf\{t \,|\, \text{for all } \epsilon > 0 : \mathcal{N}(T(t)) = \mathcal{N}(T(t+\epsilon))\}.$$

Let α denote the ascent of $\{T(t)\}$ and δ denote the ascent of the adjoint semigroup $\{T^*(t)\}$.

DEFINITION 2.1. *A solution x of equation (2.1) is called a small solution if*

$$\lim_{t\to\infty} e^{kt} x(t) = 0,$$

for all $k \in \mathbb{R}$.

In 1971 Henry [5] proved the following results

THEOREM 2.2.
(i) $\alpha \leqslant nh$;
(ii) *Small solutions are in the nullspace of* $T(\alpha)$;
(iii) $\overline{\mathfrak{M}_C} = \overline{\mathfrak{R}(T(\delta))}$.

Note that because of (iii), $\delta = 0$ implies completeness of the generalized eigenfunctions - i.e. $\overline{\mathfrak{M}_C} = C$. Also note that $\mathfrak{R}(T(t))$ 'decreases' with increasing time t because the solution becomes smoother, but that the closure of the range, $\overline{\mathfrak{R}(T(t))}$, becomes stable after finite time since, by duality, $\delta \leqslant nh$ holds as well.

The folowing questions were the motivation for a further study of the linear autonomous RFDE.

QUESTION I. Does $\alpha = \delta$ hold?

A related question is : Is completeness equivalent to the absence of non-trivial small solutions?

QUESTION II. Is there an explicit characterization of the ascents α, δ in terms of the kernel ζ such that completeness can be verified easily?

From the explicit formula for the resolvent $R(\lambda, A)$ [7] one can show that

$$\bigcap_{\lambda \in \sigma(A)} \mathfrak{R}((\lambda I - A)^{m_\lambda}) = \mathfrak{N}(T(\alpha)).$$

Therefore one might ask the following question.

QUESTION III. Can we extend the spectral decomposition over all $\lambda \in \sigma(A)$ to obtain the decomposition

$$C = \overline{\mathfrak{M}_C} \oplus \mathfrak{N}(T(\alpha))?$$

In this paper we will discuss the above questions and we will give an overview of the answers which were recently obtained.

REMARK 2.3. In this paper we use the state space C. However, all our results can be extended to the state space $M_p = \mathbb{R}^n \times L^p[-h,0]$ introduced by Delfour, Manitius and others [2]. Here we will work in the C-framework because the results of Diekmann [3] make clear that there is no need anymore to extend the state space to M_p because there is a natural extension from C to M_∞ which satisfies all our needs for a larger state space. Moreover, the semigroups on C and M_∞ have equal ascents and hence the results obtained for the C-framework can be extended without difficulties to the M_∞-framework.

3. THE RFDE VERSUS THE VOLTERRA CONVOLUTION EQUATION

The derivative of the solution of the linear autonomous RFDE (2.1) satisfies a renewal equation of the form

$$\dot{x} - \zeta * \dot{x} = h, \tag{3.1}$$

where h is defined on $\mathbb{R}_+$ and is constant on the interval $[h, \infty)$.
The convolution product is, as usual, defined by

$$h * g(t) = \int_0^t \zeta(\theta) g(t-\theta) d\theta ,$$

where $t \geqslant 0$.
Also, the adjoint semigroup $T^*(t)$ can be considered as a semigroup acting on forcing functions in the equation

$$x - \zeta^T * x = g ,$$

where g is defined on $\mathbb{R}_+$, left continuous, of bounded variation, and constant on the interval $[h, \infty)$, and ζ^T denotes the transposed kernel.

Formal calculations show that in fact equation (2.1) is equivalent to the Volterra convolution equation

$$x - \zeta * x = f, \tag{3.1}$$

where f is an element of F, the supremum-normed Banach space of all continuous functions defined on $\mathbb{R}_+$, that are constant on the interval $[h, \infty)$.

The equivalence map from an initial condition for (2.1) onto a forcing function for (3.1) is given by

$$\phi \rightarrow \phi_{-h} - \zeta * \phi_{-h} , \tag{3.2}$$

and corresponds to shifting the the solution over a distance h.

We can rescale the equation (3.1) such that the rescaled kernel $\tilde{\zeta}$ becomes an element of $L^1(\mathbb{R}_+)$. A straight forward contraction argument can be used to prove the existence and uniqueness of the rescaled solution $\tilde{x}(\cdot ; f)$ as an element of $C_0(\mathbb{R}_+) \cap L^2(\mathbb{R}_+)$. The solution $x(\cdot ; f)$ equals $\tilde{x}(\cdot ; f) e^{\gamma t}$ where γ is a fixed positive number, the scaling factor. Without loss of generality we may assume $\gamma = 0$.

Let $\{ S(t) \}$ denote the C_0-semigroup acting on forcing functions according to the induced action of $\{ T(t) \}$ on forcing functions under the equivalence mapping (3.2). Then

$$S(t) f = x_t(\cdot ; f) - \zeta * x_t(\cdot ; f), \tag{3.3}$$

and $\{ T(t) \}$ and $\{ S(t) \}$ have equal ascent.

4. Laplace transformation of the Volterra convolution equation

Since the solution $x = x(\cdot ; f)$ of the (rescaled) equation (3.1) is an element of $L^2(\mathbb{R}_+)$ we can Laplace transform the equation (3.1) to obtain for $\mathrm{Re}(z) > 0$

$$\mathcal{L}[x](z) = \Delta^{-1}(z) \left\{ f(h) + z \int_0^h e^{-zt} (f(t) - f(h)) \, dt \right\}, \tag{4.1}$$

where Δ(z), the characteristic matrix of (3.1), denotes the complex matrix valued function

$$\Delta(z) = z I - \int_0^h e^{-z\theta} d\zeta(\theta). \tag{4.2}$$

The expression (4.1) yields a meromorphic continuation of $\mathcal{L}[x]$ to the whole complex plane. We will denote this meromorphic continuation by $H(\cdot ; f)$, it is clear that the only singularities of $H(\cdot ; f)$

occur at $z = \lambda_j$, where λ_j is a root of the equation

$$\det \Delta(z) = 0. \tag{4.3}$$

The residues at these poles have the form

$$\operatorname*{Res}_{z=\lambda_j} \{ e^{zt} H(z;f) \} = p_j(t;f) e^{\lambda_j t},$$

where $p_j(\cdot\,;f)$ is a polynomial of degree $(m_\lambda - 1)$, with m_λ the multiplicity of the zero λ_j. The series

$$\sum_{j=0}^{\infty} p_j(t;f) e^{\lambda_j t},$$

is called the exponential series expansion for the solution $x(\cdot\,;f)$. Under the equivalence map of section 3 the exponential series expansion for a solution of the Volterra convolution equation is mapped onto the spectral projection series of the corresponding solution of the RFDE (2.1).

DEFINITION 4.1. *An entire function F of order 1*

$$\textit{i.e.} \quad \limsup_{r \to \infty} \frac{\log \log M(r)}{\log r} = 1,$$

where $M(r) = \max\{\, |F(r e^{i\theta})| : 0 \leq \theta \leq 2\pi \,\}$,

is of exponential type τ if

$$\limsup_{r \to \infty} \frac{\log M(r)}{r} = \tau,$$

where $0 \leq \tau \leq \infty$.

The exponential type of a vector valued entire function is defined as the maximal exponential type of the components.
If a globally defined meromorphic function is given by the quotient of two entire functions of exponential type, then the exponential growth of the meromorphic function is defined by the difference of the exponential type of the numerator and the denominator.

The following application of the Paley-Wiener theorem and the Ahlfors-Heins theorem, Boas [1], will be the key in the exponential type calculus used in this paper.

THEOREM 4.2. *Let F be an entire function which is uniformly bounded in the closed right half plane. Then F is of exponential type τ and L^2-integrable along the imaginary axis if and only if*

$$F(z) = \int_0^\tau e^{-zt} \phi(t) dt, \tag{4.4}$$

where $\phi \in L^2\,[0,\tau]$ and ϕ does not vanish a.e. in some neighborhood of τ. Moreover, the exponential growth in different directions is given by

$$\limsup_{r \to \infty} \frac{\log |F(r e^{i\theta})|}{r} = -\tau \cos\theta, \tag{4.5}$$

for θ in a dense set of $[\frac{1}{2}\pi, \frac{3}{2}\pi]$.

Functions of the form (4.4) will be called Paley-Wiener functions. Note that the exponential type of the product of two Paley-Wiener functions equals the sum of the exponential types of the factors. This property makes the exponential type calculus so useful.

Since the entries of $\Delta(z)$ are polynomials of degree 1 with coefficients constants plus Paley-Wiener functions of exponential type less than or equal to h, it follows that $\det\Delta(z)$ will be a polynomial of degree n with coefficients constants plus Paley-Wiener functions of exponential type less than or equal to nh.
The same arguments apply to the minors M_{ij} of adj $\Delta(z)$. These will be polynomials of degree (n-1) with coefficients constants plus Paley-Wiener functions of exponential type less than or equal to (n-1)h.

Define ϵ by

$$\text{exponential type } \det\Delta(z) = nh - \epsilon,$$

and σ by

$$\max_{i,j} \text{ exponential type } M_{ij} = (n-1)h - \sigma.$$

THEOREM 4.3. *If* $\epsilon > 0$ *then* $\sigma < \epsilon$.

PROOF. Suppose $\sigma = \epsilon$ then

$$(n-1)(nh-\epsilon) = \text{type } (\det \Delta(z))^{n-1} = \text{type } \det \text{adj } \Delta(z),$$

and also,

$$\text{type } \det \text{adj } \Delta(z) \leqslant n((n-1)h-\epsilon) = (n-1)(nh-\epsilon)-\epsilon,$$

which yields a contradiction if $\epsilon > 0$. □

Equation (4.1) implies that the exponential growth of $H(\cdot\,;f)$ has to be less than or equal to $(\epsilon - \sigma)$.
In case of small solutions, i.e. solutions with an entire Laplace transform, it follows that this Laplace transform has exponential type less than or equal to $(\epsilon - \sigma)$. By theorem (4.2)

$$\int_0^\infty e^{-zt}x(t)dt = \int_0^{\epsilon-\sigma} e^{-zt}x(t)dt$$

and hence the solution will be zero for $t \geqslant (\epsilon - \sigma)$, and the ascent α of $\{S(t)\}$ will be less than or equal to $(\epsilon - \sigma)$. In fact one can prove equality by constructing a forcing function f such that the right hand side of equation (4.1) is entire and has exponential type equal to $(\epsilon - \sigma)$. This has been done in [8], where we proved the following theorem

THEOREM 4.4. *The ascent* α *of the semigroup* $\{S(t)\}$ *is given by*

$$\alpha = \epsilon - \sigma.$$

Since ϵ and σ are invariant under transposing the kernel ζ we obtain as a corollary of the above theorem.

COROLLARY 4.5.

$$\alpha = \delta.$$

This means that the first two questions of section 2 are answered positively and that completeness is equivalent to 'exponential type $\det \Delta(z)$ equals nh'.

Another application of the approach introduced above is related to convergence criteria for exponential series expansions.
Let $\mathcal{M}_F$ denote the space of forcing functions f∈ F such that the solution $x(\cdot\,;f)$ can be given by a finite exponential series. Clearly, under the equivalence map of section 3, $\mathcal{M}_F$ corresponds to $\mathcal{M}_C$.

The exponential series for a solution can be obtained by applying the inverse Laplace transform theorem to the equation (4.1) and then using the Cauchy theorem of residues. From this, it follows that $C_0(\mathbb{R}_+)$-convergence of the exponential series is equivalent to

$$\Big\| \int_{C_n} e^{zt} H(z\,;f)\,dz \Big\|_{C_0(\mathbb{R}_+)} \to 0, \quad as\ n \to \infty, \tag{4.4}$$

where C_n denotes that arc of the circle with radius n which is contained in the left half plane.

If $H(\cdot\,;f)$ has positive exponential growth it is clear that (4.4) will not hold. In case $H(\cdot\,;f)$ has no exponential growth we have the following theorem [9].

THEOREM 4.6. *Let f be an absolutely continuous element of F such that*

$$\textit{exponential type}\ \operatorname{adj}\Delta(z)\,z\int_0^\infty e^{-zt} f(t)\,dt < \textit{exponential type}\ \det\Delta(z),$$

then the solution $x(\cdot\,;f)$ of (3.1) can be given as a $C_0(\mathbb{R}_+)$-convergent exponential series.

Due to the fact that the forcing function associated to an element of the $C_0(\mathbb{R}_+)$-closure of the span of the exponential series is not necessarely absolutely continuous, we have that an element of the $C_0(\mathbb{R}_+)$-closure of the span of the exponential series is not necessarely given by a $C_0(\mathbb{R}_+)$-convergent exponential series. However, for absolutely continuous forcing functions the lack of convergence of the exponential series is only in the point $t = 0$ as is shown by the following theorem [9].

THEOREM 4.7. *The solution $x(\cdot\,;f)$ is an element of the $C_0(\mathbb{R}_+)$-closure of the span of*

$$\{\, (p_j(t\,;f)e^{\lambda_j t})_{j=0}^{\infty} \,\}$$

if and only if the forcing function f satisfies the condition

$$\textit{exponential type}\ \operatorname{adj}\Delta(z)\,z\int_0^\infty e^{-zt} f(t)\,dt \leqslant \textit{exponential type}\ \det\Delta(z).$$

Moreover, if f is absolutely continuous then the solution $x(\eta+\cdot\,;f)$ can be given as a $C_0(\mathbb{R}_+)$-convergent exponential series for every $\eta > 0$.

Since all non-trivial small solutions have a Laplace transform with positive exponential type it is clear that a forcing function corresponding to a small solution can not satisfy the condition of Theorem 4.7. Consequently, we have the following corollary

COROLLARY 4.8 *There are no small solutions in the $C_0(\mathbb{R}_+)$-closure of the span of*

$$\{ (p_j(t;f)e^{\lambda_j t})_{j=0}^{\infty} \}.$$

And because of the equivalence map defined in section 3.

COROLLARY 4.9

$$\overline{\mathcal{M}_C} \cap \mathcal{N}(T(\alpha)) = \{0\}.$$

This corollary shows that $\overline{\mathcal{M}_C}$ is a closed invariant subspace of C and answers a question posed by Hale in his book on functional differential equations [4] positively.
Moreover, corollary (4.9) implies, by duality, that

$$\overline{\mathcal{M}_C} \oplus \mathcal{N}(T(\alpha))$$

decomposes at least a dense subspace of C. In general this direct sum is not closed. This is caused by the exceptional behaviour at the point $t = 0$ and is related to the lack of convergence in $t = 0$ of the spectral projection series. Detailed results will be given in [10].

5. EXAMPLE

Although the condition on f for being an element of $\overline{\mathcal{M}_F}$ are technical, they are rather simple to check in case we are dealing with differential difference equations. It turns out that $\overline{\mathcal{M}_F}$ can be explicitly determined for this type of equations.

EXAMPLE:

Consider the equation:

$$\begin{aligned} \dot{x}_1(t) &= -x_2(t) + x_3(t-1) \\ \dot{x}_2(t) &= x_1(t-1) \\ \dot{x}_3(t) &= 0 \end{aligned} \tag{5.1}$$

then the characteristic matrix becomes:

$$\Delta(z) = \begin{bmatrix} z & 1 & -e^{-z} \\ -e^{-z} & z & 0 \\ 0 & 0 & z \end{bmatrix} \tag{5.2}$$

and $\Delta^{-1}(z)$ becomes:

$$\Delta^{-1}(z) = \begin{bmatrix} z^2 & -z & ze^{-z} \\ ze^{-z} & z^2 & e^{-2z} \\ 0 & 0 & z^2 + e^{-z} \end{bmatrix} \frac{1}{z(z^2 + e^{-z})}. \tag{5.3}$$

Are there any small solutions?
Yes, $\alpha = \epsilon - \sigma = 2 - 0 = 2 > 0$.
So no completeness of the generalized eigenfunctions.

Description of the set $\overline{\mathfrak{M}_F}$:
Apply $\Delta^{-1}(z)$ to the Laplace transform of f and use Theorem 4.7, the first equation yields no information, the third equation forces f_3 to be constant, finally the second equation yields:

$$\overline{\mathfrak{M}_F} = \{f \in F : f_1(t) = f_3(1)t + c,\ c \in \mathbb{R}, \text{ and } f_3(t) = f_3(1)\}.$$

REFERENCES

[1] R. BOAS, 'Entire Functions,' Academic Press, New York, 1954.
[2] M. C. DELFOUR AND A. MANITIUS, The structural operator F and its role in the theory of retarded systems, I, II, *J. Math. Anal. Appl.* **73** (1980), 466-490; **74** (1980), 359-381.
[3] O. DIEKMANN, Perturbed dual semigroups and delay equations, Proceedings of 'Infinite Dimensional Dynamical Systems', Lisbon, 1986
[4] J. K. HALE, 'Theory of Functional Differential Equations,' Springer-Verlag, New York, 1977.
[5] D. HENRY, Small solutions of linear autonomous functional differential equations, *J. Differential Equations* **9** (1971), 55-66.
[6] D. SALAMON, 'Control and Observation of Neutral Systems,' *Research Notes in Mathematics* Vol. 91, Pitman, London, 1984.
[7] S. M. VERDUYN LUNEL, 'Linear Autonomous Retarded Functional Differential Equations: A sharp version of Henry's theorem,' Report AM-R8405, Centre for Mathematics and Computer Science, Amsterdam.
[8] S. M. VERDUYN LUNEL, A sharp version of Henry's theorem on small solutions *J. Differential Equations* **62** (1986), 266-274.
[9] S. M. VERDUYN LUNEL, Exponential series expansion for solutions of a Volterra equation of the convolution type, to appear.
[10] S. M. VERDUYN LUNEL, An invariant subspace for a linear autonomous RFDE carrying all the information, to appear.

QUALITATIVE BEHAVIOR OF THE SOLUTIONS OF PERIODIC FIRST ORDER SCALAR DIFFERENTIAL EQUATIONS WITH STRICTLY CONVEX COERCIVE NONLINEARITY

J. Mawhin
Université de Louvain
Institut Mathématique
B-1348 Louvain-la-Neuve, Belgium

1. Introduction.

It has been proved in [4] that if $f : \mathbb{R} \times \mathbb{R} \to \mathbb{R}$ is continuous, $f(., u)$ is T-periodic for each $u \in \mathbb{R}$, $f(x, .)$ is strictly convex on $\mathbb{R}$ for each $x \in \mathbb{R}$, and if $f(x, .)$ is uniformly coercive, i.e.

$$f(x, u) \to +\infty \text{ as } |u| \to \infty$$

uniformly in $x \in \mathbb{R}$, then there exists $s_1 \in \mathbb{R}$ such that the equation

$$(1) \qquad u'(x) + f(x, u(x)) = s$$

has exactly zero, one or two T-periodic solutions according to $s < s_1$, $s = s_1$ or $s > s_1$. The aim of this note is to complete the result by getting a fairly complete picture of the trajectoires of (1) under the same assumptions upon f. Our results apply in particular to the forced Bernoulli equation with periodic coefficients

$$(2) \qquad u'(x) + a_1(x)u(x) + a_2(x)u^{2k}(x) = a_0(x)$$

where the $a_i : \mathbb{R} \to \mathbb{R}$ are continuous and T-periodic, $a_0(x) > 0$ and k is a positive integer (and to its special case of the Riccati equation) and describes accurately the qualitative behavior of their solutions according to the values of $s = (1/T)\int_0^T a_0(x)dx$. Applications can be made to the equations of deterministic models for the growth of populations subject to periodic fluctuations and periodic harvesting and the reader can consult [2] and [5] for some specific contributions to this problem.

NATO ASI Series, Vol. F37
Dynamics of Infinite Dimensional Systems
Edited by S.-N. Chow, and J.K. Hale

2. The structure of the set of solutions when $s > s_1$.

Let us consider the periodic differential equation

$$(2_s) \qquad u'(x) + f(x, u(x)) = s$$

where $f : \mathbb{R} \times \mathbb{R} \to \mathbb{R}$ is continuous, $f(., u)$ is T-periodic for each $u \in \mathbb{R}$, $f(x, .)$ is strictly convex for each $x \in \mathbb{R}$ and

$$f(x, u) \to +\infty \text{ as } |u| \to \infty$$

uniformly in $\mathbb{R}$. Let $s_1 \in \mathbb{R}$ be the real such that (2_s) has exactly zero, one or two T-periodic solutions according to $s < s_1$, $s = s_1$ or $s > s_1$ (see [4] for the proof of this result). We shall assume in this section that $s > s_1$ and denote the T-periodic solutions of (2_s) by u_s and v_s respectively. Since the assumptions above imply that the Cauchy problem for (2_s) is locally uniquely solvable (see the proof in [4]), we can assume, without loss of generality that

$$u_s(x) < v_s(x)$$

for all $x \in \mathbb{R}$. It will be convenient to associate to f the function

$$R_f : \mathbb{R} \times (\mathbb{R}^2 \setminus \Delta) \to \mathbb{R},$$

$$(x, u, v) \to \frac{f(x, u) - f(x, v)}{u - v},$$

where $\Delta = \{(u, v) \in \mathbb{R}^2 : u = v\}$ is the diagonal in $\mathbb{R}^2$. The strict convexity assumption made upon f is equivalent to the fact that for each x, u or v fixed, the functions $R_f(x, ., v)$ and $R_f(x, u, .)$ are increasing on $\mathbb{R}$. We shall use the following simple observation.

<u>Lemma 1</u>. If (2) satisfies the conditions listed above and if $s > s_1$, then, for each $x_0 \in \mathbb{R}$ we have

$$\int_{x_0}^{x_0+T} R_f(x, v_s(x), u_s(x))\,dx = 0,$$

where u_s and v_s denote the two T-periodic solutions of (2_s).

Proof. We have, by T-periodicity,

$$\int_{x_0}^{x_0+T} R_f(x, v_s(x), u_s(x))\,dx = -\int_{x_0}^{x_0+T} \frac{v_s'(x) - u_s'(x)}{v_s(x) - u_s(x)}\,dx$$

$$= -\int_{x_0}^{x_0+T} \frac{d}{dx}\,[\ln(v_s(x) - u_s(x))]\,dx = 0.$$

If u is a solution of (2_s), let us denote by $I_u =]w_-(u), w_+(u)[$ its maximal existence interval. The following result describes the structure of the solution of (2_s).

Theorem 1. Let u be a solution of (2_s) with $s > s_1$, and let $x_0 \in I_u$. Then the following conclusions hold.

1) If $u(x_0) > v_s(x_0)$, one has $w_+(u) = +\infty$, $u(x) > v_s(x)$ for all $x \in]w_-(u), +\infty[$,

$$\lim_{x\to+\infty} (u(x) - v_s(x)) = 0$$

and

$$\lim_{x\to w_-(u)} (u(x) - v_s(x)) = +\infty.$$

2) If $u_s(x_0) < u(x_0) < v_s(x_0)$, one has $w_-(u) = -\infty$, $w_+(u) = +\infty$, $u_s(x) < u(x) < v_s(x)$ for all $x \in \mathbb{R}$,

$$\lim_{x\to+\infty} (v_s(x) - u(x)) = 0$$

and

$$\lim_{x\to-\infty} (u(x) - u_s(x)) = 0.$$

3) If $u(x_0) < u_s(x_0)$, one has $w_-(u) = -\infty$, $u(x) < u_s(x)$ for all $x \in]-\infty, w_+(u)[$,

$$\lim_{x\to w_+(u)} (u(x) - u_s(x)) = -\infty$$

and

$$\lim_{x\to-\infty} (u(x) - u_s(x)) = 0.$$

Proof. We shall only prove the conclusion (1), the other cases being similar. The fact that $u(x_0) > v_s(x_0)$ implies that $u(x) > v_s(x)$ for all $x \in I_u$ is a direct consequence of the uniqueness of the Cauchy problem for (2_s). Now, for $x \in I_u$, we have

$$(\ln(u(x) - v_s(x)))' = \frac{u'(x) - v_s'(x)}{u(x) - v_s(x)} = -R_f(x, u(x), v_s(x)) <$$

(3)

$$< -R_f(x, u(x), u_s(x)) < -R_f(x, v_s(x), u_s(x)).$$

as

$$u(x) > v_s(x) > u_s(x)$$

for $x \in I_u$. Now, by Lemma 1, the function

$$Z : x \to \int_{x_0}^{x} R_f(y, v_s(y), u_s(y))\, dy$$

is a T-periodic function such that $Z(x_0) = 0$. Thus, from (3), we deduce that, if $x \in I_u \cap]x_0, \infty[$, we have

$$0 < u(x) - v_s(x) < (u(x_0) - v_s(x_0))\exp(-Z(x)),$$

so that necessarily $w_+(u) = +\infty$. On the other hand,

$$\{\ln[(\exp Z(x))(u(x) - v_s(x))]\}' = [Z(x) + \ln(u(x) - v_s(x))]' =$$

$$= R_f(x, v_s(x), u_s(x)) - R_f(x, u(x), v_s(x)) < 0$$

for all $x \in I_u$. Therefore, $\ln[(\exp Z)(u - v_s)]$ is decreasing on I_u and the same is true for $(\exp Z)(u - v_s)$. Now, for all positive integers k, we have, by the T-periodicity of Z,

$$u(x_0 + kT) - v_s(x_0) = u(x_0 + kT) - v_s(x_0 + kT) =$$

$$= (\exp Z(x_0))(u(x_0 + kT) - v_s(x_0 + kT)) =$$

$$= (\exp Z(x_0 + kT))(u(x_0 + kT) - v_s(x_0 + kT)) >$$

$$> (\exp Z(x_0 + (k+1)T)(u(x_0 + (k+1)T) - v_s(x_0 + (k+1)T)) =$$

$$= u(x_0 + (k+1)T) - v_s(x_0),$$

and hence the sequence $(u(x_0 + kT) - v_s(x_0))$ is decreasing. Let $u_+ \geqslant 0$ be its limit. If $u_+ > 0$, then, denoting by $U(x; \xi, a)$ the solution of (2_s) such that

$$U(\xi; \xi; a) = a,$$

and defining the Poincaré operator P_ξ by

$$P_\xi a = U(\xi + T; \xi; a),$$

we have, by the T-periodicity of f and the uniqueness of the Cauchy problem,

$$U(\xi + 2T; \xi, a) = U(\xi + T; \xi, U(\xi + T; \xi; a)) =$$
$$= U(\xi + T; \xi, P_\xi a) = P_\xi^2 a.$$

Hence, inductively

$$U(\xi + kT; \xi, a) = U(\xi + T; \xi, U(\xi + (k-1)T, \xi, a)) =$$
$$= U(\xi + T; \xi, P_\xi^{k-1} a) = P_\xi^k a,$$

so that

$$u(x_0 + kT) = P_{x_0}^k u(x_0).$$

From

$$P_{x_0}^k u(x_0) = P_{x_0}(P_{x_0}^{k-1} u(x_0))$$

we get, for $k \to \infty$

$$u_+ + v_s(x_0) = P_{x_0}(u_+ + v_s(x_0))$$

and $U(x; x_0; u_+ + v_s(x_0))$ is a T-periodic solution greater than $v_s(x)$, a contradiction. Thus, by the monotonicity,

$$0 = u_+ = \lim_{k\to\infty} [(\exp Ƶ(x_0 + kT))(u(x_0 + kT) - v_s(x_0 + kT))] =$$
$$= \lim_{x\to+\infty} (\exp Ƶ(x))(u(x) - v_s(x))$$

so that

$$\lim_{x\to+\infty} (u(x) - v_s(x)) = 0$$

as

$$0 < u(x) - v_s(x) = \exp(-Ƶ(x))(\exp Ƶ(x))(u(x) - v_s(x)) \leqslant$$
$$\leqslant C(\exp Ƶ(x))(u(x) - v_s(x)).$$

Now, if $w_-(u) > -\infty$, we necessarily have

$$\lim_{x\to w_-(u)} u(x) = +\infty$$

and, if $w_-(u) = -\infty$, our assertion is equivalent to

$$u_- \equiv \lim_{x\to-\infty} (\exp Ƶ(x))(u(x) - v_s(x)) = +\infty$$

and we can prove it by contradiction, using an argument similar to that used in the first part of the proof.

Remark 1. Theorem 1 implies in particular that, for $s > s_1$, v_s is asymptotically stable and u_s is unstable.

Remark 2. Similar results hold if convexity is replaced by concavity when

$$f(x, u) \to -\infty \text{ as } |u| \to \infty$$

uniformly in $x \in \mathbb{R}$. Their formulation is left to the reader.

3. The structure of the set of solution when $s \leqslant s_1$.

We keep the notations of Section 2 and we denote moreover by w_{s_1} the unique T-periodic solution of (2_{s_1}). By classical results (see e.g. [1]), the functions $\lim_{s\to s_1} u_s$ and $\lim_{s\to s_1} v_s$ are T-periodic solutions of (1_{s_1}) and hence

$$(4) \qquad w_{s_1}(x) = \lim_{s\to s_1} u_s(x) = \lim_{s\to s_1} v_s(x)$$

and it is easy to show that the convergence is indeed uniform on $\mathbb{R}$. It is also shown in the proof of Theorem 1 of [4] (see part (d)) that for each $\tilde{s} > s_1$, the set of possible T-periodic solutions of (2_s) with $s \leqslant \tilde{s}$ is a priori bounded independently of s.

Lemma 2. Assume that $f : \mathbb{R} \times \mathbb{R} \to \mathbb{R}$ is continuous, $f(., u)$ is T-periodic for each $u \in \mathbb{R}$, $f(x, .)$ is strictly convex for each $x \in \mathbb{R}$ and uniformly coercive and that $D_u f$ exists and is continuous on $\mathbb{R} \times \mathbb{R}$. Let w_{s_1} be the unique T-periodic solution of (2_{s_1}). Then, for each $x_0 \in \mathbb{R}$,

$$\int_{x_0}^{x_0+T} D_u f(x, w_{s_1}(x))\,dx = 0.$$

Proof. By Lemma 1, we have, for $s > s_1$,

$$\int_{x_0}^{x_0+T} R_f(x, v_s(x), u_s(x))\,dx = 0,$$

and the result follows, using the a priori estimates upon v_s and u_s and the continuity of $D_u f$, from the dominated convergence theorem (Osgood's version for Riemann integral suffices).
We can now prove the result which corresponds to Theorem 1 when $s = s_1$.

Theorem 2. Let us assume that f satisfies the conditions of Lemma 2, let w_{s_1} the unique T-periodic solution of (2_{s_1}), let u be a solution of (2_{s_1}) and let $x_0 \in I_u =]w_-(u), w_+(u)[$, the maximal existence interval of u. Then, the following conclusions hold.

1) If $u(x_0) > w_{s_1}(x_0)$, one has $w_+(u) = +\infty$, $u(x) > w_{s_1}(x)$ for all $x \in]w_-(u), +\infty[$,

$$\lim_{x\to+\infty} (u(x) - w_{s_1}(x)) = 0$$

and

$$\lim_{x\to w_-(u)} (u(x) - w_{s_1}(x)) = +\infty.$$

2) If $u(x_0) < w_{s_1}(x_0)$, one has $w_-(u) = -\infty$, $u(x) < w_{s_1}(x)$ for all $x \in]-\infty, w_+(u)[$,

$$\lim_{x\to w_+(u)} (u(x) - w_{s_1}(x)) = -\infty$$

and

$$\lim_{x\to-\infty} (u(x) - w_{s_1}(x)) = 0.$$

Proof. It follows essentially the lines of that of Theorem 1 if one notices that, with our assumptions, the function $u \to R_f(x, u, v)$ defined if $u \neq v$ can be extended continuously at $u = v$ by $D_u f(x, v)$. Therefore, the argument of Theorem 1 holds if the function Z is now defined by

$$Z(x) = \int_{x_0}^{x} D_u f(y, w_{s_1}(y))dy.$$

The details are left to the reader.

Remark 3. Theorem 1 implies that w_{s_1} is unstable.

Remark 4. Similar results hold if convexity is replaced by concavity when

$$f(x, u) \to -\infty \text{ as } |u| \to \infty,$$

uniformly in $x \in \mathbb{R}$. Their formulation is left to the reader.

When $s < s_1$, (2_s) has no T-periodic solution and hence, by Massera's theorem [3], it has no solution bounded in the future or bounded in the past. Also, if $s_0 = \min_{\mathbb{R}\times\mathbb{R}} f$ and if (2_s) has a T-periodic solution u, then

$$s = \frac{1}{T}\int_0^T [\, u'(x) + f(x, u(x))\,]dx \geqslant s_0,$$

so that $s_0 \leqslant s_1$. Now, for $s < s_0$,

$$u'(x) = s - f(x, u(x)) < s_0 - f(x, u(x)) \leqslant 0$$

and each solution u of (2_s) is decreasing, and hence such that

$$\lim_{x\to w_-(u)} u(x) = +\infty, \quad \lim_{x\to w_+(u)} u(x) = -\infty.$$

It is indeed the case for $s < s_1$ as shown by the following result.

Theorem 3. Assume that $f : \mathbb{R} \times \mathbb{R} \to \mathbb{R}$ is continuous, $f(., u)$ is T-periodic for each $u \in \mathbb{R}$ and that $f(x, .)$ is uniformly coercive. Let $s_1 \in \mathbb{R}$ be such that (2_s) has no T-periodic solution for $s < s_1$ and let u be an arbitrary solution of (2_s) with maximal existence interval $I_u =]\,w_-(u), w_+(u)\,[$.
Then

$$(5) \qquad \lim_{x\to w_-(u)} u(x) = +\infty, \quad \lim_{x\to w_+(u)} u(x) = -\infty.$$

Proof. It follows from a remark above that u is neither bounded in the future nor in the past, i.e. that

$$\limsup_{x\to w_-(u)} |u(x)| = \limsup_{x\to w_+(u)} |u(x)| = \infty.$$

Now, by the coercivity of f(x, .), there exists R > 0 such that

$$u'(x) < 0$$

whenever $|u(x)| \geq R$, so that we necessarily have (5).

REFERENCES

1. P. HARTMAN, "Ordinary Differential Equations", Wiley, 1964.
2. A.C. LAZER and D.A. SANCHEZ, Periodic equilibria under periodic harvesting, Math. Magazine 57 (1984) 156-158.
3. J.L. MASSERA, The existence of periodic solutions of systems of differential equations, Duke Math. J. 17 (1950) 457-475.
4. J. MAWHIN, First order ordinary differential equations with several periodic solutions, to appear.
5. D.A. SANCHEZ, Periodic environments, harvesting and a Riccati equation, in "Nonlinear Phenomena in Mathematical Sciences", Lakshmikantam ed., Academic Press, 1982, 883-886.

THE SPECTRUM OF INVARIANT SETS FOR DISSIPATIVE SEMIFLOWS

Luis T. Magalhães
Departamento de Matemática
Instituto Superior Técnico
Universidade Técnica de Lisboa
1096 Lisboa Codex, PORTUGAL

1. INTRODUCTION

For a wide class of dissipative systems (see [7] for a survey) an important role is played by maximal compact invariant sets — *attractors*. As these sets contain the recurrent points throughout the evolution of the system, they are of particular importance for the discussion of qualitative dynamical properties.

The attractor sets may, in general, have a wild topologial structure. It is therefore of interest to know when they are simple. A considerable effort has been dedicated to establish conditions under which they are finite dimensional sets, are included in finite dimensional invariant manifolds, or are themselves smooth invariant manifolds [1-5, 8-13]. The concept of inertial manifolds was introduced in relation to this work. These are finite dimensional hyperbolic manifolds which are locally invariant and contain the attractor set. When inertial manifolds exist, the long time behavior of the system can be exactly described by a finite dimensional dynamical system [3-5, 12].

Hyperbolic stuctures are defined in terms of exponential rates of attraction or repulsion which can be expressed by exponential dichotomies for the linearization around the orbits they contain. A natural context for describing linearizations around solutions is that of skew-product semiflows, following the work of Sacker and Sell for flows [14, 15]. This paper describes general properties of the spectrum of a linear skew-product semiflow, in a context that is appropriate for aplications to certain retarded Functional Differential Equations (FDEs), neutral FDEs, parabolic Partial Differential Equations (PDEs), and dissipative hyperbolic PDEs (see [7] for a discussion of these types of systems). It is also remarked that, for a wide class of systems of the types just mentioned, the existence of inertial manifolds amounts to the existence of a gap in the spectrum of the linearized semiflow around the attractor set.

2. SPECTRUM OF A LINEAR SKEW-PRODUCT SEMIFLOW

Let W be a topological space. A *flow* on W is a continous mapping $\pi : R \times W \to W$ such that $\pi(0,w) = w$ and $\pi(t,\pi(s,w)) = \pi(t+s,w)$ for all $w \in W$ and $t,s \in R$. A *semiflow* on W is a continous mapping $\pi : [0,\infty) \times W \to W$ satisfying the preceding conditions for $t,s \geq 0$. When π is a semiflow, $\pi(t,w)$ are only defined for $t \geq 0$. However, they can be extended to $t < 0$ at those points w through which there is a

NATO ASI Series, Vol. F37
Dynamics of Infinite Dimensional Systems
Edited by S.-N. Chow, and J. K. Hale

backwards continuation defined for all $t \leq 0$. We consider the set B_π defined by

$$B_\pi = \{w \in W : \text{ there is a continuous function } v : (-\infty, 0] \to W \text{ such that } v(0) = w \text{ and } \pi(t, v(s)) = v(t+s) \text{ for all } s \leq 0 \text{ and } t \in [0, -s]\}.$$

A set $I \subset W$ is said to be *positively invariant* under π if $\pi(t,w) \in I$ for all $t \geq 0, w \in I$, and it is said to be *invariant* under π if $I \subset B_\pi$ and the preceding condition holds for all $t \in R$, and all possible continuations of $\pi(t,w)$ to $t \leq 0$.

Let X be a Hausdorff topological space and let E be a Banach vector bundle over X with *fiber projection* $p : E \to X$, i.e., E is a vector bundle over X with each fiber $E(x) = p^{-1}(x)$, $x \in X$, being a Banach space. Points in E can be represented by ordered pairs (x, z), with $x \in X$ and z a vector in the fiber $E(x)$. A semiflow π on E is said to be a *skew-product semiflow* on E if $B_\pi \supset \{(x, 0) : x \in X\}$, and there exists a semiflow φ on X such that the fiber projection p commutes with π and φ, i.e., π can be represented as

$$\pi(t, x, z) = (\varphi(t, w), \psi(t, x, z)),\ t \geq 0,$$

and $\psi(t,x,z)$ is in the fiber $E(\varphi(t,x))$. Such a skew-product semiflow π is a *linear skew-product semiflow* if the mapping $z \to \Psi(t,x)z = \psi(t,x,z)$ is a linear mapping from the fiber $E(x)$ to the fiber $E(\varphi(t,x))$. One defines analogously *skew-product flow* and *linear skew-product flow.*

Given a skew-product semiflow π on a vector bundle E over X, we define the *stable* and the *unstable* sets of X under π by, respectively,

$$S = \{(x,z) \in E : |\psi(t, x, z)| \to 0 \text{ as } t \to +\infty\}$$

$$U = \{(x,z) \in B_\pi : |\psi(t,x,z)| \to 0 \text{ as } t \to -\infty, \text{ for all possible continuations of } \psi(t,x,z) \text{ to } t < 0\}.$$

The sets S and U are both positively invariant under π, and the sets B_π, $S \cap B_\pi$ and U are all invariant under π. If π is a linear skew-product semiflow, it is easy to see that these sets are vector subbundles of E. For each $x \in X$, the fibers $S(x), U(x)$ are linear subspaces of $E(x)$.

Let M be any subset of the base space X. We denote by $E(M)$ the restriction of E to M, $E(M) = \{(x,z) \in E : x \in M\}$ and similarly for $S(M)$ and $U(M)$. The linear skew-product semiflow $\pi = (\varphi, \psi)$ on E is said to admit an *exponential dichotomy* over M if there exist linear projections $P(x)$ defined on $E(x)$ for $x \in M$ and depending continuously on $x \in M$, and there exist constants $K, \alpha \geq 0$ such that

1. $N(P) = \{(x,z) \in E(M) : P(x)z = 0\} \subset B_\pi$
2. $|\Psi(t,x)P(x)| \leq Ke^{-\alpha t}$, for $t \geq 0, x \in M$
3. $|\Psi(t,x)[I - P(x)]| \leq Ke^{+\alpha t}$, for $t \leq 0$ and all possible continuations of $\Psi(t,x)z$ to $t \leq 0$, with $(x,z) \in B_\pi, x \in M$.

We note that condition 3. makes sense because 1. implies that the range of the mapping $(x,z) \to (x,[I-P(x)]z)$ is contained in B_π . Whenever π admits an exponential dichotomy over M we have

$$U(M) = N(P) = \{(x,z) \in E(M) : P(x)z = 0\}$$

and

$$S(M) = R(P) = \{(x,z) \in E(M) : z = P(x)z' \text{ for some } z' \in E(x)\}.$$

Then, the stable and unstable sets of M , respectively $S(M)$ and $U(M)$, are complementary subbundles of $E(M)$.

Given a linear skew-product semiflow $\pi = (\varphi, \psi)$ on a vector bundle E , and a real number λ , we define a mapping π_λ by

$$\pi_\lambda(t,x,z) = (\varphi(t,x), e^{-\lambda t}\psi(t,x,z)).$$

It is easily seen that π_λ is also a linear skew-product semiflow on E and that the invariant sets under π and under π_λ coincide , for all $\lambda \in R$. We also define

$$\Psi_\lambda(t,x)z = e^{-\lambda t}\psi(t,x,z) = e^{-\lambda t}\Psi(t,x)z.$$

The stable and the unstable sets of X under π_λ are denoted by S_λ and U_λ , respectively. Clearly, if $\mu < \lambda$ then $S_\mu \subset S_\lambda$ and $U_\mu \supset U_\lambda$. The set of all $\lambda \in R$ for which π_λ admits an exponential dichotomy over a subset $M \subset X$ is called the ***resolvent set*** of M under π and is denoted by $\rho(M)$. The complement of this resolvent set in R is called the ***spectrum*** of M under π and is denoted by $\sum(M)$.

If the linear skew-product semiflow satisfies adequate smoothening properties then, for each λ in the resolvent set, the fibers $U_\lambda(x)$ are finite dimensional for $x \in M$. For applications to semiflows associated with equations of the types mentioned in the introduction, it is appropriate to define the required smoothening properties in terms of the concept of α-contraction (see [6, 7] for related concepts). We define the***Kuratowski measure of noncompactness***, $\alpha(B)$, of a bounded set B in a Banach space by

$$\alpha(B) = inf\{d : B \text{ has a finite open covering by sets of diameter } d\}.$$

A linear skew-product semiflow $\pi = (\varphi, \psi)$ on a vector bundle E over a base space X is said to be an ***α-contraction relative to*** X for $t \geq r$, if there exists a function $K : [0,\infty) \to [0,1)$, with $K(t) \to 0$ as $t \to \infty$, such that the linear mapping $\Psi(t,y) : E(y) \to E(\varphi(t,y))$ satisfies $\alpha(\Psi(t,y)B) \leq K(t)\alpha(B)$, for $t \geq r$, all bounded sets $B \subset E(y)$ and all $y \in X$. The linear skew-product semiflow π is said to be ***completely continuous relative to*** X for $t \geq r$ if $\Psi(t,y)B$ is a relatively compact subset of $E(\varphi(t,y))$ for $t \geq r$, all bounded sets $B \subset E(y)$ and all $y \in X$.

THEOREM 2.1

Let $\pi = (\varphi, \psi)$ be a linear skew-product semiflow on a Banach vector bundle E over a Hausdorff topological space X , and let M be a compact connected subset of X invariant under the semiflow φ . If $\lambda \in \rho(M)$, and π_λ is an α-contraction relative to M for $t \geq r$, then $U_\lambda(x)$ is finite dimensional, for all $x \in M$, with dimension independent of $x \in M$.

If the linear skew-product semiflow is completely continuous then we have available more information about the spectrum.

THEOREM 2.2

Let $\pi = (\varphi, \psi)$ be a linear skew-product semiflow on a Banach vector bundle E defined over a Hausdorff topological space X , and let M be a compact connected subset of X invariant under semiflow φ and containing a fixed point or a closed orbit of φ . If π is completely continuous relative to M for $t \geq r$, then the spectrum $\Sigma(M)$ is of one of the following forms:

1. $\Sigma(M) = \emptyset$
2. $\Sigma(M) = \cup_{i=1}^{k}[a_i, b_i]$, for some integer k
3. $\Sigma(M) = \cup_{i=1}^{\infty}[a_i, b_i]$
4. $\Sigma(M) = (-\infty, b_\infty]$
5. $\Sigma(M) = (-\infty, b_\infty] \cup \cup_{i=1}^{k}[a_i, b_i]$, for some integer k ,

where the intervals, the *spectral intervals* are nonempty and nonoverlapping and $\{a_i\}$, $\{b_i\}$ are decreasing sequences of real numbers with $a_i \leq b_i$ (an interval degenerates to a point if $a_i = b_i$) .

Associated with each spectral interval there is a *spectral subbundle* $V(M)$ of $E(M)$ such that for each $x \in M$, $V(x)$ consists of the vectors $z \in E(x)$ for which $z = 0$, or a *lim sup* , or a *lim inf* as $t \to +\infty$, or as $t \to -\infty$, of $\log|\psi(t, x, z)|/t$ lies in the particular spectral interval considered. If V_i denotes the spectral subbundle associated with a compact spectral interval $[a_i, b_i]$, then $V_i(x)$ is finite dimensional with dimension d_i independent of $x \in M$.

A useful property of the spectrum is given in the following lemma which abstracts the intuitive fact that a connecting orbit between two invariant sets decreases the number of gaps in the spectrum of each one of the sets.

LEMMA 2.3

Under the hypothesis of Theorem 2.1, if γ is a precompact orbit for the semiflow φ on M , $\alpha(\gamma)$ and $\omega(\gamma)$ are the α-limit and the ω-limit of γ , respectively, and I is an interval of real numbers, then

$$g[I \cap \Sigma(\overline{\gamma})] \leq min\ \{g[I \cap \Sigma(\alpha(\gamma))]\ ,\ g[I \cap \Sigma(\omega(\gamma))]\ \},$$

where, for a set of real numbers S , $g[S]$ denotes the number of gaps separating points of S , i.e., the number of connected components of $R - S$.

It is believed that the assumption of M containing a fixed point or a closed orbit in Theorem 2.2 is unessential, but this assumption is used in the proof we have available at this time. Since we are mostly interested in applications where the compact invariant set M is the attractor set for a system that satisfies adequate dissipation properties, and these usually guarantee the existence of a fixed point in the attractor, the theorem is useful in its present form.

The only one of the possible forms of the spectrum, indicated in Theorem 2.2, which occurs when $\pi = (\varphi, \psi)$ is a flow is 2. : a finite union of compact intervals. In that case the fibers $E(x)$ are finite dimensional. The other fours forms of the spectrum are only possible for semiflows.

EXAMPLES

We consider scalar linear retarded FDEs of the form

$$\dot{x}(t) = b(t)x(t-1) , \qquad (1)_b$$

where the function b belongs to the space X of the functions $a : R \to R$ which are continuous and bounded, taken with the topology of uniform convergence on compact subsets of R. For phase space we take $C = C([-1,0], R)$, and for each $a \in X$ and $\tau \in R$, we define the *τ-translate* of a as the function $a_\tau \in X$ such that $a_\tau(t) = a(t+\tau)$. With equation $(1)_b$, we associate the set $M = cl\{b_\tau : \tau \in R\}$, where cl denotes the closure in the topology of X. The set M is compact in X. One can define on M the flow $\varphi(t,a) = a_t$, and associate with equation $(1)_b$ the linear skew-product semiflow $\pi : R^+ \times M \times C \to C$ such that

$$\pi(t,a,\varsigma) = (\varphi(t,a), T_a(t)\varsigma) .$$

where $T_a(t)$, $t \geq 0$ denotes the solution map associated with the initial value problem at $t = 0$ for equation $(1)_a$ (for details refer to [6]). For these semiflows, we have the following examples:

1) $\Sigma(M) = \emptyset$ for

$$b(t) = \begin{cases} -2\sin^2(\pi t) & t \in [2n, 2n+1] \\ 0 & t \in (2n-1, 2n), \quad \text{for each integer } n . \end{cases}$$

2) $\Sigma(M) = (-\infty, 0]$ for

$$b(t) = \begin{cases} 0 & t \leq 0 \\ -t & 0 < t < 1 \\ -1 & t \geq 1 \end{cases}$$

3) $\Sigma(M) = (-\infty, \gamma] \cup [0, \beta]$, with $\gamma < 0 < \beta$, for

$$b(t) = \begin{cases} 0 & t \leq 0 \\ t & 0 < t < 1 \\ 1 & t \geq 1 \end{cases}$$

4) $\Sigma(M)$ is equal to an infinite union of nondegenerate compact intervals accumulating only at $-\infty$ for

$$b(t) = \begin{cases} 1-\epsilon & t \leq 0 \\ t+(1-\epsilon) & 0 < t < \epsilon \\ 1 & t \geq \epsilon, \end{cases}$$

with $0 < \epsilon < 1$.

3. ATTRACTORS FOR DISSIPATIVE SYSTEMS

In this section we review briefly a general result on atractors for dissipative systems. For more details and examples of a wide class of systems satisfying the conditions described here we refer to [7].

Let X be a Banach space and let φ be a semiflow on X. The semiflow φ is said to be *point dissipative (compact dissipative)* if there is a bounded set $B \subset X$ that attracts each point of X (each compact set of X) under φ. The semiflow φ is said to be a *conditional α-contraction* of order $K(t)$, $0 \leq t < \infty$, for $t > t_1$, if $K(t) \in [0,1]$ for $t \geq t_1$, $K(t) \to 0$ as $t \to \infty$ and, for each $t > t_1$ and each bounded set $B \subset X$ for which $\varphi(s,B)$, $0 \leq s \leq t$, is bounded, we have $\alpha(\varphi(t,B)) \leq K(t)\alpha(B)$, where α denotes the Kuratowski measure of noncompactness considered in Section 2. We say that φ is an *α-contraction* of order $K(t)$ for $t \geq t_1$ if it is a conditional α-contraction of order $K(t)$ and for each $t > t_1$ the set $\varphi(s,B)$, $0 \leq s \leq t$, is bounded if B is bounded.

THEOREM 3.1 (see [6, 7] and references therein)

If φ is an α-contraction for each $t > 0$, is point dissipative, and is such that the orbits of bounded sets are bounded then

(i) there is a compact attractor $\mathcal{A}$ which is a maximal, compact, connected, invariant set and attracts bounded sets of X. In particular, $\mathcal{A}$ is uniformly asymptotically stable.

(ii) there is an equilibrium point of φ, i.e., there is a $x_0 \in X$ such the $\varphi(t,x_0) = x_0$ for all t.

4. INERTIAL MANIFOLDS

Let X be a Banach space and φ a continuously differentiable semiflow on X that satisfies the hypothesis of Theorem 3.1. A set $Y \subset X$ is said to be an *inertial manifold* for the semiflow φ if it is a finite dimensional C^1 manifold which is locally invariant, hyperbolic, and contains the attractor set $\mathcal{A}$ whose existence is established in Theorem 3.1 (see [3-5, 12]).

In this section, the existence of inertial manifolds for dissipative systems is related to spectral properties of attractor sets.

The spectrum of a compact, connected invariant set $Y \subset X$ under a continuously differentiable semiflow φ on X can be defined using the previous theory for linear skew-product semiflows. We denote by E the subset of the tangent bundle TX defined by $E = \bigcup_{y \in Y} T_y X$. Since φ is continuously differentiable, we can define a linear skew-product semiflow π on the vector bundle E over Y by

$$\pi(t, y, z) = (\varphi(t, y), \psi(t, y, z)),$$

with

$$\psi(t, y, z) = \Psi(t, y)z = D_2\varphi(t, \varphi(t, y))z,$$

for all $t \geq 0,\ y \in Y,\ z \in E(y)$. This semiflow π is called the ***linearized skew-product semiflow around*** Y induced by the semiflow φ . The ***spectrum of*** Y under φ , denoted $\Sigma(Y)$, is defined to be the spectrum of Y under the above linearized skew-product semiflow around Y .

THEOREM 4.1

Suppose φ is a semiflow on a Banach space X which satisfies the hypothesis of Theorem 3.1 and is continuously differentiable. Denote by $\mathcal{A}$ the attractor set whose existence was established in that theorem.

A necessary and sufficient condition for the existence of an iniertial manifold containing $\mathcal{A}$ is $\Sigma(\mathcal{A}) \cap (-\infty, 0) \neq (-\infty, 0)$.

The proof of this theorem uses the spectral properties discussed in Section 2 and an argument of a type employed to establish the existence of center manifolds.

REFERENCES

1. BABIN,A.V. AND M.I. VISHIK, Attractors of partial differential equations and estimates of their dimension. *Russian Math. Surveys* **38** (1983), 151-213.

2. CONSTANTIN,P. AND C. FOIAS, Global Lyapunov exponents, Kaplan-Yorke formulas and the dimension of the attractors for 2D Navier-Stokes equations. *Comm. Pure and Appl. Math.* **38** (1985), 1-27.

3. FOIAS,C., B. NICOLAENKO, G. SELL AND R. TEMAM, Variétés inertielles pour l'equation de Kuramoto-Sivashinski. *C. R. Acad. Sci. Paris, Serie I,* **301** (1985), 285-288.

4. FOIAS C., G. SELL AND R. TEMAM, Variétés inertielles des équations différentielles dissipatives. *C. R. Acad. Sci. Paris, Serie I,* **301** (1985), 139-141.

5. GHIDAGLIA,J.M. AND R. TEMAM, Attractors for damped nonlinear hyperbolic equations. *J. Math. Pures Appl.*, to appear.

6. HALE,J.K., *Functional Differential equations.* Springer-Verlag, New York, 1977.

7. HALE,J.K., Asymptotic behaviour and dynamics in infinite dimensions, in Research Notes in Mathematics, vol. 132, 1-42, Pitman, 1985.

8. HALE,J.K., L.T. MAGALHÃES AND W.M. OLIVA, *An Introducion to Infinite Dimensional Dynamical Systems — Geometric Theory.* Springer-Verlag, New York, 1984.

9. KURZWEIL,J., Global solutions of functional differential equations, in Lecture Notes in Math., vol. 144, 134-139, Springer-Verlag, 1970.

10. MAGALHÃES,L.T., Invariant manifolds for functional differential equations close to ordinary differential equations. *Funkcialaj Ekvacioj* **28** (1985), 57-82.

11. MALLET-PARET,J., Negatively invariant sets of compact maps and an extension of a theorem of Cartwright. *J. Differential Equations* **22** (1976), 331-348.

12. MALLET-PARET,J. AND G. SELL, On the theory of inertial manifolds for reaction diffusion equations in higher space dimension. *C. R. Acad. Sci. Paris, Serie I*, ...

13. MANÉ, On the dimension of compact invariant sets of certain non-linear maps, in Lecture Notes in Mathematics, vol. 898, 230-242, Springer-Verlag, 1981.

14. SACKER,R.J. AND G.R. SELL, A spectral theory for linear differential systems. *J. Differential Equations* **27** (1978), 320-358.

15. SACKER,R.J. AND G.R. SELL, The spectrum of an invariant submanifold. *J. Differential Equations* **38** (1980), 135-160.

APPROXIMATE SOLUTIONS TO CONSERVATION LAWS VIA CONVECTIVE PARABOLIC EQUATIONS : ANALYTICAL AND NUMERICAL RESULTS *

by

Pierangelo Marcati

Dept. of Pure and Appl. Mathematics

University of L'Aquila

67100 L'AQUILA, Italy

1. INTRODUCTION

The porpuse of the present paper is to provide some results on the limiting behavior for the convective parabolic equation

(1.1) $$u_t+f(u)_x=\epsilon\psi(u)_{xx} \qquad x\in\mathbb{R}, t\geq 0$$

as the parameter ϵ goes to zero. It will be shown that the weak solutions to (1.1) converge, in the weak-star topology of L^∞, to the weak solutions of the related conservation law

(1.2) $$u_t+f(u)_x=0$$

Moreover, if there is no interval in which f is affine, then the solutions to (1.1) converge strongly in L^p_{loc}, $1\leq p<+\infty$, to the weak solutions to (1.2) satisfying the "Entropy Inequality". The theory will include also the case of degenerate diffusion (e.g. $\psi(u)=|u|^{m-1}u$, $m>1$).

As a consequence of the convergence of (1.1) to (1.2), we can approximate a shock wave (u_-,u_+,s) of (1.2) by a travellig wave of (1.1), (sec [OR]).

* Partially supported by CNR-GNAFA.

NATO ASI Series, Vol. F37
Dynamics of Infinite Dimensional Systems
Edited by S.-N. Chow, and J. K. Hale

In the degenerate diffusion case, under suitable conditions, one has that the travelling wave coincides on some region with the shock wave, then we can hope to have an approximating procedure more accurate than the usual vanishing viscosity.

Motivated by these considerations, we study the convergence of a three point monotone scheme of Lax Friedrichs type which may be regarded as the formal approximation, by using

$$u_t+f(u)_x=(\Delta t/\lambda^2)\,\psi(u)_{xx} \qquad \lambda=\Delta x/\Delta t$$

The methods employed here are inspired by [Ta 1,2] and, particularly for the convergence of the modified LF scheme, by [Di P. 1] and [CM]. The convergence will be proved by using the results on "compensated compactness" by [Ta 1,2] and [Mu 1,2] . We shall expose here some author's results. Some proofs are sketched and others omitted, for details we refer to [Ma] and to a forthcoming author's paper.

PARABOLIC APPROXIMATIONS AND CONVERGENCE

Assume that the solutions to (1.1) are absolutely continuous in x,t, u_x and u_t in $L^1(dx)$ and $\psi(u)u_x^2$ in $L^1(dxdt)$ (see [VH] and [OR]). Moreover we assume that, for any compact support initial datum $u_0(x)$ in L^∞, the solution u tends to zero as $|x|\to\infty$ for any given t, (actually in the degenerate diffusion case we may have that u is exactly zero).

Proposition (2.1) *Consider the one parameter family of convex entropies* $\{\eta_k\}_{k>0}$ *of the form* $\eta_k(u)=\{(|u|-k)^+\}^2$ *, then one has*

(2.1) $$\frac{d}{dt}\int \eta_k(u(x,t))\,dx \le 0$$

along the solutions to (1.1)

Therefore the approximating sequence $\{u^\epsilon\}$ *satisfies*

i) $\|u^\epsilon\|_{L^\infty} \le \|u_0\|_{L^\infty}$

ii) $\|u^\epsilon\|_{L^2} \le \|u_0\|_{L^2}$

iii) $\epsilon \iint \psi'(u^\epsilon)|u^\epsilon_x|^2\,dxdt \le 2\|u_0\|_{L^2}$

provided that $u_0 \in L^\infty \cap L^2$ *and* $\psi(u) > 0$, *for all* $u \neq 0$.

To apply the theory of "compensated compactness" the above estimates are not sufficent but, following [Ta 1] , we need the control of a combination of first order derivatives. This is the goal of the succeeding result.

Proposition (2.2) *For any entropy pair* (η, q), *the sequence of measures* $\{\sigma^\epsilon\}$, *where*

$$\sigma^\epsilon = \partial_t \eta(u^\epsilon) + \partial_x q(u^\epsilon),$$

belong to a compact subset of H^{-1}_{loc}.

The proof is based on the Murat lemma [Mu 1] . Namely

$$\{\text{compact } H^{-1} + \text{bounded } L^1\} \cap \text{bdd } W^{-1\infty} \hookrightarrow \text{compact } H^{-1},$$

(actually this is not the more general version of the lemma).

One has, at once

(2.2) $$\partial_t \eta(u^\epsilon) + \partial_x q(u^\epsilon) = \epsilon\partial_x[\eta'(u^\epsilon)\psi'(u^\epsilon)u^\epsilon_x] - \epsilon\eta''(u^\epsilon)\psi'(u^\epsilon)|u^\epsilon_x|^2$$

By the L^∞ bound, it follows

(2.3) $$\eta''(u^\epsilon)[\epsilon\psi'(u^\epsilon)|u^\epsilon_x|^2] \in \text{bdd } L^1$$

Moreover for any $\Omega \subset\subset \mathbb{R}\times\mathbb{R}_+$, by using the prop (2.1), one has

(2.4) $$\|\epsilon\partial_x[\eta'(u^\epsilon)\psi'(u^\epsilon)u^\epsilon_x]\|_{H^{-1}(\Omega)} = O(\epsilon^{1/2}), \text{as } \epsilon \to 0+.$$

Since the left hand side of (2.2) is in $W^{-1,\infty}$, then we are able to apply

Murat's lemma.

The detailed proofs are given in [Ma]

Combining the above results with [Ta 1,2] we can deduce the following theorem.

Theorem (2.3) *Assume that* $u_0 \in L^\infty \cap L^2$ *and* f *is a continuosly differentiable function, then the approximating sequence* $\{u^\epsilon\}$ *converges, extracting if necessary a subsequence, to a weak solution to (1.2), in the weak-star topology of* L^∞. *Moreover, if* f *is not affine on any interval, thus the* convergence *is given in the* L^p *strong topology,* $1 \leq p < +\infty$, *and the limit solution verifies the "entropy inequality".*

3. MODIFIED LAX-FRIEDRICHS SCHEME(MLF)

Let us begin by defining the difference scheme. Thus let the upper plane $t \geq 0$ be covered by a grid $t=n\,\Delta t$, $x=j\,\Delta x$, $j=0, \pm1, \pm2, ..; n=0,1,2....$

We consider the following difference scheme

$$u_j^{n+1} = u_j^n - \lambda\,\Delta^+ g(u_j^n\;,\;u_{j-1}^n) \tag{3.1}$$

where we set $\lambda = \Delta x/\Delta t$ and

$$\Delta^\pm a_j = \pm(a_{j+1} - a_j)$$

The numerical flux for MLF is given by

$$g(a,b) = \{f(a)+f(b)\}/2 - \{\psi(a) - \psi(b)\}/2\lambda \tag{3.2}$$

The scheme is consistent and conservative, moreover it is monotone,provided that

$$1 - \lambda\,\psi'(u) \geq 0 \quad , \quad \psi'(u) - \lambda\,|f'(u)| \geq 0 \tag{3.3}$$

We costruct a family of functions $\{u_h(x,t)\}$ in L^∞, starting from $\{u_j^n\}$ by defining $h=\Delta x$, $\lambda h=\Delta t$ and u_h as the solution $w(x,t,u_{j-1}^n, u_{j+1}^n)$ of the Riemann Problem in the rectangle Q_{nj}, <u>n+j even</u>, defined by

$$n\,\Delta t \le t \le (n+1)\,\Delta t\ ,\quad (j-1)h < x < (j+1)h$$

It is evident that the approximating functions have the same local structure of the Glimm's difference scheme, (see [G] and [S] for details on the Glimm's method).By using the monotonicity of the scheme, one has

<u>Lemma</u> (3.1) *The approximating functions verify*

$$\text{(3.4)}\qquad \begin{aligned} &\text{i)}\ \ \|u_h\|_{L^\infty} \le \|u_0\|_{L^\infty} \\ &\text{ii)}\ \ \sup |u_j^n| \le \|u_0\|_{L^\infty} \end{aligned}$$

The analogous role of proposition (2.1) is played here by the following result. Before to state it, we define

$$\text{(3.5)}\qquad \begin{aligned} &M=\sup\{\Psi'(u): |u| \le \|u_0\|_\infty\} \\ &N=\sup\{|f'(u)|: |u| \le \|u_0\|_\infty\} \end{aligned}$$

<u>Lemma</u>(3.2) *If* $\lambda<M/12N, M<1/2$, *then one has*

$$\text{(3.6)}\qquad \begin{aligned} &\text{i)}\ \ \sum_j |u_j^n| \le \sum_j |u_j^0| \le \|u_0\|_{L^2}^2/2 \\ &\text{ii)}\ \ \sum_j |u_j^{n+1}-u_j^n|^2 \le \|u_0\|_{L^2}^2/h \\ &\text{iii)}\ \ \sum_{n,j} |\Delta_+ u_j^n|^2 \le \text{Const}\ \|u_0\|_{L^2}^2/h \end{aligned}$$

provided that $u_0 \in L^2 \cap L^\infty$

<u>Remark</u> If $M<1/2$ does not hold, then we can modify the scheme in the following way. Replace λ by $\lambda/2M=\mu$, ψ by $\psi_M=\psi/2M$ and g by g_M, defined by

$$\text{(3.7)}\qquad g_M(a,b)=2M\{f(a)+f(b)\}/2-\{\psi_M(a)-\psi_M(b)\}/2\mu$$

Obviously this modification depends upon u_0, because of the definition of M, in (3.5)

As above, we need to prove the H^{-1}_{loc} relative compactness of

(3.8) $$\sigma_h = \partial_t \eta(u_h) + \partial_x q(u_h)$$

The next result is inspired by [DiP.1]

Proposition (3.4) *Under the above hypotheses* $\{\sigma_h\}$ *is a relatively compact subset of* H^{-1}_{loc}.

we shall give a sketch of the proof. For any smooth test function $\phi \in D(\Omega)$, $\Omega \subset\subset \mathbb{R}\times\mathbb{R}_+$, one has

(3.9) $$\iint \{\phi_t \eta + \phi_x q\} dxdt = \int \{\phi(x,t)\, \eta(x,T) - \phi(x,0)\, \eta(x,0)\} dx + $$
$$+ \sum_n \int \phi(x, n\Delta t)[\eta_n]\, dx + \sum_{n,j} \int \sigma_{n,j}(\phi) dt$$
$$= \langle M_h, \phi\rangle + \langle L_h, \phi\rangle + \langle \Sigma_h, \phi\rangle$$

where $[\eta_n] = \eta(u_h(x, n\Delta t-)) - \eta(u_h(x, n\Delta t))$ and $\sigma_{n,j}(\phi)$ is the contribution given by an eventual shock wave in $Q_{n,j}$. Since, there exists $C_1 > 0$, independent of h, such that

(3.10) $$|\langle M_h, \phi\rangle| \leq G_1 \|\phi\|_\infty$$

then $\{M_h\}$ is uniformly bounded in M (space of bounded finite measures). Moreover let us denote by

(3.11) $$E = \sup\{|\eta'(u)| : |u| \leq \|u_0\|_\infty\};$$

in every restangle $Q_{n,j}$ containing a shock wave, one has

(3.12) $$|S^n_j[\eta] - [q]| \leq 2NE|u^n_{j+1} - u^n_{j-1}|$$

Hence, by using the inequality (3.6) (iii), it follows

(3.13) $$|\langle \Sigma_h, \phi\rangle| \leq G_2 \|\phi\|_\infty$$

Thus $\{\Sigma_h\}$ lies in a bounded subset of M. Before to analyze $\{L_h\}$, we observe that there exists $C_3 > 0$ such that, for all $(j-1)\Delta x < x < (j+1)\Delta x$,

(3.14) $$\sum_n |[\eta_n]| \leq G_3 \|u_0\|_2^2 / h.$$

Decompose L_h in two addends L_h^1 and L_h^2,

$$< L_h^1, \phi > = \sum_{n,j} \phi_j^n \int_{(j-1)h}^{(j+1)h} [\eta_n] \, dx \ , \quad \phi_j^n = \phi(j\Delta x, n\Delta t)$$

Therefore $\{L_h^1\}$ is uniformly bounded in M . Moreover set $L_h^2 = L_h - L_h^1$, then

(3.15) $$| < L_h^2, \phi > | \leq \| \phi \|_{C^\alpha} \sum_{n,j} \int_{(j-1)h}^{(j+1)h} |x-jh|^\alpha \ |[\eta_n]| dx$$

where $\alpha = 1-2/p$, for all $p > 2$. Using Sobolev theorem, there exists $C_4 > 0$, such that

(3.16) $$| < L_h^2, \phi > | \leq C_4 \|\phi\|_{H_0^{1,p}} . h^{(2\alpha+1)/2}$$

and this implies that $\{L_h^2\}$ lies in a compact subset of $H^{-1,p'}(\Omega)$. A result given in [Mu1] ensure that $\{\sigma_h\}$ is relatively compact in $H_{loc}^{-1,p'}$, for all $p > 2$. But $\{\sigma_h\}$ belongs to a bounded subset of $W^{-1,\infty}$, then by standard interpolation theory, it is a relatively compact subset of H_{loc}^{-1}.

By applying again [Ta.1] we get

<u>Theorem</u> (3.5) *The* (MLF) *scheme converges* ω^*-L^∞ *to a weak solution to* (1.2) , *moreover the convergence is strong in* L^p , $1 \leq p < +\infty$, *whenever* f *has no intervals of affinity. The approximate solution verifies, in this case, the "Entropy Inequality".*

References

[Di P.] R.J. Di Perna

1. "Convergence of Approximate Solutions Conservation Laws" Arch. Rat. Mech. and Analysis 82 (1983), 27-70.
2. "Measure-valued Solutions to Conservation Laws" Arch. Rat. Mech. and Analysis 88 (1985), 223-270.
3. "Compensated Compactness and General Systems of Conservation Laws". Trans. AMS 292 (1985),383-420.

[C M] M.G. Crandall and A.Majda

1. "Monotone Difference Approximations for Scalar Conservation Laws" Mathematics of Comp.34 (1980),1-21.

[G] J.Glimm

1. "Solutions in the Large for Nonlinear Hyperbolic systems of Equations" Comm. Pure and Appl. Math. 18 (1965), 697-715.

[M] P.Marcati

1. "Convergence of Approximate Solutions to Scalar Conservation Laws by degenerate Diffusion" Preprint Univ.L'Aquila Dec. 85, (submitted).

[Mu] F. Murat

1. "L'injection du cône positif de H^{-1} dans $W^{-1,q}$ est compact pour tout q <2" J. Math.pures et appl. 60 (1981), 309-322.
2. "Compacitè par compensation" Ann. Scuola Norm. Sup. Pisa 5 (1978), 489-507; (1981),69-102.

[S] J.A.Smoller

1. "Shock Waves and Reaction Diffusion Equations" Grundlehren der mathematischen Wissenschaften 258 Springer-Verlag, Berlin, Heidelberg, New York, Tokyo 1983.

[Ta] L. Tartar

1. "Compensated Compactness and Applications to Partial Differential Equations" Res. Notes in Math. 4 (R.J.Knops, ed) Pitman Press 1979.
2. "The Compesated Compactness Method applied to Systems of

Conservation Laws". In Systems of Nonlinear PDES.

(J. Ball Ed.) D. Reidel Publishing Co.1983.

[OR] S. Osher and J. Ralston

1. "L^1 Stability of Travelling Waves with Applications to Convective Porous Media Flow". Comm. Pure and Appl. Math. 35 (1982), 737-749

[VH] A.I. Vol' pert and S.I.Hudjaev

1."Cauchy Problem, for Degenerate Second Order Quasilinear Parabolic Equations". Math. USSR Sbornik 7 (1969)365-387.

CONLEY'S CONNECTION MATRIX

Konstantin Mischaikow
Lefschetz Center for Dynamical Systems
Division of Applied Mathematics
Brown University
Providence, RI 02912

I. INTRODUCTION

An oft posed question is, given a dynamical system with fixed points, can one prove the existence of heteroclinic orbits. On a more general level one can ask whether, given a collection of invariant sets, there exist connecting solutions between them. C. Conley introduced an algebraic object called the connection matrix which is designed to answer such questions. The ideas involved are generalizations of earlier techniques based on the Conley index which allowed one to prove the existence of connections between attractor-repeller pairs. The simplest description of the connection matrix is that it organizes the information provided by the homology (or co-homology) groups of the Conley indices associated to the isolated invariant sets. Never-the-less, it is the author's belief that in its present form the connection matrix can be applied to many interesting problems by individuals with little or no training in algebraic topology. The purpose of this paper is to describe, on the most elementary level, the connection matrix and to show how strong results can be easily obtained.

II. BASIC THEORY

We begin with the underlying theory developed originally by Conley. This requires the existence of a local flow on X, a locally compact Hausdorff space, which we denote by $(x,t) \to x \cdot t$ where $x \in X$ and $t \in \mathbb{R}$ References for this are [1], [10], and [11]. Work by Rybakowski [8], [9] developed the theory in the case that X is a metric space with a semi-flow. Unless stated otherwise we shall restrict our attention to the first setting.

S is called an <u>invariant set</u> if $S \cdot \mathbb{R} = S$. A compact invariant set is said to be an <u>isolated invariant set</u> if there exists a compact neighborhood, N, of S such that S is the maximal invariant set in N. In this case N is

NATO ASI Series, Vol. F37
Dynamics of Infinite Dimensional Systems
Edited by S.-N. Chow, and J.K. Hale

called an isolating neighborhood of S. From now on S will always denote a compact isolated invariant set.

Definition 1: A pair of compact sets (N_1,N_0) is an index pair for S if $N_0 \subset N_1$ and

(i) $N_1 \setminus N_0$ is a neighborhood of S and $cl(N_1 \setminus N_0)$ is an isolating neighborhood of S.

(ii) N_0 is positively invariant in N_1, i.e. if $x \in N_0$, $t \geqslant 0$ and $x \cdot [0,t] \subset N_1$ then $x \cdot [0,t] \subset N_0$.

(iii) N_0 is an exit set for N_1, i.e. if $x \in N_1$ and $x \cdot [0,\infty) \not\subset N_1$ then there exists $t \geqslant 0$ such that $x \cdot [0,t] \subset N_1$ and $x \cdot t \in N_0$.

In the setting where X is a metric space, S is still assumed to be compact, however the index pair consists of closed but admissible sets. A set N is admissible if (a) for every sequence $\{x_n\} \subset N$ and every sequence $\{t_n\} \subset \mathbb{R}^+$ where $t_n \to \infty$ and where $x_n \cdot [0,t_n] \subset N$ then the sequence $\{x_n \cdot t_n\}$ is precompact and (b) given $x \in N$ and $x \cdot t$ defined for all $t \in [0,\omega_x)$, if $\omega_x \subset \infty$ then $x \cdot [0,\omega_x) \not\subset N$.

Given an index pair (N_1,N_0) let N_1/N_0 be the set obtained by collapsing N_0 to a point. If $p: N_1 \to N_1/N_0$ is the obvious projection map then we can topologize N_1/N_0 by letting $U \subset N_1/N_0$ be open if and only if $p^{-1}(U)$ is open in N_1. Let $[N_0]$ denote the special point in N_1/N_0 obtained from N_0.

Given S an isolated invariant set with index pair (N_1,N_0) the Conley index (or homotopy index) is the homotopy type of the pointed topological space $(N_1/N_0,[N_0])$. We denote the index by h(S). It is a standard theorem that this index is well defined. In general dealing with the homotopy equivalence classes of topological spaces is difficult. Thus to simplify matters we shall only consider the homology of the pointed topological space h(S). In particular we shall restrict our attention to the singular homology groups with $\mathbf{Z}_2$ coefficients. Thus given S we shall study the algebraic object $H_*(h(S);\mathbf{Z}_2) = H_*(N_1/N_0, [N_0];\mathbf{Z}_2)$. In fact, it can be shown [5] that one can always choose the pair (N_1,N_0) such that $H_*(N_1/N_0;[N_0];\mathbf{Z}_2) \cong H_*(N_1,N_0;\mathbf{Z}_2)$.

For the reader who has had little or no algebraic topology, the important fact is that

$$H_*(N_1,N_0;\mathbf{Z}_2) = \{H_n(N_1,N_0;\mathbf{Z}_2)\} \quad n = 0,1,2,...$$

where each $H_n(N_1,N_0;\mathbf{Z}_2)$ is a vector space over $\mathbf{Z}_2$. Thus $H_*(N_1,N_0;\mathbf{Z}_2)$ is an infinite collection of vector spaces over $\mathbf{Z}_2$ indexed by the non-negative integers. In most applications only a finite number of these are non-trivial. The following two propositions describe $H_*(h(S);\mathbf{Z}_2)$ in the case that S is a hyperbolic critical point or periodic orbit.

Proposition 2: *Let S be a hyperbolic critical point with exactly k eigenvalues having positive real part, then*

$$H_n(h(S);\mathbf{Z}_2) \cong \begin{cases} \mathbf{Z}_2 & \text{if } n=k \\ 0 & \text{otherwise}. \end{cases}$$

Proposition 3: *Let S be a hyperbolic periodic orbit with Poincare map, P. Assume DP has real positive eigenvalues exactly k of which are greater than 1. Then*

$$H_n(h(S);\mathbf{Z}_2) \cong \begin{cases} \mathbf{Z}_2 & \text{if } n=k,k+1 \\ 0 & \text{otherwise}. \end{cases}$$

Often in applications one is given a complicated invariant set S and one wishes to determine the structure of S. To do so one needs to decompose S into smaller sets and then prove the existence or non-existence of connecting orbits. Given two isolated invariant sets, S_1 and S_2, we define the set of connections from S_1 to S_2 to be

$$C(S_1,S_2) = \{x \mid \omega^*(x) \subset S_1 \text{ and } \omega(x) \subset S_2\}$$

where $\omega(x) = \bigcap_{t>0} cl(x\cdot[t,\infty))$ and $\omega^*(k) = \bigcap_{t<0} cl(x\cdot(-\infty,t])$. Let $(P,>)$ be a finite partially ordered set, i.e. P is a finite set along with a partial order relation, >, satisfying:

1) $i > i$ never holds for $i \in P$.

2) If $i > j$ and $j > k$ then $i > k$ for all $i,j,k \in P$.

Definition 4: A Morse decomposition of S is a finite collection $\mathcal{M}(S) = \{M(i) \mid i \in (P,>)\}$ of compact invariant sets in S, indexed by P, such that if $x \in S$ then either $x \in M(i)$ or $x \in C(M(i),M(j))$ where $i > j$.

The individual sets M(i) are called <u>Morse sets</u> and are isolated invariant sets. To simplify the notation we write

$$H(i) = H_*(h(M(i));\mathbf{Z}_2) \quad \text{and} \quad H_n(i) = H_n(h(M(i));\mathbf{Z}_2) .$$

III. CONNECTION MATRICES

Given S an isolated invariant set and $\mathcal{M}(S) = \{M(i) \mid i \in (P,>)\}$ a Morse decomposition associated with S, let $\Delta = [\Delta_{ij}]$, $i,j \in P$, be a matrix of matrices with coefficients in $\mathbf{Z}_2$, i.e. Δ_{ij} an entry in Δ may itself be a matrix over $\mathbf{Z}_2$.

<u>Definition 1</u>: $\Delta: \bigoplus_{i\in P} H(i) \to \bigoplus_{i\in P} H(i)$ is a <u>connection matrix</u> if:

a) $\Delta_{ij} = \mathbf{0}$ if $i \not> j$. ($\mathbf{0}$ denotes the zero matrix)

b) Δ is a <u>boundary map</u>, i.e. $\Delta^2 = \mathbf{0}$ and $\Delta_{ij} = 0$ except possibly as a map from $H_n(j)$ to $H_{n-1}(i)$.

c) Let $\Delta_n = \Delta \mid \bigoplus_{i\in P} H_n(i)$ and let $H_n\Delta = \text{Ker } \Delta_n/\text{Im } \Delta_{n+1}$, then $H_n\Delta \cong H_n(h(S);\mathbf{Z}_2)$.

<u>Theorem 2</u>: *Given S and $\mathcal{M}(S)$ there always exists at least one connection matrix. (See [2],[3]).*

It needs to be mentioned that connection matrices need not be unique. In fact, since the connection matrix satisfies certain continuation properties uniqueness is not necessarily desirable. Unfortunately, there are few satisfactory theorems concerning uniqueness or lack thereof. Reineck [6] has shown that if $\mathcal{M}(S)$ consists of hyperbolic critical points whose stable and unstable manifolds intersect transversely then Δ is unique.

The definition of the connection matrix presented here is much weaker than that of Franzosa [2]. One of the major differences being that we do not specify the isomorphism between $H_n\Delta$ and $H_n(h(S);\mathbf{Z}_2)$ whereas Franzosa does. This implies that the set of matrices which satisfies our condition may be larger than his. However, for simple applications this does not appear to be the case.

The most important property of the connection matrix is as follows.

<u>Proposition 3</u>: *Let $i < j$ and assume there does not exist k such that $i < k < j$.*

If $\Delta_{ij} \neq 0$ *then* $C(M(j),M(i)) \neq \emptyset$.

This means that in certain circumstances a 1 entry in Δ implies that a connection exists.

IV. EXAMPLES

Let S be an isolated invariant set with Morse decomposition $\mathcal{M}(S) = \{M(i) \mid i = 0,\dots,\lambda+1\}$. Furthermore assume that we have the following information about the homologies of the Morse sets.

$$H_n(i) = \begin{cases} \mathbf{Z}_2 & \text{if } n = 2i, 2i+1 \\ 0 & \text{otherwise} \end{cases} \qquad 0 \leqslant i < \lambda$$

$$H_n(\lambda) = \begin{cases} \mathbf{Z}_2 & \text{if } n = 2\lambda \\ 0 & \text{otherwise} \end{cases}$$

$$H_n(\lambda+1) = 0 \quad \text{for all } n$$

$$H_n(h(S);\mathbf{Z}_2) = \begin{cases} \mathbf{Z}_2 & \text{if } n = 0 \\ 0 & \text{otherwise .} \end{cases}$$

Then we can ask the question is $C(M(i),M(i-1))$ non-empty for $i = 1,\dots,\lambda$. (Though this may seem like an abstract example, it is modeled on results of Mallet-Paret [4] proving the existence of a Morse decomposition of the global attractor for a wide class of delay-differential equations of the form $\dot{x}(t) = f(x(t),x(t-1))$.)

Proposition 1: $C(M(i),C(M(i-1)) \neq \emptyset$ for $1 \leqslant i \leqslant \lambda$.

Proof: Rather than give a proof for the general care which might obscure the simplicity of this application of the connection matrix we shall only consider the case $\lambda = 2$.

Thus we are considering $\Delta : \bigoplus_{i=0}^{3} H(i) \to \bigoplus_{i=0}^{3} H(i)$. However $H_n(3) = 0$ for all n, so we need only consider $\Delta : \bigoplus_{i=0}^{2} H(i) \to \bigoplus_{i=0}^{2} H(i)$. Similarly, ignoring $H_n(i)$ when $H_n(i) = 0$ we have that Δ maps

$$H_0(0) \oplus H_1(0) \oplus H_2(1) \oplus H_3(1) \oplus H_4(2)$$

to itself. By III.1.a, if $i \leq j$ then $\Delta_{ij} = \mathbf{0}$. For example $\Delta_{00} : H_0(0) \oplus H_1(0) \to H_0(0) \oplus H_1(0)$ can be written as

$$\Delta_{00} = \begin{array}{c} \\ H_0(0) \\ H_1(0) \end{array} \begin{array}{c} \begin{array}{cc} H_0(0) & H_1(0) \end{array} \\ \begin{bmatrix} 0 & 0 \\ 0 & 0 \end{bmatrix} \end{array} .$$

Now consider $\Delta_{02} : H_4(2) \to H_0(0) \oplus H_1(0)$. By III.1.b, $\Delta_{02} = \mathbf{0}$. Similarly, applying III.1.b to

$$\Delta_{01} : H_2(1) \oplus H_3(1) \to H_0(0) \oplus H_1(0)$$

and $\quad \Delta_{12} : H_4(2) \to H_2(1) \oplus H_3(1)$

gives $\quad \Delta_{01} = \begin{bmatrix} 0 & 0 \\ * & 0 \end{bmatrix}$ and $\Delta_{12} = \begin{bmatrix} 0 \\ * \end{bmatrix}$

where $*$ denotes an unknown entry. Thus

$$\Delta = \begin{array}{c} \\ H_0(0) \\ H_1(0) \\ H_2(1) \\ H_3(1) \\ H_4(2) \end{array} \begin{array}{c} \begin{array}{ccccc} H_0(0) & H_1(0) & H_2(1) & H_3(1) & H_4(2) \end{array} \\ \begin{bmatrix} 0 & 0 & 0 & 0 & 0 \\ 0 & 0 & * & 0 & 0 \\ 0 & 0 & 0 & 0 & 0 \\ 0 & 0 & 0 & 0 & * \\ 0 & 0 & 0 & 0 & 0 \end{bmatrix} \end{array}$$

Finally we use III.1.c. Since $H\Delta \cong H_*(h(S);\mathbf{Z}_2)$ it must be that $\dim H\Delta = 1$, i.e. $\dim \operatorname{Ker} \Delta = \operatorname{Rank} \Delta + 1$. But clearly $\dim \operatorname{Ker} \Delta \geq 3$ thus $\operatorname{Rank} \Delta = 2$ and therefore both $*$ entries equal 1. In particular $C(M(1),M(0))$ and $C(M(2),M(1))$ are non empty by Proposition III.3. □

We finish by describing some other results which have been obtained using the connection matrix. Reineck [7] considers the qualitative behavior of solutions to a biological model of the form

$$\dot{x} = F_1(x,y)$$

$$\dot{y} = F_2(x,y)$$

where the zero sets of F_1 and F_2 can be thought of as parabolas. Depending on the directions and locations of these curves one has three distinct systems called, symbiotic, competitive, and predator-prey. In each case the possible connection matrices are determined and related to the phase plane portrait. In addition partial results are obtained for a 3-dimensional system modeling two predators and one prey.

Mischaikow [5] classifies travelling wave solutions for systems of reaction-diffusion equations of the form

$$u_t = Du_{xx} + \nabla F(u)$$

where D is a diagonal matrix with non-zero diagonal entries. These results can then be extended [6] to obtain relationships between homoclinic orbits for the Hamiltonian system

$$\dot{u} = v$$

$$\dot{v} = -\nabla F(u)$$

and heteroclinic orbits for the damped systems

$$\dot{u} = v$$

$$\dot{v} = \theta v - \nabla F(u) \qquad \theta \neq 0.$$

This research was supported in part by ARO under contract number DAAG-29-83 -K-0029 and the AFOSR under grants numbered AFOSR-81-0116-C and AFOSR-84-0376.

REFERENCES

[1] C.C. Conley, Isolated Invariant Sets and the Morse Index, CBMS Reg. Conf. Series in Math., No. 38 A.M.S., Providence, Rhode Island, 1978.

[2] R. Franzosa, Index Filtrations and Connection Matrices for Partially Ordered Morse Decompositions, Ph.D. Dissertation, Univ. of Wisconsin-Madison (1984).

[3] R. Franzosa and K. Mischaikow, Index Filtrations and Connection Matrices for Semiflows on Metric Spaces, in preparation.

[4] J. Mallet-Paret, Morse Decompositions for Delay Differential Equations, preprint.

[5] K. Mischaikow, Classification of Traveling Wave Solutions of Reaction-Diffusion Systems, LCDS No. 86-5 (1985).

[6] K. Mischaikow, Standing Waves vs. Traveling Waves, in preparation.

[7] J. Reineck, The Connection Matrix and the Classification of Flows Arising from Ecological Models, Ph.D. Dissertation, Univ. of Wisconsin-Madison (1985).

[8] K. Rybakowski, On the Homotopy Index for Infinite Dimensional Semiflows, *Trans. Amer. Math. Soc.* 269(1982), 351-382.

[9] K. Rybakowski, The Morse Index, Repeller-Attractor Pairs and the Connection Index for Semiflows on Noncompact Spaces, *J.D.E.* 47(1983), 66-98.

[10] D. Salamon, Connected Simple Systems and the Conley Index of Isolated Invariant Sets, *Trans. Amer. Math. Soc.* 291(1985), 1-41.

[11] J. Smoller, Shock Waves and Reaction-Diffusion Equations, Springer-Verlag, New York (1983).

EXISTENCE AND NON-EXISTENCE OF FINITE-DIMENSIONAL GLOBALLY ATTRACTING INVARIANT MANIFOLDS IN SEMILINEAR DAMPED WAVE EQUATIONS*

X. Mora and J. Solà-Morales

Departament de Matemàtiques, Universitat Autònoma de Barcelona
Bellaterra, Barcelona, Spain

Contents

0. Introduction

This paper is concerned with the dynamical system generated by certain semilinear damped wave equations. In §1 we reproduce a result obtained in a previous paper (Mora[1986]), which shows that, when the damping is sufficiently large this dynamical system has the property that its global attractor is contained in a finite-dimensional local invariant manifold of class C^1. In the present paper, we will show that, on the other hand, when the damping is small, it is a fairly generic fact that there is no finite-dimensional local invariant manifold of class C^1 containing the global attractor. The exact result obtained in this connection is stated in Theorem 4.1. In the way towards this result, we have developed some auxiliary results which have some interest by themselves, namely, a result giving optimal inner products for linear wave equations (Theorem 2.1), and a C^1 lineari-

* Work partially supported by the CAICYT.

NATO ASI Series, Vol. F37
Dynamics of Infinite Dimensional Systems
Edited by S.-N. Chow, and J. K. Hale

zation theorem (Theorem 3.1) .

The reason why small damping makes difficult the existence of finite-dimensional (local) invariant manifolds of class C^1 is mainly linear. When the damping is small, the linear part of the equation at a given stationary point easily has all its eigenvalues on the same vertical line of the complex plane. This immediately implies that there is no normally hyperbolic invariant manifold of class C^1 containing that point. If the eigenvalues are all simple, then we can prove that, even dropping the condition of normal hyperbolicity, one has only a countable family of finite-dimensional invariant manifolds of class C^1 . In §2 , this crucial fact is established for the linear problem. By using the C^1 linearization theorem of §3 , this fact can then be translated to the neighbourhood of a stationary state of a nonlinear problem.

Let us now consider a nonlinear problem with a heteroclinic orbit from ϕ to ψ, where ψ is a stationary state with a linearization of the type described above. Certainly, the global attractor of the system must contain the connecting orbit. But, on the other hand, it seems extremely casual that this orbit arrives at the neighbourhood of ψ by precisely one of the few finite-dimensional invariant manifolds of class C^1 which contain ψ. If actually it does not do so, then we can conclude that the global attractor is not contained in any finite-dimensional invariant manifold of class C^1. In §4 , we exhibit a family of equations depending on a parameter which varies over an open ball of a certain Banach space, for which family we have been able to prove that the property of non-existence of a finite-dimensional (local) invariant manifold of class C^1 containing the global attractor is indeed generic.

Finally, let us remark that this example of non-existence differs from the one given by Mallet-Paret, Sell [1986] for parabolic equations in that we do not require our manifolds to be normally hyperbolic.

1. The equations and a result of existence

We shall be considering evolution problems of the following form, where u is a function of $x \in (0,\pi) =: \Omega$ and $t \in \mathbb{R}$ with values in $\mathbb{R}$:

$$u_{tt} + 2\alpha u_t = u_{xx} + f(x,u) \qquad (1.1)$$

$$u_x|_{x=0} = u_x|_{x=\pi} = 0 \tag{1.2$_N$}$$

$$u|_{t=0} = u_0, \quad u_t|_{t=0} = v_0 \tag{1.3}$$

or the analogous one where $(1.2)_N$ is replaced by

$$u|_{x=0} = u|_{x=\pi} = 0 \tag{1.2$_D$}$$

Here, α is a non-negative real parameter, called the damping coefficient, and f is a function $\Omega\times\mathbb{R}\to\mathbb{R}$. The function f is assumed to satisfy the following conditions:

(F1a) $f(\cdot,u)$ is measurable for all $u\in\mathbb{R}$; $f(x,\cdot)$ is of class $C^{1+\eta}$ for almost all $x\in\Omega$ and some $\eta>0$ independent of x.

(F1b) For every bounded open interval $J\subset\mathbb{R}$, the quantity

$$\|f\|^2_{(J)} := \int_\Omega\left[\left(\sup_{u\in J}|f(x,u)|\right)^2+\left(\sup_{u\in J}|f_u(x,u)|\right)^2+\left(\sup_{\substack{u,v\in J\\ u\neq v}}\frac{|f_u(x,u)-f_u(x,v)|}{|u-v|^\eta}\right)^2\right]dx \tag{1.4}$$

is finite (i.e. the mapping $x\mapsto f(x,\cdot)$ from Ω to the Banach space $C^{1+\eta}(\bar{J})$ belongs to L_2 in the sense of Bochner)

(F2) $$\limsup_{|u|\to\infty}\frac{\operatorname{ess\,sup}_{x\in\Omega} f(x,u)}{u} < 0 \tag{1.5}$$

Thereafter, we shall consider variations in the function f. For our purpose, it will suffice to restrict our attention to variations of the following form

$$f(x,u) = f_0(x,u) + g(x,u) \tag{1.6}$$

where f_0 is a fixed function satisfying (F1a),(F1b),(F2), and the perturbation g varies within a space of the form

$$\mathcal{G}(J) := \left\{ g:\Omega\times\mathbb{R}\to\mathbb{R} \;\middle|\; \begin{array}{l} g \text{ satisfies (F1a),(F1b), and} \\ g(x,u)=0 \text{ whenever } u\notin J \end{array} \right\} \tag{1.7}$$

where J is a fixed bounded open interval of $\mathbb{R}$. One easily verifies that $\mathcal{G}(J)$ is a Banach space with norm given by (1.4).

Problem (1.1),$(1.2)_B$,(1.3) (B stands for either N or D) will be viewed as a second order evolution problem for a functional variable $u: \mathbb{R} \longrightarrow L_2 := L_2(\Omega)$, namely

$$\ddot{u} + 2\alpha\dot{u} + Au = Fu \tag{1.8}$$

$$u(0) = u_0, \quad \dot{u}(0) = v_0 \tag{1.9}$$

where A and F denote the operators on L_2 given by

$$Au := -u_{xx} \tag{1.10}$$

$$(Fu)(x) := f(x, u(x)) \tag{1.11}$$

with domains respectively equal to H^2_B and H^1_B. Here, H^k_B $(k=1,2)$ denote the closures in H^k of the set $\{u: \Omega \rightarrow \mathbb{R} \mid u \in C^\infty(\bar{\Omega})$ and satisfies $(1.2)_B\}$.

Problem (1.8),(1.9) can be rewritten as a first order evolution problem for the pair $(u, \dot{u})$ as a variable with values in $H^1_B \times L_2$. Instead of this, we shall find more convenient to use the pair

$$U := (u, v) := (u, \alpha u + \dot{u}) \tag{1.12}$$

The problem takes then the following form, where $U := (u, v)$ represents a variable with values in $H^1_B \times L_2$:

$$\dot{U} = -\alpha U + \mathbb{A}U + \mathbb{F}U \tag{1.13}$$

$$U(0) = U_0 \tag{1.14}$$

Here, $\mathbb{A}$ and $\mathbb{F}$ denote the operators on $H^1_B \times L_2$ given by

$$\mathbb{A}(u, v) := (v, \alpha^2 u - Au) \tag{1.15}$$

$$\mathbb{F}(u, v) := (0, Fu) \tag{1.16}$$

with domains respectively equal to $H^2_B \times H^1_B$ and $H^1_B \times L_2$. When necessary, the dependence of things with respect to g will be made explicit by writing g as a subindex, like in f_g, F_g, or $\mathbb{F}_g$.

It is a standard fact that the operator $\mathbb{A}$ is the generator of a group on $H^1_B \times L_2$. In fact, this will follow as a lateral result from the estimates obtained in §2 . On the other hand, the form of conditions (F1a) and (F1b) implies that, for every bounded open interval $J \subset \mathbb{R}$, the mapping $(U,g) \mapsto \mathbb{F}_g U$ goes from $(H^1_B \times L_2) \times \mathcal{G}(J)$ to $H^1_B \times L_2$, and it is of class $C^{1+\eta}$ uniformly on bounded sets. With this, the preceeding problem fits in the standard theory of semilinear evolution equations, which allows us to obtain the following result:

Theorem 1.1 . Assume that f satisfies (F1a),(F1b),(F2) . Then, the preceeding problem generates a group $T(t)$ $(t \in \mathbb{R})$ (of nonlinear operators) on $H^1_B \times L_2$ with the following properties: (i) For every $U_0 \in H^1_B \times L_2$, the mapping $T(\cdot)U_0 : \mathbb{R} \longrightarrow H^1_B \times L_2$ is continuous; if $U_0 \in H^2_B \times H^1_B$, then this mapping is continuously differentiable. (ii) For each compact interval $[T_0,T_1] \subset \mathbb{R}$, the mapping $H^1_B \times L_2 \ni U_0 \longmapsto T(\cdot)U_0 \in C([T_0,T_1], H^1_B \times L_2)$ is of class $C^{1+\eta}$ uniformly on bounded sets. (iii) There is a compact global attractor in the sense of Babin,Vishik[1983] and Hale[1985] . Assume now that f has the form (1.6) , where f_0 satisfies (F1a),(F1b),(F2) , and g varies over $\mathcal{G}(J)$, J being a fixed bounded open interval of $\mathbb{R}$. Let $T_g(t)$ $(t \in \mathbb{R})$ denote the group corresponding to a given g . Then, for each compact interval $[T_0,T_1] \subset \mathbb{R}$, the mapping $(H^1_B \times L_2) \times \mathcal{G}(J) \ni (U_0,g) \longmapsto T_g(\cdot)U_0 \in C([T_0,T_1], H^1_B \times L_2)$ is of class $C^{1+\eta}$ uniformly on bounded sets. ■

The proof of this theorem is fairly standard, so that we shall give only a summary with references.

Summary of the proof. As usual, the curves $T(\cdot)U_0 \in C([0,\infty), H^1_B \times L_2)$ are looked for as solutions of the integral equation resulting from the variation of constants formula. The local existence and uniqueness of solutions of this equation is obtained by means of a contraction mapping argument (see for instance Tanabe[1979] (Thm.6.1.4)). The proof that these solutions can be extended to the whole interval $(-\infty,+\infty)$ is based upon suitable a-priori estimates. These follow easily from the existence and properties of the Lyapunov functional

$$\Phi(u,w) := \int_\Omega \left[\tfrac{1}{2}\dot{u}^2 + \tfrac{1}{2}u_x^2 - q(\cdot,u(\cdot))\right], \quad \text{where} \quad q(x,u) := \int_0^u f(x,\xi)\,d\xi \tag{1.17}$$

(here and in the following $\dot{u}$ stands for $w - \alpha u$). It is a well-known fact that $\dot{\Phi}$, the derivative of Φ along a solution of (1.13) , is given by $\dot{\Phi} = -2\alpha\int_\Omega \dot{u}^2$. The a-priori estimates mentioned above follow from the following twoproperties of Φ , whose derivation uses the fact that f satisfies (F2) : (i) the level sets of Φ , $\{U \mid \Phi(U) \leq c\}$ $(c \in \mathbb{R})$, are bounded in $H^1_B \times L_2$; (ii) $\dot{\Phi}$ satisfies an inequality of the form $-4\alpha(\Phi + M) \leq \dot{\Phi} \leq 0$. The proof that $T(\cdot)U_0$ is continuously differentiable when $U_0 \in H^2_B \times H^1_B$ can be found for instance in Tanabe[1979] (Thm.6.1.3) . The $C^{1+\eta}$ dependence of solutions with respect to the initial state U_0 and the parameter q can be obtained by proceeding as in Henry[1981] (Thm.3.4.4) . Finally, the proof of statement (iii) can be found in Hale[1985] (§3, Thm.6.1) (see also Babin,Vishik [1983] (Thm.6.1)) .

In the following we shall refer to the group $T(t)$ $(t \in \mathbb{R})$ as the dynamical system generated by problem (1.1),$(1.2)_B$,(1.3) .

Next we reproduce a slightly generalized version of a result obtained in Mora [1986] , which establishes the fact that, for large values of α, the global attractor is contained in a finite-dimensional local invariant manifold of class C^1 . A previous result ofn this direction has been obtained by Solà-Morales, València [1986] , who, for a spatially homogeneous problem with Neumann boundary conditions, give sufficient conditions on the coefficients which ensure that all the flow is attracted by the invariant subspace formed by the spatially homogeneous states.

Theorem 1.2 . Assume that f satisfies (Fla),(Flb),(F2) and $f(x,0)=0$, and consider the dynamical system on $H^1_B \times L_2$ generated by problem (1.1), $(1.2)_B$,(1.3) . There exists a finite constant ℓ such that, for every integer n satisfying the following condition

$$2n+1 > 8\ell, \qquad \alpha^2 > (n+1)^2 + \frac{16\ell^2}{(2n+1)-8\ell} \tag{1.17}$$

there is a local invariant submanifold of class C^1 and dimension n (for $B=D$) or $n+1$ (for $B=N$) which contains the global attractor. ■

Corollary 1.3 . There exists a finite constant α^* such that, for $\alpha > \alpha^*$, the global attractor lies in a finite-dimensional local invariant submanifold of class C^1 . ■

The proof of Theorem 1.2 is a trivial generalization of the one given in

Mora [1986] . We only remark here that it is crucially based upon the use of the inner products presented in §2.1 below.

2. The linear problem

In this section we deal with an abstract linear evolution problem which includes the linear damped wave equation as well as the linearization of a semilinear equation about a stationary state. The problem under consideration has the form

$$\ddot{u} + 2\alpha\dot{u} + Au = 0 \tag{2.1}$$

$$u(0) = u_0, \quad \dot{u}(0) = v_0 \tag{2.2}$$

where u is now a variable with values in a general Hilbert space E , α is a real number (not necessarily negative), and A is a self-adjoint operator on E with numerical range bounded from below; i.e.

$$\mu_1 := \inf_{u \in \mathrm{Dom}(A)} \frac{\langle Au,u\rangle}{\langle u,u\rangle} > -\infty \tag{2.3}$$

where $\langle\cdot,\cdot\rangle$ denotes the inner product of E . As it is well-known, μ_1 coincides with the smallest element of the spectrum of A . In the following, $E^{1/2}$ will denote the Hilbert space consisting of the domain of the operator $(A+\xi I)^{1/2}$ endowed with the inner product

$$\langle u,\hat{u}\rangle_{1/2} := \langle (A+\xi I)^{1/2}u, (A+\xi I)^{1/2}\hat{u}\rangle \tag{2.4}$$

where ξ is a real number greater than $-\mu_1$. Different choices of $\xi > -\mu_1$ result in the same vector space with different but equivalent inner products.

Similarly as before, problem (2.1),(2.2) will be reconsidered as a first order evolution problem for the pair $U := (u,w) := (u, \alpha u + \dot{u})$; i.e.

$$\dot{U} = -\alpha U + \mathbb{A}U \tag{2.5}$$

$$U(0) = U_0 \tag{2.6}$$

where $\mathbb{A}$ will have the form (1.15) . Here, U will be considered as taking values in $\mathbb{E} := E^{1/2} \times E$, and $\mathbb{A}$ will be considered as an operator on

$\mathbb{E}$ with domain $\mathrm{Dom}(\mathbb{A}) := \mathrm{Dom}(A) \times E^{1/2}$.

2.1 Choice of inner products

The inner product on $\mathbb{E} := E^{1/2} \times E$ will be taken as the direct sum of one of the inner products on $E^{1/2}$ given by (2.4) and the inner product on E . Among all the possible values of ξ , we will choose a particular one which makes the numerical range of $\mathbb{A}$ to be contained in a vertical strip as small as possible. As it is shown by the following theorem, this is obtained for $\xi = \max(-\alpha^2, \alpha^2 - 2\mu_1)$, which amounts to take

$$\langle U, \hat{U} \rangle_{\mathbb{E}} := \begin{cases} \langle (A-\alpha^2 I)^{1/2} u, (A-\alpha^2 I)^{1/2} \hat{u} \rangle + \langle w, \hat{w} \rangle & \text{if } \alpha^2 < \mu_1 \\ \langle (A+(\alpha^2-2\mu_1)I)^{1/2} u, (A+(\alpha^2-2\mu_1)I)^{1/2} \hat{u} \rangle + \langle w, \hat{w} \rangle & \text{if } \alpha^2 > \mu_1 \end{cases} \tag{2.7}$$

This choice of ξ has been inspired by the work of Solà-Morales, València [1986] .

Theorem 2.1 . For $\alpha^2 < \mu_1$, the numerical range of $\mathbb{A}$ is contained in the imaginary axis. For $\alpha^2 > \mu_1$, it is contained in the strip $|\mathrm{Re}\,\lambda| \leq \sqrt{\alpha^2 - \mu_1}$. ■

Proof . We have to estimate $|\mathrm{Re}\langle \mathbb{A}U, U\rangle_{\mathbb{E}}| / \langle U, U\rangle_{\mathbb{E}}$ when $U = (u, w)$ is a general element of $\mathrm{Dom}(\mathbb{A}) = \mathrm{Dom}(A) \times E^{1/2}$.

(a) Case $\alpha^2 < \mu_1$. It suffices to notice that

$$\begin{aligned} \langle \mathbb{A}U, U \rangle_{\mathbb{E}} &= \langle (A-\alpha^2 I)^{1/2} w, (A-\alpha^2 I)^{1/2} u \rangle + \langle \alpha^2 u - Au, w \rangle \\ &= \langle w, Au - \alpha^2 u \rangle + \langle \alpha^2 u - Au, w \rangle \end{aligned}$$

which implies

$$\mathrm{Re}\,\langle \mathbb{A}U, U \rangle_{\mathbb{E}} = 0$$

(b) Case $\alpha^2 > \mu_1$. In this case we have

$$\begin{aligned} \langle \mathbb{A}U, U \rangle_{\mathbb{E}} &= \langle (A+(\alpha^2-2\mu_1)I)^{1/2} w, (A+(\alpha^2-2\mu_1)I)^{1/2} u \rangle + \langle \alpha^2 u - Au, w \rangle \\ &= \langle w, Au + \alpha^2 u - 2\mu_1 u \rangle + \langle \alpha^2 u - Au, w \rangle \\ &= 2 \langle w, (\alpha^2 - \mu_1) u \rangle + \langle w, Au - \alpha^2 u \rangle + \langle \alpha^2 u - Au, w \rangle \end{aligned}$$

which implies that

$$\mathrm{Re}\langle \mathbb{A}U,U\rangle_{\mathbb{E}} = 2\,\mathrm{Re}\,\langle w,(\alpha^2-\mu_1)u\rangle$$

From this we derive that

$$|\mathrm{Re}\langle \mathbb{A}U,U\rangle_{\mathbb{E}}| \leq 2\,|\langle w,(\alpha^2-\mu_1)u\rangle| = 2\sqrt{\alpha^2-\mu_1}\,|\langle w,\sqrt{\alpha^2-\mu_1}\,u\rangle|$$
$$\leq \sqrt{\alpha^2-\mu_1}\,[(\alpha^2-\mu_1)\langle u,u\rangle + \langle w,w\rangle]$$

where in the last step we have used Schwarz inequality. Now, we only have to verify that the quantity in square brackets is less than or equal to $\langle U,U\rangle_{\mathbb{E}}$. Indeed, we have

$$\langle U,U\rangle_{\mathbb{E}} = \langle Au + (\alpha^2-2\mu_1)u,\, u\rangle + \langle w,w\rangle$$
$$\geq (\alpha^2-\mu_1)\langle u,u\rangle + \langle w,w\rangle$$

where in the last step we have used (2.3) . Q.E.D.

From the properties of $\mathbb{A}$ one easily derives that $\mathrm{Range}(\mathbb{A}-\lambda I) = \mathbb{E}$ for any real λ with $|\lambda| > \sqrt{\alpha^2-\mu_1}$. By applying the theory of dissipative operators (see for instance Pazy [1983] (§1.4, Thm.4.3)) one then obtains the following result.

Corollary . For $\alpha^2 < \mu_1$, $\mathbb{A}$ is the generator of a group $J(t)$ $(t\in\mathbb{R})$ of unitary operators. For $\alpha^2 > \mu_1$, $\mathbb{A}$ is the generator of a group $J(t)$ $(t\in\mathbb{R})$ satisfying the bound

$$\|J(t)\| \leq \exp(\sqrt{\alpha^2-\mu_1}\,|t|) \qquad (\forall t\in\mathbb{R}) \quad \blacksquare \tag{2.8}$$

In order to see that the preceeding estimates are optimal, it suffices to notice that $\mu_1\in \mathrm{Spec}(A)$ implies $\pm(\alpha-\mu_1)^{1/2}\in \mathrm{Spec}(\mathbb{A})$, which shows that the vertical strip which contains the numerical range is the smallest possible.

2.2 The infinite-dimensional whirl

In this paragraph we assume that A has compact resolvent. In the following, e_k and μ_k $(k=1,2,\ldots)$ denote respectively a complete orthonormal system of eigenfunctions of A and the corresponding sequence of eigenvalues, which sequence is assumed to be non-decreasing. Finally, E_k will denote the one-dimensional space generated by e_k. We then have the orthogonal decomposition invariant by A $E=\bigoplus_{k=1}^{\infty} E_k$. Correspondingly, the space $\mathbb{E}:=E^{1/2}\times E$ has the orthogonal decomposition $\mathbb{E}=\bigoplus_{k=1}^{\infty}\mathbb{E}_k$, where $\mathbb{E}_k:=E_k\times E_k$. This decomposition of $\mathbb{E}$ is invariant by $\mathbb{A}$ and also by the group $J(t):=e^{\mathbb{A}t}$ $(t\in\mathbb{R})$.

Let us now assume that $\alpha^2<\mu_1$. Then the effect of $J(t)$ on $\mathbb{E}_k$ consists in a rotation of angle $\omega_k t$, where $\omega_k:=(\mu_k-\alpha^2)^{1/2}$. From this fact it follows that, for every $U\in\mathbb{E}$, the function $\mathbb{R}\ni t\longmapsto J(t)U\in\mathbb{E}$ is almost periodic.

In the following we consider the group $L(t)$ $(t\in\mathbb{R})$ generated by (2.5),(2.6). Obviously, $L(t)=e^{-\alpha t}J(t)$.

<u>Proposition 2.3</u>. Assume that A has compact resolvent, $\alpha>0$, and $\alpha^2<\mu_1$. If a positive semiorbit of $L(t)$ is contained in a submanifold M of $\mathbb{E}$ differentiable at the origin, then it is contained also in the tangent subspace of M at the origin. ■

<u>Proof</u>. Let U_0 be a point of the semiorbit which is assumed to be contained in M. Let F be the tangent subspace of M at the origin. Finally, let P denote the orthogonal projection of $\mathbb{E}$ onto F, and $Q:=I-P$. The fact that F is tangent to M at the origin means that

$$\|QU\| = o(\|PU\|) \quad \text{as} \quad U\to 0 \quad \text{on } M$$

In particular, this implies that

$$\|Q\,L(t)\,U_0\| = o(\|P\,L(t)\,U_0\|) \quad \text{as} \quad t\to+\infty$$

or, equivalently since $L(t)=e^{-\alpha t}J(t)$,

$$\|Q\,J(t)\,U_0\| = o(\|P\,J(t)\,U_0\|) \quad \text{as} \quad t\to+\infty$$

Using the fact that $J(t)\,U_0$ is an almost periodic function of t, one can then derive that $QU_0=0$, i.e. $U_0\in F$. Q.E.D.

<u>Corollary 2.4</u>. Under the hypotheses of Proposition 2.3, the only $L(t)$-invariant submanifolds of $\mathbb{E}$ differentiable at the origin are the invariant closed linear subspaces. ■

<u>Proposition 2.5</u>. Assume that A has compact resolvent, all its eigenvalues are simple, and $\alpha^2<\mu_1$. Then, the only $L(t)$-invariant closed linear

subspaces of $\mathbb{E}$ are those of the form $\mathbb{E}_K = \bigoplus_{k\in K} \mathbb{E}_k$, where K is a subset of $\mathbb{N}\setminus\{0\}$. ■

Proof. Let F be an invariant closed linear subspace. By linearity, invariance by $L(t)$ is equivalent to invariance by $J(t) = e^{\alpha t}L(t)$. On the other hand, using the fact that the operators $J(t)$ $(t\in\mathbb{R})$ are unitary, we see that the invariance of F by $J(t)$ implies the same property for $F^\perp$. In order to prove the proposition, it suffices to show that, for any $k\in\mathbb{N}\setminus\{0\}$ one has either $\mathbb{E}_k\subset F$ or $\mathbb{E}_k\subset F^\perp$. Now, since both F and $F^\perp$ are invariant and, on the other hand, the $\mathbb{E}_k$ do not contain proper invariant subspaces, the preceeding alternative is equivalent to the following statement :

for any $k\in\mathbb{N}\setminus\{0\}$, it happens that (2.9)

$$F\cap\mathbb{E}_k \neq \{0\} \text{ or } F^\perp\cap\mathbb{E}_k\neq\{0\}$$

In proving (2.9), we shall use the fact that

$$U\in\mathbb{E}_k \iff S(t)U := \tfrac{1}{2}[J(t)+J(-t)]U = (\cos\omega_k t)\,U \qquad (2.10)$$

which follows from the fact that $J(t)$ restricted to $\mathbb{E}_k$ is a rotation of frequency ω_k, and the hypothesis that the eigenvalues μ_k, and therefore the frequencies ω_k, are all different. In order to prove (2.9), we shall take an arbitrary $U\in\mathbb{E}_k$ and show that it belongs to either F or $F^\perp$. For this, we decompose $U\in\mathbb{E}_k\setminus\{0\}$ in its F and $F^\perp$ components:

$$U = V+W, \text{ where } V\in F \text{ and } W\in F^\perp$$

By applying the operators $S(t) := \tfrac{1}{2}[J(t)+J(-t)]$, and using the invariance and linearity of F and $F^\perp$, one obtains that

$$S(t)U = S(t)V + S(t)W, \text{ where } S(t)V\in F \text{ and } S(t)W\in F^\perp$$

By using (2.10), one immediately obtains that both $V\in F$ and $W\in F^\perp$ must belong to $\mathbb{E}_k$. Since V and W cannot be simultaneously zero, this proves (2.9). Q.E.D.

Corollary 2.6. Under the hypotheses of Proposition 2.5, $\mathbb{E}$ has only a countable family of finite-dimensional $L(t)$-invariant closed linear subspaces. ■

By combining the preceeding facts, we can state the following result :

Theorem 2.7 . Assume that A has compact resolvent, all its eigenvalues are simple, $\alpha > 0$, and $\alpha^2 < \mu_1$. Then the group generated by (2.5),(2.6) has only a countable family of finite-dimensional invariant submanifolds containing the origin and being differentiable at it. ∎

Remark . The result is no longer true when the condition of differentiability at the origin is dropped. If the frequencies ω_k satisfy some linear relation with integer coefficients, then one can have continuous families of finite-dimensional invariant Lipschitzian submanifolds containing the origin and being differentiable everywhere except at the origin.

3. A C^1 linearization theorem

In this section we give a C^1 linearization theorem which is applicable to certain stationary states of semilinear damped wave equations. In the finite-dimensional case, our result is included essentially in that of Hartman [1960] (Thm.(I)) , which instead of our condition (3.2) requires only that L be a contraction.

Theorem 3.1 . Let U be an open subset of a Banach space X , and T a C^1 map $U \to X$ with a fixed point p . Let L be the Fréchet derivative of T at p , i.e. $L := DT(p)$. Assume that L has a bounded inverse, and that the following properties are satisfied for some $\eta > 0$:

$$DT(p+x) - L = o(\|x\|^\eta) \quad \text{as } x \to 0 \tag{3.1}$$

$$\|L^{-1}\| \, \|L\|^{1+\eta} < 1 \tag{3.2}$$

Then, there exist V , neighbouhood of p in U , with $T(V) \subset V$, and R , a C^1 diffeomorphism onto its image, with $R(p) = 0$, $DR(p) = I$, and

$$DR(p+x) - I = o(\|x\|^\eta) \quad \text{as } x \to 0 \tag{3.3}$$

such that the following equation holds :

$$R\,T = L\,R \tag{3.4}$$

Such a map is unique in the following sense: if V' and R' satisfy also the preceeding properties, then R and R' coincide in any ball centered at p and contained in $V \cap V'$. ■

Remarks. (i) Condition (3.2) implies that L is a contraction.
(ii) The exponent η is by no means restricted to be less than 1; increasing η makes condition (3.2) less restrictive, but then condition (3.1) requires T to be closer to linear.

Proof. Without loss of generality, we assume $p = 0$. Let us rewrite equation (3.4) in the equivalent form

$$R = L^{-1} R T \tag{3.5}$$

By writing $T = L + \gamma$ and $R = I + \rho$, this equation for R transforms into the following equation for ρ:

$$\rho = L^{-1} \rho (L+\gamma) + L^{-1}\gamma \tag{3.6}$$

In the sequel, the right-hand side of (3.6) will be denoted by $K(\rho)$, and its first term will be denoted by $K_1(\rho)$:

$$K_1(\rho) := L^{-1} \rho (L+\gamma), \qquad K(\rho) := K_1(\rho) + L^{-1}\gamma \tag{3.7}$$

The existence and uniqueness of ρ satisfying (3.6) will be obtained by verifying that the transformation K is a contraction in an appropriate Banach space $\mathcal{R}$. In the following, X_δ denotes the open ball of radius δ in the space X. The Banach space $\mathcal{R}$ will consist of the mappings $\rho \in C^1(\overline{X}_\delta, X)$ satisfying $\rho(0) = 0$, $D\rho(0) = 0$, and $D\rho(x) = o(\|x\|^\eta)$ as $x \to 0$. One can check that this is a Banach space when endowed with the norm

$$\|\rho\|_{\mathcal{R}} := \sup_{x \in X_\delta \setminus \{0\}} \|x\|^{-\eta} \, \|D\rho(x)\| \tag{3.8}$$

We claim that, for δ sufficiently small, K is a contraction of $\mathcal{R}$. To have this property, we first need that $K(\rho)$ be defined on the same domain as ρ, which amounts to ask that $T = L + \gamma$ map X_δ into itself. This is true for δ sufficiently small, because L is a contraction, and γ is C^1 with $\gamma(0) = 0$ and $D\gamma(0) = 0$. In fact,

$$\|(L+\gamma)x\| \leq \left(\|L\| + \varepsilon(\delta)\right) \|x\| \qquad (\forall x \in X_\delta), \tag{3.9}$$

where

$$\varepsilon(\delta) := \sup_{x \in X_\delta} \|D\gamma(x)\|, \tag{3.10}$$

which has the property that $\varepsilon(\delta) \downarrow 0$ when $\delta \downarrow 0$. Now we can verify that K maps $\mathcal{R}$ into itself. Indeed, $L^{-1}\gamma \in \mathcal{R}$ because $\gamma \in \mathcal{R}$, and $K_1(\rho) \in \mathcal{R}$ because $\rho \in \mathcal{R}$; the fact that $D(K_1(\rho))(x) = o(\|x\|^\eta)$ as $x \to 0$ follows from the estimate

$$\begin{aligned}
\|x\|^{-\eta} \|D(K_1(\rho))(x)\| &= \|x\|^{-\eta} \|L^{-1} D\rho((L+\gamma)x)\ (L + D\gamma(x))\| \\
&\leq \|L^{-1}\| \frac{\|D\rho((L+\gamma)x)\|}{\|(L+\gamma)x\|^\eta} \frac{\|(L+\gamma)x\|^\eta}{\|x\|^\eta} \|L + D\gamma(x)\| \\
&\leq \|L^{-1}\| \left(\|L\| + \varepsilon(\delta)\right)^{1+\eta} \frac{\|D\rho((L+\gamma)x)\|}{\|(L+\gamma)x\|^\eta} \qquad (\forall x \in X_\delta \setminus \{0\})
\end{aligned} \tag{3.11}$$

where $\varepsilon(\delta)$ is the quantity defined by (3.10). Finally, from (3.11) follows that

$$\|K(\rho) - K(\sigma)\|_{\mathcal{R}} = \|K_1(\rho - \sigma)\|_{\mathcal{R}} \leq \|L^{-1}\| \left(\|L\| + \varepsilon(\delta)\right)^{1+\eta} \|\rho - \sigma\|_{\mathcal{R}} \tag{3.12}$$

which shows that, if condition (3.2) is satisfied and δ is sufficiently small, K is a contraction of $\mathcal{R}$.

In order to complete the proof, it only remains to notice that, if δ is small enough, then R will be a C^1 diffeomorphism onto its image, which follows from the inverse function theorem. The theorem is thus established with $V = X_\delta$. Q.E.D.

<u>Corollary 3.2</u>. Let X be a Banach space, and $T(t)$ $(t \in \mathbb{R})$ a group of diffeomorphisms of X with a fixed point p. Let $L(t)$ $(t \in \mathbb{R})$ be the group of bounded linear operators on X given by $L(t) := D(T(t))(p)$. Assume that, for some $\tau \in \mathbb{R}$ and some $\eta > 0$, $T := T(\tau)$ and $L := L(\tau)$ satisfy properties (3.1) and (3.2). Then there exist V, neighbourhood of p, and $R: V \to X$, a C^1 diffeomorphism onto its image, with $R(p) = 0$, $DR(p) = I$, and (3.3), such that, for every $t \in \mathbb{R}$, the equation

$$R\ T(t) = L(t)\ R \tag{3.13}$$

holds in some ball centered at p and contained in V. Such a ball can be chosen independently of t when t varies over any interval of the form $[t_0,+\infty)$ with t_0 finite. ∎

Proof . By the preceeding theorem, there exist a neighbourhood V of p and $R: V \to X$, a C^1 diffeomorphism onto its image, satisfying $R(p) = 0$, $DR(p) = I$, (3.3) , and

$$L(-\tau)\ R\ T(\tau) = R$$

Now, one easily verifies that the preceeding properties are also satisfied if V and R are replaced by $V' := T(-t)V$ and $R' := L(-t)\,R\,T(t)$, where t is any real number. By the uniqueness statement of Theorem 3.1 , this implies that equation $L(-t)RT(t) = R$ holds in any ball contained in $V \cap T(-t)V$. Finally, to establish the last statement of the corollary, it suffices to show that $\bigcap_{t \geq t_0} T(-t)V$ is a neighbourhood of p . Using the fact that $T(-\tau)V \supset V$, this reduces to see that $\bigcap_{t_0 \leq t \leq t_0+\tau} T(-t)V$ is a neighbourhood of p , i.e. there exists a $\delta > 0$ such that $T(-t)V \supset X_\delta$ for all $t \in [t_0, t_0+\tau]$, or equivalently $V \supset T(t)X_\delta$ for all $t \in [t_0, t_0+\tau]$, which follows from the joint continuity of the mapping $\mathbb{R} \times X \ni (t,x) \longmapsto T(t)x \in X$ at the points of the compact set $[t_0, t_0+\tau] \times \{p\}$. Q.E.D.

Corollary 3.3 . Let us consider problem (1.1),$(1.2)_B$,(1.3) with the hypotheses of Theorem 1.1 . Let $u = u^*(x)$ be a stationary state, and let μ_1 be the lowest eigenvalue of the differential operator $-\partial_x^2 - f_u(x, u^*(x))$ with boundary conditions $(1.2)_B$. If $\alpha^2 < \mu_1$, then, near this stationary state, the flow $T(t)\ (t \in \mathbb{R})$ is C^1-equivalent to its linearization. ∎

Proof . Let $L(t)\ (t \in \mathbb{R})$ be the group of bounded linear operators obtained by linearizing $T(t)\ (t \in \mathbb{R})$ at the stationary state $u = u^*$. This coincides with the group generated by equation (2.5) or (2.1) with A being given by the differential operator $-\partial_x^2 - f_u(x, u^*(x))$ with the boundary conditions $(1.2)_B$. Let $E = H_B^1 \times L_2$ be endowed with the inner product (2.7) . According to Corollary 2.2 , $J(t) = e^{\alpha t} L(t)\ (t \in \mathbb{R})$ is a unitary group. Corollary 3.2 can now be applied with no matter which $\tau > 0$ and small $\eta > 0$, since $T(\tau)$ is $C^{1+\eta}$ and $\|L(\tau)\| = e^{-\alpha\tau}$, $\|L(-\tau)\| = e^{\alpha\tau}$. Q.E.D.

4. Exhibiting non-existence

Our example of non-existence belongs to problem (1.1),$(1.2)_N$,(1.3) with f of the form (1.6), where f_0 will be fixed and g variable.

The fixed function f_0 will be taken to be independent of x ; accordingly, we shall write $f_0(u)$ for $f_0(x,u)$. This function f_0 is assumed to fulfil the general conditions (F1a),(F1b),(F2) , and also the following particular ones :

$$f_0(0) = f_0(1) = 0 \quad ; \quad 1 \text{ is the only positive zero of } f_0 \tag{4.1}$$

$$0 < f_0'(0) < 1 \tag{4.2}$$

$$f_0'(1) < -\alpha^2 \tag{4.3}$$

Since f_0 is independent of x, and the boundary conditions are of Neumann type, the dynamical system on $\mathbb{E} = H_B^1 \times L_2$ corresponding to $g = 0$ has a two-dimensional invariant linear subspace consisting of the states which are spatially homogeneous (i.e. constant with respect to x); on this subspace, (1.1) reduces to a second-order ordinary differential equation. In the following, $\bar{0}$ and $\bar{1}$ denote the points of this subspace given respectively by $u=0, \dot{u}=0$ and $u=1, \dot{u}=0$. Conditions (4.1)-(4.3) imply the following facts :

Both $\bar{0}$ and $\bar{1}$ are hyperbolic stationary states. (4.4)

$\bar{0}$ has a one-dimensional unstable maniflod,
and $\bar{1}$ is asymptotically stable. (4.5)

There is an heteroclinic orbit from $\bar{0}$ to $\bar{1}$. (4.6)

In fact, the heteroclinic orbit which connects $\bar{0}$ to $\bar{1}$ lies on the subspace of spatially homogeneous states. In the $(u,\dot{u})$-plane this orbit looks like shown in Fig.1 below. In the following, the corresponding solution of (1.1) (which is unique except for translations) will be denoted by $u_0(t)$ $(t \in \mathbb{R})$. As it is easily verified, there exists a time t_0^* such that

On $(-\infty, t_0^*]$, u_0 and $\dot{u}_0$ are both strictly increasing. (4.7)

On $[t_0^*, +\infty)$, u_0 remains $\geq u_0(t_0^*) =: a_0$ (4.8)

We now introduce a perturbation g which will break this special situation occurring for $g=0$. This perturbation g will be allowed to vary

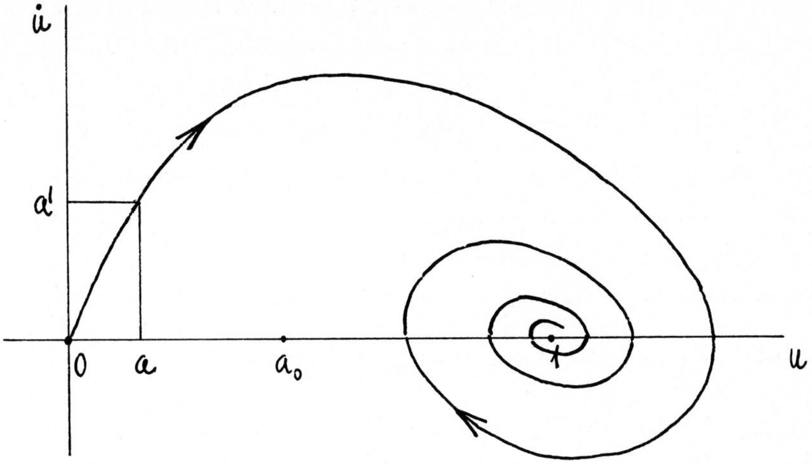

Fig.1

within a certain ball of the Banach space $\mathcal{G}(J)$ defined in (1.7), where $J := (a,b)$ is a fixed open interval with $0<a<b<1$ and $a<a_0$, where a_0 is the quantity appearing in (4.8). In the following, the space $\mathcal{G}(J)$ will be denoted simply by $\mathcal{G}$, and its ball of radius δ will be denoted by $\mathcal{G}_\delta$. Clearly, for every $g\in\mathcal{G}$, the corresponding flow still satisfies (4.4) and (4.5). In fact, these perturbed flows remain unchanged inside the open set $\mathcal{J} := \{U=(u,w) \mid u(x)\notin \bar{J}\ (\forall x\in\Omega)\} \subset \mathbb{E}$.

Let us look at the flow in the neighbourhood of the stationary state $\bar{1}$, where we know it does not depend on g. Since this stationary state is stable, every neighbourhood contains another one which is positively invariant (i.e. $T(t)$ maps it into itself for every $t\geq 0$). Condition (4.3) implies that this stationary state satisfies the hypotheses of Corollary 3.3. Therefore, we are ensured that $\mathcal{J}$ contains a positively invariant neighbourhood of $\bar{1}$ where the flow $T_0(t)\ (=T_g(t))\ (t\in\mathbb{R})$ is C^1-equivalent to its linearization at $\bar{1}$, $L(t)\ (t\in\mathbb{R})$. In the following, V denotes a small neighbourhood of $\bar{1}$ with this property. On the other hand, the group $L(t)\ (t\in\mathbb{R})$ coincides with the one generated by equation (2.5), or (2.1), with A being the operator given by $-\partial_x^2 - f_0'(1)$ with boundary conditions $(1.2)_N$, which clearly satisfies the hypotheses of Theorem 2.7. Therefore, we can conclude that V includes only a countable family of finite-dimensional local invariant manifolds of class C^1 containing $\bar{1}$.

Let us now consider the orbit that departs form $\bar{0}$ towards the positive u direction. Before leaving $\mathcal{J}$, this orbit will coincide with that corresponding to $g=0$. Therefore, by suitably choosing the time origin, we can assume that the corresponding solution of (1.1), which we shall denote by $u_g(t,x)$, satisfies the following relations :

$$u_g(t,x) = u_0(t) \quad (\forall t \leq 0), \qquad u_g(0,x) = a \tag{4.9}$$

where a is the left end of J. In the following, U_g will denote the curve $\mathbb{R} \to \mathbb{E}$ given by $U_g(t) := (u_g(t,\cdot), \hat{u}_g(t,\cdot))$, where $\hat{u}_g := \alpha u_g + \dot{u}_g$. Now, since $U_0(t) \to \bar{1}$ as $t \to +\infty$, Theorem 1.1 ensures that, if δ is small enough, then the following property will hold :

$$\exists T > 0 \text{ such that, for every } g \in \mathcal{G}_\delta, \; U_g(T) \in V \tag{4.10}$$

In particular, this implies that, for $g \in \mathcal{G}_\delta$, $U_g(t) \to \bar{1}$ as $t \to +\infty$, i.e. the corresponding dynamical system satisfies also (4.6). From now on, we assume δ small enough for property (4.10) to hold.

Let Γ denote the C^1 mapping

$$\Gamma : \mathcal{G}_\delta \ni g \longmapsto U_g(T) \in V \subset \mathbb{E} \tag{4.11}$$

where T is the quantity appearing in (4.10). Our purpose is to see that there are many $g \in \mathcal{G}_\delta$ for which $\Gamma(g) = U_g(T)$ does not belong to any of the countably many finite-dimensional local invariant manifolds of class C^1 containing $\bar{1}$. In this case, one can conclude that there is no finite-dimensional local invariant manifold of class C^1 containing the global attractor. In fact, we will prove the following result :

Theorem 4.1. Consider problem (1.1),$(1.2)_N$,(1.3) with f_0 independent of x and satisfying (F1a),(F1b),(F2),(4.1),(4.2),(4.3), and g belonging to $\mathcal{G} := \mathcal{G}(J)$, where $J := (a,b)$ is a bounded open interval with $0 < a < b < 1$ and $a < a_0$ (a_0 is the quantity appearing in (4.8), which depends on f_0). There is a $\delta > 0$ and a residual subset $\mathcal{R}$ of $\mathcal{G}_\delta$ such that if $g \in \mathcal{R}$ then there is no finite-dimensional local invariant manifold of class C^1 containing the global attractor. ∎

The proof of Theorem 4.1 will be based upon the following fact, whose proof will be given afterwards :

Proposition 4.2. Under the hypotheses of Theorem 4.1, there is a $\delta > 0$ and a dense subset $\mathcal{D}_\delta$ of $\mathcal{G}_\delta$ such that if $g \in \mathcal{D}_\delta$ then $\mathrm{Range}(D\Gamma(g))$ is infinite-dimensional. ∎

Proof of Theorem 4.1 . Let M_K (K varying among the finite subsets of $\mathbb{N}\setminus\{0\}$) be the countable family of finite-dimensional local invariant manifolds containing $\bar{\Lambda}$. We will see that, for every K , there is a open and dense subset R_K of $\mathcal{G}_\delta$ such that $g\in R_K \Longrightarrow \Gamma(g)\notin M_K$. From this the theorem will follow by a cathegory argument. The openness of R_K is an immediate consequence of the continuous dependence of solutions with respect to g (Theorem 1.1) . The denseness of R_K follows from Proposition 4.2 : Indeed, if R_K were not dense, there would be some open set $\mathcal{U}\in\mathcal{G}_\delta$ such that $\Gamma(g)\in M_K\ (\forall g\in\mathcal{U})$. But this would imply that, for every $g\in\mathcal{U}$, $\mathrm{Range}(D\Gamma(g))\subset T(M_K)_{\Gamma(g)}$, which contradicts Proposition 4.2 since $T(M_K)_{\Gamma(g)}$ is finite-dimensional. Q.E.D.

Proof of Proposition 4.2 . We will take $\mathcal{D}_\delta:=\mathcal{D}\cap\mathcal{G}_\delta$, where $\mathcal{D}:=\bigcup_{I\subset J}\mathcal{G}(I)$, which is easily verified to be dense in $\mathcal{G}:=\mathcal{G}(J)$. Let $g\in\mathcal{D}_\delta$, and assu me that it belongs to $\mathcal{G}(I)$ with $I=(a_g,b_g)$ $a<a_g<b_g<b$. In order to see that $\mathrm{Range}(D\Gamma(g))$ is infinite-dimensional, we shall see that there is a linearly independent family $h_n\ (n\in\mathbb{N})$ of elements of $\mathcal{G}$ such that the images $D\Gamma(g)h_n\ (n\in\mathbb{N})$ are linearly independent elements of $\mathbb{E}$. For every $h\in\mathcal{G}$, the value of $D\Gamma(g)h$ is given by

$$D\Gamma(g)h = Y(T) \tag{4.12}$$

where

$$Y(t) := (y(t,\cdot),\hat{y}(t,\cdot)),\quad \hat{y}=\alpha y+\dot{y} \tag{4.13}$$

and $y(t,x)$ is the solution of the first variation equation

$$y_{tt}+2\alpha y_t = y_{xx}+f'(u_g)y+g_u(x,u_g)y+h(x,u_g) \tag{4.14}$$

$$y_x|_{x=0} = y_x|_{x=\pi} = 0 \tag{4.15}$$

$$y|_{t=0} = y_t|_{t=0} = 0 \tag{4.16}$$

In the following, y_n and Y_n denote the particular y and Y that are obtained when $h=h_n$.

Our choice of the functions h_n will be based upon the following fact, whose proof is given at the end of this section:

Lemma 4.3 . If $\delta>0$ is small enough, then, for every $g\in\mathcal{D}_\delta$, we can find $c_g\in(a,a_g]$ and $t_g^*>0$ such that

$$t \leq t_g^* \Rightarrow u_g(t,x) = u_0(t) \leq a_g \quad (\forall x \in \Omega) \tag{4.17}$$

$$t \geq t_g^* \Rightarrow u_g(t,x) \geq c_g \quad (\forall x \in \Omega) \quad \blacksquare \tag{4.18}$$

We shall use this fact by taking the functions h_n in such a way that

$$h_n(x,u) = 0 \quad \text{for} \quad u \notin (a,c_g) \tag{4.19}$$

By doing so, it results that, in equation (4.14), the term $g_u(x,u_g)\,y$ vanishes for $t \leq t_g^*$, and the term $h(x,u_g)$ vanishes for $t \geq t_g^*$.

According to this fact, $D\Gamma(g)h_n$ will be given by

$$D\Gamma(g)h_n = Y_n(T) = \Phi(T,t_g^*)\,Y_n(t_g^*) \tag{4.20}$$

where $\Phi(t,s)$ $(s \leq t)$ denotes the system of evolution operators of the linear problem obtained from (4.14),(4.15) when dropping the last term of (4.14). However, these operators are isomorphisms of the Hilbert space $\mathbb{E}$. Therefore, our problem reduces to show that the h_n $(n \in \mathbb{N})$, which must satisfy (4.19), can be chosen in such a way that the resulting $Y_n(t_g^*)$ be linearly independent elements of $\mathbb{E}$.

To study this question, equation (4.14) needs to be considered only in the interval $0 \leq t \leq t_g^*$, where, by Lemma 4.3, we know that the term $g_u(x,u_g)\,y$ vanishes, and also that $u_g(t,x)$ is independent of both x and g. We shall take advantage of these facts by choosing the functions h_n in the factorial form $h_n(x,u) = c_n(x)\,\varphi_n(u)$, which makes possible to solve the equation by separation of variables. More specifically, we shall take

$$h_n(x,u) = \varphi_n(u)\cos nx \tag{4.21}$$

where, for every $n \in \mathbb{N}$, φ_n will be a function $\mathbb{R} \to \mathbb{R}$ of class $C^{1+\eta}$ with support contained in the interval $[a,c_g]$. By separating variables, we obtain that, for $0 \leq t \leq t_g^*$,

$$y_n(t,x) = Y_n(t)\cos nx \tag{4.22}$$

where Y_n is the solution of the ordinary differential equation problem

$$\ddot{Y}(t) + 2\alpha\dot{Y}(t) + n^2 Y(t) - f_0'(u_0(t))\,Y(t) = \varphi_n(u_0(t)) \tag{4.23}$$

$$Y(0) = \dot{Y}(0) = 0 \tag{4.24}$$

In the following, the expression appearing at the left-hand side of (4.23) will be denoted by $L_n(\gamma;t)$.

Obviously, to attain our purpose, it will suffice to find a family φ_n $(n\in\mathbb{N})$ of functions $\mathbb{R}\to\mathbb{R}$ of class $C^{1+\eta}$ supported in $[a,c_g]$ such that the corresponding solutions of (4.23),(4.24) satisfy

$$\gamma_n(t_g^*)\neq 0 \quad (\forall n\in\mathbb{N}) \tag{4.25}$$

If f_0 is of class $C^{2+\eta}$, this can be easily accomplished by taking

$$\varphi_n(u) := L_n(\sigma\zeta_n;\, u_0^{-1}(u)) \tag{4.26}$$

where, for each $n\in\mathbb{N}$, ζ_n is a solution of the homogeneous equation $L_n(\zeta;t)=0$ satisfying $\zeta(t_g^*)=1$, and σ is a function $\mathbb{R}\to\mathbb{R}$ of class C^∞ such that $\sigma(t)=0$ for $t\leq 0$ and $\sigma(t)=1$ for $t\geq u_0^{-1}(c_g)$. By introducing (4.26) into (4.23), one immediately obtains that $\gamma_n=\sigma\zeta_n$, which obviously satisfies $\gamma_n(t_g^*)=1$. If f_0 is not of class $C^{2+\eta}$ but merely $C^{1+\eta}$, then the resulting functions φ_n need not be of class $C^{1+\eta}$; however, any functions $\hat{\varphi}_n$ of class $C^{1+\eta}$ sufficiently near φ_n in the sup norm will serve our purpose. Q.E.D.

<u>Proof of Lemma 4.3</u>. In order to prove Lemma 4.3, we shall need the following estimate

$$|u_g(t,x)-u_0(t)| \leq K(1-e^{-\alpha t})\,\|g\|_{(J)} \qquad (\forall x\in\Omega, \forall t\in[0,T]) \tag{4.27}$$

This is an improvement of an estimate given by Theorem 1.1, namely

$$|u_g(t,x)-u_0(t)| \leq K_0\,\|g\|_{(J)} \qquad (\forall x\in\Omega, \forall t\in[0,T]) \tag{4.28}$$

In order to derive (4.27), we proceed as follows. Let us define $v(t,x) := u_g(t,x)-u_0(t)$, and $V(t) := (v(t,\cdot),\hat{v}(t,\cdot))$, where $\hat{v} := \alpha v+\dot{v}$. By the variation of constants formula, $V:\mathbb{R}\to\mathbb{E}$ is given by

$$V(t) = \int_0^t e^{-\alpha(t-s)}\, J(t-s)\left(\hat{\mathbb{F}}_g(U_g(s)) - \hat{\mathbb{F}}_0(U_0(s))\right) ds \tag{4.29}$$

where $J(t)$ $(t\in\mathbb{R})$ is the group of unitary operators generated by $\mathbb{A}$ (see § 2) when A is the operator $-\partial_x^2 - f_0'(1)$ with boundary conditions $(1.2)_N$, and $\hat{\mathbb{F}}_g$ denote the mappings $H_B^1\times L_2\to H_B^1\times L_2$ corresponding to the functions $\hat{f}_g$ defined by $\hat{f}_g(x,u) := f_g(x,u) - f_0'(1)u$.

From (4.29) one obtains that

$$\|V(t)\|_E \leq \int_0^t e^{-\alpha(t-s)} \left[\|\hat{F}_0(u_g(s)) - \hat{F}_0(u_0(s))\|_{L_2} + \|G(u_g(s))\|_{L_2} \right] ds$$

$$\leq \int_0^t e^{-\alpha(t-s)} \left[\left(\|f_0\|_{(M)} + |f_0'(1)| \right) \|u_g(s) - u_0(s)\|_{L_\infty} + \|g\|_{(J)} \right] ds \tag{4.30}$$

where $u_g(t) := u_g(t,\cdot) \in H_B^1$, $\hat{F}_0$ and G denote the functions $H_B^1 \to L_2$ corresponding respectively to $\hat{f}_0$ and g , and M is some bounded open interval of $\mathbb{R}$. From here, the desired estimate (4.27) follows by simply introducing (4.28) into the right-hand side of (4.30) .

Let us proceed with the proof of Lemma 4.3 . We start from (4.7) and (4.8) , where we remark that $t_0^* > 0$, because $a < a_0$ and we have chosen $t = 0$ when $u_0 = a$. From (4.7) , it follows that

$$0 \leq s < t \leq t_0^* \implies u_0(s) < u_0(t) - a'(t-s) \tag{4.31}$$

where $a' := \dot{u}_0(0)$. We now claim that, by virtue of (4.27) , (4.31) and (4.8) imply the following fact :

(A) If δ is small enough, then, for every $g \in \mathcal{G}_\delta$

$$0 < t \leq t_0^* \implies u_0(\tfrac{t}{2}) < u_g(t,x) \quad (\forall x \in \Omega) \tag{4.32}$$

$$t_0^* \leq t < +\infty \implies u_0(\tfrac{t_0^*}{2}) < u_g(t,x) \quad (\forall x \in \Omega) \tag{4.33}$$

Specifically, it suffices that

$$\delta \leq \frac{a'}{2K} \min\left(t_0^*, \frac{1}{\alpha}\right) \tag{4.34}$$

In fact, for $0 < t \leq t_0^*$, (4.27), (4.34), and (4.31) imply that

$$u_g(t,x) \geq u_0(t) - K(1-e^{-\alpha t})\delta > u_0(t) - K\alpha t\delta$$

$$\geq u_0(t) - a'\tfrac{t}{2} > u_0(\tfrac{t}{2})$$

On the other hand, for $t_0^* \leq t \leq T$, (4.27), (4.34), (4.8), and (4.31) imply that

$$u_g(t,x) \geq u_0(t) - K(1-e^{-\alpha t})\delta > u_0(t) - K\delta$$

$$\geq u_0(t) - a'\frac{t_0^*}{2} \geq u_0(t_0^*) - a'\frac{t_0^*}{2} > u_0(\tfrac{t_0^*}{2})$$

Furthermore, it is clear that, if the neighbourhood V is small enough, then (4.33) will also be true for $t>T$, because for such t $U_g(t,\cdot)$ remains always inside V.

Finally, the statement of Lemma 4.3 is easily obtained from (A) by taking

$$t_g^* := \begin{Bmatrix} t_0^* & \text{if } a_g \geq a_0 \\ u_0^{-1}(a_g) & \text{if } a_g \leq a_0 \end{Bmatrix}, \qquad c_g := u_0\left(\frac{t_g^*}{2}\right) \tag{4.35}$$

which quantities satisfy

$$0 < t_g^* \leq t_0^*, \qquad a < c_g \leq a_g \tag{4.36}$$

In fact, for $-\infty < t \leq t_g^*$, (4.8) implies that

$$u_g(t,x) = u_0(t) \leq u_0(t_g^*) \leq a_g$$

On the other hand, for $t_g^* \leq t \leq t_0^*$, (4.22) and (4.8) imply that

$$u_g(t,x) \geq u_0\left(\frac{t}{2}\right) \geq u_0\left(\frac{t_g^*}{2}\right) =: c_g$$

Finally, for $t_0^* \leq t < +\infty$, (4.23) and (4.8) imply that

$$u_g(t,x) > u_0\left(\frac{t_0^*}{2}\right) \geq u_0\left(\frac{t_g^*}{2}\right) =: c_g$$

Q.E.D.

Acknowledgements . We wish to thank Dan Henry for the encouraging discussions that we have had with him during the preparation of this paper.

References

1. Babin,A.V., Vishik,M.I., 1983. Regular attractors of semigroups and evolution equations. J. Math. Pures Appl. 62: 441-491.

2. Hale,J.K., 1985. Asymptotic behavior and dynamics in infinite dimensions. Res. Notes in Math. 132: 1-42.

3. Hartman,P., 1960. On local homeomorphisms of Euclidean spaces. Bol. Soc. Mat. Mexicana 5: 220-241.

4. Henry,D., 1981. Geometric Theory of Semilinear Parabolic Equations. Lecture Notes in Math. 840 .

5. Mallet-Paret,J., Sell,G.R., 1986. The principle of spatial averaging and inertial manifolds for reaction-diffusion equations. Lect. Notes in Math. (in press).

6. Mora,X., 1986. Finite-dimensional attracting invariant manifolds for damped semilinear wave equations. Contributions to Nonlinear Partial Differential Equations II (ed. by I.Díaz and P.L.Lions; Longman) (in press).

7. Pazy,A., 1983. Semigroups of Linear Operators and Applications to Partial Differential Equations. Springer.

8. Solà-Morales,J., València,M., 1986. Trend to spatial homogeneity for solutions of semilinear damped wave equations. Proc. Roy. Soc. Edinb. (in press).

9. Tanabe,H., 1979. Evolution Equations. Pitman.

SLEP METHOD TO THE STABILITY OF SINGULARLY PERTURBED SOLUTIONS WITH MULTIPLE INTERNAL TRANSITION LAYERS IN REACTION-DIFFUSION SYSTEMS

Yasumasa NISHIURA and Hiroshi FUJII
Institute of Computer Sciences, Kyoto Sangyo University
Kyoto 603, Japan

§1. INTRODUCTION

Patterns with sharp transition layers appear in various fields such as patchiness and segregation in eco-systems [2], [9], travelling and standing waves in excitable media or chemical reactions [1], [21], [3] and [12], striking patterns in morphogenesis models [7], and so on.

This paper focuses on patterns arising from a class of reaction-diffusion systems in which most of the above example fall, namely, activator-inhibitor systems with diffusions:

$$(P)^{\varepsilon,\sigma} \qquad \begin{cases} u_t = \varepsilon^2 u_{xx} + f(u,v), & (t,x) \in (0,\infty)\times I,\ I=(0,1), \\ v_t = \dfrac{1}{\sigma} v_{xx} + g(u,v), & \\ u_x = 0 = v_x & \text{on } \partial I, \end{cases}$$

where the diffusion rate ε^2 of the activator u is much smaller than that σ^{-1}, of the inhibitor v. It is noted that ε may correspond to the <u>width</u> of possible transition layers. For detailed assumptions, see §2 and Fig.2.

Concerning the construction of such patterns, whether stationary or travelling ones, the singular perturbation technique is one of the most powerful (and, sometimes the unique) tools. In fact, there are a number of works, under various situations, about existence proofs of patterns which make a use of this technique. See, for instance, [2], [4], [14], [10] and [9]. However, as may be widely known, the major problem of this technique has been the lack of a unified approach to analyse <u>stability of constructed solutions</u>. See, e.g., [2]. It is

NATO ASI Series, Vol. F37
Dynamics of Infinite Dimensional Systems
Edited by S.-N. Chow, and J. K. Hale

in [18] and [19] that the authors have firstly - at least to our best knowledge - proposed a general method, called the SLEP - Singular Limit Eigenvalue Problem - method, to resolve the stability problem for singularly perturbed solutions. Based on this idea, they have given a stability theorem to the normal SPS1, i.e., stationary singularly perturbed solutions with a single internal transition layer. (See, also [11].)

The aim of this paper is to show how nicely this SLEP idea works for the stability problem of the normal SPSn, i.e., SPS possessing n ($\in \mathbb{N}$) internal transition layers of $(P)^{\varepsilon,\sigma}$. (For the precise definition of the normal SPSn, see §2.) We note here that the normal SPSn ($n \in \mathbb{N}$) is brought into existence as a consequence of a group covariance of the system $(P)^{\varepsilon,\sigma}$. Our study leads to an interesting and rather surprising conclusion of multiple existence of stable patterns in $(P)^{\varepsilon,\sigma}$. In fact, it follows from our Main Theorem that the number of stable SPSn's tends to infinity as ε goes to zero for a fixed σ. (See, also [15].)

It should be emphasized that the stability property of an SPS with multiple internal layers is far more complicated than that of a single layered SPS, due to the presence of strong interactions between layers, thus making the behavior of the spectrum so delicate.

Let $(LP)^{\varepsilon,\sigma}$ denote the linearized eigenvalue problem at some SPSn ($n \in \mathbb{N}$):

$$(LP)^{\varepsilon,\sigma} \quad \mathcal{L}^{\varepsilon,\sigma}\begin{pmatrix} w \\ z \end{pmatrix} \equiv \begin{pmatrix} \varepsilon^2 \dfrac{d^2}{dx^2} + f_u^{\varepsilon,\sigma} & f_v^{\varepsilon,\sigma} \\ g_u^{\varepsilon,\sigma} & \dfrac{1}{\sigma}\dfrac{d^2}{dx^2} + g_v^{\varepsilon,\sigma} \end{pmatrix}\begin{pmatrix} w \\ z \end{pmatrix} = \lambda \begin{pmatrix} w \\ z \end{pmatrix}$$

The most subtle point in the spectral analysis of $(LP)^{\varepsilon,\sigma}$ is in asymptotic behaviors of critical eigenvalues as $\varepsilon \downarrow 0$. Here, by critical (or, noncritical) eigenvalues we mean those which approach (or, are bounded away from) zero as $\varepsilon \downarrow 0$. Two remarks should be made here. Firstly, as the number of internal layers increases, so does the number of critical eigenvalues, in general. In fact, there are exactly n critical eigenvalues for each normal SPSn ($n \in \mathbb{N}$). Secondly, $(LP)^{\varepsilon,\sigma}$ becomes singular in two ways as

$\varepsilon \downarrow 0$: The second derivative for w disappears, and at the same time, the coefficients $f_u^{\varepsilon,\sigma}$, $g_u^{\varepsilon,\sigma}$,... become discontinuous functions. Apparently, a formal limit of $(LP)^{\varepsilon,\sigma}$ with $\varepsilon = 0$ loses all the information from layers, thus telling us nothing about the behavior of critical eigenvalues.

The SLEP approach for the stability of a normal SPSn consists of two stages. The first one is the derivation of singular limit eigenvalue problem, and the second stage is to find the spectrum of the SLEP. The key in the first stage is that, under an appropriate scaling with respect to ε, $(LP)^{\varepsilon,\sigma}$ can be continuously extended to the limit $\varepsilon \downarrow 0$, in a function space weaker than L^2, say at least H^{-1} for one-dimensional case. This limit problem - SLEP - includes a nice invertible operator M, and a linear combination of n Dirac's point mass distributions at each layer position. Then, after an application of M^{-1}, SLEP can be converted into a finite (=n) dimensional linear algebraic eigenvalue problem with respect to the scaled critical eigenvalues - the SLEP system. In other words, in the limit of $\varepsilon \downarrow 0$, the behavior of critical eigenvalues of a normal SPSn is determined by the spectrum of an $n \times n$ matrix (see (3.16) in §3). It seems interesting to observe that this procedure is, in a sense, similar to the Lyapunov-Schmidt reduction in the bifurcation theory. Our principal result is the following. See [20] for the complete proof of it as well as that of lemmas in later sections.

Main Theorem ([20]). $\mathcal{L}^{\varepsilon,\sigma}$ at the normal SPSn has exactly n critical eigenvalues $\lambda_i^c(\varepsilon,\sigma)$ $(i=0,\ldots,n-1)$, which are real, simple, and negative for small ε, and they behave like

$$\lambda_i^c(\varepsilon,\sigma) \underset{-}{\sim} \tau_i^{*,\sigma}\varepsilon \quad \text{as} \quad \varepsilon \downarrow 0 \quad (i=0,\ldots,n-1),$$

where $\tau_i^{*,\sigma}$ $(i=0,\ldots,n-1)$ are strictly negative constants given in Theorem 3.1. See Fig. 1. All the other eigenvalues of $\mathcal{L}^{\varepsilon,\sigma}$ have strictly negative real parts independent of small ε.

It turns out that SPSn $(n \in \mathbb{N})$ are asymptotically stable if ε is taken sufficiently small. This gives an insight in understanding the phenomenon of "coexistence of multiple stable states", which are often observed in nonlinear systems in nature. (See, [5], [6], and [8].)

Final remark: The SLEP idea is, in principle, applicable to a wider range of stability problems for SPS's. It does not depend on whether I is bounded or a whole line (hence, travelling waves as well, see [16]), nor on boundary conditions, or particular type of nonlinearities. An instance is the excitable system with lateral inhibition. (The nullcline of g is changed to ——·——-line in Fig. 2. See [17].) So, although further studies are required, it seems that the SLEP idea provides us a unified approach to the stability problems for SPS's.

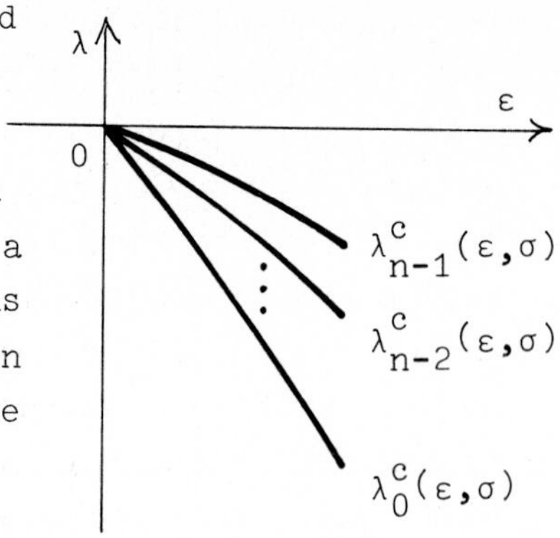

Fig. 1: Behavior of critical eigenvalues of $(LP)^{\varepsilon,\sigma}$.

This paper is divided into three parts. In the next section, the normal SPSn are constructed from the basic pattern, and several basic lemmas are prepared for later sections. In §3, we show how the linearized problem is reduced to the SLEP system. Finally, in §4, we solve the eigenvalue problem of the SLEP matrix.

We use the following notation.

$C^p_\varepsilon(\bar{I})$ = the space of p-times continuous differentiable functions with the norm $\|u\|_{C^p_\varepsilon} = \sum_{k=0}^{p} \max_{x\in I} |(\varepsilon\frac{d}{dx})^k u(x)|$,

$H^p_N(I)$ = the space of closure of $\{\cos(n\pi x)\}_{n\in\mathbb{N}}$ in $H^p(I)$.

$C^p_{c.u.}(\mathbb{R})$ = the compact uniform convergence on $\mathbb{R}$ in C^p-sense.

§2. MODEL SYSTEM, FOLDING UP PRINCIPLE, AND PRELIMINARIES TO THE SPECTRAL ANALYSIS

The stationary problem of our model system (P) takes the form:

$$(SP)^{\varepsilon,\sigma} \qquad \begin{aligned} &\varepsilon^2 u_{xx} + f(u,v) = 0 \\ &\frac{1}{\sigma} v_{xx} + g(u,v) = 0 \end{aligned} \qquad \text{in } I,$$

subject to the Neumann boundary conditions on ∂I. Assumptions for f and g are the following (see Fig. 2):

(A.1) The nullcline of f is sigmoidal and consists of three smooth curves $u = h_-(v)$, $h_0(v)$, and $h_+(v)$ with $h_-(v) < h_0(v) < h_+(v)$.

(A.2) $J(v)$ has an isolated zero at $v=v^*$ such that $dJ/dv < 0$ at $v=v^*$ where $J(v) = \int_{h_-(v)}^{h_+(v)} f(s,v)ds$.

(A.3) $f_u < 0$ on $R_+ \cup R_-$, where $R_+(R_-)$ denotes the part of the curve $u = h_+(v)(h_-(v))$ defined by $R_+(R_-) = \{(u,v) | u = h_+(v)(h_-(v))$ for $v \geqq v^*(v \leqq v^*)$, respectively.

(A.4) (a) $g|_{R_-} < 0 < g|_{R_+}$.

(b) $\det \left. \frac{\partial(f,g)}{\partial(u,v)} \right|_{R_+ \cup R_-} > 0$.

(A.5) $g_v |_{R_+ \cup R_-} \leqq 0$.

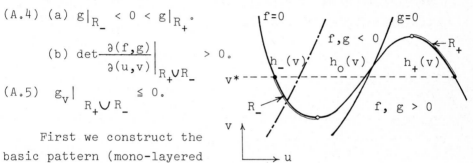

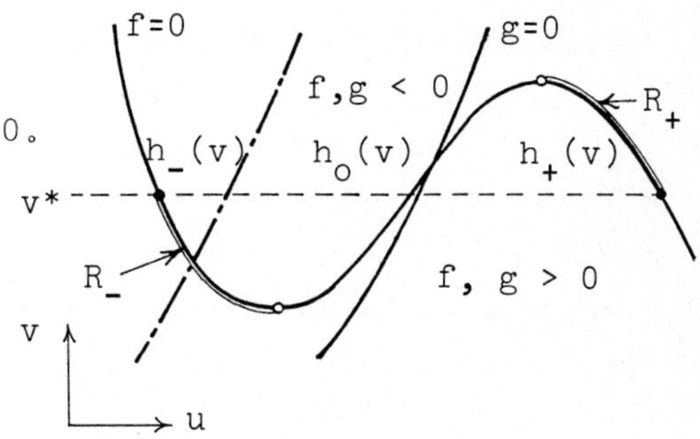

Fig.2: Functional forms of f and g, —·—·—line denotes $g=0$ for the excitable case.

First we construct the basic pattern (mono-layered solution), then we apply the Folding up principle to it to obtain the normal SPSn. The solution of the following reduced problem becomes the first approximation to the basic pattern.

$$\text{(RP.1)} \quad f(u,v) = 0, \qquad (u,v) \in L^2(I) \times \{H^2(I) \cap H^1_N(I)\},$$

$$\text{(RP.2)} \quad \frac{1}{\sigma} v_{xx} + g(u,v) = 0$$

Since we are interested in the solution of (RP) which are the limits of those of $(SP)^{\varepsilon,\sigma}$ as $\varepsilon \downarrow 0$, we take

$$\text{(2.1)} \qquad u = h^*(v) \equiv \begin{cases} h_-(v) & \text{for } v \leqq v^*, \\ h_+(v) & \text{for } v \geqq v+ \end{cases}$$

as a special solution of (RP.1). Substituting this into (RP.2), we obtain the reduced scalar equation for v:

(RSP) $\frac{1}{\sigma} v_{xx} + G^*(v) = 0, \quad v \in H^2(I) \cap H^1_N(I),$

where $G^*(v) = g(h^*(v),v)$. Note that $G^*(v)$ has a jump discontinuity at $v=v^*$.

<u>Lemma 2.1</u> There exists a uniquely determined positive constant σ_1^* such that monotone increasing(decreasing) C^1-matching solution $V_+^{*,\sigma}(x)(V_-^{*,\sigma}(x))$ of (RSP) exist for $0 < \sigma \leqq \sigma_1^*$. Moreover, $\lim_{\sigma\downarrow 0} V^{*,\sigma}(x) = v^*$ in C^1-sense.

For definiteness, we only consider the monotone increasing case and write simply $V^{*,\sigma}(x)$ instead of $V_+^{*,\sigma}(x)$. In view of (2.1), the first approximate solution to basic pattern takes the form:

(2.2) $(U^{*,\sigma}(x), V^{*,\sigma}(x)) \quad \text{for} \quad 0 \leqq \sigma \leqq \sigma_1^*,$

where $U^{*,\sigma}(x) = h^*(V^{*,\sigma}(x))$.

<u>Corollary 2.1</u> The matching point $x_1^*(\sigma)$ is well-defined by

(2.3) $V^{*,\sigma}(x_1^*(\sigma)) = v^*$

due to the monotonicity of $V^{*,\sigma}(x)$. Then $x_1^*(\sigma)$ becomes a continuous function for $0 \leqq \sigma \leqq \sigma_1^*$.

Using the singular perturbation method, we obtain

<u>Theorem 2.1</u> (<u>Existence Theorem for the Basic Pattern</u>) For any σ_0 with $0 < \sigma_0 < \sigma_1^*$, there is an $\varepsilon_0 > 0$ such that $(SP)^{\varepsilon,\sigma}$ has an (ε,σ)-family of solutions $D^1(\varepsilon,\sigma) = (u^1(x;\varepsilon,\sigma), v^1(x;\varepsilon,\sigma)) \in C^2_\varepsilon(\bar{I}) \times C^2(\bar{I})$ for $(\varepsilon,\sigma) \in Q^1 = \{(\varepsilon,\sigma) \mid 0 < \varepsilon < \varepsilon_0, 0 \leqq \sigma < \sigma_0\}$. $D^1(\varepsilon,\sigma)$ are uniformly bounded in $C^2_\varepsilon(\bar{I}) \times C^2(\bar{I})$, and satisfy

(2.4) $\lim_{\varepsilon\downarrow 0} u^1(x;\varepsilon,\sigma) = U^{*,\sigma}(x)$ uniformly on $\bar{I} \setminus I_\kappa$ for any $\kappa > 0$

and

(2.5) $\lim_{\varepsilon\downarrow 0} v^1(x;\varepsilon,\sigma) = V^{*,\sigma}(x)$ uniformly on $\bar{I}$,

where $I_\kappa = (x_1^*(\sigma)-\kappa, x_1^*(\sigma)+\kappa)$. Moreover, $D^1(\varepsilon,\sigma)$ depends continuously on $(\varepsilon,\sigma) \in Q^1$ in $C^2_\varepsilon \times C^2$-topology, and continuously on $(\varepsilon,\sigma) \in \bar{Q}^1$ in $L^2 \times C^1$-topology.

Proof. See, [4], [14] and [10].

Remark 2.1 We can introduce the matching point $x_1(\varepsilon,\sigma)$ of $D^1(\varepsilon,\sigma)$ for $\varepsilon > 0$, which tends to $x_1^*(\sigma)$ in Corollary 2.1 as $\varepsilon \downarrow 0$. See Appendix 1 in [19] for the details.

Multi-mode patterns $D^n(\varepsilon,\sigma)$ called the normal SPSn are easily constructed by applying the next proposition to the basic pattern.

Proposition 2.1 (Folding Up Principle) Suppose $W(x;\underline{d})$ is a solution of $(SP)^{\tilde{\varepsilon},\tilde{\sigma}}$ at $\underline{d} = (\tilde{\varepsilon}^2,\tilde{\sigma}^{-1})$, then $R^n(W)(x)$ is a solution of $(SP)^{\varepsilon,\sigma}$ at $\underline{d}/n^2=(\varepsilon^2,\sigma^{-1})$ for $n=1,2,\ldots,$ where $(\varepsilon,\sigma) = (\tilde{\varepsilon}/n,n^2\tilde{\sigma})$. Here

$$(2.6)\qquad R^n(W)(x) = \begin{cases} W(n(x-i/n);\underline{d}) & i=\text{even}, \\ W(n(1/n-(x-i)/n);\underline{d}) & i=\text{odd}, \end{cases}$$

for $i/n \leqq x \leqq (i+1)/n$ $(i=0,1,2..,n-1)$.

Corollary 2.2 (Existence of the Normal SPSn) Let $W = D^1(\tilde{\varepsilon},\tilde{\sigma})$ $((\tilde{\varepsilon},\tilde{\sigma}) \in Q^1)$ in Proposition 2.1, then $D^n(\varepsilon,\sigma)$ defined by $R^n(D^1(\tilde{\varepsilon},\tilde{\sigma}))$ becomes a solution of $(SP)^{\varepsilon,\sigma}$ with n interior transition layers for $(\varepsilon,\sigma) = (\tilde{\varepsilon}/n,n^2\tilde{\sigma}) \in Q^n \equiv \{(\varepsilon,\sigma) \mid 0 < \varepsilon < \varepsilon_0/n,\ 0 \leqq \sigma < n^2\sigma_0\}$. $D^n(\varepsilon,\sigma)$ is called the normal SPSn or simply SPSn. See Fig. 3.

We shall study the stability properties of the normal SPSn in the following sections. Since our model $(P)^{\varepsilon,\sigma}$ is a semilinear parabolic system, the stability is determined by the spectrum of $(LP)^{\varepsilon,\sigma}$. The following two lemmas about the Sturm-Liouville operator $L^{\varepsilon,\sigma}$ are the basic for later discussions: the first one is on the asymptotic behavior of the spectrum of $L^{\varepsilon,\sigma}$, and the second one treats the asymptotic form of eigenfunctions.

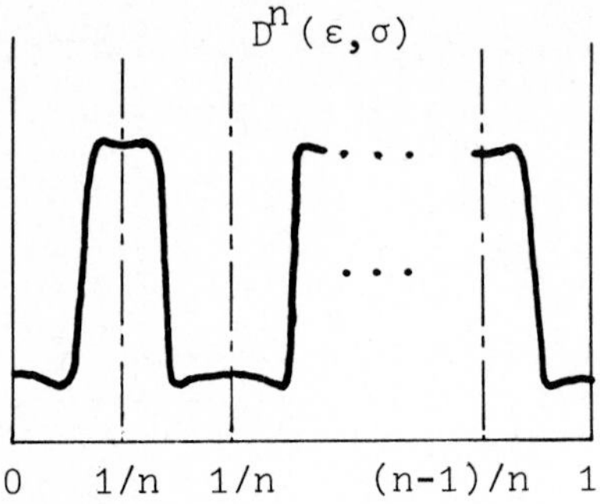

Fig. 3: The normal SPSn, $D^n(\varepsilon,\sigma) = R^n(D^1(\tilde{\varepsilon},\tilde{\sigma}))$.

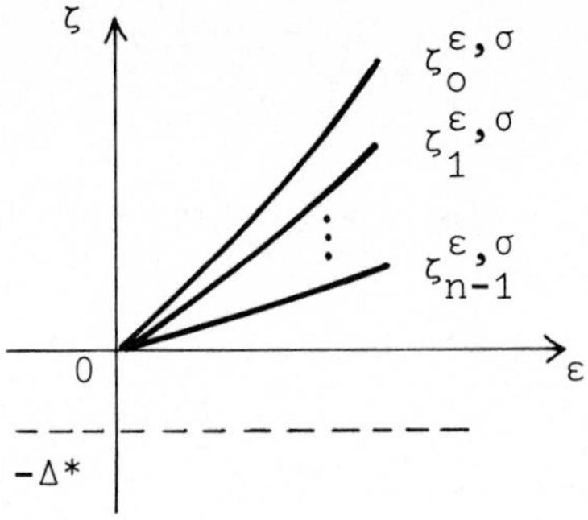

Fig. 4: Behavior of critical eigenvalues of $L^{\varepsilon,\sigma}$.

Lemma 2.2 (Asymptotic Behaviors of Eigenvalues of $L^{\varepsilon,\sigma}$) Let $\{\zeta_i^{\varepsilon,\sigma}, \phi_i^{\varepsilon,\sigma}\}_{i=0}^{\infty}$ be the complete orthonormal set of eigenvalues and eigenfunctions of $L^{\varepsilon,\sigma}$ subject to Neumann boundary conditions

$$L^{\varepsilon,\sigma}\phi = (\varepsilon^2 \frac{d^2}{dx^2} + f_u^{\varepsilon,\sigma})\phi = \zeta\phi,$$

where $f_u^{\varepsilon,\sigma}$ denotes the partial derivative f_u evaluated at the normal SPSn. The first n eigenvalues $\zeta_o^{\varepsilon,\sigma},..,\zeta_{n-1}^{\varepsilon,\sigma}$ ($\zeta_o^{\varepsilon,\sigma} > ... > \zeta_{n-1}^{\varepsilon,\sigma}$) are the critical eigenvalues of $L^{\varepsilon,\sigma}$, which are positive for $\varepsilon > 0$ and satisfy

$$\zeta_i^{\varepsilon,\sigma} = \hat{\zeta}_i(\varepsilon,\sigma)\varepsilon\sigma + \mathrm{Exp}_i(\varepsilon,\sigma), \tag{2.7}$$

where $\hat{\zeta}_i$ $(i=0,..,n-1)$ are positive continuous functions in Q^n uniquely extendable to $\varepsilon = 0$ as

$$\hat{\zeta}^{*,\sigma} \equiv \lim_{\varepsilon \downarrow 0} \hat{\zeta}_i(\varepsilon,\sigma) = \frac{1}{n}\left(\frac{\gamma^*}{c^*_\sigma}\right)^2 J'(v^*)\int_0^{x_1^*(\frac{\sigma}{n^2})} g(U^*,\frac{\sigma}{n^2}, V^*,\frac{\sigma}{n^2})dx > 0, \tag{2.8}$$

where γ^* and c^* are positive constants defined in Lemma 2.3. Here Exp_i $(i=0,..,n-1)$ are continuous functions of $(\varepsilon,\sigma) \in Q^n$ satisfying

$$|\mathrm{Exp}_i(\varepsilon,\sigma)| \leqq C \exp(-\frac{\gamma}{\varepsilon}), \tag{2.9}$$

where C and γ are positive constants independent of $(\varepsilon,\sigma) \in Q^n$ and i. See Fig.4. Note that the asymptotic slope $\hat{\zeta}^{*,\sigma}$ does not depend on i $(0 \leqq i \leqq n-1)$. The remaining eigenvalues $\zeta_i^{\varepsilon,\sigma}$ $(i \geqq n)$ are uniformly bounded away from zero with respect to small

ε, namely, it holds that

$$0 > -\Delta^* > \zeta_n^{\varepsilon,\sigma} > \zeta_{n+1}^{\varepsilon,\sigma} > \cdots \quad \text{for small} \quad \varepsilon,$$

where Δ^* is a positive constant independent of $(\varepsilon,\sigma) \in Q^n$.

Remark 2.2 Lemma 2.2 also holds under homogeneous Dirichlet boundary conditions without essential changes. In particular, the asymptotic limit $\hat{\zeta}^{*,\sigma}$ remains the same.

The characterization of the asymptotic forms of critical eigenfunctions $\phi_i^{\varepsilon,\sigma}$ $(0 \le i \le n-1)$ of $L^{\varepsilon,\sigma}$ plays a key role in the next section. Since $\phi_i^{\varepsilon,\sigma}$ $(\|\phi_i^{\varepsilon,\sigma}\|_{L^2(I)} = 1)$ has sharp peaks and valleys at layer positions, which diverge as $\varepsilon \downarrow 0$, it may be convenient to stretch it on each subinterval $I_j \equiv ((j-1)/n, j/n)$ $(j=1,\ldots,n)$ for the study of asymptotic forms. Let $\hat{\phi}_{i|j}^{\varepsilon,\sigma}(y)$ be the normalized j-th stretched function of $\phi_i^{\varepsilon,\sigma}$ on I_j defined by

$$\hat{\phi}_{i|j}^{\varepsilon,\sigma}(y) \equiv \sqrt{\varepsilon}\,\phi_i^{\varepsilon,\sigma}(x_j(\varepsilon,\sigma) + \varepsilon y) \quad \text{for} \quad y \in \tilde{I}_j, \tag{2.10}$$

where y is a stretched variable defined by $y = (x - x_j(\varepsilon,\sigma))/\varepsilon$, and $\tilde{I}_j$ is the stretched interval of I_j with center $x=x_j(\varepsilon,\sigma)$ (see Remark 2.1). It holds that

$$\int_I |\phi_i^{\varepsilon,\sigma}|^2 dx = \sum_{j=1}^{n} \int_{I_j} |\phi_i^{\varepsilon,\sigma}|^2 dx = \sum_{j=1}^{n} \int_{\tilde{I}_j} |\hat{\phi}_{i|j}^{\varepsilon,\sigma}|^2 dy. \tag{2.11}$$

Let $\tilde{u}^*$ be a unique solution (up to translation) of

$$\frac{d^2}{dy^2}\phi + f(\phi, v^*) = 0 \quad \text{with} \quad \phi(\pm\infty) = h_\pm(v^*).$$

Then, $d\tilde{u}^*/dy$ becomes a principal eigenfunction associated with zero eigenvalue of the limiting Sturm-Liouville operator L^*

$$L^*\phi = \frac{d^2}{dy^2}\phi + f_u(\tilde{u}^*, v^*)\phi = \zeta\phi, \quad \phi \in L^2(\mathbb{R}).$$

Now we have the following chracterization of $\hat{\phi}_{i|j}^{\varepsilon,\sigma}$.

Lemma 2.3 (Asymptotic Forms of Critical Eigenfunctions of $L^{\varepsilon,\sigma}$) For any fixed i $(0 \le i \le n-1)$, let $\{\hat{\phi}_i^{\varepsilon_k,\sigma}\}_{k=1}^{\infty}$ be an arbitrary sequence with $\varepsilon_k \downarrow 0$ as $k \uparrow \infty$. Then, there exists always a convergent subsequence $\{\hat{\phi}_i^{\varepsilon_{k'},\sigma}\}$ with $\varepsilon_{k'} \downarrow 0$ as $k' \uparrow \infty$ such that

$$(2.12)\qquad \lim_{k'\uparrow\infty} \hat{\phi}_{i|j}^{\varepsilon_{k'},\sigma} = \kappa_j^i \hat{w}^* \quad \text{in } C^2_{c.u.}(\mathbb{R})\text{-sense}$$

for $j=1,\ldots,n$, where $\hat{w}^*$ is the L^1-normalized principal eigenfunction of L^* defined by

$$(2.13)\qquad \hat{w}^* = \gamma^* \frac{d}{dy} \tilde{u}^* \quad \text{with } \gamma^* = 1/(h_+(v^*)-h_-(v^*)),$$

and κ_j^i $(j=1,..,n)$ are real constants. Moreover, the resulting vectors $c^*\mathbb{K}^i \equiv c^*(\kappa_1^i,..,\kappa_n^i)$ form a orthonormal set in $\mathbb{R}^n$, i.e.,

$$(2.14)\qquad (c^*)^2(\mathbb{K}^\ell,\mathbb{K}^k) = (c^*)^2 \sum_{j=1}^n \kappa_j^\ell \kappa_j^k = \begin{cases} 1 & \ell = k, \\ 0 & \ell \neq k, \end{cases}$$

for $\ell,k \in (0,\ldots,n-1)$, where the normalized constant c^* is given by

$$(c^*)^2 = \int_{-\infty}^{+\infty} |\hat{w}^*|^2 dy.$$

Remark 2.3 The orthonormal set $\{c^*\mathbb{K}^i\}_{i=1}^n$ in the above lemma may depend on the choice of the subsequence, however this does not affect the following discussions. In fact, the final form of the SLEP system in §3 does not depend on the choice of the subsequence.

§3. SLEP METHOD AND STABILITY OF THE NORMAL SPSn

The most delicate and crucial part of the linearized spectral analysis for SPSn is to find all critical eigenvalues of $(LP)^{\varepsilon,\sigma}$ and clarify their asymptotic behaviors as $\varepsilon \downarrow 0$. The SLEP method gives us a unified view point to treat this problem. The basic idea of it is to find a nice scaling which blows up the degenerate situation of $\mathcal{L}^{\varepsilon,\sigma}$ as $\varepsilon \downarrow 0$. It turns out that the study of asymptotic behaviors of critical eigenvalues is reduced to solving the linear eigenvalue problem of $n \times n$ symmetric matrix called the SLEP system. The aim of this section is to show how the linearized problem $\mathcal{L}^{\varepsilon,\sigma}$ is reduced to the SLEP system with respect to the scaled critical eigenvalues. Without loss of generality, we can restrict the region of λ to Λ_1 defined by

$$\Lambda_1 = \{\lambda \mid \operatorname{Re}\lambda > -\mu_1 > \max\{-\Delta^*, -\mu\}$$

for some fixed $\mu_1 > 0$, where $-\Delta^*$ is a negative constant appeared in Lemma 2.2, and $-\mu \equiv \sup_{x\in I} f_u^{*,\sigma}(x)$ $(f_u^{*,\sigma} \equiv f_u(U^{*,\sigma}, V^{*,\sigma})) < 0$. Also note that the first n eigenvalues $\{\zeta_i^{\varepsilon,\sigma}\}_{i=0}^{n-1}$ of $L^{\varepsilon,\sigma}$ do not belong to the spectrum of $\mathcal{L}^{\varepsilon,\sigma}$ for small ε (see [20]).

Solving the first equation of $\mathcal{L}^{\varepsilon,\sigma}$ with respect to w and substituting it into the second equation after expanding it by using the CONS of $L^{\varepsilon,\sigma}$, we have the equivalent eigenvalue problem containing only z:

$$(3.1)\quad \frac{1}{\sigma}\frac{d^2}{dx^2} z + \sum_{i=0}^{n-1} \frac{\langle -f_v^{\varepsilon,\sigma} z, \phi_i^{\varepsilon,\sigma}\rangle}{\zeta_i^{\varepsilon,\sigma} - \lambda} g_u^{\varepsilon,\sigma}\phi_i^{\varepsilon,\sigma} + g_u^{\varepsilon,\sigma}(L^{\varepsilon,\sigma}-\lambda)^{\dagger}(-f_v^{\varepsilon,\sigma} z)$$

$$+ g_v^{\varepsilon,\sigma} z = \lambda z, \quad \lambda \in \Lambda_1 \qquad (\langle\cdot,\cdot\rangle\text{: inner product in } L^2(I)),$$

where $(L^{\varepsilon,\sigma}-\lambda)^{\dagger}(\cdot) \equiv \sum_{i\geq n} \langle\cdot, \phi_i^{\varepsilon,\sigma}\rangle \phi_i^{\varepsilon,\sigma}/(\zeta_i^{\varepsilon,\sigma}-\lambda)$ which is the regular part of the resolvent $(L^{\varepsilon,\sigma}-\lambda)^{-1}$. In fact we have

<u>Lemma 3.1</u> (<u>Asymptotic Limit of $(L^{\varepsilon,\sigma} - \lambda)^{\dagger}$</u>) $(L^{\varepsilon,\sigma}-\lambda)^{\dagger}$ is a uniform L^2-bounded operator and becomes a multiplication operator in the limt of $\varepsilon \downarrow 0$. Namely,

$$\lim_{\varepsilon\downarrow 0}(L^{\varepsilon,\sigma}-\lambda)^{\dagger}(F^{\varepsilon,\sigma}h) = \frac{F^{*,\sigma}h}{f_u^{*,\sigma} - \lambda}$$

in strong L^2-sense for any bounded L^2-function h, smooth function F(u,v), and $\lambda \in \Lambda_1$, where $F^{\varepsilon,\sigma} = F(D^n(\varepsilon,\sigma))$ and $F^{*,\sigma} = F(D^n(0,\sigma))$ (i.e., evaluated at the reduced solution of $D^n(\varepsilon,\sigma)$). Moreover, if h belongs to $H^1(I)$, the above convergence is uniform on a bounded set in $H^1(I)$.

It is convenient to classify the spectrum of (3.1) into two classes; the <u>critical</u> eigenvalues which tend to zero as $\varepsilon \downarrow 0$, and <u>noncritical</u> eigenvalues which are bounded away from zero for small ε. Making use of Lemma 3.1, we can show that <u>noncritical</u> eigenvalues are <u>not</u> dangerous to the stability of $D^n(\varepsilon,\sigma)$, namely, they have strictly negative real parts independent of ε. The proof of it can be done in the same spirit of that of Proposition 2.1 in [19]. Now we can concentrate on the behavior of critical eigenvalues. Let $\lambda=\lambda(\varepsilon)$ be an arbitrary critical eigenvalue of (3.1), and assume that λ varies in the ball $B_\delta = \{\lambda \mid |\lambda| < \delta\}$ for some $\delta > 0$. In view of (2.10), (2.12), and $z \in H^1_N(I)$, we see that both the denominator and numerator of the second term of (3.1) tend to zero as $\varepsilon \downarrow 0$. Here the scaling technique comes up to convert (3.1) into more tractable and non-degenerate form. Before going into the details, we first rewrite (3.1) in the form of a <u>finite dimensional eigenvalue problem</u> by using the following operator $K^{\varepsilon,\sigma,\lambda}$.

<u>Lemma 3.2</u> (<u>Operator $K^{\varepsilon,\sigma,\lambda}$</u>) Let $\hat{B}^{\varepsilon,\sigma,\lambda}$ be a bilinear form defined by

$$\hat{B}^{\varepsilon,\sigma,\lambda}(z^1,z^2) = \frac{1}{\sigma}\langle z^1_x, z^2_x\rangle - \langle \{g_u^{\varepsilon,\sigma}(L^{\varepsilon,\sigma}-\lambda)^\dagger(-f_v^{\varepsilon,\sigma}\,\cdot) + g_v^{\varepsilon,\sigma} - \lambda\} z^1, z^2\rangle \quad \text{for } z^i \in H^1_N(I)\ (i=1,2).$$

Then, for a given $h \in H^{-1}(I)$, the equation for $z \in H^1_N(I)$

$$\hat{B}^{\varepsilon,\sigma,\lambda}(z,\psi) = \langle h,\psi\rangle \quad \text{for any } \psi \in H^1_N(I)$$

has a unique solution for small ε (including $\varepsilon=0$) and $\lambda \in B_\delta$. Define the mapping $K^{\varepsilon,\sigma,\lambda}$ by

$$K^{\varepsilon,\sigma,\lambda} h = z;\ H^{-1}(I) \to H^1_N(I).$$

$K^{\varepsilon,\sigma,\lambda}$ is a bounded operator from $H^{-1}(I)$ to $H^1_N(I)$, and depends continuously on (ε,σ) and analytically on λ in operator norm sense, respectively.

Applying $K^{\varepsilon,\sigma,\lambda}$ to (3.1), we have

$$(3.2) \qquad z = \sum_{i=0}^{n-1} \frac{\langle -f_v^{\varepsilon,\sigma} z, \phi_i^{\varepsilon,\sigma}\rangle}{\zeta_i^{\varepsilon,\sigma} - \lambda} K^{\varepsilon,\sigma,\lambda}(g_u^{\varepsilon,\sigma}\phi_i^{\varepsilon,\sigma}).$$

This shows that z is a linear combination of $K^{\varepsilon,\sigma,\lambda}(g_u^{\varepsilon,\sigma}\phi_i^{\varepsilon,\sigma})$ $(i=0,..,n-1)$ yielding

$$(3.3) \qquad z = \sum_{i=0}^{n-1} \alpha_i K^{\varepsilon,\sigma,\lambda}(g_u^{\varepsilon,\sigma}\phi_i^{\varepsilon,\sigma}),$$

where $\boldsymbol{\alpha} = (\alpha_0,\ldots,\alpha_{n-1})$ is a real vector. Note that $K^{\varepsilon,\sigma,\lambda}$

$(g_u^{\varepsilon,\sigma}\phi_i^{\varepsilon,\sigma})$ $(i=0,\ldots,n-1)$ are linearly independent. Substituting (3.3) into (3.2), we obtain a <u>n-dimensional</u> matrix eigenvalue problem:

$$M^{\varepsilon,\sigma}\begin{pmatrix}\alpha_0\\ \vdots\\ \alpha_{n-1}\end{pmatrix}=\begin{pmatrix}(\zeta_0^{\varepsilon,\sigma}-\lambda)\alpha_0\\ \vdots\\ (\zeta_{n-1}^{\varepsilon,\sigma}-\lambda)\alpha_{n-1}\end{pmatrix}, \tag{3.4}$$

where $M^{\varepsilon,\sigma}=\{<-f_v^{\varepsilon,\sigma}\phi_i^{\varepsilon,\sigma},\ K^{\varepsilon,\sigma,\lambda}(g_u^{\varepsilon,\sigma}\phi_j^{\varepsilon,\sigma})>\}_{i,j=0}^{n-1}$. This problem is highly degenerated, since all the elements of $M^{\varepsilon,\sigma}$ and $\zeta_i^{\varepsilon,\sigma}-\lambda$ $(i=0,\ldots,n-1)$ tend to zero as $\varepsilon\downarrow 0$. The following chracterization of the asymptotic form of $\phi_i^{\varepsilon,\sigma}$ by $\sqrt{\varepsilon}$-scaling plays a key role to unfold this degenerate problem.

<u>Lemma 3.3</u> (<u>Asymptotic Form of</u> $\phi_i^{\varepsilon,\sigma}/\sqrt{\varepsilon}$) Let $\{\phi_i^{\varepsilon_m,\sigma}\}_{m=1}^{\infty}$ be an arbitrary convergent sequence in the sense of Lemma 2.3 in each stretched subinterval $\tilde{I}_j$ $(j=1,\ldots,n)$ for $i\in(0,\ldots,n-1)$. Then it holds that

$$\lim_{m\uparrow\infty}-f_v^{\varepsilon_m,\sigma}\frac{\phi_i^{\varepsilon_m,\sigma}}{\sqrt{\varepsilon}}=c_1^*\,\Delta_i\equiv c_1^*\sum_{j=1}^{n}\kappa_j^i\delta(x-x_j^*(\sigma)), \tag{3.5$_a$}$$

$$\lim_{m\uparrow\infty}g_u^{\varepsilon_m,\sigma}\frac{\phi_i^{\varepsilon_m,\sigma}}{\sqrt{\varepsilon}}=c_2^*\,\Delta_i\equiv c_2^*\sum_{j=1}^{n}\kappa_j^i\delta(x-x_j^*(\sigma)), \tag{3.5$_b$}$$

both in $H^{-1}(I)$-sense, where $c_1^*\equiv-\gamma^*J(v^*)$, $c_2^*\equiv\gamma^*\{g(h_+(v^*),v^*)-g(h_-(v^*),v^*)\}$, and $\delta(x-x_j^*(\sigma))$ denotes the Dirac's δ-function at $x=x_j^*(\sigma)$. The vector $\mathbb{K}^i=(\kappa_1^i,\ldots,\kappa_n^i)$ satisfies the orthogonal relation (2.14).

Hereafter, we fix a convergent subsequence $\{\phi_i^{\varepsilon_m,\sigma}/\sqrt{\varepsilon}\}_{m=1}^{\infty}$, and for simplicity of notation, we simply write ε instead of ε_m keeping in mind that ε actually means a discrete parameter ε_m. In view of Lemma 2.2 and Lemma 3.3, we see that ε-scaling is the most suitable to blow up (3.4). In fact, dividing (3.4) by ε on both sides, we have

$$\tilde{M}^{\varepsilon,\sigma}\begin{pmatrix}\alpha_0\\ \vdots\\ \alpha_{n-1}\end{pmatrix}=\begin{pmatrix}(\zeta_0^{\varepsilon,\sigma}/\varepsilon-\lambda/\varepsilon)\alpha_0\\ \vdots\\ (\zeta_{n-1}^{\varepsilon,\sigma}/\varepsilon-\lambda/\varepsilon)\alpha_{n-1}\end{pmatrix}, \tag{3.6}$$

where $\tilde{M}^{\varepsilon,\sigma}=\{<-f_v^{\varepsilon,\sigma}\phi_i^{\varepsilon,\sigma}/\sqrt{\varepsilon},\ K^{\varepsilon,\sigma,\lambda}(g_u^{\varepsilon,\sigma}\phi_j^{\varepsilon,\sigma}/\sqrt{\varepsilon})>\}_{i,j=0}^{n-1}$.

The problem (3.6) is nondegenerate and well-defined continuously up to $\varepsilon = 0$. In fact, in the limit of $\varepsilon \downarrow 0$, (3.6) becomes

$$(3.7) \qquad \{\tilde{M}^{*,\sigma} + (\tau^{*,\sigma} - \sigma\hat{\zeta}^{*,\sigma})I\} = 0,$$

where $\tilde{M}^{*,\sigma} \equiv \lim_{\varepsilon\downarrow 0} M^{*,\sigma} = \{c_1^* c_2^* \langle\Delta_i, K^{*,\sigma}\Delta_j\rangle\}_{i,j=0}^{n-1}$, $K^{*,\sigma} \equiv K^{0,\sigma,0}$, $\tau^{*,\sigma} \equiv \lim_{\varepsilon\downarrow 0} \lambda(\varepsilon)/\varepsilon$, and $\sigma\hat{\zeta}^{*,\sigma} \equiv \lim_{\varepsilon\downarrow 0} \zeta_i^{\varepsilon,\sigma}/\varepsilon$ $(i=0,\dots,n-1)$. Here we use the fact that the critical eigenvalue $\lambda = \lambda(\varepsilon)$ can be written in the form of

$$(3.8) \qquad \lambda = \varepsilon\tau(\varepsilon,\sigma),$$

where τ is a bounded continuous function up to $\varepsilon = 0$ (see [20] and [19]). The limiting problem (3.7) is called <u>the SLEP system</u> of (3.1) with respect to the scaled eigenvalues $\tau^{*,\sigma}$. From now on, we only focus on the SLEP system (3.7), since the information on the asymptotic behaviors of critical eigenvalues for $\varepsilon > 0$ can be derived from (3.7) by using a regular perturbation (see [20] for the details). In view of the definition of $\tilde{M}^{*,\sigma}$, the problem (3.7) does not look free from the choice of the subsequence. However, by applying an appropriate change of basis to $M^{*,\sigma}$, we can obtain a normal form which is independent of the subsequence. For this purpose, we introduce the Green function $G_N = G_N(x,y;\sigma)$ associated with the operator $K^{*,\sigma}$, namely, G_N is defined by

$$(3.9) \qquad K^{*,\sigma}\phi = \langle G_N(x,y;\sigma),\phi\rangle, \quad \text{for any } \phi \in H^{-1}(I).$$

More explicitly,

$$(3.10)_a \qquad G_N(x,y;\sigma) = -\frac{\sigma}{W(h,k)} \times \begin{cases} h(x)k(y), & 0 \leqq x \leqq y \leqq 1, \\ h(y)k(x), & 0 \leqq y \leqq x \leqq 1, \end{cases}$$

where h and k satisfy the equation

$$(3.10)_b \qquad \left(-\frac{1}{\sigma}\frac{d^2}{dx^2} - \det{}^{*,\sigma}/f_u^{*,\sigma}\right)\phi = 0, \quad \phi \in H^2(I)$$

with the boundary conditions

$$(3.10)_c \qquad h(0)=1,\ h'(0)=0 \quad \text{and} \quad k(1)=1,\ k'(1)=0,$$

where $\det{}^{*,\sigma} \equiv f_u^{*,\sigma}g_v^{*,\sigma} - f_v^{*,\sigma}g_u^{*,\sigma} > 0$ (see (A.4)) and $W(h,k)$

denotes the Wronskian of h and k. Note that $h(k)$ is strictly positive and increasing (decreasing), respectively, since $-\det^{*,\sigma}/f_u^{*,\sigma}$ is strictly positive from (A.3) and (A.4). It follows from (3.9) that

$$<\delta(x-x_i^*(\sigma)),K^{*,\sigma}\delta(x-x_j^*(\sigma))> = G_N(x_i^*(\sigma),x_j^*(\sigma);\sigma).$$

Therefore, recalling that $\Delta_i = \sum_{j=1}^{n} \kappa_j^i \delta(x-x_j^*(\sigma))$, we have

$$<\Delta_i,K^{*,\sigma}\Delta_j> = {}^t\mathbb{k}^i \mathbb{G}_N \mathbb{k}^j, \tag{3.11}$$

where $\mathbb{G}_n$ is an $n\times n$ symmetric matrix with positive components defined by

$$\mathbb{G}_N \equiv \{G_N(x_i^*(\sigma),x_j^*(\sigma);\sigma)\}_{i,j=1}^{n} \tag{3.12}$$

Let us define the orthogonal matrix P by

$$P = c^*(\mathbb{k}^0,\mathbb{k}^1,\ldots,\mathbb{k}^{n-1}). \tag{3.13}$$

Then, the matrix $\tilde{M}^{*,\sigma}$ in (3.7) can be rewritten as

$$\begin{aligned}\tilde{M}^{*,\sigma} &= \{c_1^*c_2^*\ {}^t\mathbb{k}^i \mathbb{G}_N \mathbb{k}^j\}_{i,j=0}^{n-1} \\ &= \frac{c_1^*c_2^*}{(c^*)^2}\ {}^tP\mathbb{G}_N P \\ &= {}^tP\tilde{\mathbb{G}}_N P,\end{aligned} \tag{3.14}$$

where

$$\tilde{\mathbb{G}}_N \equiv \frac{c_1^*c_2^*}{(c^*)^2}\mathbb{G}_N. \quad \text{(\underline{SLEP matrix})} \tag{3.15}$$

This shows that $\tilde{M}^{*,\sigma}$ is just an orthogonal transformation of $\tilde{\mathbb{G}}_N$ which clearly does not depend on the choice of the subsequence. We reach the final form by multiplying P to (3.7):

$$\{\tilde{\mathbb{G}}_N + (\tau^{*,\sigma} - \sigma\hat{\zeta}^{*,\sigma})I\}\mathbb{\beta} = 0, \tag{3.16}$$

where $\mathbb{\beta} \equiv P\mathbb{\alpha}$. This is, what we call, <u>the normal SLEP system</u> of (3.1). Since $\tilde{\mathbb{G}}_N$ is a real symmetric matrix, it has n real eigenvalues counting their multiplicity. Especially, we are interested in the sign of the minimum eigenvalue of $\tilde{\mathbb{G}}_N$, since it determines the maximum value of the scaled critical eigenvalues $\tau^{*,\sigma}$. If the maximum value of $\tau^{*,\sigma}$ is negative (resp. positive), the normal SPSn becomes asymptotically stable (resp. unstable). We have the following result for the eigenvalue problem (3.16), the proof of which is delegated to the next section.

Theorem 3.1 (Eigenvalues of the SLEP System) Under the assumptions (A.1)-(A.5), the set of eigenvalues of $\tilde{\mathbb{G}}_N$

$$\tilde{\mathbb{G}}_N \theta = \gamma\theta \tag{3.17}$$

consists of n real distinct positive eigenvalues

$$0 < \gamma_{n-1} < \gamma_{n-2} < \cdots < \gamma_o. \tag{3.18}$$

Namely, in terms of $\tau^{*,\sigma}$, (3.16) has n distinct real eigenvalues

$$\tau_o^{*,\sigma} < \tau_1^{*,\sigma} < \cdots < \tau_{n-1}^{*,\sigma}, \tag{3.19}$$

where $\tau_i^{*,\sigma} = \sigma\hat{\zeta}^{*,\sigma} - \gamma_i$ $(0 \leq i \leq n-1)$. Moreover, it holds that

$$\tau_{n-1}^{*,\sigma} < 0. \tag{3.20}$$

Theorem 3.1 leads to the Main Theorem in §1 through a regular perturbation.

§4. EIGENVALUE PROBLEM FOR THE SLEP MATRIX

In the previous section, it was shown that the asymptotic behaviors of critical eigenvalues are completely reduced to solving the normal SLEP system. In this section, we shall show the outline of the proof of Theorem 3.1. Our strategy to solve (3.16) consists of two parts: Firstly, we switch the eigenvalue problem (3.17) to that of the inverse of $\tilde{\mathbb{G}}_N$;

$$\tilde{\mathbb{G}}_N^{-1}\theta = \gamma^{-1}\theta. \tag{4.1}$$

Here, we remark that, recalling the definition of $\tilde{\mathbb{G}}_N$, it is not difficult to show that $\tilde{\mathbb{G}}_N$ is invertible. It turns out that $\tilde{\mathbb{G}}_N^{-1}$ becomes a tri-diagonal symmetric matrix, and therefore, we can use the general result (see Lemma 4.2) to conclude Theorem

3.1 except the final result (3.20). In order to show (3.20), we need the following observation - the second point of our approach - : We introduce the matrix $\tilde{\mathbb{G}}_D$ which is similarly defined as $\tilde{\mathbb{G}}_N$ with replacing the boundary conditions in (3.10) by Dirichlet ones, namely,

$$(4.2) \qquad \tilde{\mathbb{G}}_D \equiv \frac{c_1^* c_2^*}{(c^*)^2}\, \mathbb{G}_D \equiv \frac{c_1^* c_2^*}{(c^*)^2}\{G_D(x_i^*(\sigma), x_j^*(\sigma);\sigma)\}_{i,j=1}^{n},$$

where $G_D(x,y;\sigma)$ is the Green function of the operator $(3.10)_b$ under homogeneous Dirichlet boundary conditions. Then, the minimum eigenvalue of $\tilde{\mathbb{G}}_D$ is equal to $\sigma\hat{\zeta}^{*,\sigma}$ (i.e., $\tau^{*,\sigma}=0$), since, noting that our system $(SP)^{\varepsilon,\sigma}$ is autonomous, we see that x-derivative of the normal SPSn is always an eigenfunction of $(LP)^{\varepsilon,\sigma}$ under Dirichlet boundary conditions associated with the zero eigenvalue. See Lemma 4.3. Finally, a comparison of components of $\tilde{\mathbb{G}}_N^{-1}$ and $\tilde{\mathbb{G}}_D^{-1}$ (Lemma 4.4) leads to the conclusion (3.20).

A computation of the inverse of $\tilde{\mathbb{G}}_N$ (resp. $\tilde{\mathbb{G}}_D$) implies that

Lemma 4.1 (Inverse of the SLEP Matrix) $\tilde{\mathbb{G}}_N^{-1}$ (resp. $\tilde{\mathbb{G}}_D^{-1}$) is a tri-diagonal real symmetric matrix such that

(i) All diagonal elements are equal except (1,1) and (n,n) components.

(ii) Every other off-diagonal elements is equal.

(iii) All diagonal (resp. off-) diagonal elements are positive (resp. negative).

The following result is basic for the study of the eigenvalue problem of tri-diagonal matrices. For the proof, see, for example, [22].

Lemma 4.2 (Eigenvalues and Eigenfunctions of Tri-diagonal Symmetric Matrix) Let C be a symmetric tri-diagonal matrix with non-zero off-diagonal elements of the form

$$C = \begin{pmatrix} \alpha_1 & \beta_2 & & & \\ \beta_2 & \alpha_2 & \ddots & & 0 \\ & \ddots & \ddots & \ddots & \\ & & \ddots & \ddots & \beta_n \\ 0 & & & \beta_{n-1} & \alpha_n \end{pmatrix} \qquad \beta_i \neq 0.$$

Then, C has n real, distinct, and simple eigenvalues $\lambda_1 < \lambda_2 < \cdots < \lambda_n$. The corresponding eigenvector $\kappa^k = {}^t(x_1^k,\ldots,x_n^k)$ to λ_k is explicitly written by

$$(4.3)\qquad x_1^k = 1,\quad x_r^k = (-1)^{r-1} p_{r-1}(\lambda_k)/\beta_2\beta_3\cdots\beta_r \quad (2 \leq r \leq n),$$

where $p_r(\lambda)$ denotes the leading principal minor of order r of $(C-\lambda I)$ with $p_o(\lambda) \equiv 1$. Finally, the polynomials $p_o(\lambda)$, $p_1(\lambda),\ldots,p_n(\lambda)$ satisfy the Sturm sequence property. Namely, let the quantities $p_o(\mu), p_1(\mu),\ldots, p_n(\mu)$ be evaluated for some value of μ. Then $s(\mu)$, the number of agreements in sign of consecutive members of this sequence, is the number of eigenvalues of C which are strictly greater than μ.

Since all the elements of $\tilde{G}_N$ are positive, it follows from the Perron-Frobenius Theorem that the greatest eigenvalue γ_{max} of $\tilde{G}_N^{-1}$ is real, simple, and positive with a positive eigenvector. In view of (4.3), Lemma 4.1 (iii), and the Sturm sequence property of $p_r(\lambda)$, we see that $\gamma_{max}^{-1} > 0$ is the minimum eigenvalue of $\tilde{G}_N^{-1}$, which shows the inequality $0 < \gamma_{n-1}$ of (3.18). The remaining part of Theorem 3.1 except (3.20) is a direct consequence of Lemma 4.2.

In order to show (3.20), we prepare the following two lemmas. Firstly, recalling the second part of our strategy mentioned before Lemma 4.1, and using Lemma 4.2, we can show

<u>Lemma 4.3</u> The minimum eigenvalue of $\tilde{G}_D$ is equal to $\sigma\hat{\zeta}^{*,\sigma}$. Namely, in terms of $\tau^{*,\sigma}$, the greatest eigenvalue of the problem (3.16) with replacing $\tilde{G}_N$ by $\tilde{G}_D$ is equal to zero.

In view of this lemma, we see that, in order to prove (3.20), it suffices to show that

$$(4.4)\qquad \text{Minimum eigenvalue of } \tilde{G}_N > \text{Minimum eigenvalue of } \tilde{G}_D,$$

or, equivalently,

$$(4.5)\qquad \text{Maximum eigenvalue of } \tilde{G}_N^{-1} < \text{Maximum eigenvalue of } \tilde{G}_D^{-1}.$$

For this purpose, we compare the elements of $\tilde{G}_N^{-1}$ and $\tilde{G}_D^{-1}$, and we have

<u>Lemma 4.4</u> (<u>Comparison of Elements of</u> $\tilde{\mathbb{G}}_N^{-1}$ <u>and</u> $\tilde{\mathbb{G}}_D^{-1}$) It holds that

(i) All the components of $\tilde{\mathbb{G}}_N^{-1}$ and $\tilde{\mathbb{G}}_D^{-1}$ are equal except (1,1) and (n,n) components.

(ii) For the (1,1) and (n,n) components, it holds that

$$(\tilde{\mathbb{G}}_N^{-1})_{11} < (\tilde{\mathbb{G}}_D^{-1})_{11}$$

$$(\tilde{\mathbb{G}}_N^{-1})_{nn} < (\tilde{\mathbb{G}}_D^{-1})_{nn}.$$

Namely, $\tilde{\mathbb{G}}_D^{-1}$'s components are strictly larger than those of $\tilde{\mathbb{G}}_N^{-1}$.

This lemma implies (4.5), which completes the proof of Theorem 3.1.

REFERENCES

[1] Carpenter, G.A.: A geometric approach to singular perturbation problem with application to nerve impulse equations, J. Diff. Eqs., 23 (1977) 335-367.

[2] Conway, E.D.: Diffusion and predator-prey interaction: pattern in closed systems, Research Notes in Math., 101, Pitman (1984) 85-133.

[3] Ermentrout, G.B., Hastings, S.P., and Troy, W.C.: Large amplitude stationary waves in an excitable lateral-inhibition medium, SIAM J. Appl. Math., 44 (1984) 1133-1149.

[4] Fife, P.C.: Boundary and interior transition layer phenomena for pairs of second-order differential equations, J. Math. Anal. Appl., 54 (1976) 497-521.

[5] Fujii, H., Mimura, M., and Nishiura, Y.: A picture of the global bifurcation diagram in ecological interacting and diffusing systems, Physica D, 5 (1982) 1-42.

[6] Fujii, H. and Nishiura, Y.: Global bifurcation diagram in nonlinear diffusion systems, in Nonlinear Partial Differential Equations in Applied Science, U.S.-Japan Seminar Tokyo 1982, Math. Studies 81 (1984) North-Holland.

[7] Gierer, A. and Meinhardt, H.: A theory of biological pattern formation, Kybernetik 12 (1972) 30-39.

[8] Gollub, J.P., McCarriar, A.R., and Steinman, J.F.: Convective pattern evolution and secondary instabilities, J. Fluid Mech., 125 (1982) 259-281.

[9] Hosono, Y. and Mimura, M.: Singular perturbation approach to travelling waves in competing and diffusing species models, J. Math.Kyoto Univ. 22 (1982) 435-461.

[10] Ito, M.: A remark on singular perturbation methods, Hiroshima Math. J., 14 (1985) 619-629.

[11] Ito, M. and Nishiura, Y.: The stability of interior transition layer solutions for reaction-diffusion systems, in preparation.

[12] Karfunkel, H.R. and Seelig, F.F.: Excitable chemical reaction systems, I. Definition of excitability and simulation of model systems, J. Math. Biol., 2 (1975) 123-132.

[13] Keener, J.P.: Waves in excitable media, SIAM J. Appl. Math., 39 (1980) 528-548.

[14] Mimura, M., Tabata, M., and Hosono, Y.: Multiple solutions of two-point boundary value problems of Neumann type with small parameter, SIAM J. Math. Anal., 11 (1980) 613-631.

[15] Nishiura, Y.: Every multi-mode singularly perturbed solution recovers its stability - from a global bifurcation view point, Lec. Notes in Biomath., 55 (1984) 292-301.

[16] Nishiura, Y.: Stability analysis of travelling front solutions of reaction-diffusion systems - an application of the SLEP method, to appear in the Proceedings of the IV-th International Conference on Boundary and Interior Layers: Computational and Asymptotic Methods, July 7-11, 1986, Novosibirsk, USSR.

[17] Nishiura, Y.: Stability of singularly perturbed waves in an excitable lateral-inhibition medium, in preparation.

[18] Nishiura, Y. and Fujii, H.: Stability theorem for singularly perturbed solutions to systems of reaction-diffusion equations, Proc. Japan Acad., 61, Ser A (1985) 329-332.

[19] Nishiura, Y. and Fujii, H.: Stability of singularly perturbed solutions to systems of reaction-diffusion equations, submitted for publication.

[20] Nishiura, Y. and Fujii, H.: Stability of singularly perturbed solutions with multiple internal transition layers in reaction-diffusion systems, in preparation.

[21] Rinzel, J. and Terman D.: Propagation phenomena in a bistable reaction-diffusion system, SIAM J. Appl. Math., 42 (1982) 1111-1137.

[22] Wilkinson, J.H.: The algebraic eigenvalue problem, Oxford, Clarendon Press, 1965.

ITERATED NONLINEAR MAPS AND HILBERT'S PROJECTIVE METRIC: A SUMMARY

Roger D. Nussbaum
Mathematics Department
Rutgers University
New Brunswick, New Jersey 08903/USA

Introduction:

By a cone K (with vertex at o) in a Banach space X I shall mean a closed, convex subset of X such that if $x \in K$ then $tx \in K$ for all $t > o$ and if $x \in K - \{o\}$, then $-x \notin K$. Examples of cones are provided by $K = \{x \in \mathbb{R}^n: x_i \geq o \text{ for } 1 \leq i \leq n\}$, which will be called the standard cone in $\mathbb{R}^n$, and by the set of nonnegative definite, self-adjoint bounded linear operators on Hilbert space. Every cone induces a partial ordering on X by $x \leq y$ if and only if $y - x \in K$.

If K is a cone in a Banach space X and the interior of K, $\mathring{K}$, is nonempty suppose that $f: \mathring{K} \to \mathring{K}$ is a continuous map. The map f is called "order-preserving" if for all $x, y \in \mathring{K}$ such that $x \leq y$ one has $f(x) \leq f(y)$; f is called "homogeneous of degree 1" if $f(tx) = tf(x)$ for all $x \in \mathring{K}$ and $t > o$. For simplicity assume for the moment that $f: \mathring{K} \to \mathring{K}$ is order-preserving and homogeneous of degree 1. Although I shall eventually state theorems for more general classes of maps, many of the basic difficulties are already apparent if f is order preserving and homogeneous of degree 1.

I am interested here in the following questions:

(1) Does f have an eigenvector in $\mathring{K}$, ie, does there exist $x \in \mathring{K}$ such that $f(x) = \lambda x$? Notice that if X is finite dimensional and f extends continuously to K, then a simple application of the Brouwer fixed point theorem implies that $f(y) = \lambda y$ for some $y \in K$, $||y|| = 1$. However, one does not know whether $y \in \mathring{K}$. In fact, for the maps f which will be of interest in this paper, the existence of an eigenvector in $\mathring{K}$ is frequently a delicate question.

(2) Assuming that f has an eigenvector $u \in \mathring{K}$, is this eigenvector unique (to within scalar multiples)? Notice that f may well have other

NATO ASI Series, Vol. F37
Dynamics of Infinite Dimensional Systems
Edited by S.-N. Chow, and J. K. Hale

eigenvectors in ∂K; uniqueness refers only to eigenvectors in $\overset{\circ}{K}$. The Gauss-Lagrange arithmetic-geometric mean map $f(x,y) = (\frac{x+y}{2}, \sqrt{xy})$ gives an example of a map of the standard cone in $\mathbb{R}^2$ into itself which has the unique eigenvector (1,1) in the interior but also has the eigenvector (1,0).

(3) Assuming that f has a unique normalized eigenvector u, if one defines $g(x) = \frac{f(x)}{\|f(x)\|}$, is it true that $\lim_{k\to\infty} g^k(x) = u$ for all $x \in \overset{\circ}{K}$?

(4) If, for a given $\lambda > o$, one defines $S = \{x \in \overset{\circ}{K} : f(x) = \lambda x\}$ and one assumes that S is nonempty, what can one say about the structure of S? Does it have the same homotopy type as $\overset{\circ}{K}$? If f has no eigenvectors in $\overset{\circ}{K}$, what can one say about the behavior of $g^k(x)$ as $k\to\infty$, where $g(x) = \frac{f(x)}{\|f(x)\|}$?

(5) If f has a fixed point $u \in \overset{\circ}{K}$, when is it true that for every $x \in \overset{\circ}{K}$ there exists a positive scalar $\lambda(x) > o$ such that $\lim_{k\to\infty} f^k(x) = \lambda(x)u$?

As far as I know, it has not been widely recognized that there are broad classes of nonlinear maps for which question 1 is interesting and difficult. However, there is a large literature in which questions 1-3 have been studied for linear and nonlinear maps. See, for example, (1),(3),(4),(8),(9),(20),(22),(34) and (43) for linear maps and (6),(8),(10),(21),(22),(29),(36) and (42) for nonlinear maps. Bushell's expository article (8) provides an excellent introduction. Unfortunately, most of the previous nonlinear theory provides no information in examples of interest here.

It may be worthwhile to mention a particular class of maps for which questions 1-5 are of interest. Let K be the standard cone in $\mathbb{R}^n$ and for r a real number, σ a probability vector (ie, $\sigma \in K$ and $\sum_{j=1}^{n} \sigma_i = 1$) and $x \in \overset{\circ}{K}$ define

$$M_{r\sigma}(x) = \left(\sum_{j=1}^{n} \sigma_j x_j^r\right)^{\left(\frac{1}{r}\right)}$$

If $r=o$, one defines

$$M_{o\sigma}(x) = \prod_{j=1}^{n} x_j^{\sigma_j} = \lim_{r\to o^+} M_{r\sigma}(x)$$

For each i, $1 \le i \le n$, let Γ_i be a finite collection of ordered pairs (r,σ) (r and σ as above); and for each $(r,\sigma) \in \Gamma_i$, suppose that $c_{ir\sigma}$ is a given positive real. Define $f_i:\overset{\circ}{K}\to R$ by

$$f_i(x)=\sum_{(r,\sigma)\in\Gamma_i} c_{ir\sigma}M_{r\sigma}(x)$$

and define $f:\overset{\circ}{K}\to\overset{\circ}{K}$ to be the map whose $i^{\underline{th}}$ component is $f_i(x)$. It is observed in (32) that f extends continuously to K. If $f:\overset{\circ}{K}\to\overset{\circ}{K}$ can be written in the above form, write $f\in M$. If $f\in M$ and $r\geq o$ for all $(r,\sigma)\in\Gamma_i, 1\leq i\leq n$, write $f\in M_+$; similarly, write $f\in M_-$ if $f\in M$ and $r<o$ for all $(r,\sigma)\in\Gamma_i$, $1\leq i\leq n$. Since $(x^{-1})^{-1}=x$, linear maps lie in $M_+\cap M_-$. Questions 1-5 are already nontrivial if $f\in M$. In particular, a sharp answer to question 1 can be given if $f\in M_+$; but, in general, if $f\in M_-$, it may be a subtle question to determine whether f has an eigenvector in $\overset{\circ}{K}$.

My original motivation for studying questions 1-3 came from a question in population biology posed by D. Weeks and H. Caswell (14) and solved in Section 4 of (32). If $x\in\overset{\circ}{K}$ (K the standard cone in $\mathbb{R}^n$) is considered as a "population vector", with x_i being the number of members of the population in class i, question 3 has a natural biological interpretation, and the necessity of a nonlinear map f has long been recognized (14).

However, mathematically identical questions arise in many other contexts. As one example, consider the question of "iterated means" for which the model problem is the famous Gauss-Lagrange arithmetic-geometric mean or AGM. If $f(x,y)$ is defined as in question 2 and $x>o, y>o$, Gauss and Lagrange proved $\lim_{k\to\infty} f^k(x,y)=(\lambda,\lambda)$, where $\lambda>o$ depends on x and y and λ can be explicitly written as an elliptic integral (see (11)). There have been many generalizations of the AGM: see (11),(13), (15),(19) and (26). In all cases one has the problem of proving convergence of iterates of a map f; all of these problems are subsumed by a general answer to question 5 which will be stated below.

A final motivation for studying questions 1-5 comes from so-called DAD theorems. If A is a nonnegative, nxn matrix, one can ask whether there exist positive diagonal matrices D_1 and D_2 such that D_1AD_2 is doubly stochastic, whether D_1 and D_2 are unique to within scalar multiples and how one can approximate D_1 and D_2. One can ask exactly analogous questions when $a(s,t)$ is a nonnegative, continuous kernel on $[0,1]x[0,1]$, so one seeks positive, continuous functions $y(s)$ and $x(t)$ such that if $b(s,t)=y(s)a(s,t)x(t)$, then

$$\int_0^1 b(s,\rho)d\rho=1=\int_0^1 b(\rho,t)d\rho \quad \text{for all } s,t\in[0,1]$$

These questions are the subject of an extensive literature: (6), (21),(25),(27),(28),(29),(34),(38),(39) and (40). It turns out that, to take the matrix case as an example, these questions are equivalent to studying the function $f=J\circ A^*\circ J\circ A$ (where $Ju=(u_1^{-1},u_2^{-1},\text{--},u_n^{-1})$) on $\overset{\circ}{K}$ and K is the standard cone in $\mathbb{R}^n$. Further, the answer to question 5 provides some new insight into DAD theorems.

The purpose of this note is to summarize without proof some of the theorems which are proved in (33) and which provide reasonably satisfactory answers to questions 2-5. As already remarked, the answer to question 1 may be quite delicate even for $f\in M_-$, and for reasons of length, not much will be said about question 1. Similarly, for reasons of length it will not be possible to say very much about applications. In particular I shall not discuss applications to DAD theorems, although such applications are one motivation of this paper.

1. Uniqueness of eigenvectors and convergence of iterates.

Let K be a cone in a Banach space X and assume that $\overset{\circ}{K}$, the interior of K, is nonempty. The cone K will be called "normal" if there exists a constant A such that $||x||\leq A||y||$ whenever $x,y\in K$ and $x\leq y$.

If $x,y\in\overset{\circ}{K}$, define γ to be the largest real number s that $sx\leq y$ and define β to be the smallest real number t such that $y\leq tx$. Define Hilbert's projective metric distance $d(x,y)$ by

$$d(x,y)=\log\left(\frac{\beta}{\gamma}\right)$$

One can easily prove that $d(x,y)=d(y,x)$, d satisfies the triangle inequality, $d(\lambda x,\mu y)=\lambda\mu d(x,y)$ for all positive reals λ and μ and $d(x,y)=o$ if and only if $y=sx$ for some $s>o$. If $\Sigma=\{x\in\overset{\circ}{K}:||x||=1\}$ and K is normal, then (Σ,d) is a complete metric space. This was basically proved by Birkhoff in Theorem 5 of (4), although that theorem is stated less generally. See, also (22) and (42).

As A.C. Thompson has observed in (42), one can also define a useful closely related metric $\bar{d}$ on $\mathring{K}$, namely

$$\bar{d}(x,y)=\max(\log\beta,\log\gamma^{-1}).$$

Thompson proved that $\bar{d}$ is a metric and $(\mathring{K},\bar{d})$ is complete if K is normal.

If $f:\mathring{K}\to\mathring{K}$ is order-preserving and homogeneous of degree 1, one easily can check that f is nonexpansive with respect to d and $\bar{d}$, ie, for all $x,y\in\mathring{K}$ one has

$$\bar{d}(f(x),f(y))\leq\bar{d}(x,y) \text{ and } d(f(x),f(y))\leq d(x,y)$$

The nonlinear theorems which will be stated later generalize linear theorems, so it is necessary to recall some definitions from the linear theory. If X is a Banach space and $L:X\to X$ is a bounded linear map, $r(L)$ will always denote the spectral radius of L. Define a seminorm q by

$$q(L)=\inf\{||L+C||:C \text{ is a compact linear map}\},$$

so $q(L)=o$ if L is compact. Define $\rho(L)$, the essential spectral radius of L, by

$$\rho(L)=\lim_{n\to\infty}(q(L^n))^{\frac{1}{n}}$$

See (30) for further details about $\rho(L)$. If K is a cone in X and $L(K)\subset K$ and $\rho(L)<r(L)$, it is known that L has an eigenvector in K with eigenvalue $r(L)=r$ (24,31).

If K is a cone in a Banach space X and $L:X\to X$ is a bounded linear map, L is called positive (with respect to K) if $L(K)\subset K$. If K has nonempty interior and L is positive, L is called irreducible (with respect to K) if for every $x\in K-\{o\}$ and every $\lambda>r(L)$, $(\lambda-L)^{-1}(x)\in\mathring{K}$. Irreducibility can be defined even if $\mathring{K}$ is empty. If K is the standard cone in $\mathbb{R}^n$ and L is an $n\text{x}n$ nonnegative matrix, L is irreducible if and only if for every pair (i,j) with $1\leq i,j\leq n$, there exists an integer p such that the (i,j) entry of L^p is positive.

<u>Theorem 1.1</u> Let K be a cone in a Banach space and assume that $\overset{\circ}{K}$ is nonempty. Suppose that ψ is a continuous linear functional which is positive on $\overset{\circ}{K}$ and define $\Sigma=\{x\in\overset{\circ}{K}:\psi(x)=1\}$. Let G be an open subset of $\overset{\circ}{K}$ such that $tx\in G$ if $x\in G$ and $o<t\leq 1$. Let $f:G\to\overset{\circ}{K}$ be an order-preserving map such that

$$f(tx)\geq tf(x) \text{ for all } x\in\Sigma\cap G \text{ and } o<t\leq 1. \qquad (1.1)$$

Finally, suppose that if $u\in\Sigma\cap G$ is an eigenvector of f, then f is C^1 on an open neighborhood of u, $L=f'(u)$ is irreducible and $\rho(L)<r(L)$ or $r(L)=o$ where $\rho(L)$ and $r(L)$ are as defined above. Then f has at most one eigenvector in $\Sigma\cap G$.

A more general version on Theorem 1.1 is stated in (33) so as to include certain examples which are not covered by the above theorem. In particular, the full strength of irreducibility is not needed.

If the norm on X is order-preserving or if f is homogeneous of degree 1, then Theorem 1.1 is also true with Σ replaced by

$$\Sigma_1=\{x\in\overset{\circ}{K}:||x||=1\}$$

Notice that if f is locally compact or X is finite dimensional, then the condition $r(L)=o$ or $\rho(L)<r(L)$ is automatically satisfied.

It remains to consider the convergence of normalized iterates of f to an eigenvector of f.

<u>Theorem 1.2</u> Let K,G,Σ and ψ be as in Theorem 1.1 and assume that K is normal and $G\cap\Sigma$ is connected. Suppose that $f:G\to G$ is a continuous, order-preserving map which satisfies equation (1.1). Assume that f has an eigenvector $u\in\Sigma\cap G$ and that, if $L=f'(u)$, $\rho(L)<r(L)$ and there exists an integer $m\geq 1$ such that $L^m(K-\{o\})\subset\overset{\circ}{K}$. Then it follows that u is the only eigenvector of f in $\Sigma\cap G$, and if g is defined by

$$g(x)=\frac{f(x)}{\psi(f(x))},$$

one has

$$\lim_{n\to\infty}||g^n(x)-u||=o \text{ for all } x\in\Sigma\cap G.$$

Furthermore, there exist constants $\varepsilon>o$, $o<c<1$ and $B>o$ such that for all $x\in\Sigma$ such that $d(x,u)<\varepsilon$ and for all $n\geq 1$,

$$d(g^n(x),u) \leq B\, c^n d(x,u),$$

where d denotes Hilbert's projective metric.

If f is homogeneous of degree 1, Theorem 1.2 remains true for $\Sigma=\{x \in \overset{\circ}{K} : ||x||=1\}$ and for

$$g(x)=\frac{f(x)}{||f(x)||}.$$

Notice again that the condition $\rho(L)<r(L)$ in Theorem 1.2 will automatically be satisfied if f is locally compact or X is finite dimensional.

Potter (36) and Thompson (42) study order-preserving maps $f:\overset{\circ}{K}\to\overset{\circ}{K}$ which satisfy

$$f(t,x)\geq t^{\gamma} f(x) \text{ for all } x\in \overset{\circ}{K},\ o<t\leq 1, \tag{1.2}$$

where $o<\gamma<1$ is a constant independent of t,x. If (1.2) is satisfied, $g|\Sigma$ (g,Σ as in Theorem 1.2) is a strict contraction with respect to Hilbert's projective metric d and f is a strict contraction with respect to $\bar{d}$, so one can obtain eigenvectors of f by the contraction mapping principle. However, if one adds an order-preserving, homogeneous map of degree 1 to an f which satisfies equation (1.2), one obtains a new f which is general can only satisfy equation (1.1), and one needs Theorem 1.2 to deal with such maps.

Theorems 1.1 and 1.2 deal only with order-preserving maps. To some extent, the order-preserving condition can be weakened. One can also consider, for example, order-reversing maps f. Of course, one can remark that if f is order-reversing, f^2 is order-preserving, so Theorem 1.1 gives information. However, sharper results are given in (33).

Although no details can be given, it may be worthwhile to give some idea of the proof of Theorems 1.1 and 1.2. If, for f as in Theorem 1.1 or 1.2, one defines $g(x)$ as in Theorem 1.2, it is easy to verify that g is nonexpansive with respect to d. To prove Theorem 1.1, take $x_0\in \overset{\circ}{K}\cap\Sigma$ and define $g_t(x), t\in\mathbb{R}$ by

$$g_t(x)=(1-t)g\,(x)+tx_0$$

One proves a lemma that for all $x,y\in\Sigma\cap G, x=y$,

$$d(g_t(x),g_t(y))<d(x,y) \qquad (1.3)$$

if $o<t<1$. On the other hand, for $x\in Y_1=\{z:\psi(z)=1\}$, define $h(x,t)$ by

$$h(x,t)=x-g(x,t)$$

and $h:Y_1\to Y_o=\{z:\psi(z)=o\}$. One proves that if $u\in\Sigma\cap G$ is an eigenvector of f (so $h(u,o)=o$), then the hypotheses of the implicit function theorem are satisfied, so for $o\leq t<\varepsilon$ one has a solution $u_t\in Y_1$ of

$$h(u_t,t)=o \qquad (1.4)$$

and $t\to u_t$ is C^1. If $v\in\Sigma\cap G$, $v\neq u$, were a second eigenvector of f, one would also obtain for t small

$$h(v_t,t)=o\ . \qquad (1.5)$$

For $t>o$ and t small, equation (1.4) and (1.5) contradict equation (1.3).

The proof of Theorem 1.2 is conceptually somewhat different. Under the assumptions of Theorem 1.2 one proves that if Λ is the Frechét derivative of g at $u(u\in\Sigma\cap G$ is the eigenvector of f), then $r(\Lambda)<1$. This fact depends on some delicate arguments from the theory of positive linear operators. Using $r(\Lambda)<1$ one proves that

$$\lim_{n\to\infty} d(g^n(x),u)=o \qquad (1.6)$$

for all x close to u. The full theorem is obtained by a connectedness argument in which one uses equation (1.6), the connectedness of $G\cap\Sigma$ and the fact that G is nonexpansive with respect to d.

2. Convergence of unnormalized iterates.

If $f:\mathring{K}\to\mathring{K}$ is homogeneous of degree 1, it may happen that $f(u)=u$ for some $u\in\mathring{K}$. This is precisely what occurs in the theorems about iterated means and the DAD theorems mentioned in the introduction. In such situations one might hope that for every $x\in\mathring{K}$ one has

$$\lim_{k\to\infty} f^k(x)=w,\ \text{where}\ f(w)=w\ \text{and}\ w\in\mathring{K} \qquad (2.1)$$

In particular, if positive scalar multiples of u are the only fixed points of f in $\overset{\circ}{K}$, one might hope that

$$\lim_{k\to\infty} f^k(x)=\lambda(x)u,\ \lambda(x)>o, \tag{2.2}$$

for every $x\in\overset{\circ}{K}$. Of course some further conditions on f are necessary (think of $f(x_1,x_2)=(x_2,x_1)$ on the standard cone in $\mathbb{R}^2$ or $f(x_1,x_2,x_3)=(x_3,x_2,x_1)$ in $\mathbb{R}^3$), but the theorems below give some conditions under which equation (2.1) is satisfied.

Theorem 2.1 Let K be a normal cone with nonempty interior in a Banach space X and $G\subset\overset{\circ}{K}$ an open connected set such that $tG\subset G$ for $o<t$. Let $f:G\to G$ be a map which is homogeneous of degree 1 and order-preserving. Assume that $f(u)=u$ for some $u\in G$ and that f is C^1 on an open neighborhood of u. If $L=f(u)$, assume that $\rho(L)<1$ ($\rho(L)$ is the essential spectral radius of L) and that there exists an integer m such that for each $x\in K-\{o\}$, $L^m(x)\in\overset{\circ}{K}$. Then for each $x\in G$, there exists a positive number $t=\lambda(x)$ such that

$$\lim_{k\to\infty} f^k(x)=\lambda(x)u \tag{2.3}$$

The map $x\to\lambda(x)$ is continuous, order-preserving and homogeneous of degree 1, and λ is C^1 on an open neighborhood of u. If $\lambda'(u)\in X^*$ is denoted by v^*, then v^* is nonnegative on K, $L^*(v^*)=v^*$ and $v^*(u)=1$, where L^* is the Banach space adjoint of L. If f is C^k (real analytic) on G, then λ is C^k (real analytic) on G.

Theorem 1.2, when applied to the situation of Theorem 2.1, implies that

$$\lim_{k\to\infty} f^k(x)(\psi(f^k(x)))^{-1}=u, \tag{2.4}$$

where $\psi\in X^*$ is positive on $\overset{\circ}{K}$ and $\psi(u)=1$. It is easy to check that f is actually nonexpansive with respect to Thompsor's metric $\bar{d}$, and using this fact and equation (2.4), one can prove equation (2.3). The various regularity assertions about λ can be obtained from the theory of the stable manifold of a fixed point.

Theorem 2.1 immediately implies as special cases various convergence results for iterated means, for example, the results of Everett and Metropolis in (19).

Although it is assumed in Theorem 2.1 that f is order-preserving and homogeneous of degree 1, some of the ideas of the proof still apply to functions which may not be order-preserving or even homogeneous of degree 1. For simplicity I shall describe such a result for the case of the standard cone in $\mathbb{R}^n$; a version which is valid for general cones is given in Proposition 3.1 of (33). In the following theorem, if x and u are elements of the interior of a cone K, the following notation is used:

$$m(x/u)=\sup\{\gamma>o:\gamma u\leq x\} \text{ and}$$

$$M(x/u)=\inf\{\beta>o:x\leq\beta u\}.$$

Also a matrix A is "nonnegative" if all its entries are nonnegative, and a square nonnegative matrix A is "primitive" if there exists an integer $p\geq 1$ such that A^p has all positive entries.

<u>Theorem 2.2</u> Let $K=\{x\in\mathbb{R}^n:x_i>o \text{ for } 1\leq i\leq n\}$ and suppose that $f:\overset{\circ}{K}\to\overset{\circ}{K}$ is continuous and that $f(\lambda u)=\lambda u$ for some $u\in\overset{\circ}{K}$ and all $\lambda>o$. Assume that f satisfies the following condition:

For any $x\in\overset{\circ}{K}$ and any $\gamma>o$ and $\beta>o$, the inequality $\gamma u\leq x\leq\beta u$ implies $\gamma u\leq f(x)\leq\beta u$. Let A be a nonnegative $n\times n$ matrix with the property that if $a_{ij}>o$, then $f_i(x)>m(x/u)u_i$ for all $x\in\overset{\circ}{K}$ such that $x_j>m(x/u)u_j$ ($f_i(x)$ is the ith component of $f(x)$). Let B be a nonnegative matrix such that if $b_{ij}>o$, then $f_i(x)<M(x/u)u_i$ for all $x\in\overset{\circ}{K}$ such that $x_j<M(x/u)u_j$. Assume that either A or B is primitive. Then for every $x\in\overset{\circ}{K}$ there exists a positive scalar $\lambda(x)$ such that

$$\lim_{k\to\infty} f^k(x)=\lambda(x)u. \qquad (2.5)$$

The initial hypotheses of Theorem 2.2 imply easily that for any $\lambda>o$ and $x\in\overset{\circ}{K}$, $\bar{d}(f^k(x),\lambda u)$ is a decreasing function of k. The idea of the proof is to exploit the additional structure to show that for some particular choice of λ, equation (2.5) holds.

As an immediate corollary of Theorem 2.2, one obtains the following result, which is interesting in that the maps in question are in general neither order-preserving nor homogeneous of degree 1. Recall that an $m\times n$ matrix $S=(s_{ij})$ is "row-stochastic" if $s_{ij}\geq o$ for all i,j and $\sum_{j=1}^{n} s_{ij}=1$ for $1\leq i\leq m$.

Corollary 2.1 Let K denote the standard cone in $\mathbb{R}^n$ and assume that for $x \in \overset{\circ}{K}$, $S(x)=(s_{ij}(x))$ is an $n\times n$ row stochastic matrix whose entries $s_{ij}(x)$ are continuous functions of x. For each $i,j, 1\le i,j\le n$, assume that $s_{ij}(x)$ is either identically zero for all $x \in \overset{\circ}{K}$ or strictly positive for all $x \in \overset{\circ}{K}$. Assume that for some $x \in \overset{\circ}{K}$, $S(x)$ is primitive. Write elements of $\mathbb{R}^n$ as $n\times 1$ column vectors and define $f(x)=S(x)x$. Then for every $x \in \overset{\circ}{K}$, $\lim_{k\to\infty} f^k(x)=\lambda(x)u$, where $\lambda(x)>o$ and $u \in \mathbb{R}^n$ has all entries equal to 1.

If $r \in \mathbb{R}$, σ is a probability vector in $\mathbb{R}^n$, and K is the standard cone in $\mathbb{R}^n$, define $N_{r\sigma}: \overset{\circ}{K} \to \mathbb{R}$ by

$$N_{r\sigma}(x)=\left(\sum_{j=1}^{n}\sigma_j x_j^r\right)\left(\sum_{j=1}^{n}\sigma_j x_j^{r-1}\right)^{-1}$$

Special cases of such maps have been considered by Lehmer (26). In general, for $r>1$, the maps $N_{r\sigma}$ are not order-preserving on all of $\overset{\circ}{K}$.

The next corollary also follows easily from Theorem 2.2.

Corollary 2.2 Let K denote the standard cone in $\mathbb{R}^n$. For $1\le i\le n$ let Γ_i be a finite collection of ordered pairs (r,σ), where $r \in \mathbb{R}$ and $\sigma \in K$ is a probability vector (so $\sum_{i=1}^{n}\sigma_i=1$). For each $(r,\sigma) \in \Gamma_i$, let $c_{ir\sigma}$ be a positive number and assume that

$$\sum_{(r,\sigma)\in\Gamma_i} c_{ir\sigma} = 1$$

For $1<i<n$ define $f_i: \overset{\circ}{K} \to \mathbb{R}$ by

$$f_i(x)=\sum_{(r,\sigma)\in\Gamma_i} c_{ir\sigma} N_{r\sigma}(x)$$

and define $f: \overset{\circ}{K} \to \overset{\circ}{K}$ by taking f_i to be the ith component of f. Define a nonnegative, square matrix $B=(b_{ij})$ by taking $b_{ij}>o$ if there exists $(r,\sigma) \in \Gamma_i$ such that σ_j, the jth component of σ, is positive and by writing $b_{ij}=o$ otherwise. If B is primitive, then for every $x \in \overset{\circ}{K}$ there exists a positive number $\lambda(x)$ such that

$$\lim_{k\to\infty} f^k(x)=\lambda(x)u,\ u=(1,1,\ldots,1).$$

All of the previously mentioned results concern the case in which f has exactly one normalized fixed point in $\overset{\circ}{K}$. However, equation (2.1) can be satisfied even when f has many fixed points. The following theorem (which represents joint work with Joel Cohen) is a direct generalization to bounded, self-adjoint operators of the arithmetic-geometric mean. Note that the map f below is not order-preserving.

Theorem 2.3 (Cohen and Nussbaum (16)). Let H be a Hilbert space and let C denote the set of bounded, self-adjoint nonnegative definite linear operators, so A is positive definite if $A \in \overset{\circ}{C}$. Let γ, β be given real numbers such that $0<\gamma, \beta<1$ and define $f: \overset{\circ}{C} \times \overset{\circ}{C} \to \overset{\circ}{C} \times \overset{\circ}{C}$ by

$$f(A,B)=(\gamma A+(1-\gamma)B,\ \exp(\beta \log A+(1-\beta)\log B))$$

For any $A, B \in \overset{\circ}{C}$, if $f^k(A,B)=(A_k, B_k)$, then A_k and B_k both converge in the strong operator topology to some $E \in \overset{\circ}{C}$. If H is finite dimensional, A_k and B_k converge in the operator norm to E; and if $\gamma=\beta$, the norm convergence is quadratic.

Notice that the set of fixed points of f is $\{(A,A): A \in \overset{\circ}{C}\}$.

3. Nonexpansive mappings and Hilbert's projective metric.

In most of the theorems stated so far the mapping f has been assumed to satisfy some differentiability assumptions. It is useful to have theorems about maps which are only assumed to be nonexpansive with respect to d or $\bar{d}$.

The next theorem is modeled on a result of Roehrig and Sine (37). The idea is that, although Hilbert's projective metric d is not a norm, d still has enough convexity properties that the argument in (37) can be modified.

Theorem 3.1 Let K be a cone with nonempty interior in a finite dimensional Banach space X, ψ a continuous linear functional on X which is nonnegative on K and $\Sigma=\{x \in \overset{\circ}{K}: \psi(x)=1\}$. For $x \in \overset{\circ}{K}$, define $\sigma(x)$ by

$$\sigma(x)=\inf\{||x-y||: y \in \partial K\}.$$

If $f:\Sigma \to \Sigma$ is a map which is nonexpansive with respect to Hilbert's projective metric and if f has no fixed point in Σ, then for any $x \in \Sigma$, $\lim_{k\to\infty} \sigma(f^k(x))=0$, ie, $f^k(x)$ approaches the boundary of K.

An analogous theorem holds if $f: \overset{\circ}{K} \to \overset{\circ}{K}$ is nonexpansive with respect to Thompson's metric $\bar{d}$ and f has no fixed point in K.

Theorem 3.1 treats the case in which f has no fixed points. It is also useful to have theorems which describe the structure of the fixed point set of a map f, assuming the fixed point set is nonempty.

Theorem 3.1 Let K denote the standard cone in $\mathbb{R}^n$, let ψ be a continuous linear functional which is positive on $\mathring{K}$ and $\Sigma=\{x\in \mathring{K}:\psi(x)=1\}$. Assume that $f:\Sigma\to\Sigma$ is nonexpansive with respect to Hilbert's projective metric d and that $S=\{x\in\Sigma:f(x)=x\}$ is nonempty. Then there exists a retraction r of Σ onto S such that r is nonexpansive with respect to d. In particular, S has the same homotopy type as Σ, and S is connected.

It should be noted that S need not be convex; an example of nonconvex S is given in (33). It should also be remarked that the proof of Theorem 3.1 depends strongly on special properties of the standard cone in $\mathbb{R}^n$. However, if one assumes that the map f is order-preserving and homogeneous of degree 1, then one can extend Theorem 3.2 to the case of general cones.

Theorem 3.3 Let K be a normal cone with nonempty interior in a Banach space X. Suppose that $f:\mathring{K}\to\mathring{K}$ is order-preserving,

$$f(tx)\geq tf(x) \text{ for all } x\in\mathring{K},\ o<t\leq 1,$$

and $f(B_R(x_o))$ has compact closure in $\mathring{K}$ for all $R>o$ and $x_o\in\mathring{K}$, where $B_R(x_o)=\{x\in K:e^{-R}x_o\leq x\leq e^{R}x_o\}$. For a given $\lambda>o$, let $S=\{x\in K:f(x)=\lambda x\}$ and assume that S is nonempty. Then there exists an order-preserving retraction r of $\mathring{K}$ onto S such that $r(tx)\geq tr(x)$ for all $x\in\mathring{K}$ and $o<t\leq 1$. The retraction r can be taken homogeneous of degree 1 if f is homogeneous of degree 1.

Theorem 3.3 implies that S has the same homotopy type as $\mathring{K}$, so S is connected.

4. Some examples of maps of the standard cone in $\mathbb{R}^n$.

Suppose that K is a cone in a Banach space and $f:K\to K$ is a continuous, order-preserving map which is homogeneous of degree

1. Assume that f is compact and that $f(x_o)\geq\lambda_o x_o$ for some $x_o \in K-\{o\}$ and some $\lambda_o>o$. Then a classical theorem of Krein and Rutman (see Theorem 9.1 in (24)) implies that there exists $\lambda\geq\lambda_o$ and $u\in K-\{o\}$ such that $f(u)=\lambda u$. There have been many generalizations of the above theorem, and a simple proof using the fixed point index is now available.

If K has nonempty interior, f is as above and $f(\mathring{K})\subset\mathring{K}$, define ρ by

$$\rho=\sup\{\gamma:f(x)=\gamma x \text{ for some } x\in\partial K, x\neq o\}.$$

The theorem of Krein-Rutman then implies that if $\rho<\lambda_o$, f has an eigenvector in $\mathring{K}$. In particular, if $\rho=o$, f has an eigenvector in $\mathring{K}$.

The following is a typical corollary of the above observation.

<u>Proposition 4.1</u> (Compare (32), Section 4). Let K_2 be the standard cone in $\mathbb{R}^2$ and $\phi:K_2\to(o,\infty)$ an order-preserving map, homogeneous of degree 1, such that $\phi(u,v)>o$ if u and v are positive and $\phi(u,v)=o$ if $u=o$ or $v=o$. Let K be the standard cone in $\mathbb{R}^6$ and define $f:K\to K$ by $f(x)=(\phi(x_3,x_6),ax_1,bx_2,cx_1,dx_4,ex_5)$ where a,b,c,d and e are positive constants. Then f has an eigenvector in $\mathring{K}$.

Suppose that K is the standard cone in $\mathbb{R}^n$ and $f:\mathring{K}\to\mathring{K}$ is order-preserving and homogeneous of degree 1. A nonnegative, nxn matrix $A=(a_{ij})$ is called an "incidence matrix for f" (will respect to being power bounded below) if whenever $a_{ij}>o$ there exist a positive real $c>o$ and a probability vector σ (both dependent on i and j) such that σ_j, the $j^{\underline{th}}$ component of σ, is positive and

$$f_i(x)\geq c\,x^{\sigma} \text{ for all } x\ \ K.$$

Here $f_i(x)$ is the $i^{\underline{th}}$ component of $f(x)$ and $x^{\sigma}=\prod_{k=1}^{n}x_k^{\sigma_k}$. If $g:\mathring{K}\to\mathring{K}$ is also order-preserving and homogeneous of degree 1 and g has an incidence matrix B (with respect to being power bounded below) then clearly $A+B$ is an incidence matrix for $f+g$, and it is proved in Lemma 4.1 of (32) that AB is an incidence matrix for $f\circ g$. Also, if $f\in M_+$ (M_+ defined as in the introduction) then the Jacobian matrix of f, evaluated at any

$x \in \mathring{K}$, is an incidence matrix for f. Finally, it is proved in Theorem 4.1 of (32) that if K is the standard cone in $\mathbb{R}^n$, $f:\mathring{K}\to\mathring{K}$ is order-preserving and homogeneous of degree 1 and f has an incidence matrix A which is irreducible, then f has an eigenvector in $\mathring{K}$.

Using the above results and the theorems in Section 1, one can easily obtain the following proposition.

Proposition 4.2 Let K be the standard cone in $\mathbb{R}^n$ and suppose that $f:\mathring{K}\to\mathring{K}$ is a convex combination of functions $h_i:\mathring{K}\to\mathring{K}$, $1\le i\le m$, where each h_i can be written as the composition of a finite number of functions in M_+. If $J_f(x)$, the Jacobian matrix of f at x, is irreducible for some $x\in\mathring{K}$ (and hence for all x), then f has one and exactly one eigenvector (to within scalar multiples) in $\mathring{K}$. If $J_f(y)$ is primitive for some $y\in\mathring{K}$ and u is the unique eigenvector of f in $\mathring{K}$ of norm 1, then

$$\lim f^k(x)(||f^k(x)||^{-1})=u$$

for all $x\in\mathring{K}$.

Of course a version of Proposition 4.2 is true for the class M_- if one knows a priori that f has an eigenvector in $\mathring{K}$. However , if $f\in M_-$, the question of whether f has an eigenvector in $\mathring{K}$ can be delicate. Take, for example, the following simple-looking function which is studied in (33) and generalizes a function studied by Schoen in (38):

$$f_1(x)=\alpha_1x_1+\beta_1\psi(x_1,x_2)+\gamma_1\psi(x_1,x_4)+\delta_1\psi(x_2,x_3)$$

$$f_2(x)=\alpha_2x_2+\beta_2\psi(x_1,x_2)+\gamma_2\psi(x_1,x_4)+\delta_2\psi(x_2,x_3)$$

$$f_3(x)=\alpha_3x_3+\beta_3\psi(x_3,x_4)+\gamma_3\psi(x_1,x_4)+\delta_3\psi(x_2,x_3)$$

$$f_4(x)=\alpha_4x_4+\beta_4\psi(x_3,x_4)+\gamma_4\psi(x_1,x_4)+\delta_4\psi(x_2,x_3)$$

In the above equations $\alpha_j,\beta_j,\gamma_j$ and δ_j are nonnegative and $\psi(u,v)=2uv(u+v)^{-1}$ for $u>o$, $v>o$ $(\psi(o,o)=o)$. These equations determine a map $f\in M_-$ of the standard cone in $\mathbb{R}^4$ into itself. A complete analysis of when f has an eigenvector in $\mathring{K}$ is non-trivial: see (33), where some general methods for determining when $f \in M_-$ has an eigenvector in $\overset{\circ}{K}$ are discussed.

References

1. F.L. Bauer, An elementary proof of the Hopf inequality for positive operators, Numerische Math. 7(1965), 331-337.
2. E. Beckenbach, A class of mean value functions, Amer. Math. Monthly 57(1950), 1-6.
3. G. Birkhoff, Extensions of Jentzch's theorem, Trans. Amer. Math. Soc. 85(1957), 219-227.
4. ________________, Uniformly semi-primitive multiplicative processes, Trans. Amer. Math. Soc. 104 (1962), 37-51.
5. J.B. Borwein and P.B. Borwein, The arithmetic-geometric mean and fast computation of elementary functions, SIAM Review 26(1984), 351-366.
6. R.A. Brualdi, S.V. Parter and H. Schneider, The diagonal equivalence of a non-negative matrix to a stochastic matrix, J. Math. Anal. and Appl. 16(1966), 31-50.
7. R. Bruck, Properties of fixed point sets of nonexpansive mappings in Banach spaces, Trans. Amer. Math. Soc. 179(1973), 251-262.
8. P. Bushell, Hilbert's metric and positive contraction mappings in Banach space, Arch. Rat. Mech. Anal. 52(1973), 330-338.
9. ________________, On the projective contraction ratio for positive linear mappings, J. London Math. Soc. 6(1973), 256-258.
10. ________________, On a class of Volterra and Fredholm nonlinear integral equations, Math. Proc. Camb. Phil. Soc. 79(1976), 329-335.
11. B.C. Carlson, Algorithms involving arithmetic and geometric means, Amer. Math. Monthly, 1971, 496-505.
12. ________________, A connection between elementary functions and higher transcendental functions, SIAM J. Applied Math. 17 (1969), 116-148.
13. ________________, Hidden symmetries of special functions, SIAM Review 12(1970), 332-345.
14. H. Caswell and D. Weeks, Two-sex population models: the

dynamic consequences of frequency dependent nonlinearities in structured populations, preprint.

15. J. Cohen and R.D. Nussbaum, Arithmetic-geometric means of positive matrices, submitted for publication.
16. J. Cohen and R.D. Nussbaum, The arithmetic-geometric mean for bounded linear operators, in preparation.
17. D.A. Cox, Gauss and the arithmetic-geometric mean, Notices Amer. Math. Soc. 32(1985), 147-151.
18. D.Z. Djokvic, A note on nonnegative matrices, Proc. Amer. Math. Soc. 25(1970), 80-82.
19. C.J. Everett and N. Metropolis, A generalization of the Gauss limit for iterated means, Advances in Math. 7(1971), 297-300.
20. E. Hopf, An inequality for positive linear integral operators, J. Math. Mech. 12(1963), 683-692.
21. S. Karlin and L. Nirenberg, On a theorem of P. Nowosad, J. Math. Anal. and Appl. 17(1967), 61-67.
22. M.A. Krasnoselśkii and A.V. Sobolev, Spectral clearance of a focusing operator, Functional analysis and its applications 17(1983), 58-59.
23. M.A. Krasnoselśkii, Positive Solutions of Operator Equations, Noordhoff, Groningen, 1964.
24. M.G Krein and M.A. Rutman, Linear operators leaving invariant a cone in a Banach space, Uspekhi Mat. Nauk. 3(1), 23 (1948); English transl. Amer. Math. Soc. Transl., No. 26.
25. R.S. Krupp, Properties of Kruithof's Projection Method, The Bell System Technical Journal, 58(1979), 517-538.
26. D.H. Lehmer, On the compounding of certain means, J. Math. Anal. Appl. 36(1971), 183-200.
27. M.V. Menon, Reduction of a matrix with positive elements to a doubly stochastic matrix, Proc. Amer. Math. Soc. 18(1967), 244-247.
28. M.V. Menon and H. Schneider, The spectrum of a nonlinear operator associated with a matrix, Linear Algebra and its Applications 2(1969) 321-334.
29. P. Nowosad, On the integral equation $Kf=\frac{1}{f}$ arising in a problem in communication, J. Math. Anal. Appl. 14(1966), 484-492.
30. R.D. Nussbaum, The radius of the essential spectrum, Duke

Math. J. 38(1970), 473-478.

31. ______________, Eigenvectors of nonlinear positive operators and the linear Krein-Rutman theorem, in Fixed Point Theory, Springer Verlag Lecture Notes in Math., vol. 886, 309-331.

32. ______________, Convexity and log convexity for the spectral radius, Linear Alg. and Appl. 73(1986) 59-122.

33. ______________, Iterated Nonlinear maps and Hilbert's projective metric, in preparation.

34. A.M. Ostrowski, Positive matrices and functional analysis, Recent Advances in Matrix Theory, H. Schneider, editor, Univ. of Wisconsin Press, Madison, Wisconsin, 1964, pp81-101.

35. B.N. Parlett and T.L. Landis, Methods for scaling to doubly stochastic form, Linear Alg. and Appl. 48(1982), 53-

36. A.J.B. Potter, Applications of Hilbert's projective metric to certain classes of non-homogeneous operators, Quart. J. Math. 28(1977), 93-99.

37. S. Roehrig and R. Sine, The structure of ω-limit sets of nonexpansive maps, Proc. Amer. Math. Soc. 81(1981), 398-400.

38. R. Schoen, The two-sex multi-ethnic stable population model to appear in Theoretical Population Biology.

39. R. Sinkhorn, A relationship between arbitrary positive matrices and stochastic matrices, Ann. Math. Stat. 35(1964), 876-879.

40. R. Sinkhorn and P. Knopp, Concerning nonnegative matrices and doubly stochastic matrices, Pac. J. Math. 21(1967), 343-348.

41. ______________, A note concerning simultaneous integral equations, Canadian J. Math 20(1968), 855-861.

42. A.C. Thompson, On certain contraction mappings in a partially ordered vector space Proc. Amer. Math. Soc. 14(1963), 438-443.

43. P.P. Zabreiko, M.A. Krasnosel'skii and Yu. V. Pokornyi, On a class of positive linear operators, Functional Analysis and its Applications (1972), 272-279.

JACOBI MATRICES AND TRANSVERSALITY (*)

Giorgio Fusco
Dipartimento di Metodi e Modelli Matematici
per le Scienze Aplicate
Un. di Roma - 00161 Roma, Via Antonio Scarpa, 10

Waldyr Muniz Oliva
Departamento de Matemática Aplicada
Un. de São Paulo - Caixa Postal 20570
(Agência Iguatemi) - São Paulo - Brazil

Transversality of stable and unstable manifolds of critical elements plays a central role in connection with generic results and structural stability in the theory of Dynamical Systems. That property is true, generically, for vector fields and diffeomorphisms defined on a finite dimensional manifold (see [PM] for references). Also the structural stability of Morse-Smale vector fields and diffeomorphisms on compact manifolds and the stability with respect to compact attractors in the infinite dimensional case depend strongly on transversality (see [PM] for the finite dimensional case and [HMO], [O] for infinite dimensions). In spite of this fact there is no general available method for checking transversality.

Recently, has been proved in [DH] that transversality holds for the infinite dimensional system generated, in a suitable Sobolef space H, by the Chaffee-

(*) Talk given by the second author in the NATO workshop on Dynamics of Infinite Dimensional Systems, Lisbon, May 1986.

NATO ASI Series, Vol. F37
Dynamics of Infinite Dimensional Systems
Edited by S.-N. Chow, and J. K. Hale

Infante scalar parabolic problem:

$$u_t = u_{xx} + f(u) \quad , \quad x \in [0,1], \quad t > 0 \tag{1}$$

with Dirichlet boundary conditions (see also [A]). Another class of examples in which transversality holds was presented in [MP] for delay equations.

The proofs in [DH] and [A] depend on a property which is specific to linear scalar parabolic equations, that is, the number of zeros of the solutions does not increase with time. If the number of zeros is considered as a functional $N_0: M \longrightarrow \mathbf{Z}$ defined on some dense subset M of the space H, then transversality seems to be related, in a essential way, to the following facts:

(α) the eigenvectors of the linearization of (1) around equilibria are simple and if w_k is an eigenvector corresponding to the k-th eigenvalue then $w_k \in M$ and $N_0(w_k) = k-1$;

(β) N_0 is not increasing along trajectories of the linear variational equation around solutions of (1);

(γ) N_0 has some continuity properties.

Once the number of zeros is considered as a functional with the three properties above, it seems natural to conjecture that the existence of functionals like N_0 is not a very special property of Sturm-Liouville operators but that other classes of operators may be related to some other functionals. Here we talk about families of operators rather than about a single operator because, as it happens for the functional N_0 which corresponds to the whole family of Sturm-Liouville operators, one can expect that any discrete functional N in a suitable class determines a family A of operators which is maximal with respect to properties (α) and (β).

In connection with the above conjecture it is natural to ask the following questions:

(a) What abstract properties should a functional N have in order that we can associate to N a family A of operators which have, with respect to N, the same properties that the Sturm-Liouville ones have with respect to N_0.

(b) Given a generic operator A with a simple spectrum, can we always embedd it in a class $\mathcal{A}$ corresponding to some functional N? If this is possible, in how many different ways it can be done?

(c) What are the implications of (a) and (b) for non linear problems, i.e., what is the class of non linear problems which is naturally associated to a given functional N in the sense that transversality holds?

In this talk it will be just stated, without proofs(*), some of the results obtained when one analises these questions in the finite dimensional setting.

Let E be a vector space, dim E = n. A functional N can be introduced in the following way:

Definition 1: Let $\{e_i\}$ *be a basis in* E *and* $x = \sum_{i=1}^{n} x_i e_i$ *be the representation of* $x \in E$ *in that basis. Let* $\mathcal{N} \subset E$ *be the set of all vectors such that either* $x_i \neq 0$, $i = 1,2,\ldots,n$ *or* $x_{i-1}\, x_{i+1} < 0$ *whenever* $x_i = 0$ *(we define* $x_0 = x_{n+1} = 0$*). For* $x \in \mathcal{N}$, $N(x) - 1$ *is the number of times* x_i *changes sign when* i *goes from* 1 *to* n.

We note that $\mathcal{N}$ is open and $N: \mathcal{N} \longrightarrow \{1,2,\ldots,n\}$ is a step function which is constant on each one of the connected components of $\mathcal{N}$. Therefore N is continuous in $\mathcal{N}$.

Let L(E) be the set of all linear operators in E.

Theorem 1: Given a continuous function $t \in (a,b) \longrightarrow A(t) \in L(E)$, *a necessary and sufficient condition in order that:*

(i) the flow $\phi(x,s,t)$ *of the differential equation* $\dot{x} = A(t)x$ *satisfies the condition*

$$N(\phi(x,s,t_2)) \leqq N(\phi(x,s,t_1)) \tag{2}$$

whenever $t_2 \leqq t_1$ *and both sides of (2) are defined,*

(*) The full paper with detailed proofs will appear elsewhere.

(ii) the eigenvectors of A(t) *belong to* N,

is that: for each t, *the matrix representation in* $\{e_i\}$ *of the operator* A(t) *be a positive Jacobi matrix, that is, if if* $A(t)e_j = \sum_{i=1}^{n} a_{ij} e_i$ *then* $a_{ij} = 0$ *for* $|i-j| > 1$ *and* $a_{ij} > 0$ *for* $|i-j| = 1$.

In the following proposition, which is crucial in the proof of Theorem 1, N, N, are defined as in Definition 1 with respect to the canonical basis of $\mathbb{R}^n$ (see theorem 13, page 105 of [G]).

<u>*Proposition 1:*</u> *Let* J *be a positive Jacobi matrix. Then:*

1. J *has simple eigenvalues* $\lambda_1 > \lambda_2 > \ldots > \lambda_n$.
2. *If* η_i *is an eigenvector corresponding to* λ_i *one has* $\eta_i \in N$ *and* $N(\eta_i) = i$, $i = 1,2,\ldots,n$.
3. *If* $c_h, c_{h+1}, \ldots, c_k$ *are real numbers which are not all equal to zero and* $\eta = \sum_{i=h}^{k} c_i \eta_i$ *then: either* $\eta \in N$ *and* $h \leqq N(\eta) \leqq k$ *or* $\eta \notin N$ *and there is a neighborhood* U *of* η *such that* $h \leqq N(\eta') \leqq k$ *for all* $\eta' \in U \cap N$.

The next theorem gives a sufficient condition in order that for a special class of systems the stable and unstable manifolds of hyperbolic equilibria intersect transversality. In fact:

<u>*Theorem 2:*</u> *If* $f\colon \Omega \subset E \longrightarrow E$, Ω *open, is a* C^k *function,* $k \geqq 1$, *and there is a basis* $\{e_i\}$ *in* E *such that for all* $x \in \Omega$ *the matrix representation in* $\{e_i\}$ *of the derivative* $f'(x)$ *is of positive Jacobi type, then if* $e^-, e^+ \in \Omega$ *are hyperbolic equilibria of*

$$\dot{x} = f(x) \tag{3}$$

and $\phi(t)$ *is a solution of (3) such that* $\lim \phi(t) = e^{\pm}$ *as* $t \longrightarrow \pm\infty$, *one has* $W^u(e^-) \pitchfork W^s(e^+)$.

For the proof of this theorem we needed to use some results on the assymptotic behavior of solutions of linear non autonomous equations besides the invariance of some cones defined by the functional N.

Definition 2: For any given integer $0 \leqq h \leqq n$ *let* K_h *be the set of all* $x \in E$ *such that one of the following is true:*

a) $x = 0$

b) $x \in N$ *and* $N(x) \leqq h$

c) $x \notin N$ *and there is a neighborhood* U *of* x *such that* $N(x') \leqq h$ *for* $x' \in U \cap N$.

Let K^h *be the set of* $x \in E$ *such that a) or b') or c') holds where b') and c') are like b) and c) with* $\leqq h$ *replaced by* $>h$.

The sets K_h and K^h so defined are cones. Moreover $K_h \setminus \{0\}$, $K^h \setminus \{0\}$ are open sets, $K_h \cap K^h = \{0\}$ and closure $(K_h \cup K^h) = E$.

Lemma 1. Let $a < 0 < b$ *and* $t \in (a,b) \longmapsto A(t) \in L(E)$ *be a continuous function such that for each* $t \in (a,b)$ *the matrix representation of* $A(t)$ *with respect to some fixed basis* $\{e_i\}$ *of* E *is of positive Jacobi type and let* N, K_h, K^h *the corresponding functional and cones. Then:*

(i) If $\Sigma_0 \subset K_h$ *is a linear subspace and* Σ_t *is the image of* Σ_0 *at the time* t *under the equation*

$$\dot{y} = A(t)y \tag{4}$$

then for $t \geqq 0$ *it results* $\Sigma_t \subset K_h$.

(ii) If $\Sigma^0 \subset K^h$ *is a linear subspace and* Σ^t *is the image of* Σ^0 *at the time* t *under (4), then for* $t \leqq 0$ *it results* $\Sigma^t \subset K^h$.

The idea of the proof of Theorem 2 is to apply Lemma 1 to show that if $m^- = \dim W^u(e^-)$, then $T_{\phi(t)}W^u(e^-)$ is contained in K_{m^-} and $T_{\phi(t)}W^s(e^-)$ contains a $(n-m^-)$ dimensional subspace which is contained in K^{m^-}; this implies transversality.

Another consequence of Lemma 1 is the following:

Theorem 3: Let f *be as in Theorem 2 and let* $e \in \Omega$ *be a hyperbolic equilibrium of* (3). *Then the stable and unstable manifolds* $W^s(e)$ *and* $W^u(e)$ *of* e *are imbedded submanifolds of* E. *In fact, there is a linear surjection* $P: E \longrightarrow T_e W^u(e)$ *such that the restriction* $P|W^u(e)$ *is a diffeomorphism of* $W^u(e)$, *and a similar statement holds for* $W^s(e)$.

The condition that the matrix representation of $f'(x)$ in $\{e_i\}$ is of positive Jacobi type not only implies transversality, but also put restrictions on the structure of the non wandering set which is, in this case, the set of critical points. We see then that the dynamics of (3) has many properties in common with the dynamics of a system admitting a Liapunov function. This is obviously the case when the matrix representation $A(x)$ of $f'(x)$ is a symmetric matrix and Ω is simply connected because then $f = \text{grad } U$, for some function U. In the case of a system in $\mathbb{R}^n$ given by

$$\dot{x}_i = f_i(x_i) + \sum_{j=1}^{n} a_{ij}x_j, \quad i = 1,2,\ldots,n \tag{5}$$

in which the constant matrix (a_{ij}) is of positive Jacobi type (not necessarely symmetric), it can be proved the existence of a Liapunov function for (5). We do not know if a Liapunov function always exists for a general system (3), under the hypothesis of Theorem 2.

We finish this talk with the statement of the following:

<u>Theorem 4</u>: If $A: E \longrightarrow E$ *is a linear operator with real simple eigenvalues, then there is a basis* $\{e_i\}$ *in such that the corresponding matrix representation of* A *is of positive Jacobi type. Let* $g_i: \mathbb{R}^n \longrightarrow R$ *be smooth functions with the property that* $\frac{\partial g_i}{\partial x_j}$ *is also a matrix of positive Jacobi type and* $g: E \longrightarrow E$ *be the function* $g = \sum_{i=1}^{n} g_i e_i$. *Then the system* $\dot{x} = Ax + g(x)$ *satisfies the hypothesis of Theorems 2 and 3.*

It is reasonable to believe that some extension of this result to infinite dimensions is possible. This conjecture is partially confirmed by Theorem 7.13 in [S] which states that if A is a self adjoint operator in a separable Hilbert space H and has a "simple" spectrum, then H possesses a basis such that the corresponding matrix representation of A is a infinite matrix of Jacobi type.

R E F E R E N C E S

[A] S.B.ANGENENT, "The Morse-Smale Property for a Semi-Linear Parabolic Equation", J.Diff. Equations, vol. 62, pag. 427-442 (1986).

[DH] D.B.HENRY, "Some infinite-Dimensional Morse-Smale Systems defined by Parabolic Partial Differential Equations", J. Diff. Equations, vol. 59, nº 2, Sept. 1985, pag. 165-205

[G] F.R.GANTMACHER, The Theory of Matrices. Chelsea Publ. Co., N.York, 1959, vol. 2

[H] J.K.HALE, "Infinite Dimensional Dynamical Systems", LCDS Report 81-22 (Nov. 1981), Division of Applied Math., Brown Un.

[HMO] J.K.HALE, L.T.MAGALHÃES and W.M.OLIVA, An Introduction to Infinite Dynamical Systems - Geometric Theory, Springer Appl. Math. Sciences, 47 (1984).

[MP] J.MALLET-PARET, "Morse Decompositions for delay differential equations". In preparation.

[O] W.M.OLIVA, "Stability of Morse-Smale Maps", Rel.Tec. MAP - 8301 (Jan. 1983), Inst. Mat. Est., Univ. of São Paulo.

[PM] J.PALIS & W. DE MELLO, Geometric Theory of Dynamical Systems - An Introduction, Springer-Verlag (1982).

[S] M.H.STONE, Linear Transformation in Hilbert Space and Their Applications to Analysis, AMS Coll.Publ., vol. XV, 1932.

EXAMPLES OF ATTRACTORS IN SCALAR REACTION-DIFFUSION EQUATIONS

C. Rocha
Dep. de Matemática, Instituto Superior Técnico
Universidade Técnica de Lisboa
Av. Rovisco Pais, 1096 Lisboa Codex - Portugal

1. It is well known that for a large class of nonlinearities f, the scalar reaction-diffusion equation with Neumann boundary conditions

$$\begin{aligned} u_t &= \varepsilon^2 u_{xx} + f(x,u) \quad , \quad 0 \le x \le 1 \\ u_x(0,t) &= u_x(1,t) = 0 \ , \end{aligned} \tag{1}$$

defines a well posed initial value problem in the Sobolev space $H^1(0,1)$ [12]. Moreover, this problem has a gradient-like structure which ensures that the α and ω-limit sets of any bounded orbit are equilibrium solutions [9,15,18].

If f satisfies a coerciveness condition of the form $\lim\sup_{|u|\to\infty} f(x,u)/u < 0$ uniformly in x, the set of equilibria of (1) is bounded and the dynamical system defined by (1) has a compact attractor which attracts bounded sets of $H^1(0,1)$ (see for example [8]).

Generically in f, all the equilibria of (1) are hyperbolic and, in this case, the semiflow defined on $H^1(0,1)$ has a Morse-Smale structure. This follows from the results of Henry [13] on the transversality of the stable and unstable manifolds of the equilibria (see also Angenent [2]).

To complete the study of this dynamical system, one has to characterize the attractor for the semiflow, determining all the connections between its equilibria. This problem has been solved for the special case of Chafee-Infante, where the non-linearity is of the type $f(x,u)=u-u^3$ [13], and for the general Hamiltonean case, $f(x,u)= f(u)$, some partial results are already known [4]. The general case is currently being studied

NATO ASI Series, Vol. F37
Dynamics of Infinite Dimensional Systems
Edited by S.-N. Chow, and J. K. Hale

and it is useful to have available some examples of more sofisticated attractors.

Here we deal with nonlinearities of the type

$$f(x,u) = u(1-u)(u-a(x)) \tag{2}$$

where $a:[0,1]\to(0,1)$ is a step function, and consider the characterization of the attractor of (1), (2) when ε is small and for special examples of step functions a.

For smooth functions a, this problem has been considered by several authors and using different approaches (see, for example [1,3,5,6,14]).

The attractor of (1), (2) is very simple for large values of ε [11], but for small values of ε it contains an arbitrarily large number of equilibrium solutions [14]. However, it has been shown by Angenent, Mallet-Paret and Peletier [3] that, under generic hypothesis on a, the number of stable solutions stays bounded. In fact, denoting by $0 < x_1 < \dots < x_n < 1$ the sequence of zeros of $a(x)-1/2$, they show that for functions a satisfying $a_x(x_i)\neq 0$, $i=1,\dots,n$, and for ε sufficiently small there is a one-to-one correspondence between the stable solutions and the patterns of step functions $s:[0,1]\to\{0,1\}$ with jumps at the points y_j, $j=1,\dots,m$, only if $y_j \in \{x_i, i=1,\dots n\}$ and $[s(x_i^+)-s(x_i^-)].a_x(x_i) < 0$. Furthermore, as $\varepsilon\to 0$ each stable solution approaches the corresponding step function at the points of continuity, exhibiting monotone transition layers at the jump points.

Using a phase-energy method we show that the corresponding result still holds for the case of step functions a. We consider also some special examples of step functions for which we show that the total number of equilibria stays bounded as $\varepsilon\to 0$. For these examples we determine all the equilibria, the dimension of their unstable manifolds and, through a process of parameter tunning, also the heteroclinic connections between the equilibria composing the attractor.

In this communication we only sketch the proofs of these results. The details will appear elsewhere.

2. To consider the existence and stability of equilibrium solutions of (1), (2) we use a phase-energy method developed in [7,10,17].

Denoting by E the set of equilibrium solutions, we have that $u \in E$ if and only if u is a solution of

$$\begin{aligned} &\varepsilon^2 u_{xx} + f(x,u) = 0 , \quad 0 \le x \le 1 \\ &u_x(0) = u_x(1) = 0 . \end{aligned} \tag{3}$$

From (2) one easily concludes that for u to satisfy (3), it is necessary that $u \in C^1(0,1)$ and $0 \le u(x) \le 1$ for every x in [0,1].

Associated with the boundary value problem (3) one has the following initial value problem:

$$\begin{aligned} &u_x = \frac{1}{\varepsilon^2} v , \quad v_x = -f(x,u) \\ &u(0) = u_o , \quad v(0) = 0 . \end{aligned} \tag{4}$$

We assume that the solutions of (4) are defined for all $x \in [0,1]$, since we can modify f outside the region $[0,1]\times[0,1]$ for such purpose without perturbing the attractor of (1), (2).

Then, we let M denote the solution manifold of (4), $M=\{(x,u,v): u=u(x,u_o),\ v=v(x,u_o) \text{ for } u_o \in [0,1]\}$, and let L_y be the section curve of M at x=y.

Similarly, we consider the end value problem associated to (3):

$$\begin{aligned} &u'_x = \frac{1}{\varepsilon^2} v' , \quad v'_x = -f(x,u') \\ &u'(1) = u_1 , \quad v'(1) = 0 \end{aligned} \tag{5}$$

letting M' denote its solution manifold, and L'_y the section curve of M' at x=y.

The curves L_y and L'_y provide a characterization of the equilibria of (1), (2) since, for each y in [0,1], there is a one-to-one correspondence between the set of equilibria E and the set $L_y \cap L'_y$. Moreover, it can be shown that an equilibrium is hyperbolic if and only if the corresponding intersection of the curves L_y and L'_y is transversal.

From the linear variational equation of (4) around the equilibrium solution $u= u(x,u_o)$:

$$\eta_x = \frac{1}{\varepsilon^2}\mu \ , \ \mu_x = -f_u(x,u(x,u_o)).\eta \qquad (6)$$
$$\mu(0) = 1 \ , \ \mu(0) = 0 \ ,$$

and introducing the change to polar coordinates $\eta = p \cos t$, $\mu = -p \sin t$, we obtain the differential equation satisfied by $t= t(x, u_o)$:

$$t_x = \frac{1}{\varepsilon^2}\sin^2 t + f_u(x,u(x,u_o)) \cos^2 t, \ t(0,u_o) = 0 \qquad (7)$$

In the same way, from (5) we obtain an equation for $t'=t'(x,u_o)$:

$$t'_x = \frac{1}{\varepsilon^2}\sin^2 t' + f_u(x,u(x,u_o)) \cos^2 t', \ t'(1,u_o) = 0 \ , \qquad (8)$$

and define $\varphi(y,u_o) = t(y,u_o) - t'(y,u_o)$.

Then, geometrically $\varphi=\varphi(y,u_o)$ is the angle formed by the curves L_y and L'_y at the crossing corresponding to the equilibrium solution $u=u(.,u_o)$, and we have the following stability criteria

<u>Lemma</u>: <u>The solution</u> $u=u(.,u_o)$ <u>is stable if and only if</u> $\varphi(y,u_o) < 0$ <u>for</u> $y\in[0,1]$.

In addition to this lemma, when $u=u(.,u_o)$ is not stable, one can determine from $\varphi(.,u_o)$ the dimension of the unstable manifold of the solution u, which is given by *Integer part of* $[1+\varphi/\pi]$ (see [17]).

3. One easily recognizes that, if a is not constant, the problem (1), (2) has exactly two constant solutions which are stable; u=0 and u=1.

Let us now consider a step function $a:[0,1] \to \mathbb{R}$ taking the values $a_i \in (0,1)$ in the intervals $[x_{i-1}, x_i)$, $i=1,\dots,m+1$, where $0=x_o < x_1 < \dots < x_{m+1}=1$.

Performing the change of variables $x=x_{i-1}+\varepsilon y$ in (3), we obtain

$$\begin{aligned} &u_{yy} + u(1-u)(u-a_i)=0 \quad , \quad 0 \leq y < (x_i - x_{i-1})/\varepsilon \\ &u(0)= u_{i-1} \quad , \quad u_y(0)= v_{i-1} \quad , \end{aligned} \tag{9}$$

where $u_o \in [0,1]$, $v_o=0$ and (u_j, v_j), $j=1,\dots,m$ are determined successively, corresponding to the section curves L_{x_j} after rescaled.

This equation defines a Hamiltonean system and the solutions correspond to level curves of $H(u,w)=w^2/2+\int_o^u s(1-s)(s-a_i)ds$. Then, the phase portraits depend only on the values a_i, $i=1,\dots,m+1$, changing qualitatively according to the sign of $a_i-1/2$ (see figures 1,2).

Using these phase portraits, a simple application of the inclination lemma (or λ-lemma, [16]) leads to the following

<u>Proposition</u>: <u>For any</u> $\delta > 0$ <u>there is</u> $\varepsilon_o > 0$ <u>such that for all</u> $0 < \varepsilon < \varepsilon_o$ <u>all the equilibria</u> $u=u(.,u_o) \in E$ <u>of (1), (2) with</u> $u_o \in [\delta, 1-\delta]$ <u>are unstable</u>.

This result implies that to study the stable solutions of (1), (2) we only need to consider solutions with initial conditions u_o in $[0,\delta) \cup (1-\delta,1]$.

Let U denote a small square neighborhood of the equilibrium point (1,0) in the phase portrait of (9), and V a similar neighborhood of the origin. If $a_1 > 1/2$, the phase portrait is shown in fig. 1 and we denote by Γ_o and Γ_1 the parts of the section curve L_y, $y= x_1/\varepsilon$, corresponding to the initial conditions $u_o \in [0,\delta)$ and $u_o \in (1-\delta,1]$.

Then, by the inclination lemma, Γ_o and Γ_1 get C^1 close to the unstable manifolds of the equilibria (0,0) and (1,0) respectively, as $\varepsilon \to 0$.

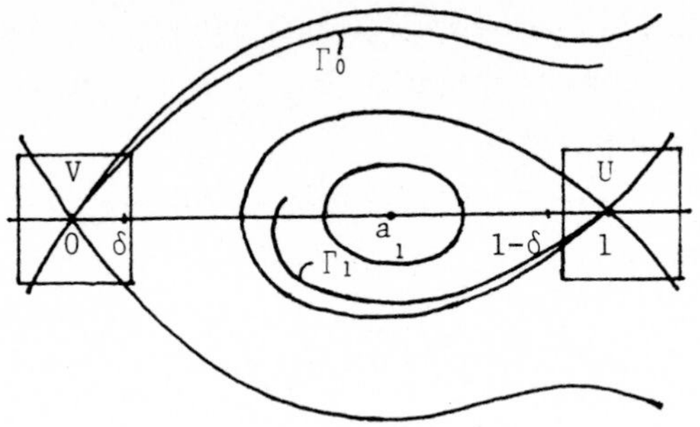

Fig. 1.

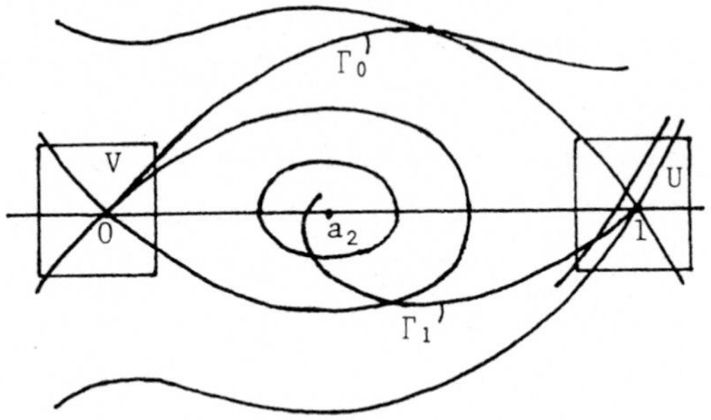

Fig. 2.

If $a_2 < 1/2$, the new phase portrait is shown in fig. 2 and we conclude that for ε small enough Γ_o will have a unique point of transversal intersection with the stable manifold of (1,0). From another application of the inclination lemma, we conclude that the image of Γ_o intersects the set U for ε small enough, and the intersection segment is C^1 close to the unstable manifold of (1,0) in U. If $x_2=1$, this segment corresponds to a segment of the section curve L_y at the end point, and is transversal to L_y' in U. The intersection corresponds to a new stable solution for (1), (2). Moreover, no other stable solutions are produced either by Γ_o or Γ_1.

Otherwise, if $a_2 > 1/2$, the new phase portrait is qualitatively the same as the one shown in Fig. 1, and we can conclude that no new stable solutions are produced.

Finally if $x_2 < 1$, we repeat the above procedure applying the inclination lemma and studying the behavior in the neighborhoods U and V (see an example in fig. 3). This way, we obtain the following result:

Theorem: If the step function $a:[0,1]\to(0,1)$ satisfies the condition $a(x)\neq 1/2$, let n denote the number of jumps of a across 1/2. Also, let $\bar{x}_1,\dots,\bar{x}_n$ define the partition of $[0,1]$ corresponding to these jumps, and $\bar{x}_o=0$, $\bar{x}_{n+1}=1$. Then, there is an $\varepsilon_o>0$ such that for every $\varepsilon\in(0,\varepsilon_o)$ the number of stable solutions of (1), (2) are exactly N_n, the n th term of the Fibonacci sequence $N_o=2$, $N_1=3$, $N_k=N_{k-1}+N_{k-2}$, $k=2,3\dots$

Furthermore, as $\varepsilon\to 0$ each stable solution u approaches 0 or 1 in each open internal $(\bar{x}_i,\ \bar{x}_{i+1})$, $i=0,1,\dots,n$, and has a monotone transition layer at $\bar{x}_j$ only if

$$[a(\bar{x}_j^+)-a(\bar{x}_j^-)]\,.\,u_x(\bar{x}_j)<0 \quad .$$

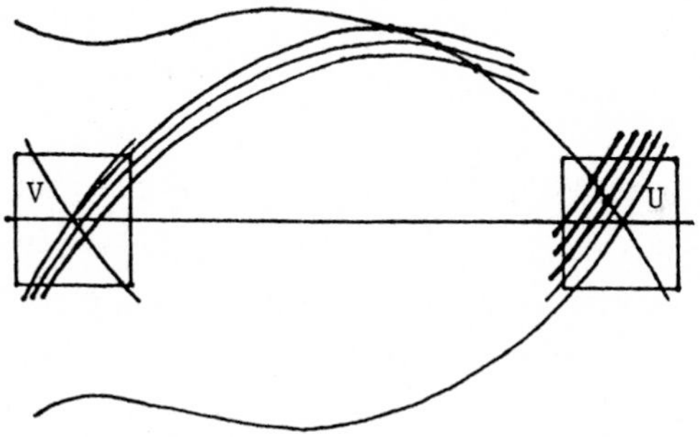

Fig. 3.

The shape of the monotone transition layers of these stable solutions is obtained by following the corresponding level curves in the phase portraits.

4. Inthis section we consider particular examples of step functions a, presenting a characterization for the attractors of problem (1), (2) for small values of ε.

Example I: As first example we consider a step function corresponding to the case n=1:

$$a(x)=\begin{cases} 1/2 + c_1 , & x < c_3 \\ 1/2 - c_2 , & x \geq c_3 \end{cases} \tag{10}$$

where c_1, $c_2 \in (0,1/2)$, $c_3 \in (0,1)$ are parameters.

This step function has a negative jump across the value 1/2. To consider an example with a positive jump one just performs a change of variables $x \to -x$.

As we have seen, the phase portrait of (9) for $a_i = 1/2-c$, $0 < c < 1/2$ has an orbit homoclinic to the origin. Then, we denote by $\gamma=\gamma(c)$ the coordinate u of the intersection point of this homoclinic orbit with the u-axis. Using the Hamiltonean of (9) we conclude that γ is a monotonically decreasing function of c in (0, 1/2). Moreover, for 1/2+c, the phase portrait can be obtained from the previous one by a reflexion around the axis u=1/2, and the corresponding homoclinic orbit intersects the u-axis at the point with coordinate $1-\gamma(c)$. Then, we have the following:

Proposition: If c_1, c_2 satisfy the condition $\gamma(c_1)+\gamma(C_2) \leq 1$, then there is an $\varepsilon_o > 0$ such that for every $0 < \varepsilon < \varepsilon_o$ the problem (1), (2), (10) has exactly 5 equilibrium solutions, 3 being stable and 2 unstable. Furthermore, each unstable equilibria have a one dimensional unstable manifold consisting of two heteroclinic orbits, one connecting to the nonconstant stable equilibrium and the other to a constant solution, either u=0 or u=1.

The condition $\gamma(c_1)+ \gamma(c_2) \leq 1$ implies that the homoclinic curve in the phase portrait for $x < c_3$ does not intersect transversally the homoclinic in the phase portrait for $x > c_3$, and tangency occurs only when equality holds.

As pointed out already, in the particular case $c_1=c_2$ the phase diagram for $x > c_3$ is the reflexion around the axis u=1/2 of the phase diagram for $x < c_3$. If we also let $c_3 = 1/2$, then the same relation holds between the section curves L_y and L'_y for $y=1/2\varepsilon$: $L'_y = \{(u,w) : (1-u,w) \in L_y\}$. We conclude that any intersection point of L_y with the axis $u=1/2$ belongs to $L_y \cap L'_y$, and the condition $\gamma(c_1) = \gamma(c_2) \leq 1/2$ implies that there exists only one such point. Studying the slope of the tangent to L_y at this point one can show that there exists a unique $\varepsilon_o > 0$ such that for $\varepsilon = \varepsilon_o$ the curves L_y and L'_y intersect nontransversally. The nonconstant solution corresponding to this intersection is unstable for $\varepsilon > \varepsilon_o$ and becomes stable for $\varepsilon < \varepsilon_o$. Moreover, two new crossings are introduced corresponding to the bifurcation of two new unstable nonconstant equilibrium solutions, with one dimensional unstable manifolds (see fig. 4). Fig. 5 presents the associated pitchfork bifurcation diagram along with the stability assignments, and fig. 6 shows a

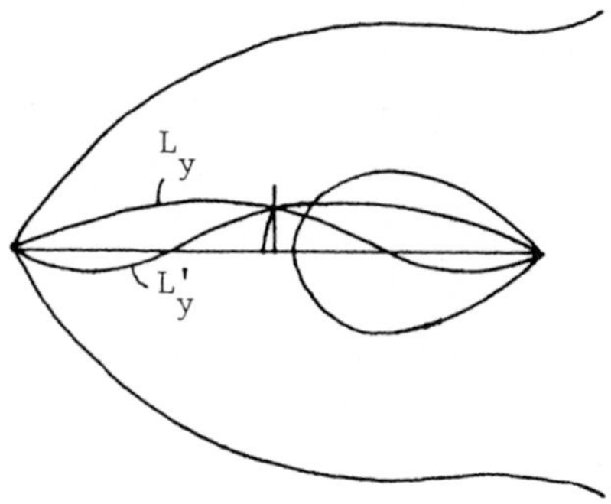

(a) $\varepsilon > \varepsilon_0$

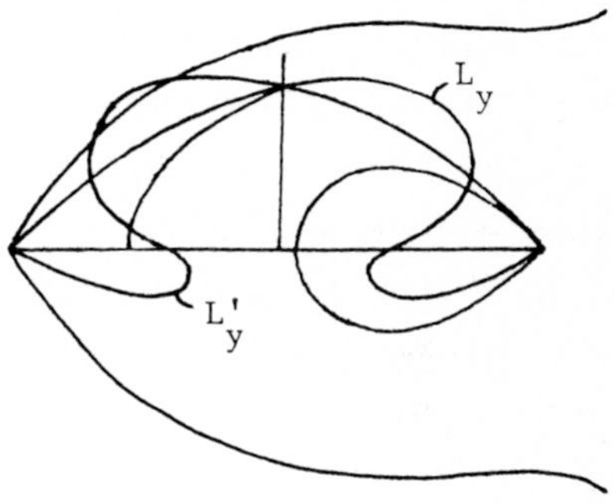

(b) $\varepsilon < \varepsilon_0$

Fig. 4.

schematic representation of the attractor for (1), (2), (10) where the heteroclinic connections follow from the center manifold theorem and the Morse-Smale structure of the flow [13].

Finally, we consider the continuous dependence of the curves L_y and L'_y on the parameters c_1, c_2 and c_3. For ε sufficiently small, a change in the parameters produces a continuous deformation of these curves preserving all intersections trans-

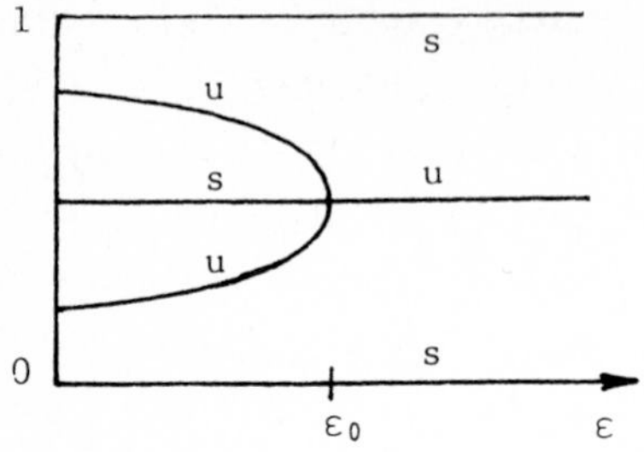

Fig. 5.

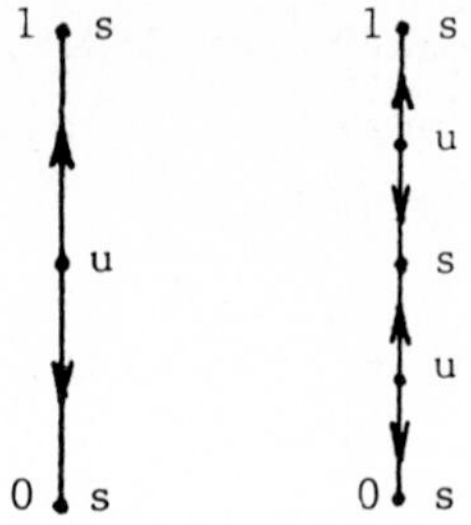

(a) $\varepsilon>\varepsilon_0$ (b) $\varepsilon<\varepsilon_0$

Fig. 6.

versal. Then, the transversality results of Henry [13] imply the proposition, yielding for (1), (2), (10) an attractor equivalent to the one in fig. 6.b.

Example II: Next, we present an example with a step function corresponding to the case n=2.

Consider the step function a of the previous example and extend it to the interval [0,2] symmetrically; that is with $a(x) = a(2-x)$. Then, rescaling the variable x back to the interval [0,1] we obtain an example of a step function with n=2 jumps across the value 1/2. For this problem, the section curve L_y at the middle point $y=1/2\varepsilon$ corresponds to the section curve at the end point for the previous problem, and, from the symmetry of a, the curve L'_y is a reflexion of L_y, $L'_y = \{(u,w):(-u,w) \in L_y\}$, (see fig. 8).

Taking the case $c_1 = c_2 > \gamma^{-1}(1/2)$, $c_3 = 1/2$, and studing the shape of the curve L_y, we conclude from the previous problem that a pitchfork bifurcation occurs for $\varepsilon=\varepsilon_o$. Furthermore, there is an ε_1 satisfying $0<\varepsilon_1<\varepsilon_o$, such that a second pitchfork bifurcation occurs at $\varepsilon=\varepsilon_1$. The first bifurcation produces one stable equilibrium and two unstable equilibria with one dimensional manifolds. From the second bifurcation arise two unstable equilibria also with one dimensional unstable manifolds, and one unstable equilibrium with a two dimensional unstable manifold. As in the first example, the heteroclinic connections between these equilibria are easily established.

Nevertheless, decreasing ε beyond ε_1, four new equilibria are introduced (see fig. 7). Two of these new equilibria are stable, making up the $N_2=5$ stable solutions referred to in the theorem of section 3, and the other two are unstable with one dimensional unstable manifolds.

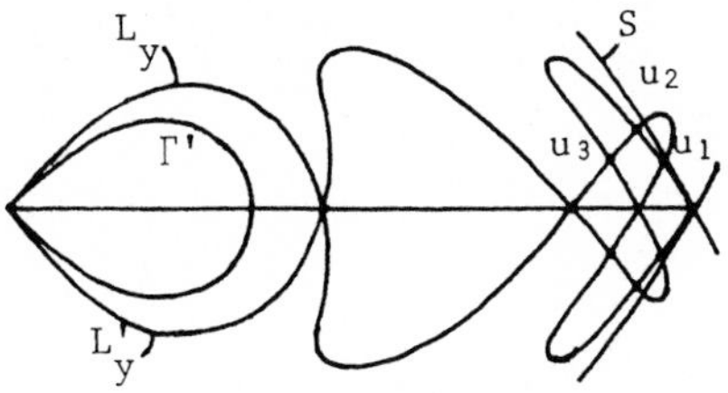

Fig. 7.

To establish the connections of these equilibria, we allow the parameters c_i to vary with ε, and consider continuous paths of the form $(\bar{\varepsilon}(\varepsilon), c_i(\varepsilon))$, with $\bar{\varepsilon}(0)=0$ and $\bar{\varepsilon}$ nondecreasing. Then, we can find two different paths along which different saddle-node bifurcations occur producing these new equilibria.

Ordering the intersections in $L_y \cap L'_y$ by their appearence along L'_y starting from $u=1$, we denote by u_1, u_2 and u_3 the three nonconstant equilibria corresponding to the three intersections immediately following $u=1$.

Then, along one of the paths, the solutions u_1 and u_2 cancel each other through saddle-node bifurcation when ε is increased, and along the other path, u_2 and u_3 cancel each other instead. The two bifurcation diagrams are shown in figure 8 with the stability assignments, where u_2 denotes unstable with a two dimensional unstable manifold. Note that, because of symmetry, two saddle-node bifurcations always occur simultaneously.

Using the transversality method of Henry as in the previous example, these diagrams justify the schematic representation of the attractor shown in fig. 9, where the connections of u_1, u_2, u_3 and the corresponding symmetric solutions are obtained

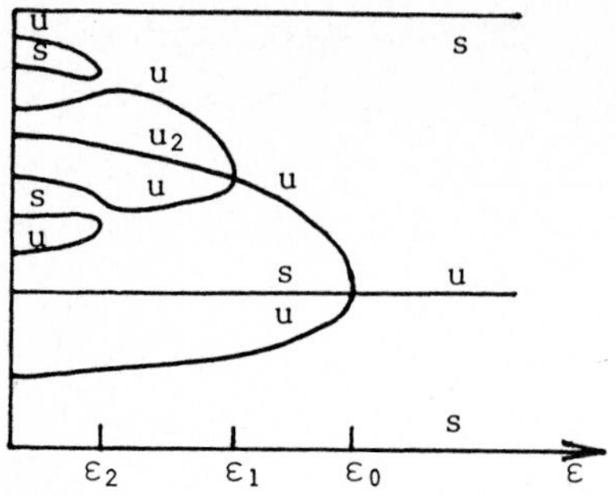

Fig. 8(a).

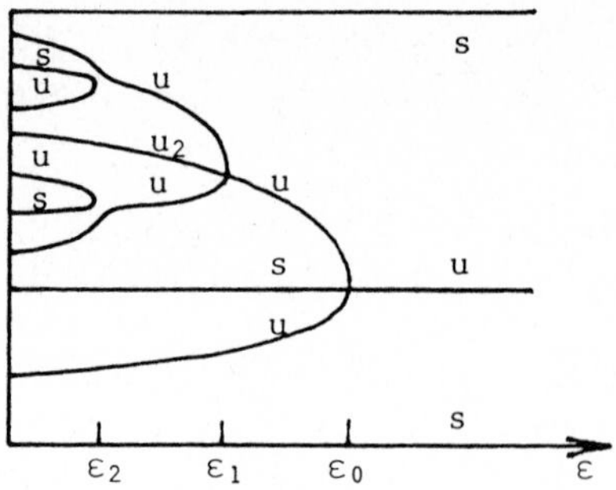

Fig. 8(b).

by following backwards (increasing the parameter ε) the two different bifurcation paths.

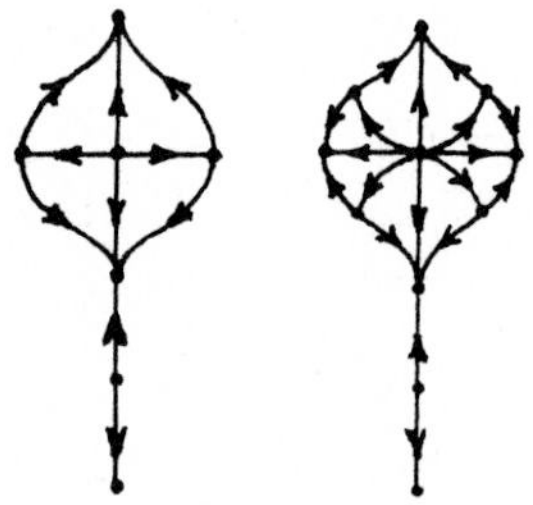

a) $\varepsilon > \varepsilon_0$ b) $\varepsilon_1 < \varepsilon < \varepsilon_0$ c) $\varepsilon_2 < \varepsilon < \varepsilon_1$ d) $\varepsilon < \varepsilon_2$

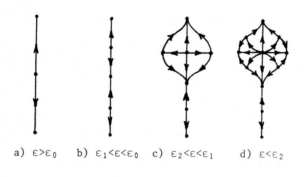

Fig. 9.

Moreover, this result still holds if we change continuously the parameters maintaining all the intersections of L_y and L'_y transversal. Then, we can consider the case of a step function a given by

$$a(x) = \begin{cases} 1/2 + c_1 , & 0 \leq x < c_3 \\ 1/2 - c_2 , & c_3 \leq x < c_5 \\ 1/2 + c_4 , & c_5 \leq x \leq 1 , \end{cases} \tag{11}$$

where c_1, c_2, $c_4 \in (0,1/2)$ and $0 < c_3 < c_5 < 1$ obtaining the following:

Proposition: *If c_1, c_2, c_4 satisfy the conditions $\gamma(c_1)+\gamma(c_2) \leq 1$ and $\gamma(c_2)+\gamma(c_4) \leq 1$ with at least one inequality being strict, there is an open set* G *in* $(0,1)\times(0,1)$ *containing the point* (1/4, 3/4) *such that if* $(c_3, c_5) \in G$ *there is an* $\varepsilon_o>0$ *such that for every* $0<\varepsilon<\varepsilon_o$ *the problem* (1), (2), (11) *has exactly* 11 *equilibrium solutions,* 5 *being stable,* 5 *unstable with one dimensional unstable manifolds and* 1 *unstable with a two dimensional unstable manifold. Moreover, the connections between these equilibria are represented in fig. 9.d.*

Using a change of variables $u \to 1-u$, one easily obtains the corresponding results for step functions with the opposite sign of the jumps.

Example III: Finally, we present some results for an example with a step function corresponding to the case n=3.

Consider again the step function of the first example with $c_1=c_2$, $c_3= 1/2$, and extend it to the interval [0,2] periodically; that is, $a(x)= a(x-1)$ for $x \in (1,2]$. Then, rescaling the variable x back to the interval [0,1] we obtain a step function with n=3 jumps across the value 1/2 at the points x=1/4, 1/2 and 3/4.

The symmetries of this problem imply that the section curve L'_y at the middle point $y=1/2\varepsilon$ is just a reflexion around u=1/2 of L_y: $L'_y= \{(u,v):(1-u,v) \in L_y\}$. Then, we can determine the intersections $L_y \cap L'_y$ for ε sufficiently small (see fig. 10), and also the corresponding stabilities.

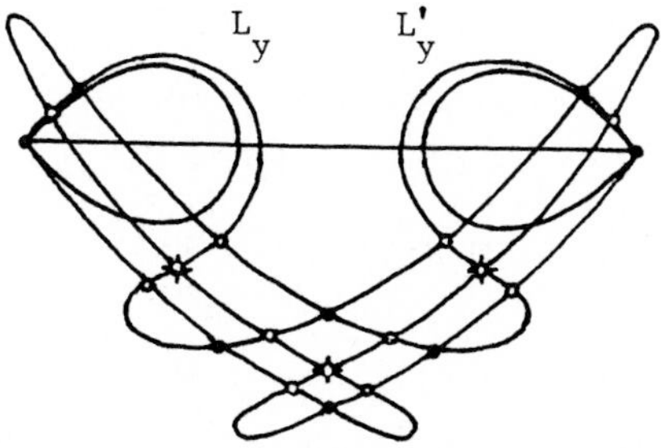

Fig. 10 .

Allowing for continuous perturbations of the parameters we can consider the example:

$$a(x)=\begin{cases} 1/2 + c_1 & , \quad 0 \leq x < c_3 \\ 1/2 - c_2 & , \quad c_3 \leq x < c_5 \\ 1/2 + c_4 & , \quad c_5 \leq x < c_7 \\ 1/2 - c_6 & , \quad c_7 \leq x \leq 1 \end{cases} \qquad (12)$$

Proposition: If c_1, c_2, c_4, c_6 satisfy the conditions $\gamma(c_1)+\gamma(c_2) \leq 1$, $\gamma(c_2)+\gamma(c_4) < 1$ and $\gamma(c_4)+\gamma(c_6) \leq 1$, there is an open set G in $(0,1)\times(0,1)\times(0,1)$ containing the point (1/4, 1/2, 3/4) such that, if $(c_3, c_5, c_7) \in G$ there is an $\varepsilon_o > 0$ such that for every $0 < \varepsilon < \varepsilon_o$ the problem (1), (2), (12) has exactly 21 solutions, 8 being stable, 10 unstable with one dimensional unstable manifolds and 3 unstable with two dimensional unstable manifolds.

With regard to the problem of determining the connections between these equilibria, it becomes quite involved and requires a rather more general method.

Aknowledgements

The author whishes to thank Professor Jack K. Hale for many helpful conversations, and also Professor J. Mallet-Paret for valuable comments.

References

[1] - N. Alikakos and K.C. Shaing, On the Singular Limit for a Class of Problems Modelling Phase Transitions, Preprint.

[2] - S.B. Angenent, The Morse-Smale Property for a Semilinear Parabolic Equation, Mathematical Institute, University of Leiden, Netherlands (1984), J. Differential Equations 62 (1986) 427-442.

[3] - S.B. Angenent, J. Mallet-Paret and L.A. Peletier, Stable Transition Layers in a Semilinear Boundary Value Problem, Mathematical Institute, University of Leiden, Netherlands (1984), Preprint.

[4] - P. Brunovsky and B. Fiedler, Connecting Orbits in Scalar Reaction Diffusion Equation, Universität Heidelberg (1986), Preprint.

[5] - P.C. Fife and L.A. Peletier, Clines Induced by Variable Selection and Migration, Proc. R. Soc. Lond. B214, 99-123 (1981).

[6] - W.H. Fleming, A Selection-Migration Model in Population Genetics, Journal of Mathematical Biology 2, 219-233 (1975).

[7] - G. Fusco and J.K. Hale, Stable Equilibria in a Scalar Parabolic Equation with Variable Diffusion, SIAM J. Math Anal, to appear.

[8] - J.K. Hale, Asymptotic Behavior in Infinite Dimensional Systems, Research Notes in Math., Vol. 132, 1-41, Pittman, 1985.

[9] - J.K. Hale and P. Massat, Asymptotic Behavior of Gradient--Like Systems, Univ. of Fla. Symp. Dyn. Systems II, Academic Press (1982).

[10] - J.K. Hale and C. Rocha, Bifurcations in a Parabolic Equation with Variable Diffusion, Nonlinear Analysis, Theory and Applications, 9 nº 5, 479-494 (1985).

[11] - J.K. Hale and C. Rocha, Varying Boundary Conditions with Large Diffusivity, J. Mat. Pures et Appl., to appear.

[12] - D. Henry, Geometric Theory of Semilinear Parabolic Equations. Lecture Notes in Math., Vol. 840, Springer-Verlag, 1981.

[13] - D. Henry, Some Infinite-Dimensional Morse-Smale Systems Defined by Parabolic Partial Differential Equations, J. Diff. Eqns. 59 (1985) 165-205.

[14] - H.L. Kurland, Monotone and Oscillatory Equilibrium Solutions of a Problem Arising in Population Genetics, Contemporary Mathematics, 17, AMS (1983).

[15] - H. Matano, Convergence of Solutions of One-Dimensional Semilinear Parabolic Equation, J. Math. Kyoto Univ. 18 (1978) 221-227.

[16] - J. Palis, On Morse-Smale Dynamical Systems, Topology, 8 (1969), 385-405.

[17] - C. Rocha, Generic Properties of Equilibria of Reaction -Diffusion Equations with Variable Diffusion, Proc. Roy. Soc. Edinb., 101A (1985), 45-55.

[18] - T.J. Zelenyak, Stabilization of Solutions of Boundary Value Problems for a Second Order Parabolic Equation with One Space Variable, Differential Eqns. 4 (1968), (Translated from Differentialniye Uravneniya).

Gauge Theory of Backlund Transformations, I

by

D.H. Sattinger[1] and V.D. Zurkowski
University of Minnesota

Abstract: Backlund transformations are obtained as gauge transformations for Hamiltonian hierarchies over $sl(2,C)$ for potentials in $so(2)$ or $su(2)$. The transformation of the scattering data is calculated, and it is shown how these transformations create or annihilate a pair of eigenvalues in the scattering data, hence create or annihilate a soliton in the potential Q. Repeated Backlund transformations are constructed which introduce higher order poles in the scattering data; and the structure of the higher order singularities is described. It is shown how an arbitrary set of poles in the scattering data may be removed by a sequence of Backlund transformations.

1. Introduction.

Consider the operator $D_x(z,Q) = \partial/\partial x - iz\sigma_0 - Q$ where Q is a 2×2 matrix with zero diagonal entries and

$$\sigma_0 = \begin{pmatrix} 1 & 0 \\ 0 & -1 \end{pmatrix}.$$

Such operators arise as isospectral operators in the theory of integrable systems. A gauge transformation G from Q_1 to Q_2 satisfies $D_x(z,Q_2)G = GD_x(z,Q_1)$, hence

[1]This research was supported in part by NSF grant DMS 85-01777.

NATO ASI Series, Vol. F37
Dynamics of Infinite Dimensional Systems
Edited by S.-N. Chow, and J.K. Hale

$$\frac{\partial G}{\partial x} = iz[\sigma_o, G] + Q_2 G - G Q_1 \quad . \tag{1.1}$$

This relationship may also be written

$$iz\sigma_o + Q_2 = \frac{\partial G}{\partial x} G^{-1} + G(iz\sigma_o + Q_1)G^{-1} \quad .$$

If ψ_1 is a solution of $D_x(z,Q_1)\psi = 0$, and $\psi_2 = G\psi_1$, then $D_x(z,Q_2)\psi_2 = 0$ also.

Backlund transformations are a specific form of gauge transformation [3], [10]. In this paper we discuss the Backlund-gauge transformations when the potential Q belongs to the Lie algebras $so(2)$ and $su(2)$, and calculate the associated transformation of the scattering data.

We consider the two cases:

$$Q = \begin{pmatrix} 0 & q \\ -q & 0 \end{pmatrix} \qquad Q^t = -Q \qquad so(2)$$

and

$$Q = \begin{pmatrix} 0 & q \\ -\bar{q} & 0 \end{pmatrix} \qquad Q^* = -Q \qquad su(2)$$

Theorem 1.1. For the $so(2)$ case the associated Backlund-gauge transformation is given as follows: Define $F(z_o,u) = (u_x + 2iz_o u)/(1+u^2)$. Then the Backlund-gauge transformation $q \to \tilde{q}$ is given by

$$q = F(z_o,u) \qquad \tilde{q} = F(z_o,-1/u) \ , \tag{1.2}$$

$$G = OAO^t \tag{1.3}$$

where O is the orthogonal matrix $O = (1+u^2)^{-1/2}\begin{pmatrix} 1 & -u \\ u & 1 \end{pmatrix}$ and $A(z) = z - z_o\sigma_o$.

In the $su(2)$ case define

$$F(z_o,\bar{z}_o,u) = \{-\bar{u}_x + \bar{u}^2 u_x + 2i\bar{u}(\bar{z}_o + z_o|u|^2)\}/(1-|u|^4). \tag{1.4}$$

The Backlund-gauge transformation for the su(2) case is:

$$q = F(z_0, \bar{z}_0, u) \qquad q = F(z_0, \bar{z}_0, -1/\bar{u}) \tag{1.5}$$

and

$$G = UAU^*$$

where U is the unitary matrix

$$U = (1+|u|^2)^{-1/2} \begin{pmatrix} 1 & -\bar{u} \\ u & 1 \end{pmatrix}$$

and $A = \begin{matrix} z-z_0 & 0 \\ 0 & z-\bar{z}_0 \end{matrix}$.

Newell [10] found a Backlund transformation for the so(2) case in a form equivalent to the one presented here. We shall discuss constructive methods for obtaining Backlund transformations in a later paper.

The scattering data for the operator $D_x(z,Q)$ transforms in a very simple way under these Backlund transformations. Following Beals and Coifman [1] we express the scattering data in terms of the singularities of a certain solution $m(x,z)$ of the equation $\partial m/\partial x - iz[\sigma_0, m] - Qm = 0$. The properties of the wave function $m(x,z)$ are described in lemma 4.1. The wave function m has in general jump conditions across the real axis and poles in the upper and lower half planes. The jump conditions of m are expressed in the form

$$m(x,\xi+i0) = m(x,\xi-i0)e^{i\xi x\sigma_0}V(\xi)e^{-i\xi x\sigma_0}$$

where V is a matrix defined on the real line. Hereafter we shall abbreviate $e^{i\xi x\sigma_0}V(\xi)e^{-i\xi x\sigma_0}$ by $V^x(\xi)$.

The analysis of the simple poles of m was carried out in [1]. That analysis has been extended to poles of arbitrary order (for the general nXn case) by Zurkowski [14] and will appear separately. At a pole z_0, m may be factored as

$$m(x,z) = \eta(x,z)V_0^x$$

where η is holomorphic in a neighborhood of z_0, and

$V_o = I+T_o$, where T_o is a strictly triangular matrix which is a polynomial in $(z-z_o)^{-1}$. T_o is upper triangular if $\mathrm{Im} z_o < 0$ and lower triangular if $\mathrm{Im} z_o > 0$. Here

$V_o^x = e^{izx\sigma_o} V_o e^{-izx\sigma_o}$. We shall call V_o the <u>principal factor</u> of m at z_o, and write $V_o = \mathcal{P}_o(m)$. In general, if $T = T(z)$ is a strictly triangular matrix with a pole of some finite order at z_o then [14] $V = I+T$ can be uniquely factored as $V = HR$ where H and R are triangular with 1's on the diagonal, H is regular at z_o, and R is a polynomial in $(z-z_o)^{-1}$ which tends to I as $z\to\infty$. We shall call R the principal factor of V at z_o, and also write $R = \mathcal{P}_o V$.

Theorem 1.2. <u>Let $Q_1 \to Q_2$ be the Backlund transformation given by (1.2)-(1.3) or (1.4)-(1.5) with $u = \psi_2/\psi_{1}$,</u>

$$\begin{pmatrix} \psi_1 \\ \psi_2 \end{pmatrix} = \begin{pmatrix} m_{11} & m_{12} \\ m_{21} & m_{22} \end{pmatrix} \begin{pmatrix} e^{iz_o x} & 0 \\ 0 & e^{-iz_o x} \end{pmatrix} \begin{pmatrix} 1 \\ a \end{pmatrix}$$

<u>The choice of the constant a will be discussed later. Assume z_o is not in the scattering data for Q_1. Then the scattering data transforms in the following manner:</u>

$$V_2(\xi) = A(\xi)V_1(\xi)A(\xi)^{-1}$$
$$V_{2j} = \mathcal{P}_j A(z)V_{1j}(z)A(z)^{-1}$$

<u>where $A(z)$ is as given in Theorem 1.1 for the so(2) and su(2) cases. In addition, new poles are introduced at $\pm z_o$ in the so(2) case and at $z_o, \bar{z}_o$ in the su(2) case with residues given by</u>

$$V_{2,z_o} = \begin{pmatrix} 1 & 0 \\ -2z_o a & 1 \end{pmatrix} .$$

In the so(2) case, $V_{2,-z_0} = V_{2,z_0}{}^t$, while $V_{2,\bar{z}_0} = (V_{2,z_0})^*$ in the su(2) case.

Repeated Backlund transformations using the same parameter z_0 may be carried out, the result being that higher order poles are introduced into the scattering data. This is in contrast to the case of the KdV equation, where only simple poles may occur (viz. simple zeroes of the function log a(k)) (cf. [7], p. 116). We shall discuss the computation of the higher order poles of m under repeated Backlund transformations with the same parameter z_0 in §5.

2. Connections and Gauge Transformations

The equation

$$\omega_{xt} = \sin \omega \ . \tag{2.1}$$

arose in the 19th century in the theory of surfaces of constant negative curvature in R^3. Today it is called the sine-Gordon equation; it models waves in nonlinear media, and is important in nonlinear optics and solid state physics [7]. Backlund obtained his transformation from a geometrical construction which took one "pseudo-spherical" surface to another. The classical Backlund transformation is

$$\begin{aligned}(v-u)_x &= 2k \sin(v+u)/2 \\ (v+u)_t &= 2k^{-1}\sin(v-u)/2 \ .\end{aligned} \tag{2.2}$$

It is an easy exercise to see that the sine-Gordon equation is a consistency condition for the Backlund transformation. That is, if u is given and one wants to find v, then a necessary condition for the existence of a function v satisfying equations (2.2) is that u satisfy (2.1); and then v must also satisfy (2.1).

Transformations akin to that found by Backlund are an intrinsic part of the subject of completely integrable systems. A comprehensive introduction to the subject may be found in several textbooks, for example [7], [11]. For our purposes it will be sufficient to formulate the class of equations in the following way. We define a connection

$$D_x(z,Q) = \frac{\partial}{\partial x} - iz\sigma_o - Q \qquad (2.3)$$
$$D_t(z,Q) = \frac{\partial}{\partial t} - A\sigma_o - B\sigma_+ - C\sigma_-$$

where

$$\sigma_o = \begin{pmatrix} 1 & 0 \\ 0 & -1 \end{pmatrix} \qquad \sigma_+ = \begin{pmatrix} 0 & 1 \\ 0 & 0 \end{pmatrix} \qquad \sigma_- = \begin{pmatrix} 0 & 0 \\ 1 & 0 \end{pmatrix}$$

The matrices $\sigma_o, \sigma_+, \sigma_-$ span the Lie algebra $sl(2,R)$ of traceless 2X2 matrices. The functions A,B, and C are functions of z and Q and its derivatives.

By making appropriate choices of A,B, and C we may construct large classes of completely integrable nonlinear evolution equations. For example, with $Q = (u_x/2)(\sigma_+ - \sigma_-)$, $A = i \cos u/4z$ and $B=C=i \sin u/4z$ we get $[D_x,D_t] = (\frac{1}{2})(u_{xt} - \sin u)\sigma_o$. Setting $[D_x,D_t] = \emptyset$ (the curvature of the connection) we get the sine-Gordon equation (2.1).

Similarly, the nonlinear Schrodinger equation is obtained by taking $Q = q\sigma_+ - \bar{q}\sigma_-$ and $A = 2iz^2 - i|q|^2$, $B = -2zq + iq_x$, and $C = 2z\bar{q} + i\bar{q}_x$, and putting $[D_x,D_t] = \emptyset$. In fact what one obtains is

$$[D_x,D_t] = \sigma_+\{q_t - iq_{xx} - 2i|q|^2 q\} - \sigma_-\{\bar{q}_t + i\bar{q}_{xx} + 2i|q|^2\bar{q}\}.$$

So setting $[D_x,D_t] = 0$ we get the equation

$$q_t - iq_{xx} - 2i|q|^2 q = 0$$

and its complex conjugate. Called the nonlinear Schrodinger equation, it was first investigated by Zakharov and Shabat in 1972.

By making other choices for Q and the functions A,B,C,

other completely integrable systems of equations can be obtained. These nonlinear partial differential equations are completely integrable: they have a Hamiltonian structure, an infinite number of conservation laws, and they can be formally integrated by the method of inverse scattering. These integrable systems are based on the semi-simple Lie algebra sl(2,C); but completely integrable systems can be constructed on any semi-simple Lie algebra. (cf. [9], [12])

The connection $\{D_x, D_t\}$ acts on the bundle $C^2 \times R^2$. We need not go deeply into the geometry of connections on vector bundles here, but we do need to explain the concept of gauge transformations. Let $\Psi(x,t)$ denote a solution of the simultaneous equations

$$D_x \Psi = 0 \qquad D_t \Psi = 0 .$$

These equations may be solved simultaneously provided the potential Q is chosen so that $[D_x, D_t] = 0$. In that case the connection is said to be flat. We shall write $D_j = \partial/\partial x_j - U_j$ where $x_1 = x$ and $x_2 = t$.

Suppose we have two potentials Q and $\tilde{Q}$, and write

$$D_j = \frac{\partial}{\partial x_j} - U_j \quad \text{and} \quad \tilde{D}_j = \frac{\partial}{\partial x_j} - \tilde{U}_j .$$

Now suppose that any two corresponding wave functions Ψ and $\tilde{\Psi}$ are simply related by a transformation $\tilde{\Psi} = G\Psi$, where G is an $n \times n$ matrix, in such a way that

$$(D_j \Psi) = \tilde{D}_j \tilde{\Psi} ,$$

i.e.

$$G D_j \Psi = \tilde{D}_j G \Psi .$$

It is easily seen that G must satisfy the differential equations

$$\frac{\partial G}{\partial x_j} + G U_j - \tilde{U}_j G = 0 \tag{2.4}$$

which also may be written as:

$$\tilde{\Psi} = G\Psi \;, \qquad \tilde{U}_j = \frac{(\partial G)}{\partial x_j} G^{-1} + GU_j G^{-1} \;.$$

Such transformations of the wave function Ψ and the potentials U_j are called gauge transformations. The gauge transformations form a representation of the gauge group: If we define

$$T_G U_j = \partial G/\partial x_j G^{-1} + GU_j G^{-1}$$

we readily find that $T_{G_1 G_2} = T_{G_1} T_{G_2}$.

We shall obtain Backlund transformations as gauge transformations G such that

$$D_x(z,Q_2)G = GD_x(z,Q_1) \qquad D_t(z,Q_2)G = GD_t(z,Q_1) \;.$$

3. Automorphisms and the projective bundle.

We now turn to the construction of the x-component of the Backlund-gauge transformations. The x-component of the transformation gives the Backlund transformation for the entire hierarchy of nonlinear evolution equations, since the x-component of the connection is the same for the entire hierarchy. The proof that the gauge transformation can also be chosen so that $D_t(z,Q_2)G = GD_t(z,Q_1)$ will be given in a subsequent paper. We shall examine the two cases in which the potential Q is skew symmetric or skew-Hermitian. We begin with the case $Q^t = -Q$, i.e. $Q = q(\sigma_+ - \sigma_-)$. This kind of potential arises in the case of the sine-Gordon equation. We shall refer to it as the $so(2)$ case, since the real 2X2 skew-symmetric matrices form the real Lie algebra $so(2)$.

The differential equations $D_x(z,Q)\Psi = 0$ can be written

$$\frac{\partial \Psi_1}{\partial x} = iz\Psi_1 + q\Psi_2$$

$$\frac{\partial \Psi_2}{\partial x} = -iz\Psi_2 - q\Psi_1 \;.$$

Setting $u = \Psi_2/\Psi_1$, one obtains the Riccati equation

$u_x = -2izu+q(1+u^2)$. One may also obtain a Riccati equation from the corresponding t-equation, $D_t(z,Q)\Psi = 0$. Chen [4] constructed Backlund transformations by considering various automorphisms of this pair of Riccati equations. The automorphisms he chose depended on the specific equation in the hierarchy of integrable equations, that is, on the choice of A,B,C. However, the x-component of the Backlund transformation depends only on the operator D_x and is the same for all equations of the hierarchy. We shall construct Backlund transformations based on the Cartan automorphisms of the algebras $sl(2,R)$ and $sl(2,C)$. In a later paper we shall show that the Backlund transformation can be chosen so that G intertwines with both D_x and D_t.

Let σ denote the automorphism of the Lie algebra $sl(2,R)$ defined by $\sigma(A) = -A^t$, and let $D_x(z)^{\sigma}=\partial/\partial x-\sigma(iz\sigma_o+Q)=\partial/\partial x+iz\sigma_o-Q$ (since $Q^t = -Q$). The Backlund transformation from Q to $\tilde{Q}$ is obtained by stipulating that

$$D_x(z_o,\tilde{Q})^{\sigma}g\Psi = D_x(z_o,Q)\Psi \qquad (3.1)$$

for some fixed z_o, wave function Ψ, and scalar function g. In other words,

$$\left(\frac{\partial}{\partial x} - iz_o\sigma_o-Q\right)\Psi = 0 \qquad \left(\frac{\partial}{\partial x} + iz_o\sigma_o-\tilde{Q}\right)g\Psi = 0 .$$

The fact that g is a scalar means that Ψ and $g\Psi$ are equivalent when considered projectively. Setting $u = \Psi_2/\Psi_1 = g\Psi_2/g\Psi_1$, we obtain the two Riccati equations:

$$u_x = -2iz_ou+q(1+u^2) \qquad u_x = 2iz_ou+\tilde{q}(1+u^2) . \qquad (3.2)$$

The Backlund transformation from q to $\tilde{q}$ is obtained by eliminating u from these two equations. For example, adding and subtracting the two equations we get

$$(\tilde{q}+q)/2 = u_x/(1+u^2) = d/dx \tan^{-1}u .$$

and

$$(\tilde{q}-q)/2 = 2iz_0u/(1+u^2) .$$

Setting $u = \tan \gamma$ we get

$$(\tilde{q}+q)/2 = \gamma_x \tag{3.3}$$

$$(\tilde{q}-q)/2 = iz_0 \sin 2\gamma .$$

In the case of the sine-Gordon equation we take $q = \omega_x/2$. Then the first equation becomes $(\tilde{\omega}+\omega)_x/2 = 2\gamma_x$. Taking the constant of integration to be zero we can assert that $(\tilde{\omega}+\omega)/2 = 2\gamma$ and

$$(\omega-\tilde{\omega})_x/2 = 2iz_0 \sin \frac{(\tilde{\omega}+\omega)}{2}$$

which is the x part of the Backlund transformation (2.2).

We shall find it more convenient, however, not to eliminate the function u, and to deal directly with the Riccati equations (3.2) themselves. We note then, for example, that (3.2) can be written as

$$q = F(z_0,u) \qquad \tilde{q} = F(z_0,-1/u) = F(-z_0,u)$$

where $F(z_0,u) = (u_x+2iz_0u)/(1+u^2)$. Thus the Backlund transformation $q\to\tilde{q}$ is achieved by the automorphism $u\to-1/u$. On the other hand, $v = -1/u$ satisfies the Riccati equation

$$v_x = 2iz_0v+q(1+v^2) .$$

If we want the new potential $\tilde{q}$ to be real then we must restrict z_0 to be purely imaginary: $z_0 = i\omega_0$, say. Then everything is real. In passing from the real vector $\{\psi_1,\psi_2\}$ to the quotients $u = \psi_2/\psi_1$ or $v = -\psi_1/\psi_2$ we are passing from the vector space R^2 to the real projective space $P_1(R)$, the space of lines through the origin in R^2. The function $u(x,t)$ may be viewed as a section of the projective bundle associated with the bundle R^2 over R^2. $P_1(R)$ may be regarded as the unit circle with antipodal points identified. The pair of coordinates u and v give a covering of this space. In dealing with these parameters one must always make sure that no singularities arise. For example, if ψ_1 vanishes at a point then u becomes infinite; but then we may use v instead. The Backlund transformation was the result of the automorphism $u \to -1/u$. The effect of this transformation in the space $P_1(R)$ is shown in the diagram below:

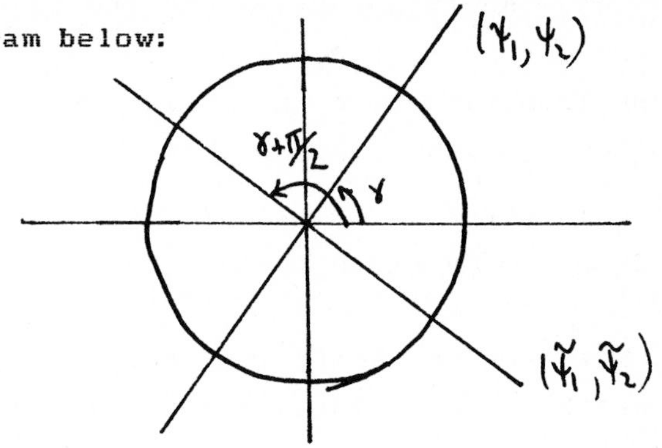

$$u = \psi_2/\psi_1 = \tan \gamma \qquad -1/u = -\cot \gamma = \tan \gamma + \pi/2$$

In terms of the parameter $\gamma = \tan^{-1} u$, the orthogonal matrix O in (1.3) is

$$O = \begin{pmatrix} \cos \gamma & \sin \gamma \\ -\sin \gamma & \cos \gamma \end{pmatrix}$$

For the case $Q = -Q^*$ (the symmetry that arises in the nonlinear Schrodinger equation; we refer to it as the su(2) case), we use the automorphism $\tau(A) = -A^*$, where A^* denotes the Hermitian conjugate of A. We define D_x^{τ} as before; and again using (3.1) to obtain the Backlund transformation, we arrive at the pair of Riccati equations

$$u_x = -2iz_o u+\bar{q}+qu^2 \qquad u_x = -2i\bar{z}_o u+\bar{\tilde{q}}+\tilde{q}u^2 .$$

To find the transformation from q to $\tilde{q}$ we adjoin the equations obtained by complex conjugation; thus for $\bar{u}$ we get the equation

$$\bar{u}_x = 2i\bar{z}_o\bar{u}+q+\bar{q}\ \bar{u}^2 \ .$$

Solving this pair of equations for q and $\bar{q}$ we get

$$q = \{\bar{u}_x-\bar{u}^2 y_x-2i\bar{u}(\bar{z}_o+z_o|u|^2)\}/(1-|u|^4) \ . \qquad (3.5)$$

The singularity $(1-|u|^4)^{-1}$ is only an apparent singularity. For if q is a smooth potential and the wave function Ψ is smooth, there can be no singularity on the right hand side. The Backlund transformation to $\tilde{q}$ is obtained by simply interchanging z_o and $\bar{z}_o$. A straightforward calculation shows that this may also be effected by the transformation $u\to -1/\bar{u}$. Thus the Backlund transformation is given by (1.5) in this case.

It is not hard to verify that the solvability equations for $\tilde{q}$ are the same of those for q if u should pass through the unit circle. In fact, if $u = e^{i\theta}$ at some point we find from the equations for q and $\bar{q}$ that at such a point we must have

$$\bar{u}_x-\bar{u}^2 u_x-2i\bar{u}(z_o+\bar{z}_o) = 0 \ .$$

But this equation is symmetric in z_o and $\bar{z}_o$, so the solvability condition for $\tilde{q}$ and $\bar{\tilde{q}}$ and is also satisfied.

This time the Backlund transformation $q\to\tilde{q}$ is effected by the automorphism $u\to -1/\bar{u}$ on $P_1(C)$. The complex manifold $P_1(C)$ is the space of all complex lines through the origin in C^2. Thus a point (z_1,z_2) in C^2 may be represented either as $(z_1/z_2,1)$ or as $(1,z_2/z_1)$. $P_1(C)$ may be identified with the extended complex plane, thus with Riemann sphere. It is an oriented manifold. The transformation $u\to -1/\bar{u}$ takes a point on the Riemann sphere to its antipodal point. In contrast to the situation in the real case, however, antipodal points on $P_1(C)$ are not identified.

Theorem 1.1 is easily verified by a direct calculation. For example, in the $so(2)$ case with $u = \tan \gamma$ we find that $O_x = BO$, where

$$B = \begin{pmatrix} 0 & 1 \\ -1 & 0 \end{pmatrix} \gamma_x$$

Then (2.1) becomes

$$G_x = [B,G] = iz[\sigma_o,G]+\tilde{Q}G-GQ$$

After some computations this reduces to

$$z_o(2\gamma_x-q-\tilde{q}) \begin{pmatrix} \sin 2\gamma & \cos 2\gamma \\ \cos 2\gamma & -\sin 2\gamma \end{pmatrix}$$

$$= z\left[-iz_o[\sigma_o,O\sigma_oO^t] + \begin{pmatrix} 0 & 1 \\ -1 & 0 \end{pmatrix} (\tilde{q}-q)\right]$$

Since this equation must hold for all z we get

$$\tilde{q}-q = -2iz_o \sin 2\gamma \quad \text{and} \quad 2\gamma_x = q+\tilde{q}$$

which are equations (3.3).

4. Evolution of the scattering data under the flow.

The transformation of the scattering data due to a Backlund transformation is easily calculated using the gauge transformation. For the KdV equation, whose isospectral problem is the Schrodinger equation, the transformation of the scattering data was computed [8], [6] using (essentially) the Darboux transformation, which can be regarded as a kind of gauge transformation for second order equations. For the 2X2 systems discussed above the scattering data used by Zakharov and Shabat and by Ablowitz et. al. was based on the Jost wave functions. For higher order problems (i.e. when Q is an nXn matrix) it is more convenient to work with the form of the scattering data introduced by Beals and Coifman. We begin with a description of this form of the scattering data. In the appendix we describe briefly the transformation between the two types of scattering data (cf. [11] for a more thorough discussion)

We consider the operator $D_x(z,Q) = \partial/\partial x - izJ - Q$ where Q and J are $n \times n$ matrices, with J diagonal and Q off-diagonal. We restrict our discussion to the case where J has real entries a_k, with $a_1 > a_2 > \ldots a_n$. If ψ is a matrix solution of $D_x(z,Q)\psi = 0$ we set $m(x,z) = \psi e^{-izxJ}$. Then m satisfies the equation

$$\frac{\partial m}{\partial x} = iz[J,m]+Qm \; . \tag{4.1}$$

Lemma 4.1: [1] There is a unique solution to the system of equations (4.1) such that

(i) $m(x,z) \to I$ as $z \to \infty$ or as $x \to -\infty$ for $\operatorname{Im} z \neq 0$.

(ii) m is meromorphic in $\operatorname{Im} z \neq 0$

(iii) m is bounded on $-\infty < x < \infty$ for regular values of z.

The poles in m arise due to the boundary condition (iii) and the normalization that $m \to I$ as $x \to -\infty$. We sometimes denote the components of m which are meromorphic in the upper and lower half planes by $m_{\pm}(x,z)$.

The singularities of m constitute the scattering data of the problem: the jump conditions of m across the real axis; the location of the poles $z_1, z_2, \ldots z_N$; and the principal factors of m at the poles. (For the case of the KdV equation it may be proved that there are finitely many discrete eigenvalues if the potential q satisfies certain growth conditions. For the case of systems it is true for a dense set of potentials in L_1[1].) These quantities are represented in the following way. The jump conditions across the real axis are written in the form

$$m_+(x,\xi) = m_-(x,\xi)V^x(\xi) \; ,$$

where V^x denotes the matrix $e^{i\xi xJ}V(\xi)e^{-i\xi xJ}$.

At an isolated pole we have [14]

Lemma 4.2 At an isolated pole z_0, m can be factored as $m(x,z) = \eta(x,z)V^x$ where η is regular at z_0, $\eta \to I$ as $x \to -\infty$, and

(i) $V = V(z) = I+T(z)$, where T is strictly upper triangular if $\mathrm{Im} z_0 < 0$ and strictly lower triangular if $\mathrm{Im} z_0 > 0$.

(ii) T is a polynomial in $(z-z_0)^{-1}$ and tends to zero as $z \to \infty$.

(iii) Here $V^x = e^{izxJ}V(z)e^{izxJ}$.

Remark: Beals and Coifman only considered the case of simple poles. Potentials for which this is true are generic, in the sense that they belong to an open dense set of all potentials. However, we shall show that multiple poles are introduced into the scattering data by repeated Backlund transformations.

The scattering data for the potential Q thus consists of

$$S(Q) = \{V(\xi);\ z_j;V_j, j=1,\dots N\} .$$

Suppose that

$$D_x(z) = \frac{\partial}{\partial x} - izJ-Q$$

$$D_t(z) = \frac{\partial}{\partial t} - iz^r K-B(z,Q)$$

where J and K are diagonal matrices, and B is a matrix polynomial in Q and its derivatives. (For the construction of B, see [9], or [12].) We assume $B \to 0$ as Q and its derivatives vanish.

Theorem 4.3: Let Q evolve in such a way that $[D_x,D_t] = 0$. Let $V(x,\xi,t)$ be the jump conditions of m across the real axis. Then $m(x,t,\xi+i0) = m(x,t,\xi-i0)V(x,\xi,t)$ where

$$V(x,\xi,t) = \exp\{i\xi^r Kt+i\xi xJ\}V(\xi)\exp\{-i\xi^r Kt-i\xi xJ\} .$$

The isolated poles $z_1,\ldots,z_N$ remain fixed. At an isolated pole z_o, m can be factored as $m(x,t,z) = \eta(x,t,z)V_o(x,t,z)$ where η is regular at z_o, and

$$\eta_x = iz[J,\eta] + Q\eta$$

$$\eta_t = iz^r[K,\eta]+B(z,Q(t))\eta$$

$$\eta \to I \quad \text{as} \quad x \to -\infty$$

and

$$V_{o,x} = iz[J,V_o] \qquad V_{o,t} = iz^r[K,V_o] .$$

The scattering data at time t is given by
$V_o(z,t) = \mathcal{P}_o \exp\{iz^r Kt\}V_o(z,0)\exp\{-z^r Kt\}$

The evolution of the scattering data for $n\times n$ systems was first given by Beals and Coifman [2] for the case of simple poles. A proof of theorem 4.3 by a different method will be given in a forthcoming article by Zurkowski and myself.

5. Transformation of the scattering data under Backlund Transformations.

The matrices V and V_j also transform in a particularly simple way under the Backlund transformations described in Theorem 1.1. We shall assume $\operatorname{Im} z_o > 0$. For the so(2) case

Theorem 5.1. Let q and $\tilde{q}$ be related by the Backlund transformation (1.2)-(1.3) with $u = \psi_2/\psi_1$, where

$$\psi = \begin{pmatrix} \psi_1 \\ \psi_2 \end{pmatrix} = \begin{pmatrix} m_{11} & m_{12} \\ m_{21} & m_{22} \end{pmatrix} \begin{pmatrix} e^{iz_o x} & 0 \\ 0 & e^{-iz_o x} \end{pmatrix} \begin{pmatrix} 1 \\ a \end{pmatrix}$$

with $m_{ij} = m_{ij}(x,z_o)$. Let $\tilde{V}$ and $\tilde{V}_j$ denote the scattering data for the potential $\tilde{q}$ and let $A(z) = zI - z_o\sigma_o$. Then

$$\tilde{V}(\xi) = A(\xi)V(\xi)A(\xi)^{-1} \qquad \tilde{V}_j = \mathcal{P}_j A(z)V_j(z)A(z)^{-1}.$$

In addition there are new poles at $z = \pm z_o$, for which the matrices $\tilde{V}_{z_o}$ and $\tilde{V}_{z_o}$ are

$$V_{z_o} = \begin{pmatrix} 1 & 0 \\ \frac{-2z_o a}{z-z_o} & 1 \end{pmatrix}$$

and $V(-z_o) = V(z_o)^t$.

Similar results are obtained in the $su(2)$ case. Notice that no new pole is introduced if $a = \emptyset$. This is similar to the situation in the case of the KdV equation. (cf. [8], [6]) In theorem 5.2 below we shall describe the extension of these results to cover the case in which m already has a pole at z_o.

Proof. We begin by showing that

$$u \to \begin{cases} 0 & \text{as} \quad x \to -\infty \\ \infty & \text{as} \quad x \to \infty . \end{cases}$$

Now $u = \Psi_2/\Psi_1 = (m_{21}e^{iz_o x} + am_{22}e^{-iz_o x})/(m_{11}e^{iz_o x} + am_{12}e^{-iz_o x}) =$ $(m_{21}+am_{22}e^{-2iz_o x})/(m_{11}+am_{12}e^{-2iz_o x})$. As $x \to -\infty$, $e^{-iz_o x} \to 0$ and $m_{12}, m_{21} \to 0$. Therefore $u \to 0$ as $x \to -\infty$.

As $x \to \infty$, m also tends to a diagonal matrix [1], [11,Chapt. 3], namely

$$m \to \begin{pmatrix} 1/d & 0 \\ 0 & d \end{pmatrix}$$

where d is an analytic function in $\text{Im} z > 0$. The zeroes of d in the upper half plane are the discrete eigenvalues $z_1, \ldots z_N$. Since $e^{iz_o x} \to 0$ as $x \to \infty$,

$$u = (m_{21}e^{2iz_o x} + am_{22})/(m_{11}e^{2iz_o x} + am_{12}) \to \infty .$$

As a consequence of this asymptotic behavior of u, $\psi = \tan^{-1}u$ tends to 0 at $-\infty$ and $\pi/2 \pmod{\pi}$ as $x\to\infty$, and the orthogonal matrix O in (1.2) goes from I at $-\infty$ to a rotation through an odd multiple of $\pi/2$ at $+\infty$. The total rotation of O over $-\infty$ to ∞ will depend on the number of zeroes of ψ_1. (Is this perhaps related to the number of solitons in Q?) In any event,

$$G \to \begin{cases} A(z,z_o) & x\to-\infty \\ A(z,-z_o) & x\to\infty \end{cases}$$

It follows that the wave function Gm satisfies $\tilde{D}_x Gm = 0$ and tends to $A(z,z_o)$ as $x\to-\infty$. Therefore the normalized wave function $\tilde{m}$ corresponding to the potential $\tilde{Q}$ is given by $\tilde{m} = GmA^{-1}$.

We now compute the transformation of the scattering data under a Backlund transformation. With $m_\pm = m(x,\xi\pm i0)$, we have

$$\begin{aligned} \tilde{V} &= (\tilde{m}_-)^{-1}\tilde{m}_+ \\ &= (Gm_-A^{-1})^{-1}Gm_+A^{-1} \\ &= Am_-^{-1}m_+A^{-1} \\ &= AVA^{-1}. \end{aligned}$$

As for the principal factor $V = V_j$ at the pole z_j of Q, we have (with $A = z-z_o\sigma_o$) $\tilde{m} = GmA^{-1}$. Factoring $m = \eta V^x$, we have

$$\tilde{m} = GmA^{-1} = G\eta A^{-1}(AVA^{-1})^x$$

But by the factorization lemma we may factor AVA^{-1} as $H\tilde{V}$ where $\tilde{V} = \mathcal{P}_j AVA^{-1}$. Hence we may write

$$\tilde{m} = G\eta A^{-1}H^x V^x$$

By the uniqueness of the factorization, $\tilde{\eta} = G\eta A^{-1}H^x$, and the transformed scattering data is $\mathcal{P}_j AVA^{-1}$.

In the case of a simple pole at z_j, for example, we have

$$AVA^{-1} = \begin{pmatrix} 1 & 0 \\ \frac{c_j}{z-z_j}\;\frac{z+z_o}{z-z_o} & 1 \end{pmatrix}$$

and

$$\mathcal{P}_j AVA^{-1} = \begin{pmatrix} 1 & 0 \\ \frac{c_j}{z-z_j}\;\frac{z_j+z_o}{z_j-z_o} & 1 \end{pmatrix}$$

Since $\tilde{m} = GmA^{-1}$, and A^{-1} has poles at $\pm z_o$, $\tilde{m}$ has additional poles at $\pm z_o$. It remains to calculate the corresponding residue matrix $V(z_o)$. Since $\tilde{m}$ has a pole at z_o we can factor it as $\tilde{\eta}(x,z)V^x$ where

$$V^x(z) = \begin{pmatrix} 1 & 0 \\ \frac{ce^{-2iz_o x}}{z-z_o} & 1 \end{pmatrix}$$

Now $\tilde{m}A(z) = \tilde{\eta}V^xA(z) = Gm$. But

$$\tilde{\eta}V^xA(z) = \tilde{\eta}(x,z) \begin{pmatrix} z-z_o & 0 \\ ce^{-2iz_o x} & 2(z+z_o) \end{pmatrix}$$

$$\underset{z \to z_o}{\to} 2z_o e^{-iz_o x} \begin{pmatrix} -\alpha\tilde{\eta}_{12}e^{-iz_o x} & \tilde{\eta}_{12}e^{iz_o x} \\ -\alpha\tilde{\eta}_{22}e^{-iz_o x} & \tilde{\eta}_{22}e^{iz_o x} \end{pmatrix}$$

where $\alpha = -c/2z_o$. Letting $z = z_o$ we get (with $S = \sin\gamma$ and $C = \cos\gamma$)

$$G(z_o)m = O\, z_o(I-\sigma_o)O^t m(x,z_o)$$

$$= 2z_o \begin{pmatrix} S(Sm_{11}-Cm_{21}) & S(Sm_{12}-Cm_{22}) \\ -C(Sm_{11}-Cm_{21}) & -C(Sm_{23}-Cm_{22}) \end{pmatrix}$$

$$= 2z_o e^{-iz_o x}\begin{pmatrix} -\alpha\tilde{\eta}_{12}e^{-iz_o x} & \tilde{\eta}_{12}e^{iz_o x} \\ -\alpha\tilde{\eta}_{22}e^{-iz_o x} & \tilde{\eta}_{22}e^{iz_o x} \end{pmatrix}$$

Comparing the elements of the top row we get

$$(S^2m_{11}-SCm_{21}) = -\alpha\,\tilde{\eta}_{12}e^{-2iz_o x}$$
$$(S^2m_{12}-SCm_{22}) = \tilde{\eta}_{12}$$

so

$$\frac{Sm_{11}e^{iz_o x} - Cm_{21}e^{iz_o x}}{Sm_{12}e^{-iz_o x} - Cm_{22}e^{-iz_o x}} = -\alpha$$

Solving for $u = \tan\gamma = S/C$ we get

$$u = \frac{m_{21}e^{iz_o x} + \alpha\, m_{22}e^{-iz_o x}}{m_{11}e^{iz_o x} + \alpha\, m_{12}e^{-iz_o x}}$$

$$= \psi_2/\psi_1$$

provided we choose $\alpha = a$.

Theorem 5.2. Suppose m has a pole of order r at $\pm z_o$. There is a one parameter family of gauge transformations $G = OAO^t$ such that $\tilde{m} = GmA^{-1}$ has poles of order $r+1$ at $\pm z_o$. The transformation of the principal factors at $\pm z_o$ and the choice of $u = \tan\gamma$ are given by (5.1) and (5.2) below.

The gauge transformation $G = O(z+z_o\sigma_o)O^t$ with $u = -m_{12}(x,z_o)/m_{22}(x,z_o)$ lowers the order of a pole by one.

(Remark: the poles of m in the upper half plane occur in the first column and those in the lower half plane occur in the second column [1].)

Proof. We construct a gauge transformation that introduces poles of order $r+1$ into the scattering data at $\pm z_o$. Again we write $\tilde{m}A = Gm$ and factor m and $\tilde{m}$ at $z = z_o$ as $m = \eta V^{x}$ and $\tilde{m} = \tilde{\eta}\,\tilde{V}^{x}$ where

$$V(z) = \begin{pmatrix} 1 & 0 \\ P_r & 1 \end{pmatrix} \quad \text{and} \quad \tilde{V} = \begin{pmatrix} 1 & 0 \\ \tilde{P}_{r+1} & 1 \end{pmatrix}$$

where

$$P_r = c_1(z-z_o)^{-1}+\ldots+c_r(z-z_o)^{-r}$$

$$\tilde{P}_{r+1} = \tilde{c}_1(z-z_o)^{-1}+\ldots\tilde{c}_{r+1}(z-z_o)^{-r-1}$$

From $\tilde{m}A = Gm$ we have

$$\begin{pmatrix} \tilde{\eta}_{11} & \tilde{\eta}_{12} \\ \tilde{\eta}_{21} & \tilde{\eta}_{22} \end{pmatrix}\begin{pmatrix} z-z_o & 0 \\ \tilde{P}_{r+1}(z-z_o)e^{-2izx} & z+z_o \end{pmatrix}$$

$$= \begin{pmatrix} \tilde{\eta}_{11}(z-z_o) + \tilde{\eta}_{12}\tilde{P}_{r+1}(z-z_o)e^{-2izx} & \tilde{\eta}_{12}(z+z_o) \\ \ldots\ldots & \ldots\ldots \end{pmatrix}$$

$$= \begin{pmatrix} (G\eta)_{11} + G(G\eta)_{12}P_r e^{-2izx} & (G\eta)_{12} \\ \ldots\ldots & \ldots\ldots \end{pmatrix}$$

Comparing the top two rows we get

$$\tilde{\eta}_{22}e^{izx}(z-z_o)+\tilde{\eta}_{12}(z-z_o)\tilde{P}_{r+1}e^{-izx} = (G\eta)_{11}e^{izx} + (G\eta)_{12}P_r e^{-izx}$$

and

$$\tilde{\eta}_{12}(z+z_o) = (G\eta)_{12} .$$

Hence

$$\eta_{11}e^{izx}(z-z_o)+\tilde{\eta}_{12}(z-z_o)\tilde{P}_{r+1}e^{-izx} = (G\eta)_{11}e^{izx}+(z+z_o)\tilde{\eta}_{12}P_r e^{-izx} .$$

This equation must hold identically in z near $z = z_o$. We write this identity as

$$\tilde{\eta}_{11} e^{izx} = (G\eta)_{11} \frac{e^{izx}}{z-z_o} + \tilde{\eta}_{12} e^{-izx} (P_r(\frac{z+z_o}{z-z_o}) - \tilde{P}_{r+1})$$

Using $(z+z_o)/(z-z_o) = 1+2z_o/(z-z_o)$ we compare the coefficients of the negative powers of $(z-z_o)$. Since the left side is regular at $z = z_o$ these all must vanish. Starting with the terms of highest order and working down we get

$$\begin{aligned} \tilde{c}_{r+1} &= 2z_o c_r \ , \\ c_j - \tilde{c}_j &= 2z_o c_{j-1}, \qquad j=r,r-1,\ldots 2 \ . \end{aligned} \tag{5.1}$$

At the lowest order we must determine $c_1-\tilde{c}_1$. It is this step that determines the function γ in the gauge transformation. The coefficient of $(z-z_o)^{-1}$ is

$$(G\eta)_{11}(z_o)e^{iz_o x} + \tilde{\eta}_{12} e^{-iz_o x}(c_1-\tilde{c}_1) = 0 \ . \tag{5.2}$$

Again,

$$G(z_o) = 2z_o \begin{pmatrix} S^2 & -SC \\ -SC & C^2 \end{pmatrix}$$

where $S = \sin \gamma$ and $C = \cos \gamma$. Hence

$$G(z_o)\eta = 2z_o \begin{pmatrix} S^2\eta_{11}-SC\eta_{21} & S^2\eta_{12}-SC\eta_{22} \\ -CS\eta_{11}+C^2\eta_{21} & -CS_\eta+C^2\eta_{22} \end{pmatrix}$$

At $z = z_o$, $\tilde{\eta}_{12} = (G\eta)_{12}/2z_o$; so equation (5.1) becomes

$$(S\eta_{11}-C\eta_{21})e^{iz_o x}+(S\eta_{12}-C\eta_{22})e^{-iz_o x}\Delta = 0$$

where $\Delta = (c_1-\tilde{c}_1)/(2z_o)$. Setting $u = \tan \gamma = S/C$ we can solve this equation for u , obtaining $u = \psi_2/\psi_1$, where ψ is the wave function

$$\Psi = \begin{pmatrix} m_{11}\, e^{iz_o x} & m_{12} e^{-iz_o x} \\ m_{21}\, e^{iz_o x} & m_{22} e^{-iz_o x} \end{pmatrix} \begin{pmatrix} 1 \\ \Delta \end{pmatrix} \qquad (5.3)$$

In particular, when we introduce the first pole at $z = z_o$, the residue is related to Δ by $\tilde{c}_1 = -2z_o\Delta$. The family of Backlund transformations is parametrized by Δ.

We now want to show how to remove a pole, or more generally to reduce the order of a pole at $z = z_o$. We have $\tilde{m} = GmA^{-1}$, but this time we take $A = z+z_o\sigma_o$ and $G = O(z+z_o\sigma_o)O^t$. If m has a pole of order r at z_o we can factor it as

$$m = \eta \begin{pmatrix} 1 & 0 \\ P_r e^{-2izx} & 1 \end{pmatrix}$$

where P_r is a polynomial of order r in $(z-z_o)^{-1}$. Now

$$GmA^{-1} = (G\eta) \begin{pmatrix} (z+z_o)^{-1} & 0 \\ \dfrac{P_r e^{-2izx}}{z-z_o} & (z-z_o)^{-1} \end{pmatrix}$$

must have a pole of order $r-1$ in the first column and be regular in the second column at $z = z_o$. This means that the second column of $G\eta$ must vanish at $z = z_o$. A simple computation shows that

$$G(z_o)\eta = 2z_o \begin{pmatrix} C^2\eta_{11} + CS\eta_{12} & C^2\eta_{12}+CS\eta_{22} \\ SC\eta_{11} + S^2\eta_{21} & SC\eta_{12}+S^2\eta_{22} \end{pmatrix}$$

so we need

$$C\eta_{12}+S\eta_{22} = 0$$

hence

$$u=S/C=\tan\gamma=\sin\gamma/\cos\gamma=-\eta_{12}/\eta_{22}=-m_{12}/m_{22}.$$

This is the choice of the function $u(x) = \tan\gamma(x)$ we must make in order to remove a bound state, or reduce the order of a pole at z_o.

In a later paper we shall give a constructive derivation of the gauge transformations of Theorem 1.1 as well as proofs that the Backlund-gauge transformations preserve the flows in the Hamiltonian hierarchy. A proof of this latter fact was given by Newell [10] for the 2X2 case using Kac-Moody algebras. We shall, in fact, construct "Backlund" type gauge transformations for hierarchies modeled on a general semisimple algebra $\underset{\sim}{g}$.

Appendix

Let us explain briefly how the Jost scattering data is related to the scattering data we have used above. For a discussion of the nXn case, see [11, Chapter 3]. For real ξ let $\phi(x,\xi)$ and $\psi(x,\xi)$ be solutions of $D_x(x,Q)\psi(x,\xi) = 0$ such that

$$\phi(x,\xi)e^{-i\xi xh} \to 0 \quad \text{as} \quad x\to-\infty$$
$$\psi(x,\xi)e^{-i\xi xh} \to 0 \quad \text{as} \quad x\to\infty .$$

These two fundamental solution matrices are related by $\phi = \psi S$, where S is an nXn matrix. In the 2X2 case [7,11]

$$\phi = \|\varphi_-,\varphi_+\| \qquad \psi = \|\psi_+,\psi_-\|$$

$$S = \begin{pmatrix} a & b \\ c & d \end{pmatrix}$$

It happens that φ_- and ψ_- have analytic continuations into the lower half plane, and φ_+, ψ_+ have analytic continuations into the upper half plane. The matrices ϕ,ψ, and S have determinant one, and it is easily found that

$$a = \det\|\varphi_-,\psi_-\| \qquad d = \det\|\varphi_+,\psi_+\| .$$

The eigenvalues of $D_x(z)$ are given by the zeroes of a in $\mathrm{Im}\,z<0$ and those of d in $\mathrm{Im}\,z>0$. At a zero, say z_j, of a, the wave functions φ_- and ψ_- are linearly dependent, hence $\varphi_-(x,z_j) = c_j\psi_-(x,z_j)$. The scattering data in this case consists of the functions b/a and c/d on the real line, the zeroes z_j of a and d, and the corresponding coupling constants c_j. The evolution of this data when the potential $Q(x,t)$ is such that $[D_x,D_t] = 0$ has been worked out, and the Gelfan'd-Levitan-Marchenko integral equation for the inverse scattering problem is usually written in terms of this data [7,11].

For the nXn inverse scattering problems, however, it is more convenient to use the Beals-Coifman scattering data we have introduced above. The transformation of S to this data is obtained via triangular factorizations of S [13,11]:

$$S^{+} = SS^{-} \qquad \text{and} \qquad R^{-} = SR^{+} ,$$

where $^{+}$ denotes an upper triangular matrix and $^{-}$ denotes a lower triangular matrix. Such factorizations are not unique, but are determined up to multiplication of $S^{\pm}$ and $R^{\pm}$ on the right by diagonal matrices. There is, however, a unique solution in which the entries of $S^{\pm}$ and $R^{\pm}$ polynomials in the entries of S.

For the 2x2 case we find

$$S^{+} = \begin{pmatrix} 1 & b \\ 0 & d \end{pmatrix} \quad S^{-} = \begin{pmatrix} d & 0 \\ -c & 1 \end{pmatrix} \quad R^{-} = \begin{pmatrix} a & 0 \\ c & d \end{pmatrix} \quad R^{+} = \begin{pmatrix} 1 & -b \\ 0 & a \end{pmatrix}$$

Having obtained these triangular factorizations of S we form

$$\Omega_{+} = \phi S^{-} = \psi S^{+} \qquad \text{and} \qquad \Omega_{-} = \phi R^{+} = \psi R^{-} .$$

Then $\Omega_{\pm}$ is analytic in $\pm \mathrm{Im} z > 0$. For the 2x2 case we easily find that $\Omega_{+} = \| \psi_{+}, \varphi_{+} \|$ and $\Omega_{-} = \| \varphi_{-}, \psi_{-} \|$. For $\mathrm{Im} z > 0$ the asymptotic behavior of Ω_{+} as $x \to \pm\infty$ is:

$$\Omega_{+} \to \begin{pmatrix} 1 & 0 & 0 & 0 \\ 0 & \Delta_{n1}^{-} & 0 & \\ 0 & 0 & \Delta_{n-2}^{-} . & \\ 0 & 0 & & \\ . & & & \\ 0 & 0 & .. & \Delta_{1}^{-} \end{pmatrix} \qquad x \to \infty$$

and

$$\Omega_{+} \to \begin{pmatrix} \Delta_{n-1}^{-} & 0 & 0 & 0 \\ 0 & \Delta_{n-2}^{-} & 0 & 0 \\ 0 & 0 & \Delta_{n-3}^{-} & 0 \\ \vdots & & & \\ 0 & 0 & & 1 \end{pmatrix} \qquad x \to -\infty$$

where Δ_{j}^{-} are the jth lower minors of the matrix S. Thus, in the 2X2 case $\Delta_i^- = d$ and $\Delta_2^- = \det S = 1$, while $\Delta_i^+ = a$ etc. For the nXn case the lower minors are analytic in Im z>0 and the upper minors are analytic in Imz<0. Moreover [11] the matrices $\Omega_{\pm}(x,z)$ are analytic in ±Imz>0.

Following Beals and Coifman we have chosen the unique factorizations of S in which the diagonal entries of S_- and R^+ are 1. This factorization introduces poles into the wave function at the zeroes of a and d. For the 2X2 case,

$$m_+ = \| \Psi_+/d,\ \varphi_+ \| e^{-izx\sigma_0}$$

$$m_- = \| \varphi_-, \Psi_-/a \| e^{-izx\sigma_0} .$$

References

1. Beals, R. and Coifman, R., "Scattering and inverse scattering for first order systems, Comm. Pure and Applied Math. 37 (1984), 39-90.

2. ______ "Inverse scattering and evolution equations," Comm. Pure and Applied Math. 38 (1985), 29-42.

3. Boiti, M., and Tu, G. "Backlund transformations via gauge transformations," Il Nuovo Cimento, 71B, (1982), 253-264.

4. Chen, H. "Relation between Backlund transformations and inverse scattering problems," in Backlund Transformations, ed. R.M. Miura, Springer Lecture Notes in Mathematics, #515, Heidelberg, 1976.

5. Darboux, G. C.R. Acad. Sciences, Paris, 94 (1882), p. 1456.

6. Deift, P. and Trubowitz, E. "Inverse scattering on the line," Comm. Pure Applied Mathematics, 32, (1979), 121-251.

7. Dodd, Eilbeck, Morris, and Gibbons, Solitons and Nonlinear Equation, Academic Press, 1982.

8. Flaschka, H. and McLaughlin, D. Some comments on Backlund transformations, and the

inverse scattering method," in Backlund Transformations, loc. cit.

9. Flaschka, H. Newell, A., and Ratiu, T.
"Kac-Moody Lie algebras and soliton equations," Physica 9D (1983), 300-323.

10. Newell, A.
Solitons, CBMS, Siam, 1985.

11. Novikov, S., Manakov, S.V., Pitaevskii, L.B., and Zakharov, V.E.,
Theory of Solitons, Plenum Publishing, New York, 1984.

12. Sattinger, D.
"Hamiltonian hierarchies on semisimple Lie algebras," Studies in Appl. Math. 72 (1985), 65-86.

13. Shabat, A.B.
"An inverse scattering problem," Diff. Equations, 15, (1979), 1299-1307.

14. Zurkowski, V.D.
Ph.D thesis, University of Minnesota, June 1987.

RECENT DEVELOPMENTS IN THE THEORY OF NONLINEAR SCALAR FIRST AND SECOND ORDER PARTIAL DIFFERENTIAL EQUATIONS

P.E. Souganidis(*)

Lefschetz Center for Dynamical Systems
Division of Applied Mathematics
Brown University
Providence, Rhode Island 02912

In this note we review some recent applications of the notion of *viscosity solution* of first and second order scalar nonlinear partial differential equations of the form

$$F(y,u,Du) = 0 \quad \text{in} \quad \Omega \subset \mathbb{R}^M \tag{1.1}$$

and

$$F(y,y,Du,D^2u) = 0 \quad \text{in} \quad \Omega \subset \mathbb{R}^M \tag{1.2}$$

Here Ω is an open subset of $\mathbb{R}^M$, F and u are continuous functions of their arguments and Du, D^2u denote the first and second derivatives of u respectively. Equations (1.1) and (1.2) arise in several areas of applications; in particular the calculus of variations, deterministic and stochastic optimal control theory and differential games. (See for example [22],[23],[26],[29],[49],[50],[51],[52] etc.) It is well known that (1.1) and (1.2) do not have global smooth solutions in general; moreover, solutions which satisfy the equations almost everywhere are not unique. Recently M.G. Crandall and P.-L. Lions [9] (also see M.G. Crandall, L.C. Evans and P.-L. Lions [7]) introduced the notion of viscosity solution of (1.1). We recall the definition.

<u>**Definition 1([7],[9]):**</u> A continuous function u is a viscosity subsolution (resp. supersolution) of (1.1) if for every smooth function ϕ and every local maximum (resp. minimum) y of u-ϕ we have

$$F(y,u(y),D\phi(y)) \leqslant 0 \quad (\text{resp. } F(y,u(y),D\phi(y)) \geqslant 0).$$

A continuous function u is a viscosity solution if it is both sub- and supersolution.

It turns out that viscosity solutions are the correct class of generalized solutions of (1.1). Indeed, the original work of M.G. Crandall and P.-L. Lions [9] as well as subsequent work of M.G. Crandall and P.-L. Lions [10], H. Ishii [39],[40],[41] and M.G. Crandall, H. Ishii and P.-L. Lions [8] imply, that under quite general

(*)Partially supported by the NSF under grant No. DMS-8401725 and the ONR under contract No. N00014-83-K-0542.

NATO ASI Series, Vol. F37
Dynamics of Infinite Dimensional Systems
Edited by S.-N. Chow, and J.K. Hale

assumptions on F, viscosity solutions of (1.1) are unique. The definition and the proofs have also been extended to apply to discontinuous functions (H. Ishii [42]) and to infinite dimensional analogues of (1.1) (M.G. Crandall and P.-L. Lions [11],[12],[13],[14]). The existence of viscosity solutions has been established in a series of papers (P.-L. Lions [47],[48], P.E. Souganidis [59], G. Barles [4],[5], H. Ishii [43]). For a review of the general theory we refer to the book by P.-L. Lions [47], and the review papers by M.G. Crandall and P.E. Souganidis [15] and P.-L. Lions [53]. In the next section we briefly describe in the form of examples some of the many applications of the notion of viscosity solutions. These include optimal control, differential games, porous-medium equation, large deviations, reaction-diffusion equations, asymptotic expansions etc. In the last section of this note we touch upon equation (1.2).

2. Example 2.1: Differential games.

A two-person zero sum differential game consists of the *dynamics*, i.e. an ordinary differential equation of the form

$$\begin{cases} \dot{x}(s) = f(x(s),y(s),z(s)), & t \leqslant s \leqslant T \\ x(t) = x \end{cases} \tag{2.1}$$

and the *payoff*

$$P_{x,t}(y,z) = \int_t^T h(x(s),y(s),z(s))ds + g(x(T)) \tag{2.2}$$

where $y(\cdot) \in Y$ and $z(\cdot) \in Z$ with Y,Z compact sets. The underlying idea is that there are two players I and II. Player I controls $y(\cdot)$ and wants to maximize P. Player II controls $z(\cdot)$ and wants to minimize P. This introduces some difficulties concerning how to define a value function. R. Isaacs [38] provided with a heuristic treatment of differential games. Consequently, W. Fleming [22],[23],[24],[25],[26], A. Friedman [35],[36] and R. Elliott and N. Kalton [17] were successful in giving a rigorous treatment. In particular, they defined upper and lower value functions and showed, via involved probabilistic arguments, that these value functions exist and are equal if the Isaacs condition is satisfied. It turns out that all of the above can be considerably simplified by the use of the viscosity solutions. P.-L. Lions [47] first observed the relation between the value function of an optimal control problem and the viscosity solution of the underlying Bellman equation. This observation generalizes to differential games (N. Barron, L.C. Evans and R. Jensen [6], L.C. Evans and P.E. Souganidis [19], P.-L. Lions and P.E. Souganidis [55],[56], P.E. Souganidis [60]). We conclude with a result from L.C. Evans and P.E. Souganidis [19].

<u>Theorem: ([19])</u>. Let U (resp. V) be the upper (resp. lower) value of the differential game (2.1),(2.2). Then U is the unique viscosity solution of

$$\begin{cases} U_t + H^+(x,DU) = 0 \text{ in } \mathbb{R}^N\times[0,T) \\ U(x,T) = g(x) \text{ in } \mathbb{R}^N \end{cases} \quad \text{resp.} \quad \left[\begin{array}{l} V_t + H^-(x,DV) = 0 \text{ in } \mathbb{R}^N\times[0,T) \\ V(x,T) = g(x) \text{ in } \mathbb{R}^N \end{array}\right]$$

where $H^+(x,p) = \min_{z\in Z} \max_{y\in Y} \{f(x,y,z)\cdot p+h(x,y,z)\}$

$$(\text{resp. } H^-(x,p) = \max_{y\in Y} \min_{z\in Z} \{f(x,y,z)\cdot p+h(x,y,z)\}).$$

If $H^+(x,p) = H^-(x,p)$ for every $(x,p) \in \mathbb{R}^N\times\mathbb{R}^N$

(Isaacs condition), then U=V; i.e. the differential game has value.

<u>Example 2.2:</u> Porous medium and eikonal equations.
The porous medium equation

$$\begin{cases} V_{mt} = (m-1)V_m\Delta V_m + |DV_m|^2 \\ V_m(\cdot,0) = V_{m0} \geqslant 0 \end{cases} \tag{2.3}$$

arises naturally as a mathematical model in several areas of applications (e.g. percolation of gas through porous media, radiative heat transfer in ionized plasma, etc.). As far as its mathematical properties are concerned, (2.3) exhibits both parabolic and hyperbolic behavior (for details see J.L. Vasquez [62]). In view of this observation, a natural question is whether V_m converges as $m \downarrow 1$ to the solution of the eikonal equation

$$\begin{cases} V_t = |DV|^2 \\ V(\cdot,0) = V_0 \end{cases} \tag{2.4}$$

which is of main interest in optimal control and geometrical optics ([47]) where it describes the propagation of wave fronts. The relation between (2.3) and (2.4) was examined by D.G. Aronson and J.L. Vasquez [1] and P.-L. Lions, P.E. Souganidis and J.L. Vasquez [58], where we refer for more details concerning the significance of this relation. Here we only state a sample theorem. We have:

Theorem: ([1],[58]) If, as $m \downarrow 1, V_{m0} \to V_0$, then $V_m \to V$ where V is the unique viscosity solution of (2.4). Moreover, $\{x:V_m(x,t) > 0\} \to \{x:V(x,t) > 0\}$ where the convergence holds in the sense of sets.

Example 2.3: Large deviation problems of Freidlin-Ventcel type.
A typical large deviations problem of Freidlin-Ventcel type considers a stochastic differential equation

$$\begin{cases} dX_s^\epsilon = b(X_s^\epsilon)ds + \sqrt{\epsilon}\, c(X_s^\epsilon)dW_s \\ X_s^\epsilon = x \end{cases} \tag{2.5}$$

where W_s is a standard Brownian motion and a sample path functional u^ϵ which goes to 0 as $\epsilon \to 0$. The question is to try to estimate how fast does u^ϵ converge to zero; in particular, to find I such that $u^\epsilon = \exp(\frac{-I+0(1)}{\epsilon})$ as $\epsilon \to 0$.

Typical u^ϵ are $E[\exp(-\lambda\tau_x^\epsilon)]$, $P(X^\epsilon_{\tau_x^\epsilon} \in \Omega)$: $t \leq s \leq T$) etc., where E(f) is the expected value of f, P(A) is the probability of A, $\lambda > 0$, τ_x^ϵ is the exit time of X^ϵ from a smooth domain Ω, Γ is a part of the boundary of Ω etc. Such problems were treated by S.R.S. Varadhan [61], W. Fleming [27], A. Friedman [37] and M.I. Freidlin and A.D. Ventcel [34] via probabilistic and stochastic control methods. It turns out, however, that the notion of viscosity solutions and the techniques that are associated with them provide with purely analytic proofs (L.C. Evans and H. Ishii [18], W. Fleming and P.E. Souganidis [30], S. Kamin [45], M. Bardi [3], A. Eizenberg [16], P.-L. Lions [54]). To illustrate the PDE-viscosity solution techniques, we consider a simple case ([18]) where $u^\epsilon(x) = E[\exp(-\lambda\tau_x^\epsilon)]$ with $b \equiv 0$ and $c \equiv 1$ in (2.5). It is easy to check that u^ϵ solves the equation

$$\begin{cases} \lambda u^\epsilon - \frac{\epsilon}{2}\Delta u^\epsilon = 0 \text{ in } \Omega \\ u^\epsilon = 1 \text{ on } \partial\Omega \end{cases}$$

Let $V^\epsilon = -\epsilon \log u^\epsilon$. A simple calculation shows that V^ϵ solves the PDE

$$\begin{cases} -\frac{\epsilon}{2}\Delta V^\epsilon + \frac{1}{2}|DV^\epsilon|^2 = \lambda \quad \text{in} \quad \Omega \\ \\ V^\epsilon = 0 \quad \text{on} \quad \partial\Omega \end{cases} \tag{2.6}$$

The structure of (2.6) is such that one can easily obtain estimates on $|V^\epsilon|$ and $|DV^\epsilon|$ which are independent of ϵ. Thus, along a subsequence V^ϵ, converges to V which is a viscosity solution of the PDE

$$\begin{cases} \frac{1}{2}|DV|^2 = \lambda \quad \text{in} \quad \Omega \\ \\ V = 0 \quad \text{on} \quad \partial\Omega \end{cases} \tag{2.7}$$

Equation (2.7) has, however, unique viscosity solution which in general can be identified via the underlying optimal control or differential game problem. (For (2.7) it is more or less immediate that $V(x) = \sqrt{2\lambda}\ d(x,\partial\Omega)$([47].)). Thus the whole family V^ϵ converges as $\epsilon \to 0$ and $I = V$.

<u>Example 2.4.</u> . Geometric optics approach to reaction-diffusion equations.

For the sake of a better exposition we consider here the simplest version of a reaction-diffusion equation

$$u_t = \frac{1}{2}\Delta u + f(u) \tag{2.8}$$

One of the most interesting properties of (2.8) is the existence of travelling wave solutions $u = q(x-\alpha t)$, and the convergence of the solutions of the Cauchy problem for (2.8) to a travelling wave solutions as $t \to \infty$ (see R. Fisher [21], A. Kolmogorov, I. Petrovskii and N. Piskunov [46], D. Aronson and H. Weinberger [2]) , under certain assumptions on f. Another way of looking at this problem is to introduce the scalling $u^\epsilon(x,t) = u(\frac{x}{\epsilon},\frac{t}{\epsilon})$ and to investigate what happens as $\epsilon \to 0$ (see M. Freidlin [32],[33]). For definiteness let us consider the problem

$$\begin{cases} u^\epsilon_t = \frac{\epsilon}{2}\Delta u^\epsilon + f(u^\epsilon) \quad \text{in} \quad \mathbb{R}^N \times (0,T) \\ u^\epsilon(x,0) = \chi_G(x) \quad \text{in} \quad \mathbb{R}^N \end{cases} \tag{2.9}$$

where $f(0) = f(1) = 0$, $f(u) > 0$ for $0 < u < 1$, $f(u) < 0$ for $u \notin [0,1]$,

$f'(0) = \sup\limits_{0<u<1} u^{-1}f(u)$ and χ_G the characteristic function of a smooth set G. Either probabilistic methods (M. Freidlin [32],[33]) or the PDE-viscosity solutions methods explained in Example 2.3 (L.C. Evans and P. E. Souganidis [20]) yield the following theorem (which is stated without all the necessary technical assumptions).

Theorem: [32],[20] Let V be the unique viscosity solution of $V_t = \frac{1}{2}|DV|^2 + f'(0)$ in $\mathbb{R}^n\times(0,T)$ with Cauchy data 0 on G and $-\infty$ on $\mathbb{R}^N\backslash G$. Then $u^\epsilon \to 0$ in $\{V(x,t) < 0\}$ and $u^\epsilon \to 1$ in $\{V(x,t) > 0\}$.

The idea of the PDE-viscosity solution approach is to introduce the function $V^\epsilon = \epsilon \log u^\epsilon$ which satisfies the equation

$$\begin{cases} V_t^\epsilon = \frac{\epsilon}{2}\Delta V^\epsilon + \frac{1}{2}|DV^\epsilon|^2 + \dfrac{f(u^\epsilon)}{u^\epsilon} \\ V^\epsilon(x,0) = \begin{cases} 0, & x \in G \\ -\infty, & x \notin G \end{cases}, \end{cases}$$

to obtain a priori estimates and to pass to the limit.

Example 2.5. Asymptotic expansions.

An important question in stochastic control for small noise intensities (W. Fleming [28]) is whether we can approximate the value functions and the optimal policies by the value function and optimal policy of the corresponding deterministic control problem. A related question is whether in case of the large deviations problem (see Example 2.3) we can improve the estimate $\exp-(\frac{I+0(1)}{\epsilon})$ to a real WKB approximation, i.e. to write u^ϵ as an asymptotic series with respect to ϵ. Some results in this direction have been obtained by W. Fleming and P.E. Souganidis [31] via the use of PDE-viscosity solution techniques. We give here a sample result.

Theorem: [31] For $\epsilon \geq 0$ let u_ϵ be the solution of the problem $-\epsilon\Delta u_\epsilon + |Du_\epsilon|^2 + \lambda u_\epsilon = L(x)$ in Ω with $u_\epsilon = 0$ on $\partial\Omega$ where $L \geq c > 0$, $\lambda \geq 0$ and Ω is a domain with smooth boundary. In compact subsets of the regions where u_0 is classical for every $m \geq 1$ we have:

$$u_\epsilon = u_0 + \epsilon v_1 + \epsilon^2 v_2 + \dots + \epsilon^m v_m + 0(\epsilon^{m+1})$$

where v_i can be found by formally differentiating the equation for u_ϵ i times and letting $\epsilon = 0$.

3. We briefly discuss here how to extend the notion of viscosity solution in order to apply to problems of the form (1.2). Such equations, i.e. fully nonlinear second-order elliptic or parabolic equations, arise in the theory of stochastic control and stochastic differential games. For details see P.-L. Lions [49],[50],[52], P.-L. Lions and P.E. Souganidis [57], R. Jensen [44]. We have:

__Definition:__ A continuous function u is a viscosity subsolution (resp. supersolution) of (1.2) if for every smooth function ϕ and a local maximum y (resp. minimum) of u-ϕ we have

$$F(D^2\phi(y),D\phi(y),u(y),y) \leq 0 \text{ (resp. } F(D^2\phi(y),D\phi(y),u(y),y) \geq 0).$$

A continuous function u is a solution if it is both sub- and supersolution of (1.2).

The question of uniqueness of viscosity solution has not yet been settled completely, but only in special cases. In particular, P.-L. Lions [49],[50],[51] has developed a complete theory in the case where F is convex with respect to (D^2u,Du) (stochastic control case). On the other hand, R. Jensen [44] recently obtained the first uniqueness results for the general problem. These results do not allow, however, very general spatial dependence on F. (For more details see P.-L. Lions and P.E. Souganidis [57].) A complete analysis of uniqueness problem will shed light on the theory of stochastic differential games.

4. References

[1] Aronson, D.G. and J.L. Vasquez, The Porous Medium Equation as a Finite Speed Approximation to a Hamilton-Jacobi Equation, J. d'Analyse Nonlineaire, to appear.

[2] Aronson, D.G. and H. Weinberger, Nonlinear diffusion in population genetics, combustion and nerve propagation, Lecture Notes in Math. 446, Springer-Verlag,Berlin-Heidelberg-New York, 1975, 5-49.

[3] Bardi, M., An asymptotic formula for the Green's function of an elliptic operator, to appear.

[4] Barles, G., Existence results for first order Hamilton-Jacobi equations, Nonlinear Analysis, Annales de l'Institut Henri Poincare, 1(1984), 325-340.

[5] Barles, G., Remarques sur des resultats d'existence pour les equations de Hamilton-Jacobi du premier ordre, Annales del' Institut H. Poincare, Analyse Nonlineaire, 2(1985), 21-32.

[6] Barron, E.N., L.C. Evans and R. Jensen, Viscosity solutions of Isaacs' equations and differential games with Lipshitz controls, J. of Diff. Eq., 53 (1953), 213-233.

[7] Crandall, M.G., L.C. Evans and P.-L. Lions, Some properties of viscosity solutions of Hamilton-Jacobi equations, Trans. AMS. 282 (1984), 487-502.

[8] Crandall, M.G., H. Ishii and P.-L. Lions, Uniqueness of viscosity solutions revisited, to appear.

[9] Crandall, M.G. and P.-L. Lions, Viscosity solutions of Hamilton-Jacobi equations, Trans. AMS, 277 (1983), 1-42.

[10] Crandall, M.G. and P.-L. Lions, On existence and uniqueness of solutions of Hamilton-Jacobi equations, Journal of Nonlinear Analysis, TMA 10 (1986), 353-370.

[11] Crandall, M.G. and P.L. Lions, Solutions de viscosite non bornees des equations de Hamilton-Jacobi du premier ordre, C.R. Acad. Sci. Paris 298 (1984), 217-220.

[12] Crandall, M.G. and P.-L. Lions, Hamilton-Jacobi equations in infinite dimensions, Part I, J. Funct. Analysis, 63 (1985), 379-396.

[13] Crandall, M.G. and P.-L. Lions, Hamilton-Jacobi equations in infinite dimensions, Part II, J. Funct. Anal., 65 (1986), 368-405.

[14] Crandall, M.G. and P.-L. Lions, Hamilton-Jacobi equations in infinite dimensions, Part III, J. Funct.. Anal., to appear.

[15] Crandall, M.G. and P.E. Souganidis, Developments in the theory of nonlinear first-order partial differential equations, Proceedings of International Symposium on Differential Equations, Birmingham, Alabama (1983), Knowles and Lewis, eds., North-Holland Math. Studies 92, North Holland, Amsterdam, 1984.

[16] Eizenberg, A., The vanishing viscosity method for Hamilton-Jacobi equations with a singular point of attracting type, to appear.

[17] Elliott, R.J. and N.J. Kalton, The existence of value in differential games, Mem. AMS #126 (1972).

[18] Evans, L.C. and H. Ishii, A PDE approach to some asymptotic problems concerning random differential equations with small noise intensities, Ann. Inst. H. Poincare Analyse Nonlineaire, 2 (1985), 1-20.

[19] Evans, L.C. and P.E. Souganidis, Differential games and representation formulas for solutions of Hamilton-Jacobi-Isaacs equations, Indiana U. Math. J., 33 (1984), 773-797.

[20] Evans, L.C. and P.E. Souganidis, in preparation.

[21] Fisher, R., The advance of advantageous genes, Ann. Eugenics, 7 (1937), 355-369.

[22] Fleming, W.H., The convergence problems for differential games, J. Math. Analysis and Applications, 3 (1961), 102-116.

[23] Fleming, W.H., The convergence problem for differential games II, Advances in Game Theory, Ann. Math. Studies 52, Princeton U.

[24] Fleming, W.H., The Cauchy problem for degenerate parabolic equations, J. Math. Mech. 13 (1964), 987-1008.

[25] Fleming, W.H., The Cauchy problems for a Nonlinear First-Order Partial Differential Equation, J. of Dif. Equations, 5 (1969), 515-530.

[26] Fleming, W.H., Nonlinear partial differential equations, probabilistic and game theoretic methods, Problems in Nonlinear Analysis, CIME, Ed. Cremonese, Roma, 1971.

[27] Fleming, W.H., Stochastic control for small noise intensities, SIAM J. Control 9 (1971), 473-517.

[28] Fleming, W.H., Exit Probabilities and Optimal Stochastic Control, Appl. Math. Optim. 4 (1978), 329-346.

[29] Fleming, W.H. and R.W. Rishel, Deterministic and Stochastic Optimal Control, Springer-Verlag, New York 1975.

[30] Fleming, W.H. and P.E. Souganidis, Asymptotic series and the method of vanishing viscosity, Indiana U. Math. J., 35 (1986), 425-447.

[31] Fleming, W.H. and P.E. Souganidis, PDE-viscosity solution approach to some problems of large deviations, An. Scuola Norm. Sup., to appear.

[32] Freidlin, M.I., Limit theorems for large deviations and reaction-diffusion equations, Annals of Prob., 13 (1985), 639-675.

[33] Freidlin, M.I., Geometric optics approach to reaction-diffusion equations, SIAM J. Appl. Math., 46 (1986), 222-232.

[34] Freidlin, M.I. and A.D. Wentzell, Random Perturbations of Dynamical Systems, Springer-Verlag, New York (1984).

[35] Friedman, A., Differential Games, Wiley, New York, 1971.

[36] Friedman, A., Differential Games, CBMS 18, AMS, Providence, 1974.

[37] Friedman, A., Stochastic Differential Equations and Applications, Vol. II, Academic Press, New York, 1976.

[38] Isaacs, R., Differential Games, Wiley, New York, 1965.

[39] Ishii, H., Remarks on Existence of Viscosity Solutions of Hamilton-Jacobi Equations, Bull. Facul. Sci. & Eng., Chuo Univ., 26 (1983), 5-24.

[40] Ishii, H., Uniqueness of unbounded solutions of Hamilton-Jacobi equations, Indiana Univ. Math. J., 33 (1984), 721-748.

[41] Ishii, H., Existence and uniqueness of solutions of Hamilton-Jacobi equations, Funkcial. Ekvar., to appear.

[42] Ishii, H., Viscosity solutions of Hamilton-Jacobi equations with discontinuous Hamilton and differential games, in preparation.

[43] Ishii, H., Perron's method for Hamilton-Jacobi equations, to appear.

[44] Jensen, R., in preparation.

[45] Kamin, S., Singular perturbation problems and the Hamilton-Jacobi equation, Comm. in PDE, 9 (1984), 197-213.

[46] Kolmogorov, A., I. Petrovskii and N. Piskunov, Etude de l'equation de la quantite de la matiere et son application a un probleme biologique, Moscow Univ. Bull. Math., 1 (1937), 1-25.

[47] Lions, P.-L., Generalized Solutions of Hamilton-Jacobi Equations, Pitman, Boston, 1982.

[48] Lions, P.-L., Existence results for first order Hamilton-Jacobi equations, Ricerche Mat. Napoli, 32 (1983), 3-23.

[49] Lions, P.-L., Optimal control of diffusion processes and Hamilton-Jacobi-Bellman equations, Part 1, Comm. P.D.E. 8, (1983), 1101-1174.

[50] Lions, P.-L., Optimal control of diffusion processes and Hamilton-Jacobi-Bellman equations, Part 2, Comm. P.D.E. 8 (1983), 1229-1276.

[51] Lions, P.-L., Optimal control of diffusion processes and Hamilton-Jacobi-Bellman equations, Part 3, Nonlinear PDE's and their Applications, College de France Seminar, Vol. V, Pitman, London, 1983.

[52] Lions, P.-L., Some recent results in the optimal control of diffusion processes, Stochastic Analysis, Proceedings of the Tanigachi International Symposium on Stochastic Analysis, Katata and Kyoto, 1982, Kinokuniya, Tokyo, 1984.

[53] Lions, P.-L., Viscosity solutions of Hamilton-Jacobi equations and boundary conditions, Proceedings of Conference in PDE's, L'Aquila, February 1986, to appear.

[54] Lions, P.-L., Grandes deviations, Calcul des variations et solutions de viscosite, to appear.

[55] Lions, P.-L. and P.E. Souganidis, Differential games, optimal control and directional derivatives of viscosity solutions of Bellman's and Isaacs' equations, SIAM J. of Control and Optimization, 23 (1985), 566-583.

[56] Lions, P.-L. and P.E. Souganidis, Differential games, optimal control and directional derivatives of viscosity solutions of Bellman's and Isaacs' equations II, SIAM J. of Control and Opt., to appear.

[57] Lions, P.-L. and P.E. Souganidis, Viscosity solutions of second-order equations, stochastic control and stochastic differential games, Proceedings of Workshop on Stochastic Control and PDE's, IMA, June 1986, to appear.

[58] Lions, P.-L., P.E. Souganidis and J.L. Vasquez, The porous medium and eikonal equations, to appear.

[59] Souganidis, P.E., Existence of viscosity solutions of Hamilton-Jacobi equations, J. Diff. Equations, 56 (1985), 345-390.

[60] Souganidis, P.E., Approximation schemes for viscosity solutions of Hamilton-Jacobi equations with applications to differential games, J. of Nonlinear Analysis, Theory Methods and Applications 9 (1985), 217-257.

[61] Varadhan, S.R.S., On the behavior of the fundamental solution of the heat equation with variable coefficients, Comm. Pure Appl. Math., 20 (1967), 431-455.

[62] Vasquez, J.L., Hyperbolic aspects in the theory of the porous medium equation, Proceedings of Workshop on Metastability, IMA 1985, to appear.

Hopf Bifurcation for an Infinite Delay Functional Equation

Olof J. Staffans
Institute of Mathematics
Helsinki University of Technology
SF-02150 Espoo 15, Finland

1. Introduction

In this lecture we discusse a Hopf bifurcation problem for the functional equation

$$x(t) = F(\alpha, x_t), \quad t \in \mathbf{R}. \tag{1.1}$$

Here x is a vector in $\mathbf{R}^n$, the parameter α is an element of a finite dimensional real Banach space $\mathcal{A}$, and F is a mapping from $\mathcal{A} \times BUC(\mathbf{R};\mathbf{R}^n)$ into $\mathbf{R}^n$. Moreover, $F(\alpha, 0) = 0$, so $x \equiv 0$ is a solution of (1.1). In addition we suppose that the linearization of (1.1) has a one-parameter family of nontrivial periodic solutions at a critical value α_0 of the parameter. Our aim is to show that also the nonlinear equation has a one-parameter family of nontrivial periodic solutions for some values of α close to α_0.

Most of the existing Hopf bifurcation results for (1.1) require F to have a finite delay, i.e., the term $F(\alpha, x_t)$ is only allowed to depend on the values of $x(s)$ for $t - r \leq s \leq t$, where r is some fixed finite number (see e.g. [1], [3], [4], [11], [12], and [16]). Some resent results do exists (see [5] and [20]) where F is allowed to include an infinite delay, but then it is still assumed that F has an "exponentially fading" memory, i.g. that $F(\alpha, x_t)$ are well defined also for functions x which grow exponentially at minus infinity. In [7] a Hopf bifurcation result is proved for a certain quadratic scalar integral equation, in which F does not have an exponentially fading memory. In all the infinite delay results listed above it is assumed that F is "smooth" in the sense that it is well defined also for discontinuous functions, and even functions which are locally unbounded. For example, this means that functionals F which evaluate the function x a some point, such as $F(\alpha, x_t) = f(\alpha, x(t-r))$, are prohibited.

The purpose of this lecture is to describe a general Hopf bifurcation result for (1.1) which uses minimal assumptions on the rate of decay of the built-in memory of F. This result is proved in [19], together with an analogous result for the functional differential equation

$$\tfrac{d}{dt}\left(x(t) - F(\alpha, x_t)\right) = G(\alpha, x_t), \quad t \in \mathbf{R}. \tag{1.2}$$

Here we only outline the proofs, and refer the reader to [19] for more details.

The basic result is that the problem of the existence of nontrivial small periodic solutions of (1.1) can be studied through a *bifurcation equation*, the zeros of which correspond in a one-to-one way to periodic solutions of (1.1). In the case when F is a causal operator with exponentially fading memory, the stability properties of the bifurcation

NATO ASI Series, Vol. F37
Dynamics of Infinite Dimensional Systems
Edited by S.-N. Chow, and J. K. Hale

periodic solutions can be determined directly from the the bifurcation equation (see [20]). In the case discussed here the question of the stability of solutions seems to be rather difficult.

2. The Hopf Bifurcation Theorem

As we already mentioned above, the function F in (1.1) is supposed to map $\mathcal{A} \times BUC(\mathbf{R};\mathbf{R}^n)$ into $\mathbf{R}^n$, where the parameter space $\mathcal{A}$ is a finite dimensional real Banach space, and $BUC(\mathbf{R};\mathbf{R}^n)$ is the space of uniformly continuous functions from $\mathbf{R}$ into $\mathbf{R}^n$. The function x in (1.1) is required to belong to $BUC(\mathbf{R};\mathbf{R}^n)$, and $x_t(s) = x(t+s)$ for all $s \in \mathbf{R}$. Under these assumptions (1.1) makes sense as an equation in $BUC(\mathbf{R};\mathbf{R}^n)$.

The hypotheses which we use can be devided into three parts:

- size and smoothness assumptions on F;
- existence of a one-parameter family $ce^{i\nu_0 t}\xi$, $c \in \mathbf{C}$, of periodic solutions of the linearized version of (1.1) for some critical value α_0 of the parameter, and nonexistence of any resonant periodic solutions.
- existence of a "pseudo-inverse" of the linear part of the operator $x \mapsto x(t) - F(\alpha, x_t)$ for α close to α_0.

Under these assumptions we have proved the existence of a bifurcation function $g(\alpha, c)$, where α is the original parameter and $c \in \mathbf{R}^+ = [0, \infty)$, such that there is a one-to-one correspondence between small positive zeros $c(\alpha)$ of the bifurcation equation $g(\alpha, c(\alpha)) = 0$ and small periodic solutions of (1.1) with a period close to $2\pi/\nu_0$. The zero $c(\alpha)$ is roughly proportional to the square root of the amplitude of the corresponding periodic solution of (1.1). The bifurcation function can be given any desired degree of smoothness. In the sequel we fix an integer $k \geq 1$, and require g to be a C^k-function from $\mathcal{A} \times \mathbf{R}^+$ into $\mathbf{R}$.

Before we state our smoothness assumptions, let us look at a specific example which we want to be able to treat. If we let $F(\alpha, x)$ be the function $F(\alpha, x) = G(x(-\alpha))$, where G is a smooth mapping from $\mathbf{C}^n$ into itself and α represents a (real) variable delay, then (1.1) becomes

$$x(t) = G(x(t-\alpha)), \quad t \in \mathbf{R}. \tag{2.1}$$

The right hand side of this equation is smooth with respect to variations in the function x, but not smooth with respect to variations in the parameter α, because it is impossible to take derivatives with respect to α unless x is differentiable. The smoothness assumptions which we use below have been modeled after this example, with a function G which is C^{2k+1} from $\mathbf{C}^n$ into itself.

Let us state the first set of assumptions, i.e. the size and smoothness assumptions on F.

To obtain a reasonably simple result, we feel that it is more or less necessary to assume that F has a "fading memory", i.e., to assume that F does not depend too much on values close to $\pm\infty$. To materialize this statement we introduce an (arbitrary) positive continuous weight function η satisfying $\eta(t) \to 0$ as $|t| \to \infty$, $\eta(0) = 1$ and $\eta(s+t) \geq \eta(s)\eta(t)$ for $s, t \in \mathbf{R}$, and define

$$\eta_j(t) = \frac{\eta(t)}{(1+|t|)^j}, \qquad t \in \mathbf{R}, \quad j \geq 0.$$

For example, one can take η to be $\eta(t) = \big(1+|t|\big)^{-\gamma}$ or $\eta(t) = \big(1+\log(1+|t|)\big)^{-\gamma}$ for some $\gamma > 0$. Next we define $BC_0(\mathbf{R};\mathbf{R}^n;\eta_j)$ to be the space of continuous functions x from $\mathbf{R}$ into $\mathbf{R}^n$ satisfying $\eta_j(t)x(t) \to 0$ as $|t| \to \infty$, with norm $\|x\| = \max_{s\in\mathbf{R}} \eta(s)|x(s)|$, and require $F(\alpha, x_t)$ to be well defined for all functions $x \in BC_0(\mathbf{R};\mathbf{R}^n;\eta_k)$.

The exact set of conditions stated below looks rather complicated, but it has a fairly simple interpretation: We start our with a total of k moments (i.e., $F(\alpha, x)$ maps $\mathcal{A} \times BC_0(\mathbf{R};\mathbf{R}^n;\eta_k)$ into $\mathbf{R}^n$). Every time we differentiate with respect to α we are willing to give up one moment, one time derivative, and two x-derivatives. The exact formulation is the following (here $BC_0^j(\mathbf{R};\mathbf{R}^n;\eta_{k-j})$ is the space of j times differentiable functions on $\mathbf{R}$ which together with its first j derivatives belong to $BC_0(\mathbf{R};\mathbf{R}^n;\eta_{k-j})$):

(F1) F maps $\mathcal{A} \times BC_0(\mathbf{R};\mathbf{R}^n;\eta_k)$ continuously into $\mathbf{R}^n$, and it has partial derivatives of total order up to $2k+1$ and of partial order with respect to α up to k in the following sense: For each j, $0 \leq j \leq k$, $(\frac{\partial}{\partial\alpha})^j F(\alpha, x)$ has $2k+1-2j$ locally bounded, simply continuous derivatives with respect to x as a mapping from $BC_0^j(\mathbf{R};\mathbf{R}^n;\eta_{k-j})$ into $\mathbf{R}^n$, and for each $x \in BC_0^j(\mathbf{R};\mathbf{R}^n;\eta_{k-j})$, $(\frac{\partial}{\partial x})^m F(\alpha, x)$ has $\min\{j, k-(m-1)/2\}$ if m is odd or $\min\{j, k-m/2\}$ if m is even locally bounded, simply continuous derivatives with respect to α as a mapping from $\mathcal{A}$ into $\mathbf{R}^n$. Moreover, $\frac{\partial}{\partial x}F(\alpha,x) - \frac{\partial}{\partial x}F(\alpha,0) \to 0$ in the total variation norm in $M(\mathbf{R};\mathbf{R}^{n\times n})$ as $\alpha \to \alpha_0$ in $\mathcal{A}$ and $x \to 0$ in $BUC(\mathbf{R};\mathbf{R}^n)$.

Here the words "simply continuous" refer to the notion of simple convergence (sometimes also called "strong" or "pointwise" convergence) in a space of linear or multilinear operators. In equation (2.1) the partial derivatives will exist as Frechet derivatives, but it is actually enough if they exist as Gateaux derivatives in a "simple" (pointwise) sense, i.e., before one differentiates a j-th order derivative to get a $(j+1)$-th order derivative one is allowed to apply the j-th order derivative to j-tuple of vectors to get a function from $\mathcal{A} \times BC_0^j(\mathbf{R};\mathbf{R}^n;\eta_{k-j})$ into $\mathbf{R}^n$, which is then differentiated. In particular, in equation (2.1) we do not have joint Frechet differentiability with respect to the pair (α, x) on the indicated spaces.—The final requirement about convergence in the total variation norm makes sense, because it follows from the earlier assumptions with $j = 0$ that $\frac{\partial}{\partial x}f(\alpha,x)$ maps $BC_0(\mathbf{R};\mathbf{R}^n;\eta_k)$ continuously into $\mathbf{R}^n$, hence by Riesz representation theorem, it induces a measure in $M(\mathbf{R};\mathbf{R}^{n\times n})$ (cf. the discussion of the measure $\mu(\alpha)$ below).

The remainder of our assumptions refer to the linear part of F with respect to x. Before we state our assumptions on this linear part, let us rewrite (1.1) in such a way that the linear part of F becomes visible. It follows from (F1) and from the Riesz representation theorem that the linar part of F is induced by an α-dependent $\mathbf{R}^{n\times n}$-valued measure $\mu(\alpha)$. In other words, it is possible to write (1.1) in the form (recall that we assume that $F(\alpha, 0) = 0$)

$$x(t) = (\mu(\alpha) * x)(t) + H(\alpha, x_t), \quad t \in \mathbf{R}, \tag{2.2}$$

where H is of second order and satisfies (F1) with F replaced by H, $\mu(\alpha)$ is a $\mathbf{R}^{n\times n}$-valued measure satisfying

$$\int_{\mathbf{R}}(1+|s|)^k|\mu(\alpha,ds)|<\infty, \tag{2.3}$$

and $(\mu(\alpha)*x)(t)=\int_{\mathbf{R}}\mu(\alpha,ds)x(t-s)$. Replacing H by zero we get the linearized version

$$x(t)=(\mu(\alpha)*x)(t),\quad t\in\mathbf{R}, \tag{2.4}$$

of (2.2).

In particular, (2.3) implies that the Fourier transform

$$\tilde{\mu}(\alpha,\omega)=\int_{\mathbf{R}}\mu(\alpha,ds)e^{i\omega s},\quad \omega\in\mathbf{R},$$

of $\mu(\alpha)$ is k times continuously differentiable with respect to ω. Moreover, it follows from (F1) and the fact that μ is the linear part of F that $\tilde{\mu}(\alpha,\omega)$ is k times continuously differentiable, jointly in α and ω. The so called characteristic function $\Delta(\alpha,\omega)$ of (2.2) is defined by

$$\Delta(\alpha,\omega)=I-\tilde{\mu}(\alpha,\omega),\quad \omega\in\mathbf{R}. \tag{2.5}$$

Clearly also Δ is k times continuously differentiable with respect to (α,ω).

We suppose that for some critical value $\alpha=\alpha_0$, the linear equation (2.4) has a periodic solution $e^{i\nu_0 t}\xi$, where $\nu_0\in\mathbf{R}$, $\nu_0\neq 0$, and $\xi\in\mathbf{C}^n$, $\xi\neq 0$. This implies that

$$\Delta(\alpha_0,\nu_0)\xi=0,$$

i.e. in the terminology of e.g. [14], ν_0 is an eigenvalue and ξ is an eigenvector of the matrix-valued function $\omega\mapsto\Delta(\alpha_0,\omega)$. We suppose that this eigenvalue is simple, and that there are no resonant eigenvalues, i.e., we assume that

(L1) the function $\omega\mapsto\det\Delta(\alpha_0,\nu_0)$ has a first order zero at ν_0, and $\det\Delta(\alpha_0,k\nu_0)\neq 0$ for all integers $k\neq\pm 1$.

Here "first order" means that $\lim_{\omega\to\nu_0}\det\Delta(\alpha_0,\omega)/(\omega-\nu_0)$ exists and is nonzero.

One final assumption remains to be stated, namely the assumption that the linear part of F has a pseudo-inverse in the following sense:

(L2) There is a α-dependent $\mathbf{R}^{n\times n}$-valued measure $\lambda(\alpha)$ with the same growth and smoothness properties as $\mu(\alpha)$ and a constant $\Omega>0$ such that $\tilde{\lambda}(\alpha,\omega)\Delta(\alpha,\omega)=I$ for $|\omega|\geq\Omega$ and all α in a neighborhood of $\alpha_0\in\mathcal{A}$.

We are now able to state our Hopf bifurcation theorem for (1.1).

Theorem 2.1. *Let F satisfy the growth and smoothness assumption* (F1), *and let* $F(\alpha,0) = 0$ *for all* $\alpha \in \mathcal{A}$. *Let* $\mu(\alpha)$ *be the measure induced by* $\frac{\partial}{\partial x}F(\alpha,0)$, *define the characteristic function* $\Delta(\alpha,\omega)$ *by (2.5), and suppose that* (L1) *and* (L2) *hold. Then there exist* C^k*-functions* $g(\alpha,c)$, $\nu(\alpha,c)$ *and* $u(\alpha,\nu,b)$ *with the following properties: The functions* g *and* ν *map* $\{ \mathcal{A} \times [0,\epsilon] \,|\, |\alpha - \alpha_0| \leq \epsilon \}$ *into* $\mathbf{R}$, *with*

$$g(\alpha_0,0) = 0, \quad \nu(\alpha_0,0) = \nu_0.$$

The function u *maps* $\{ \mathcal{A} \times [\nu_0 - \epsilon, \nu_0 + \epsilon] \times [0,\epsilon] \,|\, |\alpha - \alpha_0| \leq \epsilon \}$ *into the space of* 2π*-periodic continuous functions on* $\mathbf{R}$ *with values in* $\mathbf{R}^n$, *it as derivatives up to order* $2k+1$ *with respect to* b, *and*

$$u(\alpha_0,\nu_0,0) = 0.$$

These functions describe the small periodic solutions of (1.1) with a period close to $2\pi/\nu_0$ *in the sense that a nontrivial* $2\pi/\nu$*-periodic function* x *with*

$$\sup_{t\in\mathbf{R}}|x(t)| \leq \epsilon, \quad |\nu - \nu_0| \leq \epsilon,$$

is a solution of (1.1) if and only if

$$\nu = \nu(\alpha,(c(\alpha)),$$

and, modulo a phase shift,

$$x(t) = u(\alpha,\nu,\sqrt{c(\alpha)})(\nu t),$$

where $c(\alpha)$ *is a solution of the bifurcation equation*

$$g(\alpha,c(\alpha)) = 0$$

in the interval $(0,\epsilon]$. *In particular, in the so called generic case when both* $\frac{\partial}{\partial\alpha}g(\alpha_0,0)$ *and* $\frac{\partial}{\partial c}g(\alpha_0,0)$ *are nonzero (these two numbers are given in formulas (4.1) and (4.2) below), the equation* $g(\alpha,0) = 0$ *defines a* C^k*-manifold with codimension one in a neighborhood of the point* α_0 *in* $\mathcal{A}$ *such that on one side of this manifold (1.1) does not have any small nontrivial periodic solution with period close to* $2\pi/\nu_0$, *and on the other side (1.1) has exactly one small nontrivial periodic solution with period close to* $2\pi/\nu_0$ *(if we ignore phase shifts).*

The function $c(\alpha)$ above has a direct physical interpretation, namely, it is asymptotically (as $c \to 0$) proportional to the square of the norm of the corresponding periodic solution x of (1.1).

In the nongeneric case when either $\frac{\partial}{\partial\alpha}g(\alpha_0,0) = 0$ or $\frac{\partial}{\partial c}g(\alpha_0,0) = 0$ one must compute higher derivatives of g at $(\alpha_0,0)$ in order to determine the bifurcation behavior of (1.1). This can be done through a set of recursive formulas given in [19]. For a fairly detailed discussion of higher order bifurcations we refer the reader to [20] and [21].

3. Outline of Proof

The proof which we outline below looks quite similar to the proofs given for the cases when one has exponentially decaying memories. This is largely due to the fact that we have omitted most of the technical details. The major part of the proof given in [19] deals with difficulties which are either easy to solve, or which do not exist at all in the case of an exponentially fading memory.

To prove Theorem 2.1 we use of a Liapunov-Smith method in the space $P_{2\pi}(\mathbf{R};\mathbf{R}^n)$ of 2π-periodic continuous functions from $\mathbf{R}$ into $\mathbf{R}^n$. However, before we can do this, we have to transform (1.1) into a more suitable form.

In general already a small perturbation in (1.1) leads to a change of frequency of its periodic solutions. Therefore we change the time scale, i.e., we define $u(t) = x(\nu t)$, and look for a 2π-periodic solution u of a modified version of (1.1). Let us define two operators τ_h and σ_ν, which act on $BUC(\mathbf{R};\mathbf{R}^n)$, by

$$(\tau_h x)(s) = x(s+h), \quad (\sigma_\nu x)(s) = x(s/\nu), \quad h \in \mathbf{R}, \quad \nu \in (0,\infty), \quad s \in \mathbf{R}.$$

Then the function u defined above can be written in the form $u = \sigma_\nu x$. Applying σ_ν to (1.1), and replacing x by $\sigma_{1/\nu} u$, we get

$$u(t) = F(\alpha, \tau_{t/\nu}(\sigma_{1/\nu} u)), \quad t \in \mathbf{R}.$$

This equation can also be written in the form

$$u(t) = F_\nu(\alpha, \tau_t u), \tag{3.1}$$

where F_ν is the mapping $F_\nu(\alpha,u) = F(\alpha, \sigma_{1/\nu} u)$. This equation is of the same form as (1.1). The only difference is that we have replaced the original parameter α by a pair of parameters (α,ν).

One can show that the right hand side $F_\nu(\alpha,\tau_t u)$ of (3.1) maps $BUC(\mathbf{R};\mathbf{R}^n)$ into itself, and that it behaves in a reasonably good way with respect to all the variables:

Lemma 3.1. *The operator which maps (α,ν,u) into the function $t \mapsto F_\nu(\alpha,\tau_t u)$ is continuous from $\mathcal{A} \times (0,\infty) \times BUC(\mathbf{R};\mathbf{R}^n)$ into $BUC(\mathbf{R};\mathbf{R}^n)$. Moreover, if we denote the parameter pair (α,ν) bu β, the for each $j \le k$ and $m \le 2k+1-2j$, the operator which maps u into the function $t \mapsto (\frac{\partial}{\partial\beta})^j F_\nu(\alpha,\tau_t u)$ has $2k+1-2j$ locally bounded, simply continuous derivatives as a mapping from $BUC^{j+m}(\mathbf{R};\mathbf{R}^n)$ into $BUC^m(\mathbf{R};\mathbf{R}^n)$, and for and for each $u \in BUC^j(\mathbf{R};\mathbf{R}^n)$, the operator which maps β into the function $t \mapsto (\frac{\partial}{\partial u})^m F_\nu(\alpha,\tau_t u)$ has $\min\{j, k-(m-1)/2\}$ if m is odd or $\min\{j, k-m/2\}$ if m is even locally bounded, simply continuous derivatives with respect to β as a mapping from $\mathcal{B} = \mathcal{A} \times (0,\infty)$ into $BUC(\mathbf{R};\mathbf{R}^n)$. Moreover, $\frac{\partial}{\partial u} F_\nu(\alpha,u) - \frac{\partial}{\partial u} F_\nu(\alpha,0) \to 0$ in the total variation norm in $M(\mathbf{R};\mathbf{R}^n)$ as $\alpha \to \alpha_0$ in $\mathcal{A}$ and $u \to 0$ in $BUC(\mathbf{R};\mathbf{R}^n)$, uniformly for ν in some neighborhood of* 1.

Now that we have settled the question of the smoothness of equation (3.1) with respect to α, ν and u, let us use Hypothesis (L2) together with the second part of Hypothesis (L1) to reduce the question of the existence of small periodic solutions of (3.1) to a finite dimensional fixed point problem.

It follows from (L2) that for all ν sufficiently close to ν_0 and all α sufficiently close to α_0,

$$\tilde{\lambda}(\alpha, j\nu)\Delta(\alpha, j\nu) = I, \quad |j| > K,$$

where K is some sufficiently large number. By convolving λ with another suitable scalar real measure we may further assume that

$$\tilde{\lambda}(\alpha, j\nu)\Delta(\alpha, j\nu) = 0, \quad |j| \leq K.$$

Then, for every function $u \in P_{2\pi}(\mathbf{R}; \mathbf{R}^n)$,

$$(\lambda(\alpha) * u)(t) = u(t) - \sum_{j=-K}^{K} [u]_j e^{ijt},$$

where we have used the notation $[u]_j$ for the j-th Fourier coefficient

$$[u]_j = \frac{1}{2\pi} \int_0^{2\pi} e^{-ijt} u(t) dt.$$

of u. In particular, if we write (3.1) in the form

$$u(t) = (\mu_\nu(\alpha) * u)(t) + H_\nu(\alpha, \tau_t u),$$

where $\mu_\nu(\alpha, ds) = \mu(\alpha, ds/\nu)$ and H_ν is defined analogously to F_ν, and convolve this equation with $\lambda_\nu(\alpha, ds) = \lambda(\alpha, ds/\nu)$, then we get

$$u(t) - \sum_{j=-K}^{K} [u]_j e^{ijt} = \int_R \lambda_\nu(\alpha, ds) H_\nu(\alpha, \tau_{t-s} u).$$

The next step in the proof is to use a fixed point theorem (given in [19]) to solve this equation modulo the function

$$v(t) = \sum_{j=-K}^{K} [u]_j e^{ijt},$$

i.e., we assume that we know v, and solve the equation for the remainder $w = u - v$. The function $u = u(\alpha, \nu, v)$ which we get in this way is 2π-periodic and solves (3.1) modulo a finite number of Fourier coefficients. More specifically, we have

$$u(\alpha, \nu, v)(t) - F_\nu(\alpha, \tau_t u(\alpha, \nu, v)) = \sum_{j=-K}^{K} (\Delta(\alpha, j\nu)[u]_j - [h]_j) e^{ijt},$$

where $[h]_j$ is the j-th Fourier coefficeint of the function $t \mapsto H_\nu(\alpha, \tau_t u(\alpha, \nu, v))$, i.e.,

$$[h]_j = \frac{1}{2\pi} \int_0^{2\pi} e^{-ijt} H_\nu(\alpha, \tau_t u(\alpha, \nu, v)).$$

This means that in order to complete the proof of Theorem 2.1, we have to show that the set of equations

$$\Delta(\alpha, j\nu)[u]_j - [h]_j = 0, \quad |j| \leq K, \tag{3.2}$$

can be reduced to one scalar equation of the type given in Theorem 2.1.

We observe immediately that by using (L1) together with the implicit function theorem we can solve the equations above with $k \neq \pm 1$ for the function v as a function of ν, α, and of the first Fourier coefficient $[u]_1$ of u (note that $[u]_{-1} = \overline{[u]_1}$). Then we are left with an equation in $\mathbf{C}^n$, i.e., with the equation

$$\Delta(\alpha,\nu)[u]_1 - [h]_1 = 0. \tag{3.3}$$

The next step is to reduce the equation (3.3) in n complex dimensions to an equation in one complex dimension. For this we use Condition (L1).

Under our smoothness assumptions, the zero of $\Delta(\alpha_0,\omega)$ at ν_0 will automatically be a "locally analytic zero" in the sense of Definition 3.1 in [13] (with respect to the weight function $\rho \equiv 1$). Moreover, $\Delta(\alpha_0,\omega)$ has a first order Smith decomposition

$$\Delta(\alpha_0,\omega) = P(\omega)\begin{pmatrix} I & 0 \\ 0 & (\omega-\nu_0)\end{pmatrix} Q(\omega)$$

in a neighborhood of ν_0, where I is the identity matrix in $\mathbf{R}^{(n-1)\times(n-1)}$, and P and Q are k times differentiable matrices which are invertible at ν_0. In particular,

$$\begin{pmatrix} I & 0 \\ 0 & (\omega-\nu_0)\end{pmatrix} = P^{-1}(\omega)\Delta(\alpha_0,\omega)Q^{-1}(\omega),$$

so if we let ς be the last row vector in $P^{-1}(\nu_0)$ and ξ the last column vector in $Q^{-1}(\nu_0)$, then

$$\varsigma\Delta(\alpha_0,\nu_0) = 0, \quad \Delta(\alpha_0,\nu_0)\xi = 0, \quad \varsigma\frac{\partial}{\partial\omega}\Delta(\alpha_0,\nu_0)\xi = 1. \tag{3.4}$$

Let us decompose $\mathbf{C}^n$ in two different ways:

$$\begin{aligned}\mathbf{C}^n &= \text{span}[\xi] \oplus X_1 = X_0 + X_1,\\ \mathbf{C}^n &= Y_0 \oplus \text{range}[\Delta(\alpha_0,\nu_0)] = Y_0 + Y_1,\end{aligned}$$

where X_0 is the null space of $\Delta(\alpha_0,\nu_0)$, Y_1 is the range of $\Delta(\alpha_0,\nu_0)$, and the complementary spaces X_1 and Y_0 can be chosen e.g. in the following way: As $\varsigma\frac{\partial}{\partial\omega}\Delta(\alpha_0,\nu_0)\xi \neq 0$, it follows that the set $\{x \in \mathbf{C}^n \mid \varsigma\frac{\partial}{\partial\omega}\Delta(\alpha_0,\nu_0)x = 0\}$ is a $(n-1)$-dimensional subspace which is complementary to the span of ξ. We choose this space to be X_1. Similarily, we have $\varsigma y = 0$ for every y in the range of $\Delta(\alpha_0,\nu_0)$, and therefore $\frac{\partial}{\partial\omega}\Delta(\alpha_0,\nu_0)\xi$ does not belong to the range of $\Delta(\alpha_0,\nu_0)$. We choose Y_0 to be the span of $\frac{\partial}{\partial\omega}\Delta(\alpha_0,\nu_0)\xi$.

The preceding choices of X_1 and Y_0 makes it easy to construct complementary projections P_0 and P_1 which split $\mathbf{C}^n$ into $X_0 \oplus X_1$, and complementary projections Q_0 and Q_1 which split $\mathbf{C}^n$ into $Y_0 \oplus Y_1$. One simply defines

$$\begin{aligned} P_0 &= \xi\varsigma\frac{\partial}{\partial\omega}\Delta(\alpha_0,\nu_0), \quad P_1 = I - P_0,\\ Q_0 &= \frac{\partial}{\partial\omega}\Delta(\alpha_0,\nu_0)\xi\varsigma, \quad Q_1 = I - Q_0.\end{aligned} \tag{3.5}$$

It is easy to check that these operators are projections (use the fact that $\varsigma\frac{\partial}{\partial\omega}\Delta(\alpha_0,\nu_0)\xi = 1$), that the ranges of P_0 and P_1 are X_0 and X_1, and that the ranges of Q_0 and Q_1 are Y_0 and Y_1.

The n dimensional complex equation (3.3) which we want to solve can be written in the form

$$\Delta(\alpha_0,\nu_0)a = \big(\Delta(\alpha_0,\nu_0) - \Delta(\alpha,\nu)\big)a + [h]_1.$$

We split this equation into two by applying Q_0 and Q_1 to both sides. This gives us

$$\begin{aligned}\Delta(\alpha_0,\nu_0)P_1 a &= Q_1\big(\Delta(\alpha_0,\nu_0) - \Delta(\alpha,\nu)\big)a + Q_1[h]_1,\\ 0 &= Q_0\big(\Delta(\alpha_0,\nu_0) - \Delta(\alpha,\nu)\big)a + Q_0[h]_1,\end{aligned} \tag{3.6}$$

The operator $\Delta(\alpha_0,\omega_0)$ is invertible as a mapping from X_1 into Y_1. This means that we can use the implicit function theorem to solve $P_1 a$ a function of ν, α, and $P_0 a$ from the first of these two equations.

We are left with the second of the two equations above. In this equation we replace Q_0 by $\frac{\partial}{\partial\omega}\Delta(\alpha_0,\nu_0)\xi\varsigma$, use the fact that $Q_0\Delta(\alpha_0,\nu_0) = 0$, denote $P_0 a$ by $b\xi$, and split the result into its real and imaginary parts to get

$$\begin{aligned}\Re\Big\{\varsigma\Delta(\alpha,\nu)(P_1 a + b\xi) - \varsigma[h]_1\Big\} &= 0,\\ \Im\Big\{\varsigma\Delta(\alpha,\nu)(P_1 a + b\xi) - \varsigma[h]_1\Big\} &= 0.\end{aligned} \tag{3.7}$$

It follows from (3.4) that one can use the implicit function theorem to solve ν as a function of b from the first of these two equations.

After one final change of variable we arrive at the conclusion of Theorem 2.1. We define $c = b^2$, or equivalently, $b = \sqrt{c}$. The bifurcation equation $g(\alpha,c) = 0$ comes from the second equation in (3.7). We define G by

$$G(\alpha,\nu,b) = \varsigma\Delta(\alpha,\nu)a - \varsigma[h]_1 = \varsigma\Delta(\alpha,\nu)[u]_1 - \varsigma[h]_1, \tag{3.8}$$

and then the second of the two equations in (3.7) becomes $\Im G(\alpha,\nu,b) = 0$. The function G is odd in b, so we get a function $g(\alpha,c)$ with the required properties by defining

$$g(\alpha,c) = \frac{\Im G(\alpha,\nu(\alpha,c),\sqrt{c})}{\sqrt{c}}. \tag{3.9}$$

4. The Bifurcation Formulas

It is shown in [19] how to compute the derivatives of the bifurcation function g which are needed when one wants to determine the bifurcation behavior of (1.1). Below we only give the formulas which are needed to determine if we are in the generic case or not, i.e., the formulas which determine $\frac{\partial}{\partial\alpha}g(\alpha_0,0)$ and $\frac{\partial}{\partial c}g(\alpha_0,0)$. The first of these derivatives is given by

$$\frac{\partial}{\partial\alpha}g(\alpha_0,0) = \Im\Big\{\varsigma\frac{\partial}{\partial\alpha}\Delta(\alpha_0,\nu_0)\xi\Big\}. \tag{4.1}$$

The second derivative can be computed from the formulas

$$
\begin{aligned}
u_{2,2} &= \left[\Delta(\alpha_0, 2\nu_0)\right]^{-1} H_2(\alpha_0)\left(\xi e^{i\nu_0\cdot}\right)^2, \\
u_{0,2} &= 2\left[\Delta(\alpha_0, 0)\right]^{-1} H_2(\alpha_0)\left(\xi e^{i\nu_0\cdot}\right)\left(\overline{\xi} e^{-i\nu_0\cdot}\right), \\
h_{1,3} &= 3H_3(\alpha_0)\left(\xi e^{i\nu_0\cdot}\right)^2\left(\overline{\xi} e^{-i\nu_0\cdot}\right) + 2H_2(\alpha_0)\left(u_{2,2}e^{2i\nu_0\cdot}\right)\left(\overline{\xi} e^{-i\nu_0\cdot}\right) \\
&\quad + 2H_2(\alpha_0)\left(u_{0,2}\right)\left(\xi e^{i\nu_0\cdot}\right), \\
\frac{\partial}{\partial c} g(\alpha_0, 0) &= -\Im\{\varsigma h_{1,3}\}.
\end{aligned}
\tag{4.2}
$$

As a side product of the computation above we get

$$
\frac{\partial}{\partial c}\nu(\alpha_0, 0) = -\Re\{\varsigma h_{1,3}\},
\tag{4.3}
$$

and we also get the two first terms in the expansion of the periodic function u, namely

$$
u(\alpha_0, \nu_0, b) = 2\Re\{\xi e^{it}\}b + \left[2\Re\{u_{2,2}e^{2it}\} + u_{0,2}\right]b^2 + O(b^3).
\tag{4.4}
$$

References

1. N. Chafee, A bifurcation problem for a functional differential equation of finitely retarded type, J. Math. Anal. Appl. 35 (1971), 312–348.
2. S-N. Chow and J. K. Hale, Methods of Bifurcation Theory, Springer-Verlag, Berlin and New York, 1982.
3. J. R. Claeyssen, Effect of delays on functional differential equations, J. Differential Equations 20 (1976), 404–440.
4. J. R. Claeyssen, The integral-averaging bifurcation method and the general one-delay equation, J. Math. Anal. Appl. 78 (1980), 429–439.
5. O. Diekmann and S. A. van Gils, Invariant manifolds for Volterra integral equations of convolution type, J. Differential Equations 54 (1984), 139–180.
6. R. R. Goldberg, Fourier Transforms, Cambridge University Press, London, 1970.
7. G. Gripenberg, Periodic solutions of an epidemic model, J. Math. Biology 10 (1980), 271–280.
8. G. Gripenberg, On some epidemic models, Quart. Appl. Math. 39 (1981), 317–327.
9. G. Gripenberg, Stability of periodic solutions of some integral equations, J. reine angew. Math. 331 (1982), 16–31.
10. J. K. Hale, Theory of Functional Differential Equations, Springer-Verlag, Berlin and New York, 1975.
11. J. K. Hale, Nonlinear oscillations in equations with delays, in Nonlinear Oscillations in Biology, Lectures in Applied Mathematics, Vol. 17., 157–189, American Mathematical Society, Providence, 1978.
12. J. K. Hale and J. C. F. de Oliveira, Hopf bifurcation for functional equations, J. Math. Anal. Appl. 74 (1980), 41–58.
13. G. S. Jordan, O. J. Staffans and R. L. Wheeler, Local analyticity in weighted L^1-spaces and applications to stability problems for Volterra equations, Trans. Amer. Math. Soc. 274 (1982), 749–782.

14. G. S. Jordan, O. J. Staffans and R. L. Wheeler, Convolution operators in a fading memory space: The critical case, SIAM J. Math. Analysis, to appear.
15. G. S. Jordan, O. J. Staffans and R. L. Wheeler, Subspaces of stable and unstable solutions of a functional differential equation in a fanding memory space: The critical case, to appear.
16. J. C. F. de Oliveira, Hopf bifurcation for functional differential equations, Nonlinear Anal. 4 (1980), 217–229.
17. J. C. F. de Oliveira and J. K. Hale, Dynamic behavior from bifurcation equations, Tôhoku Math. J. 32 (1980), 577–592.
18. O. J. Staffans, On a neutral functional differential equation in a fading memory space, J. Differential Equations 50 (1983), 183–217.
19. O. J. Staffans, Hopf bifurcation of functional and functional differential equations with infinite delay, to appear.
20. H. W. Stech, Hopf bifurcation calculations for functional differential equations, J. Math. Anal. Appl. 109 (1985), 472–491.
21. H. W. Stech, Nongeneric Hopf bifurcations in functional differential equations, SIAM J. Math. Anal. 16 (1985), 1134–1151.

A NUMERICAL ANALYSIS OF THE STRUCTURE OF PERIODIC ORBITS IN AUTONOMOUS FUNCTIONAL DIFFERENTIAL EQUATIONS

Harlan W. Stech
Department of Mathematics and Statistics
University of Minnesota, Duluth
Duluth, Minnesota 55812
and
Department of Mathematics
Virginia Polytechnic Institute
Blacksburg, Virginia 24060

1. Introduction

Understanding the structure of periodic solutions in nonlinear, autonomous functional differential equations is a problem that often arises when such equations are used in the mathematical modeling of "real-world" phenomena. Knowledge of the existence, stability, and parameter dependence of such periodic solutions provides valuable insight into the general dynamics of the system. Stable steady states and periodic orbits are of particular interest since they correspond to observable states in the system being modeled. However, unstable steady states and periodic orbits are of importance as well since (through variation of parameters in the model) these solutions can themselves change stability and therefore, become "observable".

Numerical simulation of the associated initial value problem often provides evidence of the existence of stable equilibria and stable periodic orbits. However, it is of limited value in the study of unstable solutions.

Linearization provides a straight-forward means of analyzing equilibria and their stability types. A careful study of the associated characteristic equation ideally leads to the identification of the subset of parameter space at which a variation of the system parameters can

NATO ASI Series, Vol. F37
Dynamics of Infinite Dimensional Systems
Edited by S.-N. Chow, and J. K. Hale

induce a qualitative change in the nature of solutions near the equilibrium. Generically, system parameters corresponding to the existence of the characteristic value $\lambda=0$ correspond to branch points of equilibria, while the existence of a complex conjugate characteristic root pairs $\lambda=\pm i\omega$ correspond to the existence of small-amplitude periodic orbits.

At parameter values of this last type (Hopf bifurcation) there is now a straight-forward technique for the determination of the stability and parameter dependence (i.e., direction of bifurcation) of such orbits [7]. Fixing all but one system parameter, it is natural to ask how variation of the remaining parameter effects the periodic orbit, inducing changes of stability and secondary bifurcations. Towards this end, the theory of global Hopf bifurcation is valuable in identifying the available alternatives [2].

This paper concerns the use of numerical methods (other than simple simulation studies) to aid in the analysis of both the local and global natures of periodic solutions to parameter-dependent autonomous periodic orbits. Section 2 discusses a numerical implementation ot the Hopf bifurcation algorithm of [7]. Section 3 outlines the use of numerical tracking techniques to determine certain information concerning the global bifurcation picture in one-parameter problems. The usefulness of such techniques is illustrated in Section 3, where the result of the analysis of a model of nerve firing are described.

2. Local Analysis

Consider the differential equation

$$x'(t) = f(\alpha; x_t) \qquad [2.1]$$

in which it is assumed that $x=0$ is an equilibrium for all values of the system parameters $\alpha \in R^k$. Somewhat arbitrarily, we have chosen $f : R^k \times C \to R^n$, where $C = C([-r,0], R^n)$ is the usual Banach space of continuous R^n-valued functions on $[-r,0]$; other phases spaces can be used as well. Given adequate smoothness (which we, henceforth, assume without mention), we may expand the right hand side in series form

$$x'(t) = L(\alpha)x_t + H_2(\alpha;x_t,x_t) + H_3(\alpha;x_t,x_t,x_t) \cdots , \qquad [2.2]$$

where $L(\alpha)$ is bounded and linear on C, and $H_2(\alpha;\cdot,\cdot)$ and $H_3(\alpha;\cdot,\cdot,\cdot)$ are, respectively, bounded symmetric bilinear and trilinear forms on C.

The linearized problem reads

$$y'(t) = L(\alpha)y_t, \qquad [2.3]$$

which has exponential solutions $y(t) = \varsigma\, e^{\lambda t}$ if and only if

$$[\lambda I - L(\alpha)\, e^{\lambda\,\cdot}]\, \varsigma \equiv \Delta(\alpha;\lambda)\varsigma = 0. \qquad [2.4]$$

We assume the existence of a critical parameter $\alpha = \alpha_c$ at which [2.4] has a nontrivial solution (i.e., $\det\Delta(\alpha_c;\lambda) = 0$) with $\lambda = \pm i\omega$ a purely imaginary root pair. Assuming simplicit of the root $i\omega$, it is known that the corresponding characteristic vector ς is uniquely defined up to a scalar multiple. Furthermore, the implicit function theorem shows that there exists a unique smooth family $\lambda(\alpha)$ of characteristic roots defined in a neighborhood of α_c in R^k and satisfying $\lambda(\alpha_c) = i\omega$. For α near α_c, we write $\lambda(\alpha) = \mu(\alpha) + i\omega(\alpha)$ and $\varsigma = \varsigma(\alpha)$. For simplicity, we assume that at α_c there are no other purely imaginary root pairs.

Define $\varsigma^* = \varsigma^*(\alpha) \neq 0$ to be any solution of

$\varsigma^*(\alpha)\Delta(\alpha;\lambda(\alpha))=0$ for α near α_c. For λ near $\lambda(\alpha)$, let $\hat{\varsigma}=\hat{\varsigma}(\alpha;\lambda)\equiv\varsigma^*/[\varsigma^*\Delta'(\alpha;\lambda)\varsigma]$, where $\Delta'=\partial\Delta/\partial\lambda$. By simplicity of the characteristic value $\lambda(\alpha)$ the denominator above is nonzero.

The following theorem, whose proof may be found in [7], reduces the problem of analyzing the existence of small periodic solutions with frequence near ω to that of considering a scalar "bifurcation function".

Theorem 2.1: Under the above hypotheses, there are smooth functions $G(\alpha;c,\nu)$ (C–valued) and $x(t,\alpha;c,\nu)$ (R^n–valued and $2\pi/\nu$–periodic in t) defined in a neighborhood of $(\alpha_c,0,\omega)$ in $R^k\times R\times R$ such that [2.1] has a small $2\pi/\nu$–periodic solution $x(t)$ with (α,ν) near (α_c,ω) if and only if $x(t)=x(t,\alpha;c,\nu)$ up to phase shift, and (α,c,ν) solves the bifurcation equation

$$G(\alpha;\ c,\ \nu) = 0. \qquad [2.5]$$

Moreover,

$$x(t,\alpha;c,\nu) = 2\,\mathrm{Re}\{\varsigma(\alpha)e^{\omega it}\}c + O(c^2), \qquad [2.6]$$

G is odd in c, and has the expansion

$$G(\alpha;\ c,\ \nu) = (\lambda - \nu i)c + M_3(\alpha;\ \nu,\ \lambda)c^3 + O(c^5), \qquad [2.7]$$

where $\lambda=\lambda(\alpha)$, $M_3(\alpha;\nu,\lambda)=\hat{\varsigma}(\alpha;\lambda)\cdot N_3(\alpha;\nu)$,

$$N_3(\alpha;\ \nu) \equiv 3H_3(\varphi^2,\ \overline{\varphi}) + 2H_2(\overline{\varphi},A_{2,2}e^{2\nu i}) + 2H_2(\varphi,\ A_{2,0}),$$

with $\varphi(s)=\varsigma(\alpha)e^{i\nu s}$ for $s\leq 0$ and $A_{2,2}$, $A_{2,0}$ the unique solutions of

$$\Delta(\alpha;\ 2\nu i)A_{2,2} = H_2(\varphi^2),$$

$$\Delta(\alpha;0)A_{2,0} = 2H_2(\varphi,\overline{\varphi}), \tag{2.8}$$

respectively.

The imaginary part of [2.5] can be easily solved (e.g., by iteration) to obtain $\nu=\omega(\alpha)+O(c^2)$. Upon substituting this into the real part of [2.5], one obtains the "reduced" bifurcation equation

$$0 = g(\alpha;\ c) \equiv \mu(\alpha)c + K_3(\alpha)c^3 + O(c^5). \tag{2.9}$$

Given $K_3(\alpha_c)\neq 0$, (the so-called "generic" case), one can show the existence of nonzero solutions $c=c^*(\alpha)$ for values of α near α_c for which $\text{sgn}\{\mu(\alpha)\}=-\text{sgn}\{K_3(\alpha_c)\}$. If in the case k=1 $\mu(\alpha)$ increases with α and $K_3(\alpha_c)<0$, the solution of [2.9] near $c=0$ requires $\mu(\alpha)>0$ (supercritical bifurcation). Similarly $K_3(\alpha_c)>0$ corresponds to subcritical bifurcation.

Concerning the stability of the associated periodic orbits, it is known [7] that if all other characteristic roots have negative real parts, then the periodic orbits posses the same stability type as that of c^* when viewed as an equilibrium solution of the scalar ordinary differential equation

$$c' = g(\alpha;\ c) \tag{2.10}$$

Thus, $K_3(\alpha_c) < 0$ corresponds to an orbitally asymptotically stable periodic orbit, while $K_3(\alpha_c)>0$ corresponds to an unstable periodic orbit.

The above algorithm, although usually too involved to allow an algebraic determination of the structure of Hopf bifurcations, does lend itself to numerical evaluation. This has been recently implemented in the FORTRAN code BIFDE by A. Sathaye [6]. It is there presumed that one can obtain from the linearized equation [2.3] analytic expressions for $\Delta(\alpha;\lambda)$, as well as the partial derivatives of $\Delta(\alpha;\lambda)$ with respect to λ and some (use-chosen) coordinate of α. It is also assumed that one is able to identify critical values of the parameter α_c for which a simple purely imaginary root pair $\lambda=\pm\omega i$ exists, and the other spectral assumptions listed above hold. Finally, it is expected that $H_2(\alpha;\varphi_1,\varphi_2)$ and $H_3(\alpha;\varphi_1,\varphi_2,\varphi_3)$ can be evaluated, where each of the arguments φ_j are of the form $\varphi_j(s)=we^{zs}$ for complex values z and complex n-vectors w.

Given the above data, BIFDE coordinates the calculation of the left and right characteristic vectors ς^* and ς (by inverse iteration), identification and solution of the linear systems [2.8] (by Gauss elimination with implicit pivoting) and the evaluation of N_3 (hence, M_3 and K_3). The program uses the partial derivatives of $\Delta(\alpha;\lambda)$ to compute $\mu'(\alpha_c)$ the partial derivative of μ with respect to a user-chosen coordinate of α, and thereby determine the direction of bifurcation with respect to that coordinate of α.

The program is complementary to BIFDD of Hassard [5] in that BIFDD assumes [2.1] to be of delay-difference form, yet identifies the required higher order terms numerically. In [5], rather than making use of Theorem 2.1, the stability and direction of bifurcation is determined by center manifold approximation techniques and the Poincaré normal form.

Remark 2.2: Given BIFDE or BIFDD, a principle difficulty lies in the determination of the bifurcation data. That is, the critical value(s) of the system parameters α and the associated frequency ω of bifurcating periodic orbits. For one-parameter problems (k=1), solution of

$$\det \Delta(\alpha; i\omega) = 0 \qquad [2.11]$$

can be obtained by standard rootfinding techniques (e.g., Newton on Quasi-Newton methods) provided the size of the system n is not prohibitively large, and a sufficiently accurate approximation to the bifurcation data is known in advance.

For k large, one can seek bifurcation data by considering the associated nonlinear minimization problem:

Minimize $|\Delta(\alpha;i\omega)|^2$, subject to the constraint that α and ω lie within a compact interval of ω values and α lies within a compact subset of admissable system parameters.

Precise approximations for this minimization problem, although useful in specifying the above constraints, are not necessary.

For k=2, one expects the underdetermined system [2.11] to have a one-parameter family of bifurcation data. Given one set of bifurcation data (perhaps by considering the above minimization problem) one can apply now standard continuation techniques to identify curves in parameter space (a subset of R^2) along which [2.11] has a solution. Indeed, in many instances, this family of critical values can be parameterized in terms of ω itself. A simple example serves to illustrate the point.

Example 2.3: Consider an ordinary differential equation

$$x'(t) = f(x(t)); \qquad x \in R^n \qquad [2.12]$$

in which one coordinate x_j of x is thought to act as a feedback in one of the n equations in [2.12]. In studying the effects of time delay in this feedback, one replaces the appropriate term $x_j(t)$ with $x_j(t-r)$. To determine the stabilizing/destabilizing effect on equilibria, one encounters a characteristic equation of the form

$$p(\lambda) + q(\lambda)se^{-r\lambda} = 0, \qquad [2.13]$$

where p and q are polynomials, r corresponds to the length of the time delay, and s represents a measure of the strength and type (positive or negative) of the feedback. One can algebraically solve for r (then s) in terms of λ by considering [2.13] and its conjugate. The details are elementary and omitted. Observe that this provides a convenient reparameterization of [2.1] in terms of λ rather than r and s in which (generically) with $\mathrm{Re}\{\lambda\}=0$, $\mathrm{Im}\{\lambda\}$ determines the location of α_c on the imaginary root curves in R^2, and with $\mathrm{Im}\{\lambda\}$ fixed, $\mathrm{Re}\{\lambda\}$ determines the stability of the equilibrium.

3. Global Analysis

Consider [2.1] in the special case when $k=1$, and suppose that at some critical value of the parameter α_c the equation has been shown to satisfy the hypotheses of Theorem 2.1. Equation [2.6] provides an asymptotic estimate of the resulting one-parameter family of periodic orbits bifurcating from the equilibrium. We discuss in this section numerical methods for continuation of this one-parameter family away from the equilibrium, calculation of the stability of the orbits, and identification of secondary bifurcation points.

The numerical approximation of periodic orbits must not rely on the stability type of the orbits if a complete global bifurcation picture is to be obtained. For that reason, periodic solutions are viewed as solutions of a boundary value problem of the form

$$F(\alpha,\ T,\ x_t,\ \dot{x}_t) = 0 \qquad [3.1]$$

$x(t+1)=x(t)$, where $F : R\times R^+\times C\times C \rightarrow R^n$. The independent variable t has been scaled so that T-periodic solutions of [2.1] correspond to 1-periodic solutions of [3.1].

For periodic solutions x(t) of [3.1] we introduce the finite dimensional approximation

$$x^{(N)}(t) = \sum_{j=1}^{N} c_j\ \phi_j(t), \qquad [3.2]$$

where the ϕ_j represent appropriate scalar 1-periodic basis functions and the c_j are in R^n. Both truncated Fourier series and k^{th} order periodic B-splines are examples of approximations of this type. See [1] and [3].

Collocation provides one means of computing the coefficients c_j. That is, for N distinct nodes t_j chosen from [0.,1.) one considers the nN equations

$$F(\alpha,\ T,\ x^{(N)}(t_j),\ \dot{x}^{(N)}(t_j)) = 0 \qquad [3.3]$$

in the nN+2 unknowns c_j; j=1,...,N, α, and T.

We adjoin to these equations a scalar phase constraint to remove the indeterminacy due to the fact that the phase shift of any periodic solution of [3.1] is also a periodic solution. In the case of trucated Fourier series, this corresponds to simply setting one of the coordinates of the primary Fourier coefficients equal to zero. However, there are more sophisticated methods of instituting such a constraint.

Finally, we adjoin a scalar equation in order to (in a sense) specify which of the one-parameter family of periodic solutions is to be computed. More precisely, given that a solution $(\alpha_i, T_i, x^{(N)}_i)$ to [3.3] has been obtained, one seeks a solution $(\alpha_{i+1}, T_{i+1}, x^{(N)}_{i+1})$ that lies a given arclength away.

Having obtained two points on the one-parameter family of periodic orbits, one can linearly extrapolate ("predict") an initial approximation to the next desired member of the family, then iteratively improve that approximation ("correct") by solving the above nN+2 simultaneous nonlinear equations by some Newton-like scheme.

It should be remarked that in the case of ordinary differential equations, the use of B-splines has an advantage over truncated Fourier series in that the Jacobian matrices encountered are sparse; the precise structure being dependent only on the order of the splines in use. For functional differential equations such sparsity is lost, with the structure of the Jacobian dependent on the form of the equation [3.1] as well as the parameters T and α. Despite this fact, splines possess certain numerical characteristics which speak in favor of their use over truncated Fourier series.

Having computed an approximation $x^{(N)}$ to [3.1] at the parameter values α and T, one determines the stability of the orbit (and identifies secondary bifurcation points) by computing approximations to the orbit's Floquet multipliers. For equations with finite delay, some iterate of the (linearized) Poincaré map is compact [4]. The Floquet multipliers are, therefore eigenvalues of finite multiplicity with zero as their only cluster point.

Let $X^{(M)}$ denote an M dimensional approximation to the phase space X in use. We assume $X^{(M)} \subseteq X$ and let $P^{(M)} : X \to X^{(M)}$ denote a projection of X onto $X^{(M)}$. The approximate (linearized) Poincaré map is defined to be

$$\mathcal{P}^{(M)} = P^{(M)} \circ \Pi \Big|_{X^{(M)}}, \qquad [3.4]$$

where Π is the period 1 map defined by the linearized equation associated with [3.1]. The eigenvalues of $\mathcal{P}^{(M)}$ serve as approximations to the Floquet multipliers of the periodic solution to [3.1].

Finally, we remark that due to the autonomous nature of [3.1], 1 is always a Floquet multiplier [4]. This fact provides a useful monitor of the overall accuracy of the periodic solution approximation $x^{(N)}$ and the multiplier approximation scheme described above.

4. An Example

We conclude with a brief description of the results of applying the methodology described in the previous sections to a model in physiology. We refer the reader to [1] for details, and seek only to indicate the kinds of information that can be obtained when applying these ideas to a particular mathematical model.

The two dimensional delay-difference system

$$v' = h(v) - w + \mu[v(t-\tau) - v_0]$$

$$w' = \rho[v + a - bw] \qquad [4.1]$$

arises as a model of recurrent neural feedback. Here, $h(v)=v-v^3/3$, $\rho>0$ is small, $0<b<1$, $1-2b/3<a<1$, and v_0 is the v coordinate of the unique equilibrium (v_0,w_0) that exists for [4.1] when $\mu=0$. Fixing ρ, a and b, one can consider the associated Hopf bifurcation problem in the two remaining parameters τ and μ, which are restricted to be positive and negative, respectively.

Linearizing about the equilibrium of [4.1], one obtains as

characteristic equation

$$0 = \lambda^2+(\rho b-\sigma)\lambda+\rho(1+b\sigma)-(\mu\lambda+\mu\rho b)e^{-\lambda\tau}, \qquad [4.2]$$

where $\sigma=h'(v_0)$. As indicated in Section 2, one expects for this two parameter problem that there should be curves of critical parameters at which [4.2] possesses purely imaginary root pairs $\lambda=\pm\omega i$. Since [4.2] has the form [2.13], one expects these "imaginary root" curves to be parameterized by frequency ω.

Figure 4.1 shows a few of these curves for $\rho=.08$, $a=.7$ and $b=.8$. Along them one is able to determine the stability-determining constant K_3 by numerically implementing the algorithm discussed in Section 2. Solid lines correspond to $K_3<0$, while dashed lines correspond to $K_3>0$. One can show that at nonintersection points of the imaginary root curves the required spectral hypotheses hold, and that for small μ that all characteristic roots must have negative real parts. Thus, [4.1] supports both stable and unstable periodic orbits.

If one additionally fixes τ, one can apply numerical tracking techniques similar to those discussed in Section 3 to the resulting one-parameter problem. Figure 4.2 depicts the global bifurcation diagram with $\tau=25$. Solid lines indicate stable periodic orbits, while dashed lines correspond to unstable periodic orbits. See [1] for details.

5. Acknowledgement

This work was partially sponsored by the Air Force Office of Sponsored Research under Grant #86-0071.

6. References

1. Castelfranco, A. and H.W. Stech, Periodic solutions in a model of recurrent neural feedback, SIAM Journal of Applied Mathematics, to appear.

2. Chow, S. - N. and J. Mallet-Paret, The Fuller index and global Hopf bifurcation, Journal of Differential Equations, 29(1978), 66-85.

3. Doedel, E.J. and P.C. Leung, Numerical techniques for bifurcation problems in delay equations, Cong. Num., 34(1982), 225-237.

4. Hale, J.K., Functional Differential Equations, Applied Math. Sciences Vol. 3, Springer-Verlag, N.Y., 1971.

5. Hassard, B., Numerical Hopf Bifurcation Computation for Delay Differential Systems, Proceedings Adelphi University 1982, ed J. P.E. Hodgson, Springer-Verlag Lecture Notes in Biomathematics #51, 1983.

6. Sathaye, A., BIFDE: A Numerical Software package for the Analysis of the Hopf Bifurcation Problem in Functional Differential Equations, Master's Thesis, Virginia Polytechnic Institute, 1986.

7. Stech, H.W., Hopf bifurcation calculations for functional differential equations, Journal of Math. Analysis and Applications, 109(1985), 472-491.

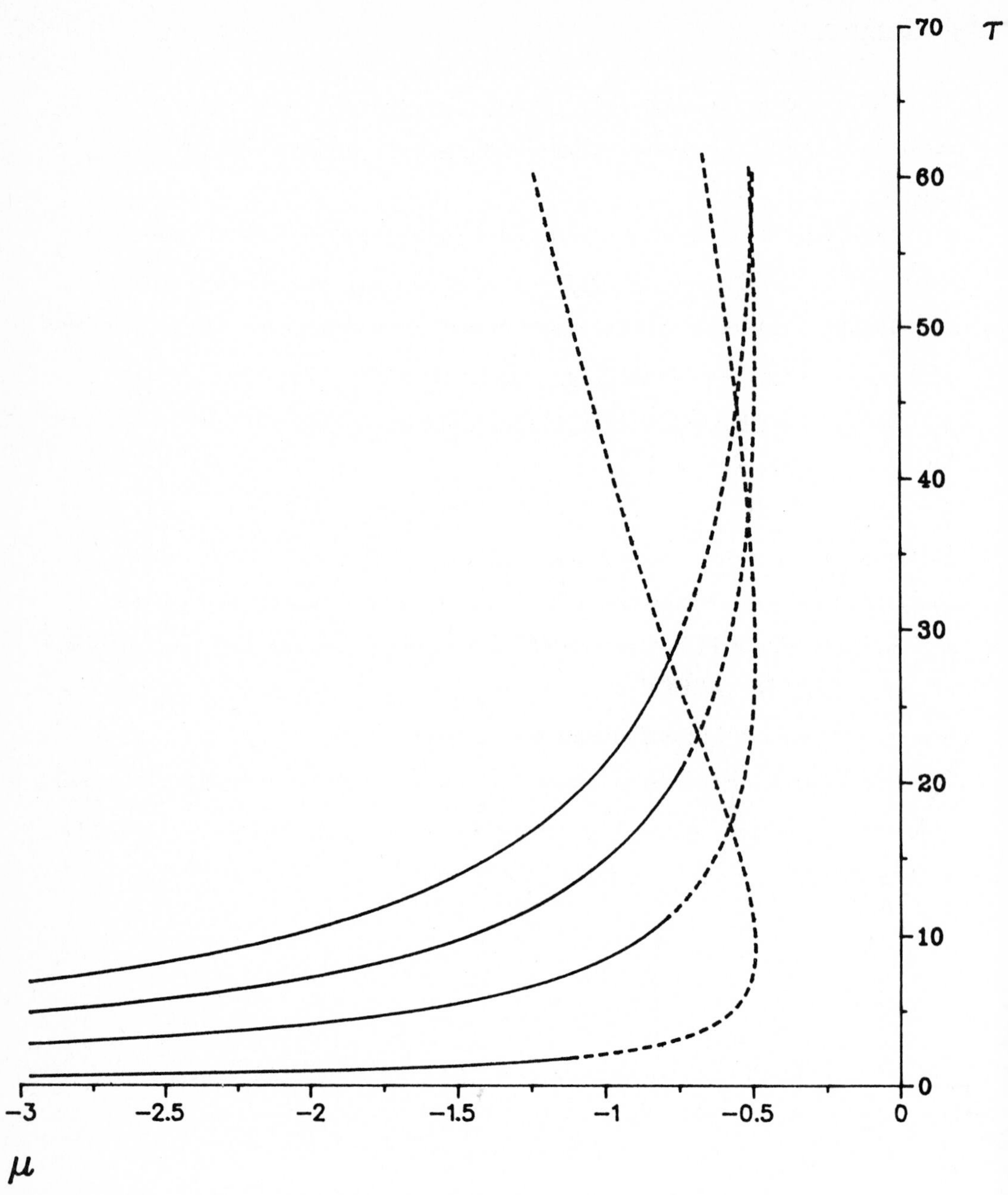

Figure 4.1

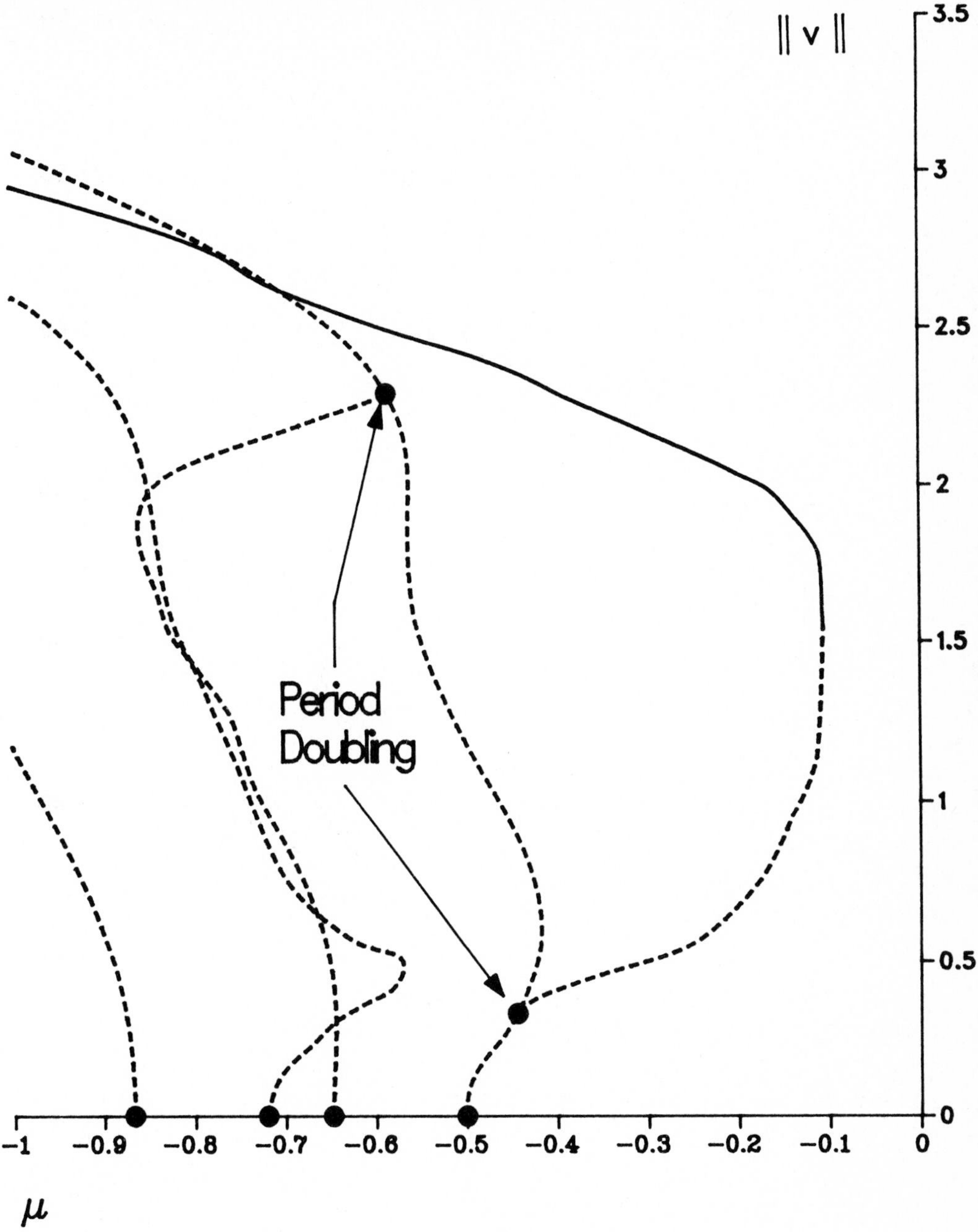

Figure 4.2

OSCILLATIONS AND ASYMPTOTIC BEHAVIOUR

FOR TWO SEMILINEAR HYPERBOLIC SYSTEMS

Luc TARTAR
Centre d'Etudes de Limeil-Valenton
DMA/MCN. BP.27
94190 Villeneuve Saint Georges, FRANCE

We want to describe here properties of two semilinear hyperbolic systems in one space variable, which have been associated to the kinetic theory of gases : the "Carleman" model and the "Broadwell" model.

The Carleman model is given by

$$u_t + u_x + u^2 - v^2 = 0 \tag{C1}$$

$$v_t - v_x - u^2 + v^2 = 0 \tag{C2}$$

and the Broadwell model is the system

$$u_t + u_x + uv - w^2 = 0 \tag{B1}$$

$$v_t - v_x + uv - w^2 = 0 \tag{B2}$$

$$w_t \qquad - uv + w^2 = 0 \tag{B3}$$

Although the first one is attributed to Carleman because it appears in the appendix of a posthumous work edited by Carleson and Frostman (Carleman [4], 1957), it should not be considered as a model of kinetic theory because it fails to conserve momentum. It describes the density of two families of particles, propagating with speed ±1 along the axis, which interact in a self- destructive way : when two particles of the same family "meet" they may decide to change their direction. This model satisfies an H-theorem which is its only similarity with the Boltzmann equation.

NATO ASI Series, Vol. F37
Dynamics of Infinite Dimensional Systems
Edited by S.-N. Chow, and J.K. Hale

In spite of the fact that Maxwell introduced a similar discrete velocity model (see Gatignol [7]) the second system has been named after Broadwell who used it in its fluid dynamical limit (Broadwell [2], 1964). It is the simplest model satisfying conservation of mass and momentum together with an H-theorem. It describes the density of four families of particles moving in the plane with velocities ±1 parallel to one of the axes, where the two families propagating in the y direction have the same density (this is indeed possible if no dependence in y occurs). If y is restricted to a unit interval then the mass density is $\rho=u+v+2w$ and the momentum is $q=u-v$.

We describe some features here that show how different these two equations are, e.g. although they scale in the same way, they have a completely different asymptotic behaviour since one has self-similar solutions and the other does not.

Firstly for the Carleman model and then for the Broadwell model, we examine properties of global existence and asymptotic behaviour, discuss different estimates and study propagation and interaction of oscillations.

I. The Carleman model

We consider the Cauchy problem on the real line for the system

$$u_t + u_x + u^2 - v^2 = 0 \qquad \text{(C1)}$$

$$v_t - v_x - u^2 + v^2 = 0 \qquad \text{(C2)}$$

with the initial data

$$u(x,0) = u_o(x) \ ; \ v(x,0) = v_o(x). \qquad \text{(I.1)}$$

The data u_o and v_o are assumed to be nonnegative. Existence and uniqueness of a global solution were investigated first by Kolodner ([10], 1963) who saw the importance of the order relation.

<u>Proposition 1</u> : Assume that

$$0 \leq u_o(x) \leq A_o \text{ a.e } ; \ 0 \leq v_o(x) \leq B_o \text{ a.e.} \qquad (I.2)$$

Then there exists a unique global solution satisfying

$$0 \leq u(x,t),\ v(x,t) \leq \max(A_o,B_o) \text{ a.e for } t \geq 0. \qquad (I.3)$$

If $\underline{u}_o$, $\underline{v}_o$ verify

$$0 \leq u_o \leq \underline{u}_o \leq M \text{ a.e } ; \ 0 \leq v_o \leq \underline{v}_o \leq M \text{ a.e} \qquad (I.4)$$

then the solution $\underline{u}$, $\underline{v}$ corresponding to initial data $\underline{u}_o$, $\underline{v}_o$ are such that

$$0 \leq u(x,t) \leq \underline{u}(x,t) \leq M \ ; \ 0 \leq v(x,t) \leq \underline{v}(x,t) \leq M \text{ a.e for } t \geq 0. \qquad (I.5)$$

<u>Remark 1</u> : Using $\underline{u} = A(t)$, $\underline{v} = B(t)$, one can improve (I.3) into

$$0 \leq u(x,t) \leq A(t) = \frac{A_o+B_o}{2} + \frac{A_o-B_o}{2} \exp(-2(A_o+B_o)t) \qquad (I.6)$$

$$0 \leq v(x,t) \leq B(t) = \frac{A_o+B_o}{2} + \frac{B_o-A_o}{2} \exp(-2(A_o+B_o)t). \qquad (I.7)$$

<u>Remark 2</u> : The order preserving property is certainly very useful from a mathematical point of view but it is not shared by reasonable models of kinetic theory.

<u>Remark 3</u> : The existence of bounded invariant regions follows from (I.3). Indeed if $0 \leq u_o(x)$, $v_o(x) \leq M$ a.e. on R then one has $0 \leq u(x,t)$, $v(x,t) \leq M$ a.e. on $R \times [0,\infty[$. Once again this property is not satisfied by reasonable models.

With integrable data one has the following

<u>Proposition 2</u> : Assume that (u_o, v_o) and $(\underline{u}_o, \underline{v}_o)$ are nonnegative integrable and bounded data. Then one has the estimate

$$\|u(.,t)-\underline{u}(.,t)\| + \|v(.,t)-\underline{v}(.,t)\| \leq \|u_o-\underline{u}_o\| + \|v_o-\underline{v}_o\| \text{ for } t \geq 0 \qquad (I.8)$$

where $\|.\|$ denotes the norm in $L^1(R)$.

<u>Remark 4</u> : This property was noticed by Liggett (see Crandall [5], 1971) and it shows that the Carleman model defines a contraction semigroup on nonnegative integrable functions.

This is also a property which is not shared by some models. Indeed, under conservation of mass the L^1 contraction property is equivalent to the order preserving property (Crandall & Tartar [6]).

The first result on asymptotic behaviour was obtained by Illner & Reed ([8], 1981).

<u>Proposition 3</u> : Let $0 \leq u_o, v_o$ and let $\int (u_o(x) + v_o(x))dx = m$ be finite. Then one has the estimate

$$0 \leq u(x,t),\ v(x,t) \leq \frac{C(m)}{t}. \qquad (I.9)$$

<u>Remark 5</u> : If the data have compact support then the support can grow with time and has a length $O(t)$ at time t. Thus, due to conservation of mass, a uniform decay in $1/t$ is the best possible.

<u>Remark 6</u> : The initial proofs (Illner & Reed [8], Tartar [15]) gave an exponential bound for $C(m)$. From the self similar solution I suggested in [17] that for large m one could expect that $C(m)=O(m^2)$ (one has $C(m) \geq 1$ for positive m). I will sketch a proof based on the use of the generalized invariant method.

<u>Remark 7</u> : The inequality (I.9) expresses irreversibility : if initial data has a finite mass m one cannot go backward in time and keep nonnegative data. Change in sign will certainly occur before time $T = \frac{-\ C(m)}{\sup_x (u_o(x),\ v_o(x))}$.

In order to see how the mass spreads one has to rescale.

<u>Proposition 4</u> : Let u_o, v_o be nonnegative integrable data with $m = \int(u_o+v_o)dx$ and let (u,v) be the solution of (C1-C2) corresponding to these initial data. Define

$$u_n(x,t) = nu(nx,nt)\ ;\ v_n(x,t) = nv(nx,nt). \qquad (I.10)$$

Then (u_n,v_n) converges as n goes to ∞ to the self-similar solution of mass m which has the form

$$u_\infty(x,t) = \frac{1}{t} U_m(\frac{x}{t}) \; ; \; v_\infty(x,t) = \frac{1}{t} V_m(\frac{x}{t}). \qquad (I.11)$$

where (U_m, V_m) have the following explicit form

$$U_m(\sigma) = \frac{1+\sigma}{2+C(1-\sigma^2)} \text{ in } (-1,1), \text{ 0 outside,} \qquad (I.12)$$

$$V_m(\sigma) = \frac{1-\sigma}{2+C(1-\sigma^2)} \text{ in } (-1,1), \text{ 0 outside,} \qquad (I.13)$$

with the constant C related to m by the relation

$$\int_{-1}^{+1} U_m(\sigma)d\sigma = \frac{m}{2}. \qquad (I.14)$$

Remark 8 : If one believes that all the sequence (u_n, v_n) converges in a strong topology to a limit then this limit will inherit the self-similar property and the mass m. However one knows only that u_n and v_n are uniformly bounded by $C(m)/t$ and the objective of the first part of the proof is to rule out oscillations for a subsequence. The second part of it consists in showing that the only solutions of (C1-C2) having support in $-t \leq x \leq t$ are the self-similar solutions and a precise form of the L^1 contraction property is used there.

Remark 9 : The equation for self-similar solution (U_m, V_m) is

$$\frac{d}{d\sigma}((1-\sigma)U_m(\sigma)) = \frac{d}{d\sigma}((1+\sigma)V_m(\sigma)) = V_m^2(\sigma)-U_m^2(\sigma) \qquad (I.15)$$

which easily yields (I.12)-(I.13). When m goes to ∞, C goes to -2 and one can check that $\sup_\sigma U_m(\sigma) = O(m^2)$.

Remark 10 : The solution (I.12)-(I.13) corresponds to initial data $u_o = v_o = \frac{m}{2}\delta_o$. There is no special reason to rescale the solution as in (I.10) using a particular origin in space and time but it is an open question how to obtain precise error estimates or to discuss the asymptotic interaction of different masses m_j bursting at time t_j out of a point x_j.

Remark 11 : There are no solutions of (C1-C2) corresponding to initial data which are Dirac measures $(u_o = \alpha\delta_o \; ; \; v_o = \beta\delta_o)$ if $\alpha \neq \beta$. If a sequence of nonnegative uniformly integrable initial data (u_ν, v_ν) converges to $(\alpha\delta_o, \beta\delta_o)$ in the sense of measures, then the corresponding sequence of solutions will converge to the self-similar solution for mass $m = \alpha + \beta$. There is a boundary layer near $t = 0$ where the nonlinear terms play an important role and equilibrates the coefficients of the Dirac measures before the transport terms can come into play.

Remark 12 : In order to obtain a good bound for $C(m)$, one should compare u and v not to functions which are constant in x and depend only upon t but with functions having an x dependance similar to the one shown by self-similar solutions.

Using the conservation of mass, expressed by the equation $(u + v)_t + (u - v)_x = 0$, one introduces a potential function W satisfying

$$W_x = (u + v) \; ; \; W_t = (v - u). \qquad (I.16)$$

We normalized it by setting $W(-\infty, t) = 0$ in order to have $W \geqslant 0$ and $W(+\infty, t) = m$. The generalized invariant region method consists in seeking inequalities of the type

$$0 \leqslant u(x,t) \leqslant A(t,W(x,t)) \; ; \; 0 \leqslant v(x,t) \leqslant B(t,W(x,t)). \qquad (I.17)$$

From the form of the self-similar solutions it is then natural to try to obtain

$$0 \leqslant u(x,t) \leqslant \frac{1}{t}A(W(x,t)) \; ; \; 0 \leqslant v(x,t) \leqslant \frac{1}{t}B(W(x,t)). \qquad (I.18)$$

This leads to looking for functions A and B satisfying the following inequalities

$$2B\frac{dA}{dW} - A \geqslant B^2 - A^2 \text{ and } A \geqslant 1 \text{ for } 0 \leqslant W \leqslant m \qquad (I.19)$$

and

$$-2A\frac{dB}{dW} - B \geqslant A^2 - B^2 \text{ and } B \geqslant 1 \text{ for } 0 \leqslant W \leqslant m. \qquad (I.20)$$

The explicit formula for self-similar solutions (corresponding to a mass bigger than m) can be used to generate functions A and B satisfying (I.19) and (I.20) together with a uniform bound of order m^2 for large m.

As mentioned in remark 8, one of the difficulties in proving the convergence of the rescaled solutions is to show that no oscillations occur. A uniform bound for nonnegative solutions only enables us to extract a subsequence converging in a weak * topology and the limit may not be a solution of the system (C1-C2). This is due to the fact that oscillations are compatible with the equations because of their hyperbolic nature. They can propagate and interact in a way that we will describe later, but we notice firstly that they cannot be created, as shown by the following result.

Proposition 5 : Let (u_ν, v_ν) be a uniformly bounded sequence of nonnegative solutions of (C1-C2) defined on an open set Ω of the x,t plane. Assume that, as ν goes to ∞, this sequence converges in the weak * topology of $L^\infty(\Omega)$ to (u_∞, v_∞) and that the sequence (u_ν^2, v_ν^2) converges to $(u_\infty^2 + S_u^2, v_\infty^2 + S_v^2)$. This will define S_u and S_v which measure the strength of the oscillations in the sequences u_ν and v_ν. Then, the fact that oscillations propagate and cannot be created is expressed by the differential inequalities

$$(S_u)_t + (S_u)_x + u_\infty(S_u) \leqslant 0 \text{ a.e in } \Omega \qquad (I.21)$$

and

$$(S_v)_t - (S_v)_x + v_\infty(S_v) \leqslant 0 \text{ a.e in } \Omega. \qquad (I.22)$$

Remark 13 : Without nonnegativity, inequalities (I.21)-(I.22) have to be modified replacing u_∞ by $u_\infty + M$ and v_∞ by $v_\infty + N$ where M and N are respectively lower bounds for $u_\nu(x,t)$ and $v_\nu(x,t)$ in Ω.

When analysing a system of partial differential equations the study of general oscillating solutions should be added, I believe, to the usual questions of existence, uniqueness and asymptotic behaviour. A simple case consists in taking initial data which are periodically modulated. One starts with a function f(x,y) having period 1 in y and one considers the sequence $f_\varepsilon(x) = f(x, \frac{x}{\varepsilon})$ for which all the weak limits of functions of f_ε can be expressed explicitly : if f is bounded and G is continuous then $G(f_\varepsilon)$ converges in L^∞ weak * to $\int_0^1 G(f(x,y))dy$.

When possible one should try to understand general oscillations (and even questions of correlations of oscillations for which little is known) and with our model this can be done.

Such results for periodically modulated functions were obtained first in joint work with G. Papanicolaou (see Tartar [17] and McLaughlin & Papanicolaou & Tartar [11])

Proposition 6 : Let $a(x,y)$, $b(x,y)$ be nonnegative bounded functions with period 1 in y and considers the solution $(u_\varepsilon, v_\varepsilon)$ of (C1-C2) corresponding to initial data $(a_\varepsilon, b_\varepsilon)$ as above then

$$u_\varepsilon(x,t) - A(x,t,\frac{x-t}{\varepsilon}) \to 0 \qquad (I.23)$$

$$v_\varepsilon(x,t) - B(x,t,\frac{x+t}{\varepsilon}) \to 0 \qquad (I.24)$$

where $A(x,t,y)$, $B(x,t,y)$ are the functions of period 1 in y solution of the the system

$$A_t + A_x + A^2 - \int_0^1 B^2(x,t,z)dz = 0 \qquad (I.25)$$

$$B_t - B_x - \int_0^1 A^2(x,t,z)dz + B^2 = 0 \qquad (I.26)$$

with initial data

$$A(x,0,y) = a(x,y) \; ; \; B(x,0,y) = b(x,y). \qquad (I.27)$$

Remark 14 : The convergence in (I.23)-(I.24) holds, for example, in L^∞ weak * and L^p_{loc} strong for finite p. Then the weak * limit of $G(u_\varepsilon, v_\varepsilon)$ can be expressed for any continuous function G by

$$G(u_\varepsilon,v_\varepsilon) \to \int_0^1\int_0^1 G(A(x,t,y),B(x,t,z))dydz \text{ in } L^\infty \text{ weak } *. \qquad (I.28)$$

Remark 15 : The system (I.25)-(I.26) possesses the following property similar to gauge invariance in some equations used by physicists : if g is any application from (0,1) onto itself which is measure preserving then from a solution (A, B) one can construct another one (A^g, B^g) given by

$$A^g(x,t,y)=A(x,t,g(y)) \; ; \; B^g(x,t,y)=B(x,t,g(y)). \qquad (I.29)$$

Moreover the solution $(u^g_\varepsilon, v^g_\varepsilon)$ is such that the weak * limit of $G(u^g_\varepsilon, v^g_\varepsilon)$ does not change, as can be seen immediately from (I.28).

In some way, the preceding remark says that no knowledge

of any correlation for oscillations is needed to describe the evolution of these oscillations. Therefore, it is not surprising that for the model (C1-C2) the case of general oscillating data can be deduced from the case of the periodically oscillating case (I had obtained general results before, [16], 1981).

Proposition 7 : Let $(u_\varepsilon, v_\varepsilon)$ be a sequence of nonnegative uniformly bounded solutions of (C1-C2) corresponding to initial data $(u_{o\varepsilon}, v_{o\varepsilon})$ and assume that

$$u_\varepsilon(x,0)^m \to \int_0^1 a(x,y)^m dy \text{ in } L^\infty \text{ weak * for all m} \quad (I.30)$$

$$v_\varepsilon(x,0)^m \to \int_0^1 b(x,y)^m dy \text{ in } L^\infty \text{ weak * for all m.} \quad (I.31)$$

Then (I.28) holds where (A, B) is defined by (I.25)-(I.26)

Remark 16 : If for every integer m the sequence $u_\varepsilon(x,0)^m$ has a weak * limit $a_m(x)$ then one can construct a function a(x,y) which has the moments a_m. Also, one can choose a to be nonincreasing in y on (0,1) and this property will be conserved by A for $t \geq 0$.

II. The Broadwell model

We consider the Cauchy problem on the real line for the system

$$u_t + u_x + uv - w^2 = 0 \quad (B1)$$

$$v_t - v_x + uv - w^2 = 0 \quad (B2)$$

$$w_t - uv + w^2 = 0 \quad (B3)$$

with the initial data

$$u(x,0) = u_o(x) \ ; \ v(x,0) = v_o(x) \ ; \ w(x,0) = w_o(x). \quad (II.1)$$

The data u_o, v_o and w_o are assumed to be nonnegative.

The first result of global existence was obtained by Mimura & Nishida ([12], 1975).

Proposition 8 : There exist m_o and κ such that, if the data are nonnegative bounded and satisfy

$$\int(u_o(x)+v_o(x)+2w_o(x))dx \leq m_o, \qquad (II.2)$$

then the solution exists for all $t \geq 0$ and satisfies

$$\sup_{x,t}(u(x,t),v(x,t),w(x,t)) \leq \kappa \sup_x(u_o(x),v_o(x),w_o(x)). \qquad (II.3)$$

Remark 17 : There is no bounded invariant region for this system. If instead one looks for regions of the type

$$0\leq u(x,t)\leq A(t) \ ; \ 0\leq v(x,t)\leq B(t) \ ; \ 0\leq w(x,t)\leq C(t) \qquad (II.4)$$

while asking that if (II.4) is satisfied at time t_o then it stays true for $t \geq t_o$, this leads to impose the following inequalities

$$\frac{dA}{dt} \geq C^2 \ ; \ \frac{dB}{dt} \geq C^2 \ ; \ \frac{dC}{dt} + C^2 \geq AB \qquad (II.5)$$

which have no global solution on $(0,\infty)$ except 0.

Using finite propagation speed and the H-theorem, which says that $\int$ (uLogu + vLogv + 2wLogw)dx is nonincreasing in t, it was shown by Crandall and myself how to obtain global existence from a result like proposition 8 (see Tartar [13], 1976 and [14]).

Proposition 9 : There exists a continuous function F(M,t) such that if the data satisfy

$$0 \leq u_o(x),v_o(x),w_o(x)\leq M \qquad (II.6)$$

then the solution exists for $t \geq 0$ and satisfies

$$0 \leq u(x,t),v(x,t),w(x,t) \leq F(M,t). \qquad (II.7)$$

Asymptotic behaviour was understood first for data with small mass (Tartar [15], 1980) using suitable functional spaces in (x,t) instead of the usual semigroup approach. The idea is to introduce the following functional spaces (equipped with natural norms)

$$V_c = \{f(x,t) \; ; \; \int \|f_t + cf_x\| dxdt + \int \|f(x,0)\| dx \text{ is finite}\} \quad (II.8)$$

which is continuously imbedded in

$$W_c = \{f(x,t) \; ; \; \|f(x,t)\| \leq g(x-ct) \text{ with } g \text{ integrable}\}. \quad (II.9)$$

The first important fact is that if a function f belongs to V_c then there exists an integrable function f_∞ such that

$$\int \|f(x,t) - f_\infty(x-ct)\| dx \to 0 \text{ as } t \text{ tends to } \infty. \quad (II.10)$$

The second one is that if two functions f_a and f_b belong respectively to spaces W_a and W_b with $a \neq b$, then $f_a f_b$ is integrable in (x,t). Then the goal is to prove existence of a solution with (u,v,w) belonging to $(V_1 \times V_{-1} \times V_o)$.

Proposition 10 : For (u_o, v_o, w_o) nonnegative integrable data there exists a unique solution (u,v,w) in $(V_1 \times V_{-1} \times V_o)$. The asymptotic behaviour is given by

$$u(x,t) \approx u_\infty(x-t) \; ; \; v(x,t) \approx v_\infty(x+t) \; ; \; w(x,t) \approx 0. \quad (II.11)$$

Remark 18 : As it was noticed by Caflish the proof showed also that w^2 is integrable and so the limit of w had to be 0.

My initial proof only applied to small integrable data (by a different approach Illner obtained also similar results in [9], 1984). T. Beale ([1], 1985) showed that the same result was true for finite mass (it was in order to simplify his argument that I derived the method of generalized invariant regions sketched already in remark 12 for the Carleman system).

From conservation of mass and momentum it is easy to compute the mass contained in the two signals u_∞ and v_∞,

$$\int u_\infty(x) dx = \int (u_o(x) + w_o(x)) dx \quad (II.12)$$

$$\int v_\infty(x) dx = \int (v_o(x) + w_o(x)) dx. \quad (II.13)$$

Then, it is more natural to write the conservation laws in the form

$$(u + w)_t + u_x = 0 \quad (II.14)$$

$$(v + w)_t - v_x = 0 \quad (II.15)$$

and to introduce the two potential functions satisfying

$$U_x = u + w \ ; \ U_t = -u \qquad \text{(II.16)}$$

$$V_x = -(v + w) \ ; \ V_t = -v \qquad \text{(II.17)}$$

and normalized by $U(-\infty,t)=0$ and $V(+\infty,t)=0$.

The generalized invariant region method consists in seeking inequalities of the type

$$0\leq u\leq A(t,U,V) \ ; \ 0\leq v\leq B(t,U,V) \ ; \ 0\leq w\leq C(t,U,V). \qquad \text{(II.18)}$$

The condition to impose on A,B,C can be computed easily but, owing to our knowledge of the asymptotic behaviour, it is natural to try the simpler case

$$0\leq u(x,t)\leq A(U(x,t)) \ ; \ 0\leq v(x,t)\leq B(V(x,t)) \ ; \ 0\leq w(x,t)\leq C \qquad \text{(II.19)}$$

which yields the conditions

$$\frac{dA}{dU} \geq C \ ; \ \frac{dB}{dV} \geq C \ ; \ C^2 \geq AB. \qquad \text{(II.20)}$$

Proposition 11 : Given three positive numbers α, β, λ consider the following inequalities

$$\sup_x (\alpha + U(x))(\beta + V(x)) \leq 1 \qquad \text{(II.21)}$$

and

$$\sup_x \max\left\{\frac{u(x)}{\alpha+U(x)} , \frac{v(x)}{\beta+V(x)} , w(x)\right\} \leq \lambda. \qquad \text{(II.22)}$$

If they are satified at time 0, then they are satisfyed for all $t > 0$, i.e. they define a generalized invariant region.

Remark 19 : If the total mass is less than 1 then (II.21) is true for α and β small and we obtain again the result of proposition 8. The advantage of (II.21) is that it measures a bound of the future interaction, in the same spirit than the functional introduced by J. Glimm for quasilinear systems of conservation laws.

Remark 20 : As it was noticed by S.R.S. Varadhan one can obtain quickly a basic estimate for proposition 10 by considering the functional

$$I(t) = \iint_{x\leq y} (u+w)(x,t)(v+w)(y,t)dxdy \qquad \text{(II.23)}$$

whose derivative is

$$\frac{dI}{dt} = - \int (2uv + uw + vw)dx. \qquad (II.24)$$

Remark 21 : Although I had not seen its implications, I found earlier an estimate for the integral of (2uv+uw+vw) by applying an argument of compensated compactness to the system (II.14)-(II.15). Multiplying (II.14) by V and integrating by parts on a strip $0 \leq t \leq T$, one obtains the same information as in (II.23)-(II.24). Its advantage is that this estimate is independent of the mean free path (for simplicity I have considered here the mean free path equal to 1). I did these computations in order to obtain a long time version of the result of Caflish & Papanicolaou on the fluid dynamical limit ([3], 1979). Because of the lack of estimates for the case where the mean free path goes to zero one may expect the presence of oscillations ; if oscillations were present the formal procedure of the Hilbert or Chapman & Enskog expansions could be wrong.

Remark 22 : The asymptotic behaviour for the Broadwell model is then quite different than the decay in 1/t found for the Carleman model. As suggested by R. Caflish one could expect such a decay in 1/t for w. For finite mass I conjecture that, on each bounded interval, w^2 and uv do decay in $1/t^2$.

Remark 23 : For the Carleman model there exists self-similar solutions and the asymptotic behaviour for the general solution is described by a one parameter family. For the Broadwell model there are no self-similar solutions with finite mass (except the zero solution) and the asymptotic behaviour is described by functions with arbitrary shape. Indeed when $w_o=0$, if u_o and v_o are any nonnegative functions with compact support such that the support of u_o is entirely to the right of the support of v_o, then the solution is given by $u(x,t) = u_o(x-t)$, $v(x,t) = v_o(x+t)$ and $w(x,t) = 0$.

Let us consider now the description of oscillations of solutions for (B1-B2-B3). As for the Carleman model, a uniform bound for nonnegative solutions enables us to extract a subsequence converging in a weak * topology and the limit may not be a solution of the system. Oscillations are also compatible with the equations but they propagate and interact in a different way : oscillations on u and v cannot be created but here a new phenomenon occurs because the interaction of oscillations of u and oscillations of v can generate oscillations on w. This result is expressed by

Proposition 12 : Let (u_ν, v_ν, w_ν) be a uniformly bounded sequence of nonnegative solutions of (B1-B2-B3) defined on an open set Ω of the x,t plane. Assume that, as ν goes to ∞, this sequence converges in the weak * topology of $L^\infty(\Omega)$ to $(u_\infty, v_\infty, w_\infty)$. Let S_u, S_v and S_w, which measure the strength of the oscillations in the sequences u_ν, v_ν and w_ν, be such that the sequence $(u^2_\nu, v^2_\nu, w^2_\nu)$ converges to $(u^2_\infty + S^2_u, v^2_\infty + S^2_v, w^2_\infty + S^2_w)$ in the weak * topology. Then

$$(S_u)_t + (S_u)_x + v_\infty(S_u) = 0 \text{ a.e in } \Omega \qquad \text{(II.25)}$$

$$(S_v)_t - (S_v)_x + u_\infty(S_v) = 0 \text{ a.e in } \Omega \qquad \text{(II.26)}$$

which express the fact that oscillations propagate and cannot be created for u and v. Also the propagation and the possibility of creation of oscillations for w is given by

$$(S_w)_t + w_\infty(S_w) \leqslant (S_u)(S_v) \qquad \text{(II.27)}$$

Effective creation of oscillations can be shown in an explicit form in the simple case where initial data are periodically modulated.

Proposition 13 : Let a(x,y), b(x,y), c(x,y) be nonnegative bounded functions with period 1 in y and consider the solution $(u_\varepsilon, v_\varepsilon, w_\varepsilon)$ of (B1-B2-B3) corresponding to the periodically modulated initial data $(a_\varepsilon, b_\varepsilon, c_\varepsilon)$. Then

$$u_\varepsilon(x,t) - A(x,t,\frac{x-t}{\varepsilon}) \to 0 \qquad \text{(II.28)}$$

$$v_\varepsilon(x,t) - B(x,t,\frac{x+t}{\varepsilon}) \to 0 \qquad \text{(II.29)}$$

$$w_\varepsilon(x,t) - C(x,t,\frac{x}{\varepsilon}) \to 0 \qquad \text{(II.30)}$$

where A(x,t,y), B(x,t,y), C(x,t,y) are the functions of period 1 in y solution of the system

$$A_t + A_x + A \int_0^1 B(x,t,z)dz - \int_0^1 C^2(x,t,z)dz = 0 \quad \text{(II.31)}$$

$$B_t - B_x - B \int_0^1 A(x,t,z)dz - \int_0^1 C^2(x,t,z)dz = 0 \quad \text{(II.32)}$$

$$C_t - \int_0^1 A(x,t,y-z)B(x,t,y+z)dz + C^2 = 0 \quad \text{(II.33)}$$

with initial data

$$A(x,0,y)=a(x,y) \; ; \; B(x,0,y)=b(x,y) \; ; \; C(x,0,y)=c(x,y). \quad \text{(II.34)}$$

Remark 24 : Unlike the case of the Carleman model the situation of general oscillating data is not entirely understood for the Broadwell model. Some knowledge of correlations of oscillations is necessary to describe their evolution.

References

1. Beale J.T., "Large-time behaviour of the Broadwell model of a discrete velocity gas," Comm. Math. Phys. 102, (1985) 217-235
2 Broadwell J.E., "Shock structure in a simple discrete velocity gas", Phys. Fluids 7, (1964) 1243-1247
3. Caflish R. & Papanicolaou G., "The fluid dynamical limit of a nonlinear model Boltzmann equation" Comm. Pure Appl. Math. 32, (1979) 589-619
4. Carleman T., Problèmes Mathématiques dans la Théorie Cinétique des Gaz, Publications Scientifiques de l'Institut Mittag-Leffler, Almqvist & Wiksells (Uppsala) 1957.
5. Crandall M.G., "Semigroups of nonlinear transformations in Banach spaces" in Contributions to Nonlinear Functional Analysis (Zarantonello E.H. ed) Academic Press (New-York) 1971. 157-179
6. Crandall M.G. & Tartar L.C., "Some relations between nonexpansive and order preserving mappings," Proc. Amer. Math. Soc. 78 3 (1980) 385-390.

7. Gatignol R., Théorie Cinétique des Gaz à Répartition Discrète de Vitesses Lecture Notes in Physics 36 Springer (Berlin) 1975

8. Illner R. & Reed M.C., "The decay of solutions of the Carleman model", Math. Meth. in Appl. Sci. 3, (1981) 121-127

9. Illner R., "The Broadwell model for initial values in $L^1_+(R)$" Comm. Math. Phys. 93, (1984) 351-353

10. Kolodner I., "On the Carleman's model for the Boltzmann equation and its generalizations", Ann. di Mat. Pura Appl. Ser. IV, LXIII (1963) 11-32

11. McLaughlin D. & Papanicolaou G. & Tartar L.C., "Weak limits of semilinear hyperbolic systems with oscillating data," Macroscopic Modelling of Turbulent Flows (Frisch U, Keller J.B, Papanicolaou G, Pironneau O. eds) Lecture Notes in Physics 230 Springer (Berlin) 1985. 277-289.

12. Mimura M. & Nishida T., "On the Broadwell's model for a simple discrete velocity gas", Proc. Japan Acad. 50, (1974) 812-817

13. Tartar L.C., "Existence globale pour un système hyperbolique semilinéaire de la théorie cinétique des gaz," Goulaouic- Schwartz Seminar (1975-1976) exp.I. Ecole Polytechnique (Palaiseau) 1976.

14. Tartar L.C., "Quelques remarques sur les systèmes semilinéaires," Etude Numérique des grands Systèmes, (Lions J.L, Marchouk G.I. eds) Méthodes Mathématiques de l'Informatique, Dunod (Paris) 1978. 198-212.

15. Tartar L.C., "Some existence theorems for semilinear hyperbolic systems in one space variable," Report 2164 Mathematics Research Center, University of Wisconsin (Madison) 1980.

16. Tartar L.C., "Solutions oscillantes des équations de Carleman," Goulaouic-Meyer-Schwartz Seminar (1980-1981), exp.XII. Ecole Polytechnique (Palaiseau) 1981.

17. Tartar L.C., "Etude des oscillations dans les équations aux dérivées partielles non linéaires," dans Trends and Applications of Pure Mathematics to Mechanics (Roseau M. and Ciarlet P.G. eds) Lecture Notes in Physics 195 Springer (Berlin) 1984. 384-412.

AN APPLICATION OF THE CONLEY INDEX TO COMBUSTION

David Terman
Department of Mathematics
Michigan State University
East Lansing, Michigan 48824

I. Introduction

This paper is concerned with the existence of laminar flames with complex chemistry. Mathematically, the problem reduces to proving the existence of traveling wave solutions of a reaction-diffusion system. The reaction-diffusion system takes the form

$$U_t = DU_{xx} + F(U) \tag{1.1}$$

where $U = (T, Y_1, \cdots, Y_{n-1}) \in \mathbb{R}^n$ and D is a positive, diagonal matrix. The components of U specify the dimensionless temperature and the concentration of the reactants. For a background of the physical motivation of these equations, see [2], [8].

By a traveling wave solution of (1.1) we mean a nonconstant, bounded solution of the form $U(x,t) = U(z)$, $z = x + \theta t$. These correspond to solutions which appear to be traveling with constant shape and velocity. Note that a traveling wave solution satisfies the system of ordinary differential equations

$$DU'' - \theta U' + F(U) = 0 . \tag{1.2}$$

We shall be interested in the sequential, two step reaction

$$A \longrightarrow B \longrightarrow C . \tag{1.3}$$

Let T be the dimensionless temperature, Y_1 the concentration of A, and Y_2 the concentration of B. In order to understand what sort of phenomena may arise from this reaction, we first imagine (1.3) as taking place in two steps, and thus producing two flames.

The first reaction, $A \longrightarrow B$ in (1.3), will convert the given unburned state to a partially burned state. This will produce a flame, F1, with speed, say θ_1. The second reaction, $B \longrightarrow C$ in (1.3), will then act on

NATO ASI Series, Vol. F37
Dynamics of Infinite Dimensional Systems
Edited by S.-N. Chow, and J. K. Hale

the product, or final state, of F1 and convert the partially burned state to a completely burned state. We denote this second flame by F2, and its velocity by θ_2.

The above flames will proceed at different velocities. For example, if the speed of F2, which is built on the products of F1, is slower than F1 itself, then one can imagine two flames both existing, but the distance between them ever increasing. Another, more interesting, phenomenon is if F2 is faster than F1. In this case, the rear flame approaches the forward one. As it does so, its effect is to heat up the forward one. One expects that the speed of the forward wave to then increase. One may then expect that what eventually evolves in a single configuration moving with constant velocity. In [7] we prove that such a third wave does indeed exist. In this paper we present the essential features of the proof of this result.

In the next section we state our main result precisely. An outline of the proof is given in Section 3. A key ingredient on the proof of the main theorem is the Conley Index. In Section 4 we briefly describe how the Conley index is used in the proof.

2. Precise Statement of the Problem and the Main Result

The system of reaction-diffusion equations associated with (1.3) is

$$\begin{aligned} T_t &= d_0 T_{xx} + Q_1 Y_1 f_1(T) + Q_2 Y_2 f_2(T) \\ Y_{1t} &= d_1 Y_{1xx} - Y_1 f_1(T) \\ Y_{2t} &= d_2 Y_{2xx} + Y_1 f_1(T) - Y_2 f_2(T) . \end{aligned} \tag{2.1}$$

Here d_0, d_1, d_2, Q_1, and Q_2 are positive constants. The nonlinearities, $f_1(T)$ and $f_2(T)$, will be defined shortly. We look for a traveling wave solution of (2.1); that is, a solution of the form

$$(T(z), Y_1(z), Y_2(z)), \quad z = x + \theta t .$$

A traveling wave solution satisfies the system of ordinary differential equations

$$d_0T'' - \theta T' = -Q_1Y_1f_1(T) - Q_2Y_2f_2(T)$$
$$d_1Y_1'' - \theta Y_1' = Y_1f_1(T) \qquad (2.2)$$
$$d_2Y_2'' - \theta Y_2' = -Y_1f_1(T) + Y_2f_2(T) \ .$$

We assume that the unburned state

$$(T,Y_1,Y_2)(-\infty) = (T_-,Y_{1-},Y_{2-})$$

is prescribed.

The assumptions we make on $f_1(T)$ and $f_2(T)$ are the following. For $i = 1$ and 2 assume that there exists a positive constant T_i such that

$$\begin{aligned} &a) \quad f_i(T) = 0 \quad \text{for} \quad T < T_i \\ &b) \quad f_i(T) > 0 \quad \text{for} \quad T > T_i \qquad (2.3) \\ &c) \quad f_i(T) \ \text{is continuous for} \ T > T_i \ . \end{aligned}$$

We think of each T_i as an ignition temperature. It is introduced to avoid the so called "cold boundary difficulty". Details are given in [7] and the references sited there.

For convenience, we assume in this paper that

$$\begin{aligned} &a) \quad (T_-,Y_{1-},Y_{2-}) = (0,1,1) \\ &b) \quad T_1 < Q_1 < T_2 < Q_1 + 2Q_2 \ . \end{aligned} \qquad (2.4)$$

We make these assumptions here only to describe, as simply as possible, the basic ideas in [7]. Our methods also apply when (2.4) is not satisfied. As we now show, (2.4b) is needed to guarantee the existence of simple flames.

We first discuss the simple flame F1. Notice that if $T < T_2$, then $f_2(T) = 0$ and (2.2) reduces to the two equations

$$\begin{aligned} d_0T'' - \theta T' &= -Q_1Y_1f_1(T) \\ d_1Y_1'' - \theta Y_1' &= Y_1f_1(T) \ . \end{aligned} \qquad (2.5)$$

We seek a solution of this reduced system which satisfies

$$(T,Y_1)(-\infty) = (0,1) \quad \text{and} \quad (T,Y_1)(+\infty) = (T^1,0) \qquad (2.6)$$

for some T^1. By integrating the equations in (2.5) for $-\infty < z < \infty$ we find that $T^1 = Q_1$. Berestycki, Nicolaenko, and Scheurer [1] prove that there exists a solution of (2.4), (2.5) for some value of θ. They also prove that $T(z)$ is monotonically increasing. Hence, by (2.4b),

$$T(z) < Q_1 < T_2$$

for all z, and the flame F1 is well defined. We do not know if the solution of (2.5), (2.6) is unique. Let

$$\theta_1 = \sup\{\theta : \text{there exists a solution of (2.5), (2.6) with speed } \theta\} .$$

Notice that if we integrate (2.2) for $-\infty < z < \infty$ then we find that along F1,

$$Y_2(+\infty) = 2 .$$

We next consider flame F2. Along this flame we have that $Y_1(z) = 0$. If this is the case, then (2.2) reduces to the lower order system

$$\begin{aligned} d_0 T'' - \theta T' &= -Q_2 Y_2 f_2(T) \\ d_2 Y_2'' - \theta Y_2' &= Y_2 f_2(T) . \end{aligned} \tag{2.7}$$

For boundary conditions we have

$$(T, Y_2)(-\infty) = (Q_1, 2) \quad \text{and} \quad (T, Y_2)(+\infty) = (T_+, 0) . \tag{2.8}$$

Integrating the equations in (2.7) for $-\infty < z < \infty$, we find that

$$T_+ = Q_1 + 2Q_2 .$$

As before, there exists a monotone solution of (2.7), (2.8). We do not know if the solution of (2.7), (2.8) is unique. Let

$$\theta_2 = \inf\{\theta : \text{there exists a solution of (2.7), (2.8) with speed } \theta\} .$$

We are interested in the case $\theta_1 < \theta_2$. In [7], we prove

Theorem: Assume (2.3), (2.4), and $\theta_1 < \theta_2$. Then there exists a solution of (2.2) for some speed θ^* which satisfies

$$(T,Y_1,Y_2)(-\infty) = (0,1,1) \quad \text{and} \quad (T,Y_1,Y_2)(+\infty) = (Q_1 + 2Q_2,0,0) \; .$$

Moreover, T, Y_1, and Y_2 are positive for all z, $T(z)$ is monotone increasing, and $Y_1(z)$ is monotone decreasing.

The proof of the Theorem is quite geometrical. In the next section we describe the basic geometrical features of the proof. A key ingredient in the proof of the Theorem is the Conley index. In Section 4 we briefly describe how the Conley index is used in the proof.

3. An Outline of the Proof

The first step in the proof of the Theorem is to reduce (2.2) to a first order system. Let $T' = q$, $Y_1' = p_1$, and $Y_2' = p_2$. Then (2.2) is equivalent to the system

$$\begin{aligned}
T' &= q \\
d_0 q' &= \theta q - Q_1 Y_1 f_1(T) - Q_2 Y_2 f_2(T) \\
Y_1' &= p_1 \\
d_1 p_1' &= \theta p_1 + Y_1 f_1(T) \\
Y_2' &= p_2 \\
d_2 p_2' &= \theta p_2 - Y_1 f_1(T) + Y_2 f_2(T) \; .
\end{aligned} \tag{3.1}$$

Let

$$\gamma(z) = (T(z),q(z),Y_1(z),p_1(z),Y_2(z),p_2(z)) \; ,$$

$$A = (0,0,1,0,1,0), \quad B = (Q_1,0,0,0,2,0) \; ,$$

$$\text{and} \quad C = (Q_1 + 2Q_2,0,0,0,0,0) \; .$$

Then F1 corresponds to a solution $\gamma_1(z)$ of (3.1), for some θ, which satisfies

$$\lim_{z\to-\infty} \gamma(z) = A \quad \text{and} \quad \lim_{z\to+\infty} \gamma(z) = B \, .$$

F2 corresponds to a solution $\gamma_2(z)$ of (3.1), for some θ, which

$$\lim_{z\to-\infty} \gamma(z) = B \quad \text{and} \quad \lim_{z\to+\infty} \gamma(z) = C \, .$$

For convenience we assume, in this section, that these two waves are uniquely determined. In particular, the speeds of the waves, say θ_1 and θ_2, are uniquely determined. We wish to prove that if $\theta_1 < \theta_2$, then there exists a solution $\gamma_0(z)$ of (3.1) which satisfies

$$\lim_{z\to-\infty} \gamma_0(z) = A \quad \text{and} \quad \lim_{z\to+\infty} \gamma_0(z) = C \, .$$

In each case the traveling wave solution corresponds to a trajectory in phase space which connects two critical points.

One of the key ideas in the proof of the Theorem is to attach to (3.1) the equation

$$\theta' = \varepsilon(\theta^2 - \theta_0^2) \tag{3.2}$$

where $\theta_0 >> 1$ and $0 < \varepsilon << 1$. Let

$$A_1 = (A,\theta_0) \, , \quad B_1 = (B,\theta_0) \, , \quad C_1 = (C,\theta_0) \, ,$$

$$A_2 = (A,-\theta_0) \, , \quad B_2 = (B,-\theta_0) \, , \quad C_2 = (C,-\theta_0) \, ,$$

$$\ell_A = \{(\gamma,\theta) \; : \; \gamma = A \, , \quad |\theta| \le \theta_0\} \, ,$$

$$\ell_B = \{(\gamma,\theta) \; : \; \gamma = B \, , \quad |\theta| \le \theta_0\} \, ,$$

$$\ell_C = \{(\gamma,\theta) \; : \; \gamma = C \, , \quad |\theta| \le \theta_0\} \, .$$

First consider the case $\varepsilon = 0$. When $\varepsilon = 0$, ℓ_A, ℓ_B, and ℓ_C correspond to lines of critical points of (3.1), (3.2). Then $(\gamma_1(z),\theta_1)$ and $(\gamma_2(z),\theta_2)$ trace out curves in phase space as shown in Figure 1. In Figure 1A we have shown the case $\theta_1 > \theta_2$, while in Figure 1B we have shown $\theta_1 < \theta_2$.

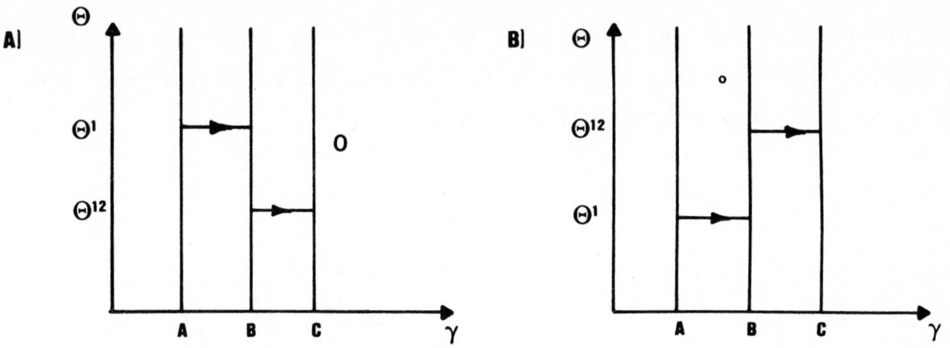

Figure 1

Now suppose that $\varepsilon > 0$. The lines ℓ_A, ℓ_B, and ℓ_C now correspond to trajectories which connect the six critical points A_1 and A_2, B_1 and B_2, C_1 and C_2. Crucial to the proof of the Theorem is the following

Proposition 3.1: For each $\varepsilon > 0$ there exists a solution $(\gamma^\varepsilon(z),\theta^\varepsilon(z))$ of (3.1), (3.2) which satisfies

$$\lim_{z\to-\infty} (\gamma^\varepsilon(z),\theta^\varepsilon(z)) = A_1 \quad \text{and} \quad \lim_{z\to+\infty} (\gamma^\varepsilon(z),\theta^\varepsilon(z)) = C_2 .$$

Remark: This proposition is true in both cases, $\theta_1 < \theta_2$ and $\theta_1 > \theta_2$.

Proposition 3.1 is proved in [7]. We comment on the proof shortly. First we discuss what happens to $(\gamma^\varepsilon(z),\theta^\varepsilon(z))$ as $\varepsilon \to 0$. We shall have apriori bounds so it will be clear that at least some subsequence of $\{(\gamma^\varepsilon(z),\theta^\varepsilon(z))\}$ converges to something.

Let us first consider the case $\theta_1 > \theta_2$. It is possible that if $0 < \varepsilon << 1$, then $(\gamma^\varepsilon(z),\theta^\varepsilon(z))$ is as shown in Figure 2A. That is, $(\gamma^\varepsilon(z),\theta^\varepsilon(z))$ lies close to ℓ_A for $\theta_1 \leq \theta \leq \theta_0$, then lies close to $(\gamma_1(z),\theta_1)$, then lies close to ℓ_B for $\theta_2 < \theta < \theta_1$, then lies close to $(\gamma_2(z),\theta_2)$, and finally lies close to ℓ_C for $-\theta_0 \leq \theta < \theta_2$. In the limit $\varepsilon \to 0$, the curves traced out by $(\gamma^\varepsilon(z),\theta^\varepsilon(z))$ converge to the union of the curves ℓ_A for $\theta_1 \leq \theta \leq \theta_0$, $(\gamma_1(z),\theta_1)$, ℓ_B for $\theta_2 \leq \theta \leq \theta_1$, $(\gamma_2(z),\theta_2)$, and ℓ_C for $-\theta_0 \leq \theta \leq \theta_2$. In this case, $(\gamma^\varepsilon(z),\theta^\varepsilon(z))$ does not converge to a traveling wave solution.

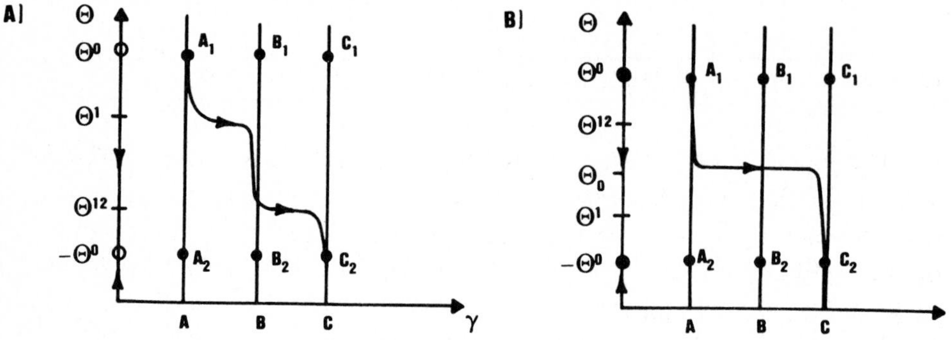

Figure 2

In the above paragraph we strongly used the assumption that $\theta_1 > \theta_2$. Since $\theta' < 0$, it follows that if $\theta_1 < \theta_2$, then $(\gamma^\varepsilon(z),\theta^\varepsilon(z))$ cannot converge to a curve which consists of pieces of ℓ_A, ℓ_B, ℓ_C, $(\gamma_1(z),\theta_1)$, and $(\gamma_2(z),\theta_2)$. In [7], we prove that the only other possibility is that there exists θ^*, and a sequence $\{\varepsilon_n\}$ such that $\varepsilon_n \to 0$ as $n \to \infty$, and $(\gamma^{\varepsilon_n}(z),\theta^{\varepsilon_n}(z))$ converges to a curve which consists of three pieces. These are:

a) ℓ_A for $\theta^* < \theta < \theta_0$

b) a trajectory $(\gamma_0(z),\theta^*)$ which satisfies (3.1), (3.2) for all z and $\varepsilon = 0$, and satisfies

$$\lim_{z\to-\infty} \gamma_0(z) = A \quad \text{and} \quad \lim_{z\to+\infty} \gamma_0(z) = C$$

c) ℓ_C for $-\theta_0 < \theta < \theta^*$.

Then $\gamma_0(z)$ is the desired solution.

We conclude this section with a few remarks concerning the proof of Proposition 3.1. The proof of this result consists of three main steps. The first step is to obtain apriori estimates on both the variables (T,q,Y_1,p_1,Y_2,p_2) and the wave speed θ^*. These estimates are derived in

[7]. The second step in the proof of the Proposition is to prove the result in the special case $d_0 = d_1 = d_2$. The proof in this case is based on the Conley index. In the next section we describe how the Conley index is used in the proof. The third step in the proof of Proposition 3.1 is to continue the solution from the case $d_0 = d_1 = d_2$ to the case of general, positive diffusion constants. Again, this step is carried out in [7].

4. Outline of the Proof of Proposition 3.1 for $d_0 = d_1 = d_2$

Throughout this section we assume that $d_0 = d_1 = d_2 = 1$. The proof of Proposition 3.1 in this case is based on the Conley index. For the proof we will need such notions as isolated invariant set, the Conley index, and Morse decomposition. Those readers unfamiliar with these notions are referred to [3], [5], and [6].

If we multiply the second equation in (2.2) by $Q_1 + Q_2$, the third equation by Q_2, add the resulting equations to the first, and let

$$Z = T + (Q_1 + Q_2)Y_1 + Q_2Y_2 ,$$

we find that

$$Z'' - \theta Z' = 0 .$$

Because $Z'(-\infty) = Z'(\infty) = 0$ it follows that

$$T + (Q_1 + Q_2)Y_1 + Q_2Y_2 = \text{constant} .$$

By assumption $Z(-\infty) = Q_1 + 2Q_2$. Hence,

$$Y_2 = \frac{1}{Q_2}[Q_1 + 2Q_2 - T - (Q_1 + Q_2)Y_1] . \tag{4.1}$$

Plugging (4.1) into (2.2) we obtain

$$\begin{aligned} T'' - \theta T' &= -Q_1Y_1f_1(T) - [Q_1 + 2Q_2 - T - (Q_1 + Q_2)Y_1]f_2(T) \\ Y_1'' - \theta Y_1' &= Y_1f_1(T) . \end{aligned} \tag{4.2}$$

The traveling wave equations (3.1) become, after substitution of (4.1),

$$
\begin{aligned}
T' &= q \\
q' &= \theta q - Q_1 Y_1 f_1(T) - [Q_1 + 2Q_2 - T - (Q_1 + Q_2)Y_1]f_2(T) \\
Y_1' &= p_1 \\
p_1' &= \theta p_1 + Y_1 f_1(T) \ .
\end{aligned}
\tag{4.3}
$$

We seek a solution of (4.3), (3.2) which satisfies

$$
\begin{aligned}
(T,q,Y_1,p_1,\theta)(-\infty) &= (0,0,1,0,\theta_0) \equiv \alpha_1 \\
(T,q,Y_1,p_1,\theta)(+\infty) &= (Q_1 + 2Q_2,0,0,0,-\theta_0) \equiv \gamma_2 \ .
\end{aligned}
\tag{4.4}
$$

The first step in proving the existence of a solution of (4.3), (4.4) is to analyse the rest point set of (4.2). This set is very large. If we let

$$\mathcal{P} = \{(T,Y_1) : T \geq 0,\ Y_1 \geq 0\} \ ,$$

then the rest points of (4.2) in $\mathcal{P}$ are

$$\{(T,Y_1) : 0 \leq T \leq T_1,\ Y_1 \geq 0\} \cup \{(T,Y_1) : Y_1 = 0,\ 0 \leq T \leq T_1\} \cup$$
$$\cup \ \{(Q_1 + 2Q_2,0)\} \ .$$

This set is much too large to work with. For this reason we perturb the vector field given by the right hand side of (4.2) to a new vector field, which we denote by $(g_\delta(T,Y_1),h_\delta(T,Y_1))$, $0 < \delta << 1$. As $\delta \to 0$, (g_δ,h_δ) approaches the vector field given by the right side of (4.2). Moreover, (g_0,h_δ) has only five rest points. The flow given by

$$
\begin{aligned}
T' &= g_\delta(T,Y_1) \\
Y_1' &= h_\delta(T,Y_1)
\end{aligned}
\tag{4.5}
$$

is as shown in Figure 3.

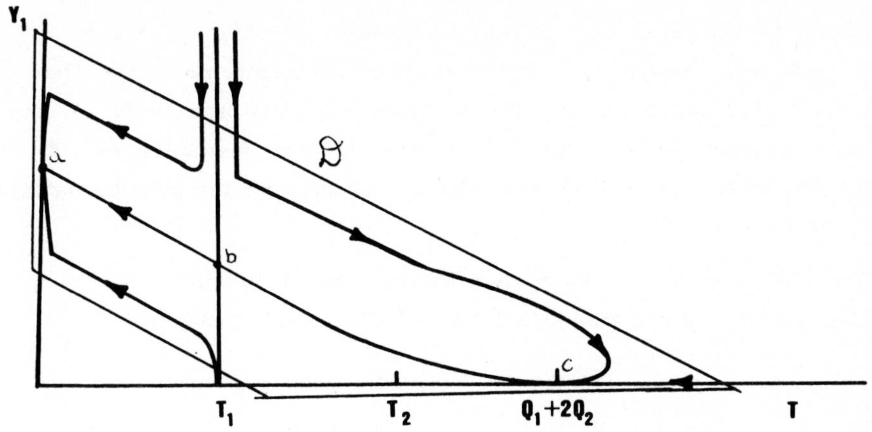

Figure 3

The precise definition of (g_δ, h_δ) is given in [7].

For $0 < \delta << 1$ we replace (4.3), (3.2) by

$$
\begin{aligned}
T' &= q \\
q' &= \theta q - g_\delta(T, Y_1) \\
Y_1' &= p_1 \\
p_1' &= \theta p_1 - h_\delta(T, Y_1) \\
\theta' &= \varepsilon(\theta^2 - \theta_0^2) .
\end{aligned}
\tag{4.6}
$$

The proof of Proposition 2.1 is now broken up into two steps. We first prove

<u>Proposition 4.1</u>: There exists $\delta_0 > 0$ such that for each $\delta \in (0, \delta_0)$ and $\varepsilon > 0$, there exists a solution of (4.6), (4.4).

To complete the proof of Proposition 2.1 we show that as $\delta \to 0$, the solutions of (4.6), (4.4) approach a solution of (4.3), (4.4). In the remainder of this section we outline the proof of Proposition 4.1.

We wish to apply the Conley index theory. To do this we begin by defining some isolating neighborhoods. The first step is to construct a set $\mathcal{D}$ in $\mathcal{P}$ as shown in Figure 3. $\mathcal{D}$ has the properties that it is convex, it contains only the three rest points a, b, and c shown in Figure 3, and it is positively invariant. By positively invariant we mean that any solution of (4.5) which lies in $\mathcal{D}$ for some z_0 must remain in $\mathcal{D}$ for all $z > z_0$. Another way to say this is that on the boundary of $\mathcal{D}$, the vector field given by (4.5) points into $\mathcal{D}$. Of course, the precise definition of $\mathcal{D}$ is given in [7].

The next step is to prove that there exists a constant V such that if (T,q,Y_1,p_1,θ) is a bounded solution of (4.6), such that

$$(T(z),Y_1(z)) \in \mathcal{D} \quad \text{for all} \quad z ,$$

then

$$\|(q,p)\| < V \quad \text{for all} \quad z .$$

This is not very hard to do. Then let

$$N = \{(T,q,Y_1,p_1,\theta) \; : \; (T,Y_1) \in \mathcal{D}, \; \|(q,p)\| \leq V, \; |\theta| \leq \theta_0 + 1\} .$$

In [7] we prove that N is an isolating neighborhood; that is, any trajectory which lies on the boundary of N must leave N in forwards or backwards time. It is here that we use the fact that $\mathcal{D}$ is convex and all of the diffusion constants are equal. Of course, we used the fact that $d_0 = d_1 = d_2$ before to reduce the order of the system, but that was mainly for convenience - so that the pictures are easier to draw.

Let S equal to the maximal invariant in N. We shall define a Morse decomposition of S. To do this it will first be necessary to split N into three subsets. These subsets will be denoted by N_1, N_2, and N_3, and are illustrated in Figure 4. To define these sets, we first introduce sets G_1 and G_2 which are, respectively, neighborhoods of the local unstable manifold at α_1 and the local stable manifold at γ_2. The precise definitions of these subsets are given in [7]. Let

$$N_1 = \{(T,q,Y_1,p_1,\theta) \; : \; |\theta - \theta_0| \leq 1\} \setminus G_1 ,$$

$$N_2 = \{(T,q,Y_1,p_1,\theta) \; : \; |\theta| \leq \theta_0 - 1\} \cup G_1 \cup G_2 ,$$

$$N_3 = \{(T,q,Y_1,p_1,\theta) \; : \; |\theta + \theta_0| \leq 1\} \setminus G_2 ,$$

and, for $i = 1,2,3$,

$$S_i = \text{maximal invariant set in } N_i .$$

The sets G_1 and G_2 are chosen so that each N_i is an isolated invariant set. Moreover, any trajectory which lies on the border of N_1 and N_2 must enter N_2, and any trajectory on the border of N_2 and N_3 must enter N_3. This is shown in Figure 4. An immediate consequence of this fact is that (S_3,S_2,S_1) forms a Morse decomposition of S.

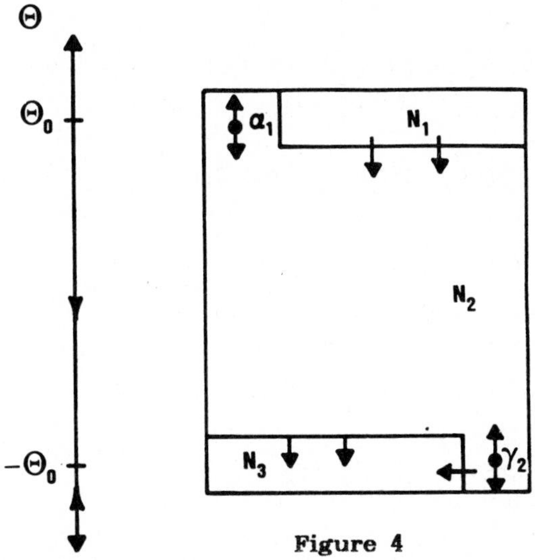

Figure 4

Recall that we are trying to prove that there is a trajectory which connects α_1 and γ_2. To prove this we will show that there is a bounded trajectory in N_2 besides α_1 and γ_2. It is clear the $\theta(z)$ is decreasing on any nonconstant, boundary trajectory in N_2. Hence, $\theta(z)$ serves as a Liapunov function in N_2. An immediate consequence of this is that any bounded trajectory in N_2 must be either a rest point or a trajectory which connects two rest points. Since α_1 and γ_2 are the only rest points in N_2, it follows that if we can prove that

$$S_2 \neq \{\alpha_1\} \cup \{\gamma_2\} , \tag{4.7}$$

then Proposition 4.1 is true; that is, there exists a trajectory which connects α_1 and γ_2.

We prove (4.7) by showing that

$$h(S_2) = \bar{0} . \tag{4.8}$$

Here $h(S_2)$ is the Conley index of S_2 and $\overline{0}$ is the index of the empty set. The reason that (4.7) follows from (4.8) is because $h(\alpha_1) \neq \overline{0}$ and $h(\gamma_2) \neq \overline{0}$. In fact, because $\dim W^u_{\alpha_1} = 3$ and $\dim W^u_{\gamma_2} = 2$ it follows that

$$h(\alpha_1) = \Sigma^3 \text{ and } h(\gamma_2) = \Sigma^2 ,$$

where Σ^n is the pointed n-sphere. If (4.7) were not true, then we must have (see Conley [3]),

$$h(S_2) = h(\alpha_1) \vee h(\gamma_2) .$$

Since $\overline{0} \neq \Sigma^2 \vee \Sigma^2$, this is impossible.

So it remains to prove (4.8). This is very hard to do directly. We prove this by using the fact that (S_3,S_2,S_1) forms a Morse decomposition of S. We will prove that

$$h(S) = h(S_1) = h(S_3) = \overline{0} . \tag{4.9}$$

It is not hard to prove (4.8) from (4.9). To do this one constructs an exact sequence which relates the homologies of the indices in S, S_1, S_2, and S_3. See [7] for details, or [5] for the general theory.

We conclude by outlining the proof of (4.9). We prove that $h(S) = \overline{0}$ by continuing the given flow, (4.6), to one in which the maximal invariant set in N is the empty set. We then apply the basic continuation theorem, see [3], to conclude that $h(S) = \overline{0}$. The continuation is defined as follows. Replace the last equation in (4.6) by

$$\theta' = \varepsilon(\theta^2 - \theta_0^2) + \lambda . \tag{4.10}$$

If λ is sufficiently large, then $\theta' > 0$ for all z. This implies that there are no bounded trajectories in N. Of course, one must check that N is an isolated invariant set for all λ, but this is not very hard to do. We have thus continued S to the empty set, proving that $h(S) = \overline{0}$.

We next describe why $h(S_1) = \overline{0}$. It is clear that

$$S_1 \subset \{(T,q,Y_1,P_1,\theta) : \theta = \theta_0\} .$$

Let $h_1(S_1)$ equal to the index of S_1 considered as an isolated invariant set of the flow defined by (4.6) restricted to the set $\{\theta = \theta_0\}$. We prove that $h_1(S_1) = \bar{0}$. This immediately implies that $h(S_1) = \bar{0}$.

The first step is to note that for θ_0 large, the bounded solutions of (4.6) have projection onto (T,Y_1) space lying close to the bounded solutions of (4.5). This is proved in Conley and Gardner [4]. Moreover, their proof shows that if P is the projection of N_1 onto (T,Y_1) space and M is the maximal invariant set in P, then

$$h_1(M_1) = h(M) \wedge \Sigma^2 ,$$

the smash product of the index of M, as an invariant set of (4.5), and Σ^2, the pointed two sphere. Because $\bar{0} = \bar{0} \wedge \Sigma^2$, it suffices to prove that $h(M) = \bar{0}$.

Now P is equal to $\mathcal{D}$, as shown in Figure 3, minus a small neighborhood of (0,1). Note that (0,1) is an attractor for the flow defined by (4.5). Hence, we can find a compact neighborhood H of (0,1) such that $H \subset \mathcal{D}$, $M \cap H = \emptyset$, and on ∂H the vector field defined by the right hand side of (4.5) points into H.

It follows that $\mathcal{D}\backslash H$ is an isolating neighborhood, and the maximal invariant set in $\mathcal{D}\backslash H$ is M. Note that trajectories enter $\mathcal{D}\backslash H$ along its outward boundary, since $\mathcal{D}$ is positively invariant, and leave $\mathcal{D}\backslash H$ along ∂H. From this one can easily compute $h(M)$ directly. One finds that $h(M) = \bar{0}$.

In a similar manner one proves that $h(S_3) = \bar{0}$.

REFERENCES

1. Berestycki, H., B. Nicolaenko, and B. Scheurer, Traveling wave solutions to reaction-diffusion systems modelling combustion, Nonlinear Partial Differential Equations (J. Smoller, ed.), Contemporary Mathematics, Vol. 17, American Mathematical Society, 1983, 189-207.

2. Buckmaster, J. and G.S.S. Ludford, "Theory of Laminar Flames", Cambridge University Press (1982).

3. Conley, C., Isolated Invariatn Sets and The Morse Index, CBMS Regional Conference Series in Mathematics, No. 38, American Mathematical Society, 1978.

4. Conley, C. and R. Gardner, An application of the generalized Morse index to traveling wave solutions of a competitive reaction-diffusion model, Indiana Univ. Math. J., 33, 319-343 (1984).

5. Conley, C. and E. Zehnder, Morse-type index theory for flows and periodic solutions for Hamiltonian equations, Comm. or Pure and Applied Math., 37, 207-253 (1984).

6. Smoller, J., "Shock Waves and Reaction-Diffusion Equations", Springer-Verlag (1983).

7. Terman, D., Traveling wave solutions arising from a combustion model, IMA Preprint Series #216, University of Minnesota (1986).

8. Williams, F., "Combusion Theory", Addision-Wesley (1963).

PATH CONTINUATION - A SENSITIVITY ANALYSIS APPROACH

Klaus Ulrich
Institut für Angewandte Mathematik
Universität Hannover
Welfengarten 1
D-3000 Hannover 1

Introduction

Determining the zero set of nonlinear mappings is a most delicate task. As is often the case, especially for problems in applications like nonlinear structural mechanics, biological, chemical, or physical systems, analytical solutions are not available. Instead, more or less appropriate approximate versions of the original problems have to be studied numerically.

Principally, this task - referred to as path continuation or numerical bifurcation analysis - is an ill-posed problem; for slight perturbations in the model equations may completely change the solution set, in quality and nature. Moreover it is not hard to imagine (as indeed can be shown by theoretical reasoning as well as by numerical examples) the influence of various effects on the computational solution, effects due to discretization, approximation, iteration, interpolation, integration processes, limited machine accuracy, and so on as well as their mutual interaction.

Quite powerful methods have been developped, both by mathematicians and engineers, to handle a wide spectrum of nonlinear problems, and various algorithms based on these methods have been designed. The interested reader is referred to Kearfott /15/, Keller /16,17/, Weber /49/, Chow/Hale /5/, Küpper/Mittelmann/Weber /22/, Mittelmann/Weber /25/, Rabinowitz /34/, Rhein-

NATO ASI Series, Vol. F37
Dynamics of Infinite Dimensional Systems
Edited by S.-N. Chow, and J. K. Hale

boldt /36,37/, Wacker /46/; Gerardin/Idelsohn/Hogge /10/, Winters/Cliffe/Jackson /50/, and Hinton/Owen/Taylor /13/, for a survey of methods and trends in bifurcation analysis and some applications in engineering sciences; and to Aluko/Chang /1/, Bank/Chan /3/ and Mittelmann /24/, Doedel /8/, Kubicek /21/ and Holodniok/Kubicek /14/, Rheinboldt/Burkardt /38/, Seydel /41/, and Watson/Fenner /47/, for algorithms and algorithmic comments.

Surprisingly, discretization of the underlying continuous model and following the solution curve(s) generally appear as distinct components in solution strategies, hardly yielding feedback abilities on tracing the discretized model curve itself. Having this in mind, a problem-oriented self-adaptive quasi-dynamical path continuation strategy using optimal design formulations and sensitivity analysis techniques is suggested, and the future use of such procedures is highly recommended.

State of the problem

The general problem can be formulated as follows:

Given a nonlinear mapping $F: X \to Y$, where X and Y denote suitable Banach spaces, determine the (local) zero set of F, $\ker_{loc} F$ (respectively $\ker F = \bigcup \ker_{loc} F$, the union being taken over suitable decompositions).

Conveniently X may be thought of as a product $X_1 \times X_2$ where X_1 denotes the so-called state space and X_2 the space of "parameters". X_2 may represent the domain of some "natural" parameter (i. e. one inherent to the problem, with some physical meaning like load or deflection etc.) or of an "artificial" parameter (i. e. one introduced after modelling the problem), or simply denote the complement of X_1 in X. In order to make the general problem stated above accessible to proper mathematical treatment additional assumptions e. g. on the differentiability properties of F have to be made. In what follows F is assumed to be at least once Fréchet differentiable and its first Fréchet derivative, DF, to be a Fredholm operator of positive index (i. e. both ker DF and coker DF are finite dimensional, and dim ker DF is greater than dim coker DF) thus allowing for some kind of Lyapunov-Schmidt reduction. (For further details, and general bifurcation analysis methods, the reader is referred to the excellent book of Chow/Hale /5/.)

Most frequently analytical solutions are not available, or properties of solutions have to be known within a large region of physical significance; consequently interest centres on numerical methods for determining $\ker_{loc} F$. More precisely, generally appropriate approximations $\tilde{F}: \tilde{X} \to \tilde{Y}$, $\tilde{X}$ and $\tilde{Y}$ finite dimensional, to F have to be studied. (This of course may lead to significant changes in quality and nature of the solution set; this problem shall not be discussed here.) There is an extensive literature on this subject (see /5/, /22/, /25/, /34/, /37/, /46/ and the references quoted there); still a lot of work is to be done. Among others, higher order bifurcations, hysteresis phenomena, or so-called snap-through effects are not readily accessible to proper numerical treatment; also in the field of design of algorithms further improvement is to be achieved.

Conceptual comments

To give a more concrete understanding of the strategy which has been announced above the main steps on the way from the original problem to its possible computational solution will be reviewed. Emphasis is made on the genuine constituents to be dealt with, and on feedback and sensitivity analysis abilities.

0) Modelling.

Here a-priori knowledge may lead to better handling; more sophisticated models (with e. g. symmetry breaking properties) are desirable.

1) Discretization.

(Here the main features are discussed for the finite element method, in a rather general way.) There are three basic approaches to obtain improved quality of computational results based on finite element discretizations: the so-called h-, p-, and r-methods. In its pure form the h-method suggests the use of finer grids (not necessarily fixing initial nodal points), whereas the order of the element shape functions remains unchanged. The p-method requires the use of element shape functions of increased order while the initial grid remains unchanged. Finally, the r-method aims at generating a new mesh configuration mainly by suitably rearranging (inner) nodal points; thus the initially given finite elements may change in size and shape, whereas the order of the element shape functions as well as the number of the finite elements used remain unchanged.

Considerable progress has been achieved, for the h-method, in using adaptive mesh refinement and multigrid techniques (see e. g. Babuska /2/ and Zienkiewicz/Gago/Kelly /51/). For the p-method no such progress can be reported. Very recently quite successful attempts have been made for the r-method using optimal design formulations for the finite element grid adaptation procedures (see Demkowicz/Oden /6/, Diaz/Kikuchi/Taylor /7/, Ghia/Ghia /11/, Komkov /19/, Komkov/Irwin /20/, Lions /23/, Pedersen /32/, Rabitz /35/, Russell/Christiansen /39/, Siu/Turcke /42,43/, Saltzman/Brackbill /40/, Bathe/Sussman /4/, Sussman/Bathe /44/, Szefer /45/ for evolutionary lines and further background; and Mota Soares /26/, Haug/Choi/Komkov /12/, Kikuchi /18/ for recent advances and trends in optimal design both theory and applications). Though pure refinement techniques are powerful tools they may fail (e. g. when dealing with problems with sharp, or moving, stress concentrations in nonlinear structural mechanics, fluid dynamics problems for specific domains). Properly rearranged, "optimized" meshes (for which the relocation of the nodal points has been obtained by solving a suitable optimal control problem), however, did not only lead to surprisingly well improved quality, but were also capable to give relevant results for problems where refined meshes would fail (see Demkowicz/Oden /6/ for a striking example). Best (and mathematically justified) results have been obtained by means of sensitivity analysis techniques using e. g. the finite element error indicator. As in principle optimization procedures are quite costly (and of course there is no uniquely determined optimal mesh, for a given constellation) it would be wise to work with somewhat suboptimal meshes rather than optimal ones. Numerous numerical experiments still have to be carried out. Moreover, according to actual requirements, optimization could be restricted to certain subdomains where it really leads to significant grid modifications. It is noted, however, that a combination of optimization with adaptive mesh refinement or multigrid techniques is strongly recommended, for two main reasons: First, in order to let nodal points move it may turn out necessary to refine. Second, some numerical experience has shown that properly optimized meshes look similar for different levels of refinement. So the power and efficiency of these two methods should be brought together.

2) Path continuation.

Here homotopy methods and incremental (predictor-corrector) methods are most popular and cover a wide range of applications. Their evolutionary lines go back to both mathematicians (primarily for homotopy methods) and engineers (for incremental continuation procedures); see e. g. Kearfott /15/, Watson/Fenner /47/, Keller /16,17/, Wacker /46/, and Hinton/Owen/Taylor /13/ for some of the main aspects of constructive bifurcation theory. Algorithms based on these methods have been mentioned earlier. The behaviour of continuation procedures "near" possible critical points may be quite delicate, for several reasons (like e. g. numerical instabilities, spurious critical points, etc.) - if ever they are able to detect such points. For incremental procedures, for example, a more accurate line-search and stepsize control is required. It seems that the facilities of homotopy methods have not yet been fully utilized; in many applications these methods could be of some advantage when compared with incremental methods. Principally, there is no preference for either these methods.

3) Local bifurcation analysis.

Apart from following a solution curve within a certain range, (relevant) critical points have to be detected or in addition accurately located; upper bounds for the number of bifurcating branches are of great interest; (approximate) directions of bifurcating branches have to be computed; curves have to be continued beyond critical points; bifurcating branches have to be followed. Particular interest centres on higher order bifurcations ("follow the folds") and "invisible" bifurcations (for which there is no proper (e. g. experimental) a-priori knowledge and which may easily be passed by without having been detected). Spurious bifurcations (which can occur due to effects of discretization, but which are numerically irrelevant) have to be handled appropriately (at present this means successive refinement until they hopefully vanish...). As can be seen once more, a better link between path continuation and local bifurcation analysis modules and the discretization module is required. Finally the use and hence the implementation of new, more sophisticated (nonlinear) methods resulting from applied algebraic geometry (like constructive blowing up, see Rabier /33/) seems to be indispensable in order to achieve further progress. To show all their effectiveness these methods may in turn require a special design of algorithms.

4) Computerization.

Here parallel rather than vector processing abilities should be provided, to take full advantage of the discretized model's structure.

Sketch of the strategies and further developments

In view of what has been pointed out above the use of sensitivity analysis techniques is suggested to design problem-oriented self-adaptive path continuation and local bifurcation analysis procedures:

Based on an initial finite element grid an optimized (suboptimal rather than optimal) grid is computed, by means of an appropriate optimal control process. This procedure is to be combined with adaptive refinement or multigrid techniques to overcome difficulties which have been outlined and to achieve further improvement in quality and efficiency. On following a curve, various strategies are at one's disposal, according to the actual purposes: "simple" path continuation (away from critical points) (in one, possibly several parameters), detecting or in addition accurately locating critical points, branch switching (with or without previously computing approximate directions of bifurcated branches). The nonlinear response of the discretized system to changes in control (respectively perturbation) parameters is being measured directly <u>on</u> following the curve, to provide feedback facilities.

Moreover, different levels of accuracy may be chosen for different segments of a curve under consideration, according to the actual response (or to some a-priori knowledge); grids may be re-optimized at any stage of the computational process. The underlying domain may be split up into "static" and "dynamic" subdomains for which different strategies may be applied. In addition, the use of so-called reduction methods may lead to considerably increased efficiency and may even provide further information on the system's response. Such methods were apparently presented first by Nagy /27/ and have subsequently been improved and studied by Noor /28,29/, Noor/Peters /30,31/, Fink/Rheinboldt /9/ and Rheinboldt /37/. Indeed, they have not yet revealed their full power, more numerical experiments are needed.

Possible applications of such strategies for quasi-dynamical path continuation are in static as well as in originally time-dependent problems (e. g. reaction-diffusion systems, hysteresis phenomena, even shock waves). Improved quality and efficiency in numerical bifurcation analysis is hopefully achieved, the implementation of far advanced techniques facilitated, and the future use of more sophisticated models inspired. It is about time.

Acknowledgement: The work reported in this paper has been supported by Stiftung Volkswagenwerk.

Bibliography

/ 1/ Aluko,M.;Chang,H.-C.: PEFLOQ: An algorithm for the bifurcational analysis of periodic solutions in autonomous systems. Manuscript, Dept. Chem. Nucl. Eng., UCSB, 1984

/ 2/ Babuska,I.: Feedback, adaptivity, and a posteriori estimates in finite elements: Aims, theory, and experience. Univ. Maryland, Inst. Phys. Sci. Tech., Tech. Note BN-1022(1984)

/ 3/ Bank,R.E.;Chan,T.F.: PLTMGC: A multigrid continuation program for parametrized nonlinear elliptic systems. SIAM J. Sci. Stat. Comp. 3(1982), 173-194

/ 4/ Bathe,K.J.;Sussman,T.D.: An algorithm for the construction of optimal finite element meshes in linear elasticity. A.S.M.E., A.M.D.; in: Recent Developments in Computing Methods for Nonlinear Solid and Structural Mechanics, June 1983

/ 5/ Chow,S.-N.;Hale,J.K.: Methods of Bifurcation Theory. Grundl. der math. Wiss. 251, Springer-Verlag, New York, 1982

/ 6/ Demkowicz,L.;Oden,J.T.: On a mesh optimization method based on a minimization of interpolation error. Int. J. Engng. Sci. 24,1(1986), 55-68

/ 7/ Diaz,A.;Kikuchi,N.;Taylor,J.E.: Optimal design formulations for finite element grid adaptation. In: Komkov,V.(ed.): Sensitivity of Functionals with Applications to Engineering Sciences. Lecture Notes in Mathematics 1086, Springer-Verlag, Berlin, 1984, 56-76

/ 8/ Doedel,E.J.: AUTO: A program for the automatic bifurcation analysis of autonomous systems. Congr. Num. 30(1981), 265-284

/ 9/ Fink,J.P.;Rheinboldt,W.C.: On the error behaviour of the reduced basis technique for nonlinear finite element approximations. Z. angew. Math. u. Mech., Berlin, 63,1(1983), 21-28

/10/ Gerardin,M.;Idelsohn,S.;Hogge,M.: Computational strategies for the solution of large nonlinear problems via quasi-Newton methods. Comp. & Struct. 13(1981), 73-81

/11/ Ghia,K.;Ghia,U.: Advances in Grid Generation. FED-vol.5, Amer. Soc. Mech. Eng., New York, 1983

/12/ Haug,E.J.;Choi,K.K.;Komkov,V.: Design Sensitivity Analysis of Structural Systems. Academic Press, New York, 1986

/13/ Hinton,E.;Owen,D.R.J.;Taylor,C.(eds.): Recent Advances in Nonlinear Computational Mechanics. Pineridge Press, Swansea, 1982

/14/ Holodniok,M.;Kubicek,M.: DERPER - An algorithm for the continuation of periodic solutions in ordinary differential equations. J. Comp. Phys. 55,2(1984), 254-267

/15/ Kearfott,R.B.: A derivative-free arc continuation method and a bifurcation technique. In: Allgower,E.L.;Glashoff,K.; Peitgen,H.O.(eds.): Numerical Solution of Nonlinear Equations. Springer-Verlag, New York, 1981
/16/ Keller,H.B.: Global homotopies and Newton methods. In: De Boor,C.;Golub,G.H.(eds.): Recent Advances in Numerical Analysis. Academic Press, New York, 1978, 73-84
/17/ Keller,H.B.: The bordering algorithms and path-following near singular points of higher nullity. SIAM J. Sci. Stat. Comp. 4(1983), 4
/18/ Kikuchi,N.: Adaptive grid design for finite element analysis in optimization. Part 1: Review of finite element error analysis. Part 2: Grid optimization. Part 3: Shape optimization. In: Mota Soares,C.A.(ed.): Computer Aided Optimal Design. Pre-proceedings Conf. Tróia, Portugal, June 29 - July 11, 1986. CEMUL, Lisboa, 1986, 307-363
/19/ Komkov,V.(ed.): Sensitivity of Functionals with Applications to Engineering Sciences. Lecture Notes in Mathematics 1086, Springer-Verlag, Berlin, 1984
/20/ Komkov,V.;Irwin,C.: Uniqueness for gradient methods in engineering optimization. In: Komkov,V.(ed.): Sensitivity of Functionals with Applications to Engineering Sciences. Lecture Notes in Mathematics 1086, Springer-Verlag, Berlin, 1984, 93-118
/21/ Kubicek,M.: Algorithm 502: Dependence of solutions of nonlinear systems on a parameter. ACM Trans. Math. Software 2(1976), 98-107
/22/ Küpper,T.;Mittelmann,H.D.;Weber,H.(eds.): Numerical Methods for Bifurcation Problems. ISNM 70, Birkhäuser Verlag, Basel, 1984
/23/ Lions,J.L.: Contrôle optimale de systèmes distribués singuliers. Dunod, Paris, 1983
/24/ Mittelmann,H.D.: An efficient algorithm for bifurcation problems of variational inequalities. Math. Comput. 41 (1983), 473-485
/25/ Mittelmann,H.D.;Weber,H.(eds.): Bifurcation Problems and Their Numerical Solution. ISNM 54, Birkhäuser Verlag, Basel, 1980
/26/ Mota Soares,C.A.(ed.): Computer Aided Optimal Design. Pre-proceedings Conf. Trôia, Portugal, June 29 - July 11, 1986. CEMUL, Lisboa, 1986
/27/ Nagy,D.A.: Model representation of geometrically nonlinear behaviour by the finite element method. Comp. & Struct. 10 (1977), 683-688
/28/ Noor,A.K.: Recent advances in reduction methods for nonlinear problems. Comp. & Struct. 13(1981), 31-44
/29/ Noor,A.K.: Recent advances in the application of variational methods to nonlinear problems. In: Kardestuncer,H.(ed.): Unification of Finite Element Methods. Elsevier Science Publishers B. V. (North-Holland), New York, 1984, 275-302
/30/ Noor,A.K.;Peters,J.M.: Reduced basis technique for nonlinear analysis of structures. AIAA J. 18(1980), 455-462
/31/ Noor,A.K.;Peters,J.M.: Bifurcation and post-buckling analysis of laminated composite plates via reduced basis technique. Comp. Meth. Appl. Mech. Eng. 29(1981), 271-295

/32/ Pedersen,P.: Sensitivity analysis for non-selfadjoint problems. In: Komkov,V.(ed.): Sensitivity of Functionals with Applications to Engineering Sciences. Lecture Notes in Mathematics 1086, Springer-Verlag, Berlin, 1984, 119-130

/33/ Rabier,P.: Topics in One-Parameter Bifurcation Problems. Tata Institute Lecture Notes 76, Springer-Verlag, Berlin, 1985

/34/ Rabinowitz,P.(ed.): Applications of Bifurcation Theory. Academic Press, New York, 1977

/35/ Rabitz,H.: Sensitivity methods for mathematical modelling. In: Komkov,V.(ed.): Sensitivity of Functionals with Applications to Engineering Sciences. Lecture Notes in Mathematics 1086, Springer-Verlag, Berlin, 1984, 77-92

/36/ Rheinboldt,W.C.: Numerical methods for a class of finite dimensional bifurcation problems. SIAM J. Numer. Anal. 15 (1978), 1-11

/37/ Rheinboldt,W.C.: Numerical Analysis of Parametrized Nonlinear Equations. Wiley, New York, 1986

/38/ Rheinboldt,W.C.;Burkardt,J.V.: A locally parameterized continuation process. ACM Trans. Math. Software 9,2(1983), 215-235

/39/ Russell,R.D.;Christiansen,J.: Adaptive mesh selection strategies for solving boundary value problems. SIAM J. Numer. Anal. 15(1978), 59-80

/40/ Saltzman,J.;Brackbill,J.: Applications and generalizations of variational methods for generating adaptive meshes. In: Thompson,J.F.(ed.): Numerical Grid Generation. Elsevier Science Publishers B. V. (North-Holland), Amsterdam New York, 1982, 865-884

/41/ Seydel,R.: Numerical computation of periodic orbits that bifurcate from stationary solutions of ordinary differential equations. Appl. Math. Comp. 9(1981), 257-271

/42/ Siu,D.;Turcke,D.J.: Optimum grids in nonlinear finite element analysis. Symp. on Math. Modelling in Structural Eng., NASA-Langley Research Center, Oct. 1979

/43/ Siu,D.;Turcke,D.J.: Optimal finite element discretization for nonlinear constitutive relations. Comp. & Struct. 13 (1981), 83-87

/44/ Sussman,T.;Bathe,K.J.: The gradient of the finite element variational indicator with respect to nodal point co-ordinates: an explicit calculation and applications in fracture mechanics and mesh optimization. Int. J. Numer. Meth. Engng. 21(1985), 763-774

/45/ Szefer,G.: Optimierung und Sensibilität elastischer Strukturen. Vorlesungsausarbeitung, Inst. für Mechanik (Bauwesen), Univ. Stuttgart, Dez. 1985

/46/ Wacker,H.J.(ed.): Continuation Methods. Academic Press, New York, 1978

/47/ Watson,L.T.;Fenner,D.: Chow-Yorke algorithm for fixed points or zeros of C^2-maps. ACM Trans. Math. Software 6,2(1980), 252-259

/48/ Watson,L.T.;Yang,W.H.: Methods for optimal engineering design problems based on globally convergent methods. Comp. & Struct. 13(1981), 115-119

/49/ Weber,H.G.: Multigrid bifurcation iteration. SIAM J. Numer. Anal. 22(1985), 262-279

/50/ Winters,K.H.;Cliffe,K.A.;Jackson,C.P.: The Prediction of Instabilities Using Bifurcation Theory. Preprint AERE-TP. 1172(1986) (to appear in Numer. Meth. in Transient and Coupled Systems, John Wiley, Chichester, England)

/51/ Zienkiewicz,O.C.;De S.R. Gago,J.P.;Kelly,D.W.: The hierarchical concept in finite element analysis. Comp. & Struct. 16,1-4(1983), 53-65

Confinor and Anti-confinor in Constrained "Lorenz" System

Shigehiro Ushiki† **and René Lozi**
CNRS, Institut de Mathématiques et Sciences Physiques,
Université de Nice, Parc Valrose, 06034 Nice, France
† Permanent address : Institute of Mathematics,
Yoshida College, Kyoto University, Kyoto 606 Japan

Abstract The bifurcation of possibly chaotic oscillatory patterns of a family of differential equations constrained on the cusp surface is studied. Notions of "confinor" and "anti-confinor" are introduced to get a rough diagram of bifurcations of this family. "Anti-bifurcation" and "rainbow-bifurcation" are found as bifurcations of confinors.
Keywords bifurcation, confinor, constrained equation, dynamical system, Lorenz attractor, singular perturbation

§0. Introduction

Paper-sheet models (J.Guckenheimer[5], J.Guckenheimer & R.Williams[6]) and slow-fast equations (O.E.Rössler[14], F.Takens[16,17], R.Lozi[8,9]) with slow manifolds have been employed to understand the dynamics and the bifurcation of "strange attractors". Geometrically constructed models permits us to study the bifurcation to only a limited extent. We shall construct a family of ordinary differential equations constrained on the cusp surface starting from a versal family of vector fields which "unfolds" a degenerate singular point of vector field. We employ the notion of confinors and anti confinors in place of attractors and repellers, since the notion of attractor is too strict and too complicated when one is interested in the global bifurcation diagrams.

The existence of some type of confinor can imply the existence of orbits with specific patterns of oscillations. In particular, oscillatory patterns consisting of small amplitude oscillations and large amplitude relaxation-oscillations can be related to confinors.

O.E.Rössler[14] gave many equations which produce strange attractors. When he discovered these formulae, he was guided by his intuition based on slow-fast type ordinary differential equations. H.Oka and H.Kokubu[12] found an "attractor" which looks like the Lorenz attractor. Their "attractor" is constrained on the cusp surface. Our study on the bifurcation of oscillatory patterns is done for their family of constrained equations.

§1. A family of differential equations constrained on the cusp surface

Let us consider a vector field defined in a neighborhood of the origin, O, of $\mathbf{R}^3$ of the following form :

$$du/dt = L\,u + F(u), \text{ where } u = {}^t(x,y,z),\ L = \begin{bmatrix} 0&0&0\\ 0&0&0\\ 0&1&0 \end{bmatrix} \text{ and } F(O) = 0. \tag{1.1}$$

NATO ASI Series, Vol. F37
Dynamics of Infinite Dimensional Systems
Edited by S.-N. Chow, and J. K. Hale

In order to obtain a sufficiently degenerate but not too complicated systems, we suppose (1.1) is equivariant under the symmetry $s : (x,y,z) \to (-x,-y,z)$.

By a differentiable change of coordinates around the origin, we can always transform (1.1) into the form :

$$\begin{cases} dx/dt = y + \alpha xz + a' x^3 \\ dy/dt = \beta xz + b' x^3 + c' yz^2 + e' xz^2 \\ dz/dt = \gamma x^2 + \delta z^2 + d' z^3 + f' x^2 z, \end{cases} \tag{1.2}$$

up to higher order terms. Here, coefficients $\alpha, \beta, \gamma, \delta, a', b', c', d', e', f'$ can be computed from the third order Taylor expansion of the vector field at the origin. Two of these eight parameters can be normalized by an appropriate linear scaling of coordinates. We shall employ such a rescaling later, for the renormalization of the cusp surface. Family of ordinary differential equations (1.2) can be considered as a family of normal forms in the truncated sense under the symmetry s, in the sense of F.Takens[15].

We suppose that

$$\beta \neq 0 \quad \text{and} \quad \gamma \neq 0. \tag{1.3}$$

Then we can execute a further transformation of coordinates to suppress the coefficients e' and f'. Thus we obtain the following family up to higher order terms :

$$\begin{cases} dx/dt = y + \alpha xz + a x^3 \\ dy/dt = \beta xz + b x^3 + c yz^2 \\ dz/dt = \gamma x^2 + \delta z^2 + d z^3 . \end{cases} \tag{1.4}$$

The choice of these terms in the family of normal forms was taken by H.Oka & H.Kokubu[12]. They selected these terms in order to get a Lorenz-attractor-like "strange attractor" almost constrained on the cusp surface (observed in numerical experiments).

We suppose that the origin is always singular. By adding the linear versal family due to V.I.Arnold[2], we obtain a "versal family" in the truncated sense :

$$\begin{cases} dx/dt = y + \alpha xz + a x^3 \\ dy/dt = Ax + By + \beta xz + b x^3 + c yz^2 \\ dz/dt = Cz + \gamma x^2 + \delta z^2 + d z^3 . \end{cases} \tag{1.5}$$

Note that the right hand side of the first line, if equated to zero, could give a cusp surface. Let us restrict ourselves to considering only a small subfamily :

$$\begin{cases} dx/dt = y + \alpha xz + a x^3 \\ dy/dt = Ax + By + \beta xz \\ dz/dt = Cz + \gamma x^2 . \end{cases} \tag{1.6}$$

by setting $b = c = d = \delta = 0$, in order to facilitate the computation and numerical experiments.

Next, consider a family of ordinary differential equations

$$\begin{cases} dx/dt = (\tau/\varepsilon)(y + \alpha xz + a x^3) \\ dy/dt = \tau(Ax + By + \beta xz) \\ dz/dt = \tau(Cz + \gamma x^2) . \end{cases} \tag{1.7}$$

with extra parameters ε and τ with $\varepsilon > 0$ and $\tau > 0$. The parameter τ does not change the phase portrait of (1.7), since it gives nothing but the change of time scale.

In order to obtain a family with a normalized cusp surface as a "slow manifold", we execute a linear rescaling of coordinates :

$$x = pX,\ y = qY,\ z = rZ,\ \ p \neq 0,\ q \neq 0,\ r \neq 0, \tag{1.8}$$

which transforms (1.7) into

$$\begin{cases} dX/dt = (\tau/\varepsilon)((q/p)Y + r\,\alpha\,XZ + p\,a\,X^3) \\ dY/dt = \tau((p/q)AX + BY + (pr/q)\,\beta\,XZ) \\ dZ/dt = \tau(CZ + (p^2/r)\,\gamma\,X^2)\,. \end{cases} \tag{1.9}$$

We suppose $\alpha \neq 0$ and $a < 0$. Set the scaling parameters p, q, and r as

$$p = 1/\sqrt{-a\tau}\,,\ \ q = 1/\tau\sqrt{-a\tau},\ \ r = 3/4\alpha\tau \tag{1.10}$$

The time scale parameter τ will be determined later as a function of parameters α, β, B, and C, hence independent of ε. We obtain a family :

$$\begin{cases} \varepsilon\ dX/dt = Y + (3/4)\,XZ - X^3 \\ dY/dt = \tau^2 AX + \tau\,BY + (3\tau/4\alpha)\,\beta\,XZ \\ dZ/dt = \tau\,CZ - (4\alpha\tau/3a)\,\gamma\,X^2, \end{cases} \tag{1.11}$$

which has a cusp surface $Y = X^3 - (3/4)\,XZ$ as a constrained surface when $\varepsilon \to 0$. The equation (1.11) defines a family of so-called "constrained" or "slow-fast" equations. For more general definition of constrained equations, we refer to F.Takens[16] and E.Benoit[3]. We adopt the Takens' definition of a **solution** or an **integral curve** of a constrained equation. Fix parameters A, B, C, a, α, β, γ, τ and rewrite (1.11) as

$$\begin{cases} \varepsilon\,dX/dt = f(X,Y,Z) \\ dY/dt = g(X,Y,Z) \\ dZ/dt = h(X,Y,Z). \end{cases} \tag{1.12}$$

Projections $\pi : \mathbf{R}^3 \to \mathbf{R}^2$ and $\sigma : \mathbf{R}^3 \to \mathbf{R}^1$ are defined by $\pi(X,Y,Z) = (Y,Z)$ and $\sigma(X,Y,Z) = X$. The **slow surface** M and the **stable part** S of the slow surface are defined as

$$M = \{\ (X,Y,Z) \mid f(X,Y,Z) = 0\ \} \text{ and } S = \{\ (X,Y,Z) \mid f(X,Y,Z) = 0,\ \partial f/\partial X(X,Y,Z) \leq 0\ \}.$$

A map $\phi :]t_1, t_2[\to S$ is a **solution** of the constrained equation (1.12) if :

i) $\pi \circ \phi$ is continuous;

ii) for each $t_0 \in]t_1, t_2[$, $\phi(t_0) = \lim_{t \downarrow t_0} \phi(t)$;

iii) $\partial_r(\pi \circ \phi(t)) = (g(\phi(t)), h(\phi(t)))$, where ∂_r denotes the right derivative operator;

iv) for each $t_0 \in]t_1, t_2[$, $\phi^-(t_0) = \lim_{t \uparrow t_0} \phi(t)$ exists;

v) whenever $\phi^-(t_0) \neq \phi(t_0)$, for all η on the open segment $]\phi^-(t_0), \phi(t_0)[$, $f(\eta)\cdot(\sigma(\phi(t_0)) - \sigma(\phi^-(t_0))) > 0$.

As we treat only constrained equations (1.11), such a solution can be approximated by solutions of (1.11) for sufficiently small $\varepsilon > 0$. For detailed analysis of such "singular perturbation" problems, we refer to E.Benoit[3].

Roughly speaking, the solution curve is given by "integral curves" of the "slow vector field" on the stable part S. When the orbit reaches a boundary point of S, then by jumping, along the "fast" vector field, to another sheet of S, orbit continues along the "slow" vector field defined there (see fig.1.1).

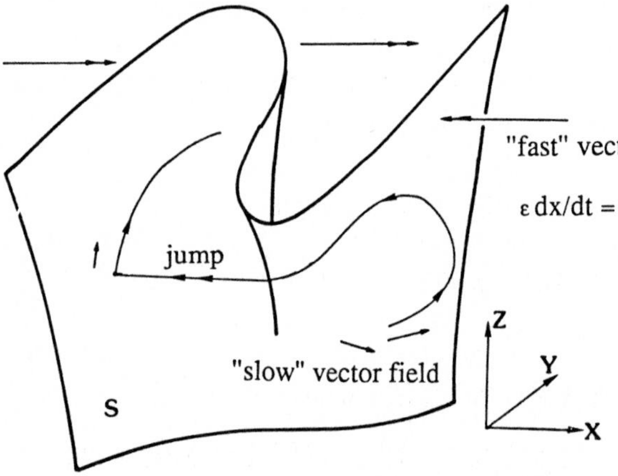

fig. 1.1

§2 "Slow" vector field on the cusp surface

The "Slow" vector field on the stable part S of the cusp surface is the vector field defined on S, tangent to the cusp surface and whose (Y,Z)-component is given by

$$dY/dt = g(X,Y,Z) \quad \text{and} \quad dZ/dt = h(X,Y,Z). \tag{2.1}$$

Let us adopt $(x,z) \in \mathbf{R}^2$ as a coordinate parametrizing the cusp surface $\{ f(X,Y,Z) = 0 \}$ by

$$X = x, \quad Y = x^3 - (3/4)\, xz, \quad Z = z. \tag{2.2}$$

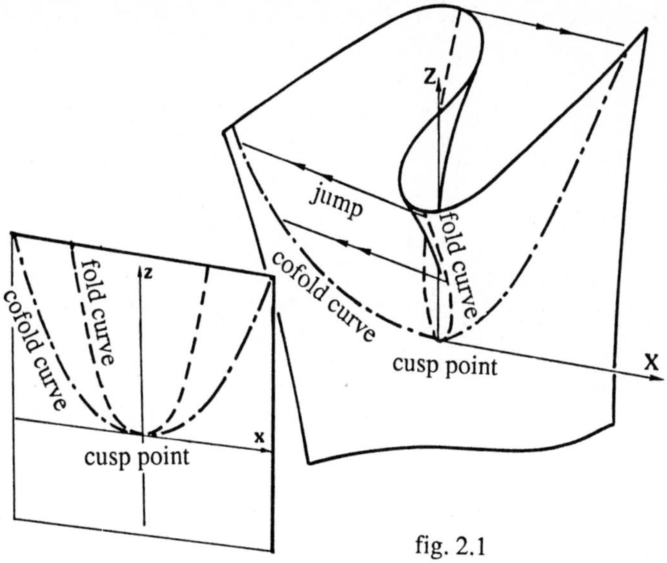

fig. 2.1

Definition A point $(X,Y,Z) \in \mathbf{R}^3$ is a **fold point** if

$$f(X,Y,Z) = \partial f/\partial X(X,Y,Z) = 0, \quad \text{and} \quad \partial^2 f/\partial X^2(X,Y,Z) \neq 0. \tag{2.3}$$

The fold points form **fold curves.**

Definition A point $(X,Y,Z) \in \mathbf{R}^3$ is a **cusp point** if

$$f(X,Y,Z) = \partial f/\partial X(X,Y,Z) = \partial^2 f/\partial X^2(X,Y,Z) = 0,$$
$$\text{and} \quad \partial f^3/\partial X^3(X,Y,Z) \neq 0 \quad (\text{and} \quad \partial^2 f/\partial X \partial Z(X,Y,Z) \neq 0\ ). \tag{2.4}$$

Let J denote the fold curve. The points of S, to which orbits arrive after a jump at fold points form a manifold in S. This manifold consists of curves which are components of $(\pi|_S)^{-1} \circ \pi(J)$. Such a curve is called a **cofold** curve (see J.Argemi[1]).

In our case, the fold curve is given by

$$(X,Y,Z) = (\ s, -2s^3, 4s^2\ ), \quad s \in \mathbf{R} \setminus \{0\}. \tag{2.5}$$

The origin, $(X,Y,Z) = (0,0,0)$ is a cusp point. The cofold curve is given by

$$(X,Y,Z) = (\ -2s, -2s^3, 4s^2\ ), \quad s \in \mathbf{R} \setminus \{0\}. \tag{2.6}$$

Using the coordinate (x,z) of the cusp surface M defined by (2.2), these curves are given by the following equations :

$$z = 4\,x^2 \qquad (\text{ fold curve }), \tag{2.7}$$

$$z = x^2 \qquad (\text{ cofold curve }). \tag{2.8}$$

When an orbit of the "slow" vector field on S arrives at a fold point, say (x_0,z_0), it jumps to $(-2x_0,z_0)$ on the cofold curve (see fig. 2.2).

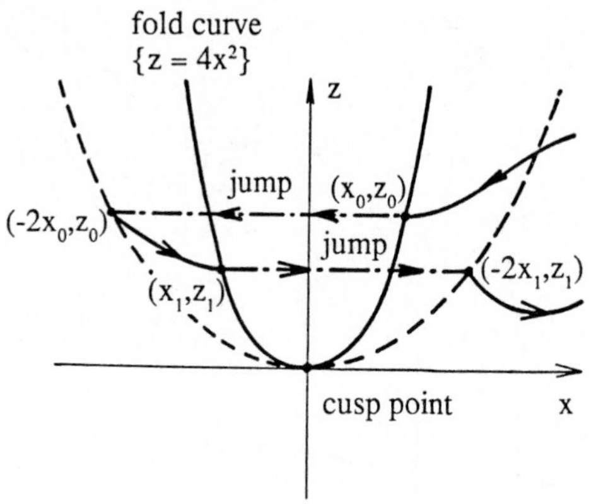

fig. 2.2

The "slow" vector field on the cusp surface is computed as follows. It is tangent to S and its (Y,Z)-component is given by (2.1). Differenciating the equation of the cusp surface f(X,Y,Z) = 0 with respect to the time, we obtain

$$\partial f/\partial X \, dX/dt + \partial f/\partial Y \, dY/dt + \partial f/\partial Z \, dZ/dt = 0. \tag{2.9}$$

Hence, for $(X,Y,Z) \in S$, we have

$$\begin{cases} dX/dt = -(\partial f/\partial Y \, g(X,Y,Z) + \partial f/\partial Z \, h(X,Y,Z))/(\partial f/\partial X) \\ dY/dt = g(X,Y,Z) \\ dZ/dt = h(X,Y,Z), \end{cases} \tag{2.10}$$

which is called the **induced vector field**. This vector field is, generally, not bounded near the fold curve. By multiplying $-\partial f/\partial X$ to (2.10), we obtain the **reduced vector field** (see F.Takens[16], E.Benoit[3]) :

$$\begin{cases} dX/dt = \partial f/\partial Y \, g(X,Y,Z) + \partial f/\partial Z \, h(X,Y,Z) \\ dY/dt = -g(X,Y,Z) \, \partial f/\partial X \\ dZ/dt = -h(X,Y,Z) \, \partial f/\partial X, \end{cases} \tag{2.11}$$

Since $-\partial f/\partial X > 0$ in the interior of the stable part S, the phase portrait of the slow vector field is not affected. In our case, the reduced vector field in the (x,z) coordinate on S takes the form :

$$\begin{cases} dx/dt = \partial f/\partial Y(x,Y,z) \, g(x,Y,z) + \partial f/\partial Z(x,Y,z) \, h(x,Y,z) \\ dz/dt = -h(x,Y,z) \, \partial f/\partial X(x,Y,z) \end{cases} \tag{2.12}$$

where $Y = x^3 - (3/4) xz$. It turns out to be :

$$\begin{cases} dx/dt = x\,(\tau^2 A + \tau\,(B - \alpha\gamma/a)\,x^2 + (3\tau/4)\,(\beta/\alpha + C - B)\,z) \\ dz/dt = -(3/4)\,C\,\tau\,(z - 4x^2)\,(z - (4\,\alpha\gamma/3aC)\,x^2). \end{cases} \tag{2.13}$$

We suppose

$$\beta / \alpha + C - B < 0, \tag{2.14}$$

that is, we only look at the parameter region satisfying this inequality. The time scale parameter τ is determined here by

$$\tau = -(4/3)/(\beta/\alpha + C - B) \tag{2.15}$$

in order to simplify the reduced slow vector field into the form :

$$\begin{cases} dx/dt = -x\,(z - U - V x^2) \\ dz/dt = D\,(z - 4x^2)\,(z - W x^2), \end{cases} \tag{2.16}$$

where we set

$$U = \tau^2 A, \quad V = \tau\,(B - \alpha\gamma/a), \quad D = -3C\tau/4, \quad \text{and} \quad W = -4\,\alpha\gamma/3aC. \tag{2.17}$$

Further, we suppose

$$U > 0, \ 1 > W > V > 0 \ \text{ and } \ D > 0. \tag{2.18}$$

The phase prtrait of the reduced vector field (2.16) in this case is depicted in fig. 2.3.

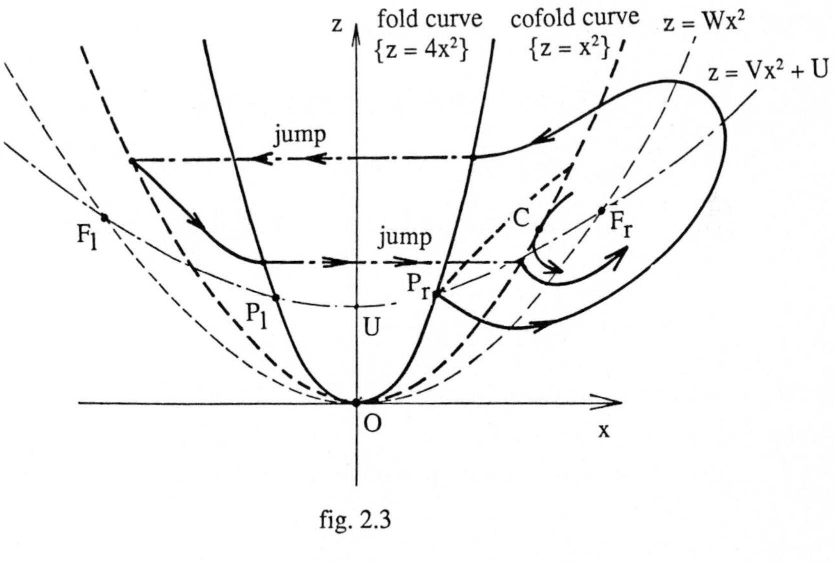

fig. 2.3

The stable part S is represented by the region $\{ z \leq 4x^2 \}$. The z-component of the reduced vector field vanishes on the fold curve $\{ z = 4x^2 \}$ and on parabola $\{ z = Wx^2 \}$. The x-component vanishes on the z-axis and parabola $\{ z = Vx^2 + U \}$. Besides the origin, there are four singular points P_r, P_l, F_r and F_l. They are symmetric with respect to the z-axis. Singular points P_r and P_l are singular points on the fold curve. They are called **pseudo-singular-points** (see J.Argemi[1]). Points F_r and F_l are singular points of the reduced vector field. The coordinate of these singular points are given by the followings :

$$P_r, P_l = (\pm \sqrt{U/(4-V)}, 4U/(4-V)), \quad \text{and} \quad F_r, F_l = (\pm \sqrt{U/(W-V)}, WU/(W-V)).$$

Note that when orbit of the reduced vector field arrives at the fold curve, the orbit is continued onto the branch of cofold curve on the opposite side by a jump. In order to facilitate the calculations we identify the right and left half planes of the phase portrait by setting

$$\xi = x^2. \tag{2.19}$$

This identification might be confusing, since we lose some information. For example, we cannot distinguish two non-symmetric orbits which are symmetric with respect to the z-axis. However, (2.19) simplifies the reduced vector field (2.16) into the form :

$$\begin{cases} d\xi/dt = -2\xi(z - U - V\xi) \\ dz/dt = D(z - 4\xi)(z - W\xi). \end{cases} \tag{2.20}$$

We consider only the region $\xi \geq 0$. This coordinate system permits us a further reduction of parameters. By changing the time scale, $t \to kt$, and by a homothetic change of coordinates $(\xi, z) \to (\xi/k, z/k)$, we see that the qualitative properties of the phase portrait does not depend upon U. We set $U = W - V$ so that the singular point is given by

$$F = (1, W), \tag{2.21}$$

in the (ξ, z) coordinate. The coordinate of the pseudo singular point P is given by :

$$P = (\xi_P, z_P) = (\xi_P, 4\xi_P), \quad \text{where } \xi_P = (W - V)/(4 - V). \tag{2.22}$$

After all, the family of reduced vector fields is simplified in the following form :

$$\begin{cases} d\xi/dt = -2\xi(z - (W-V) - V\xi) \\ dz/dt = D(z - 4\xi)(z - W\xi) \end{cases} \tag{2.23}$$

with fold line $\{ z = 4\xi \}$ and the cofold line $\{ z = \xi \}$. The jump from the fold line to the cofold line is given by $(\xi, z) \to (4\xi, z)$.

In the (ξ, z) coordinates, the Jacobian matrix, J_F, at the singular point F is given by

$$J_F = \begin{bmatrix} 2V & -2 \\ 2EW & -2E \end{bmatrix}, \tag{2.24}$$

where we set $E = D(4-W)/2$. We have

$$\text{trace}(J_F) = 2(V-E) \quad \text{and} \quad \det(J_F) = 4E(W-V). \tag{2.25}$$

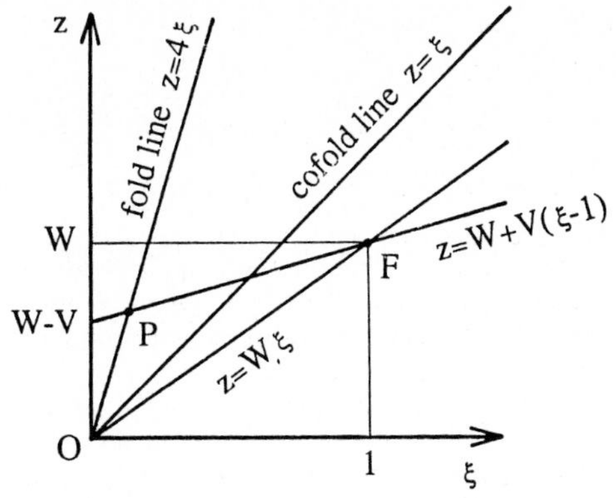

fig. 2.4

If $V > E$ then F is an unstable singular point. The eigenvalues of J_F are

$$\lambda = V - E \pm \sqrt{(V+E)^2 - 4EW}, \qquad \text{if } (V+E)^2 \geq 4EW, \tag{2.26}$$

$$\lambda = V - E \pm \sqrt{4EW - (V+E)^2}\, i, \qquad \text{if } (V+E)^2 \leq 4EW. \tag{2.27}$$

At the pseudo-singular point P, the Jacobian matrix is given by

$$J_P = \begin{bmatrix} 2V\,\xi_P & -2\,\xi_P \\ -8E\,\xi_P & 2E\,\xi_P \end{bmatrix} . \tag{2.28}$$

We have

$$\det(J_P) = \xi_P{}^2 4E(V-4).$$

As we consider in the parameter region given by (2.18), P is a saddle point. Vectors $(1, V - \lambda_{\pm}/2\,\xi_P)$ are the eigenvectors for eigenvalues $\lambda_{\pm} = \xi_P\,(V+E\pm\sqrt{(V-E)^2 + 16E}\,)$.

A point $C = (\xi_C, z_C)$ is a **co-pseudo-singular point** if the reduced vector field is tangent to the cofold curve at C. Co-pseudo-singular point is given by the following equations

$$z = \xi, \qquad D(z-4\,\xi\,)(z-W\,\xi\,) + 2\,\xi\,(z-(W-V)-V\,\xi\,) = 0. \tag{2.29}$$

Besides the degenerate singular point $O = (0, 0)$, there exists a co-pseudo-singular point $C = (\xi_C, z_C)$ with :

$$\xi_C = z_C = (W-V)/(1-V-3D(1-W)/2) \tag{2.30}$$

if $2(1-V) > 3D(1-W)$. Note that $z_C > z_P$ holds in our parameter region (2.18).

§3 Confinor and Anti-confinor

Definition A **confinor** is a compact subset C of the stable part S, which is positively invariant under the semiflow defined by the reduced vector field and the "jump" from fold points into cofold curves.

In this note, we consider confinors whose boundary consists of a finite number of arcs of integral curves, arcs of fold curves, arcs of cofold curves, and singular points (pseudo-singular point, co-pseudo-singular point). Especially, the unstable manifold of the hyperbolic pseudo-singular point and the integral curve starting from the co-pseudo-singular point are important as the boundary of the confinor.

The choice of a confinor might be quite arbitrary. But some types of confinors give information about the semi-flow. For example, if there is a confinor as depicted in fig.3.1, then one can, at least, conclude that there exists an oscillatory orbit. If necessary, one can study the behavior of the orbits in the confinor by looking at some of its sub-confinors.

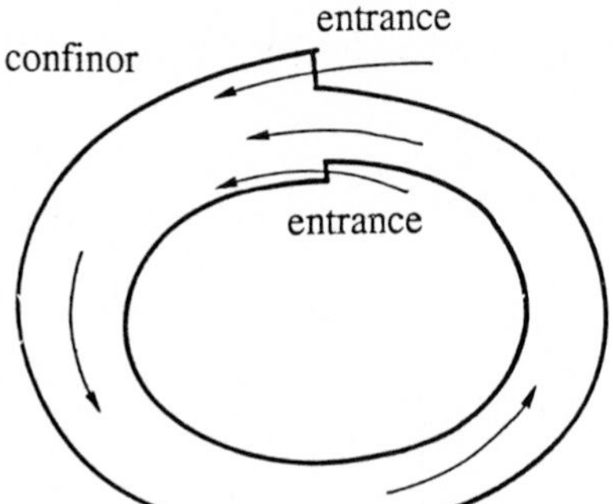

fig. 3.1

Proposition 3.1 Assume $0 < V < W < 1$ and $D > 0$. If the trajectory of (2.23) starting (at $t = -\infty$) from the pseudo-singular point P, which is contained in the branch of the unstable manifold of P emanating into the stable part S, intersects the fold line above P, $\{ z = 4\xi, z > z_P \}$, then there exists a confinor whose boundary consists of the arc of trajectory starting from P and a segment of the fold line.

Proof The unstable manifold of P emanates into the triangle POF, in which we have $d\xi/dt > 0$ and $dz/dt < 0$. Hence, the unstable manifold intersects the cofold line. We denote this first intersection point by P_0. The stable manifold of P emanates into the sector $\{ 4\xi > z > W + V(\xi - 1) \}$. In the region $\{ 4\xi > z > W + V(\xi - 1)$ and $z > W\xi \}$, we have $d\xi/dt < 0$ and $dz/dt < 0$. We denote by P_* the first intersection point (time direction reversed) of the separatrix curve and the cofold line. If the trajectory starting from P intersects the fold line above P, it must pass above the separatix curve (i.e. the stable manifold) of P. Let P_1 denote the first intersection point of the trajectory, starting from P, and the cofold line above the co-pseudo-singular point C. Let $Q = (z_P, z_P)$. Let P_1' denote the first intersection point, with the fold line, of the trajectory after passing by P_1. Let P_2 denote the point on the cofold line onto which the trajectory jumps from P_1'. The "re-injection interval" QP_2 is included in P_0P_1. Hence the region surrounded by the arc of the unstable manifold $PP_*P_1P_1'$ and the segmant PP_1' in the fold line is a confinor.

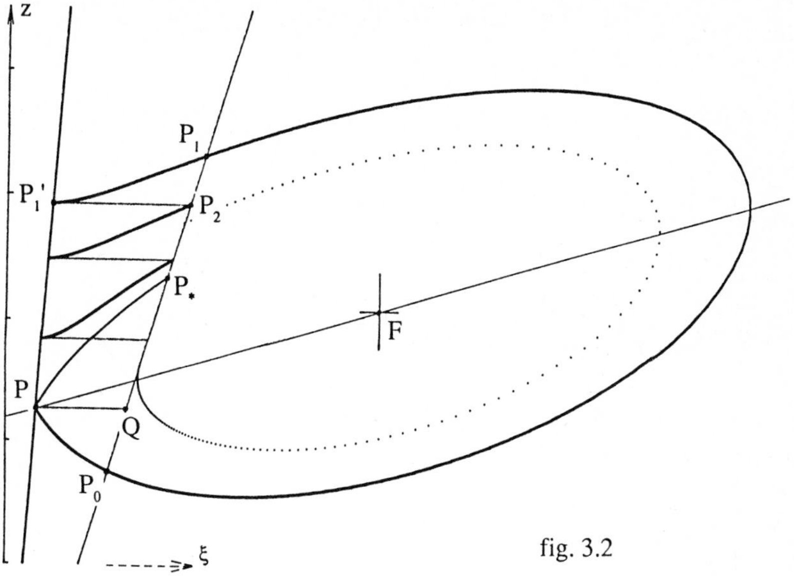

fig. 3.2

We denote this confinor by C_0. On the cofold line, $\{ z = \xi \}$, we define an order by :

$$(z_1, z_1) \leq (z_2, z_2) \text{ if and only if } z_1 \leq z_2.$$

Definition A confinor is said to be a **trapping region** if it has a neighborhood in S such that all trajectories starting from points in this neighborhood enter into the confinor in finite time.

Proposition 3.2 The confinor C_0 is a trapping region.

Proof In region $\{ 4\xi > z > W + V(\xi -1)$ and $z > W\xi \}$, we have $d\xi/dt < 0$ and $dz/dt < 0$. Hence $P_2 < P_1$. Take a sufficiently small neighborhood of C_0 in S so that all the trajectories outside of C_0 jump into the segment P_2P_1.

Definition An **anti-confinor** is an open subset, A, of a trapping region, which is negatively invariant under the semi-flow restricted to the trapping region.

We consider anti-confinors whose frontiers ($= \bar{A} \setminus A$) consists of finite number of arcs of trajectories, fold curves, cofold curves and singular points.

Note that a confinor can include anti-confinors as well as sub-confinors. And an anti-confinor can include confinors and sub-anti-confinors.

Proposition 3.3 If the trajectory starting from the co-pseudo-singular point C intersects the segment CP_1, then there exists an ant-confinor whose frontier is the union of an arc of the trajectory starting from C and a segment in the cofold line.

We set $C_1 = C$. Let C_2 denote the first intersection point of the trajectory after passing by C_1. The open region surrounded by the arc of trajectory from C_1 to C_2 and the segment C_1C_2 is an anti-confinor. We denote this region by A_0.

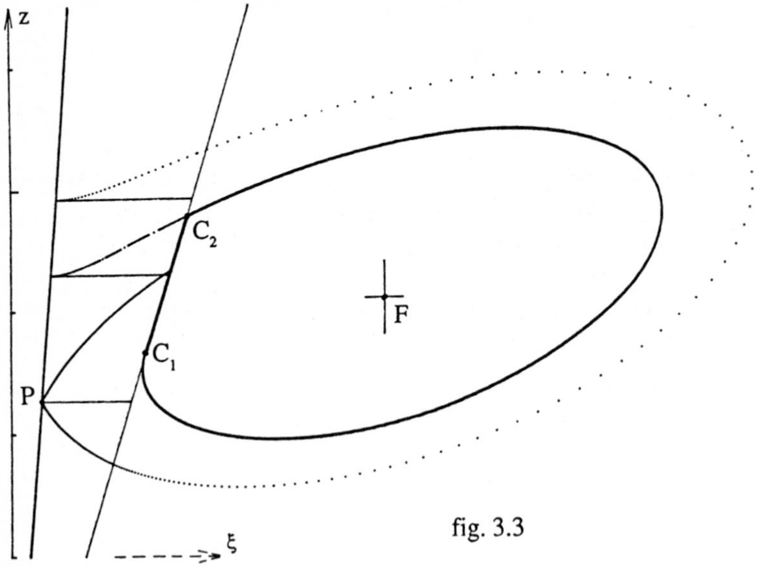

fig. 3.3

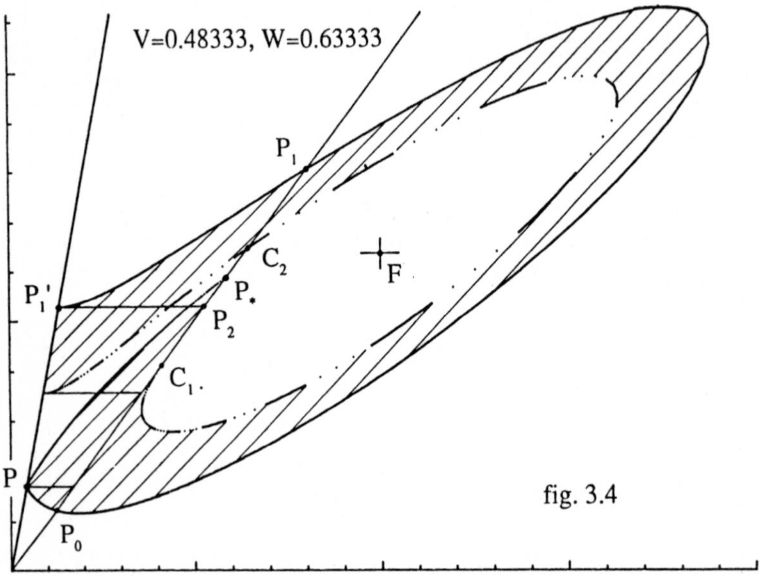

fig. 3.4

Proposition 3.4 The set $C_1 = C_0 \setminus A_0$ is a confinor.

Note that if the conditions of propositions 3.1 and 3.3 are satisfied, then $C_1 \neq \emptyset$. Hence in this case, there exists an oscillatory solution for (2.23).

Figure 3.4 shows an example of confinor. The condition of this case is given by

$$P_2 < P_* < C_1.$$

This confinor contains an oscillatory orbit which makes a turn around F and a jump for each turn. Since we have identified x and -x by $\xi = x^2$, the oscillation in the original constrained equation should be described as follows. First, it makes a turn around F_r and by a jump it goes to the other sheet of S to make a turn around F_l before jumping back to the right hand side sheet. It could be even chaotic since the Poincaré's return map is given by a one dimensional continuous map on the interval $[C_2, P_1]$. Let us forget the distinction $x > 0$ and $x < 0$ and denote this oscillation pattern by S^1L^1. The symbol S^1 represents a turn around F and L^1 represents a jump.

In our study of bifurcation of confinors, we found that the development of the bifurcation series seen in the family of one dimensional mappings (e.g. period doublings, cascade of chaotic bands, windows of peridic attractice orbits, chaos, etc) is often interrupted by a "bifurcation of confinors".

Figure 3.5 shows an example of non-trivial confinor which has S^1L^3, S^1L^2, and S^1L^1 as ocsillatory patterns.

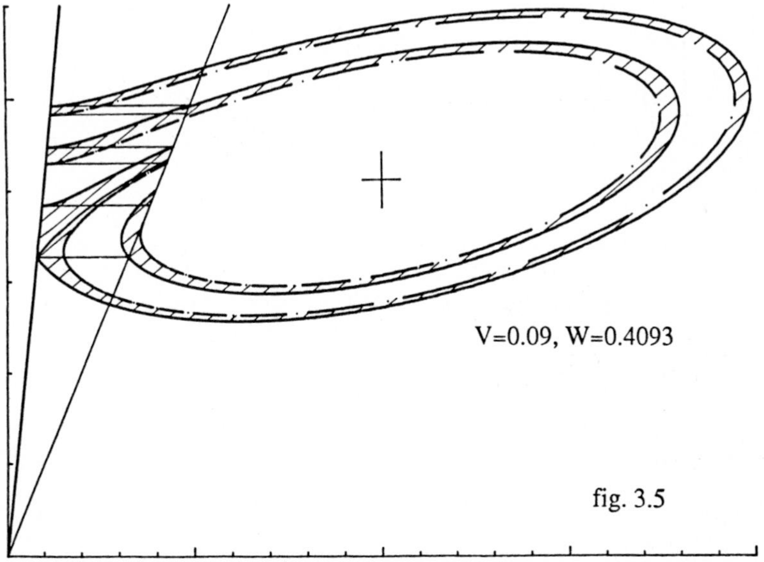

fig. 3.5

The possible succesion patterns of such oscillations are given by the following scheme.

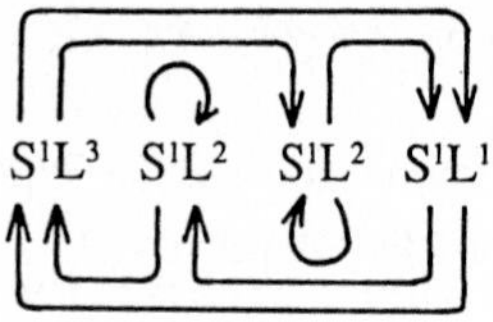

fig. 3.6

The reduced vector field must verify several conditions to have confinors of such type. We denote the successive intersection points of trajectories starting from the pseudo-singular point and the co-pseudo-singular point as follows.

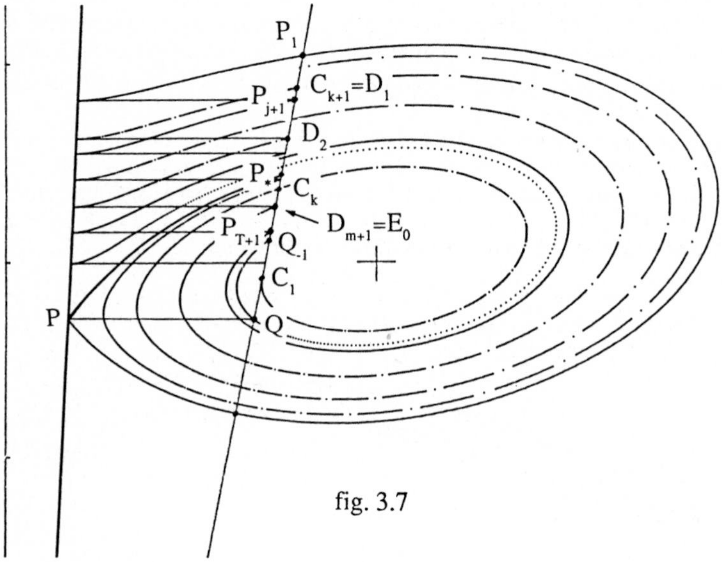

fig. 3.7

i) Let $C_1 = C$ and $C_2, \ldots, C_k$ be the successive intersection point of the trajectory of the co-pseudo-singular point with the segment CP_1 of the cofold line. The number k is determined so that

$$C_1 < C_2 < \ldots < C_k < P_* \leq C_{k+1}.$$

ii) Let $D_1 = C_{k+1}$ and $D_2, \ldots, D_m, D_{m+1} = E_0$ be the successive arrival points after jumps with

$$D_1 > D_2 > \ldots > D_m \geq P_* > D_{m+1} = E_0.$$

iii) Compute the trajectory starting from P_1 to have successive arrival points of jumps $P_2, \ldots, P_j, \ldots, P_{j+n} = P_T, P_{T+1}$ with

$$P_1 > \dots > P_j > D_1 \geq P_{j+1} > \dots > P_{j+n} = P_T > P_* \geq P_{T+1}.$$

The conditions to have a confinor of the fig. 3.5 type are given by :

(A) $k = 1, m = 1, j = 2, T = 3, \ P_{T+1} > E_0, E_1 > P_2, E_j > D_1,$

and

(B) (a) $P_{T+1} \leq Q_{-1}$ and $\{ \ Q_{-1} > R_{m+2}$ or $P_j \geq R_1 \ \}$

or

(b) $P_{T+1} > Q_{-1}$ and $P_j \geq R_1,$

where Q_{-1} is the preceding intersection point of the trajectory of Q, $R_0 = \max(P_{T+1}, Q_{-1}),$ and $R_1, \dots , R_{m+2}$ are its successors.

§4 Bifurcations of Confinors and Anti-confinors

Definition An **anti-bifurcation** is a "bifurcation" of anti-confinors, i.e., the qualitative change of anti-confinors in varying the parameters continuously.

Figure 4.1 shows an example of anti-bifurcation, where a simple confinor with pattern S^1L^2 is transformed into a confinor described in the previous section. In this case, the anti-confinor loses the separatrix curve of P. The confinor gets a new "route". To the opening of this new route corresponds a "bifurcation" of schemes as the aquisition of new arrows and new patterns.

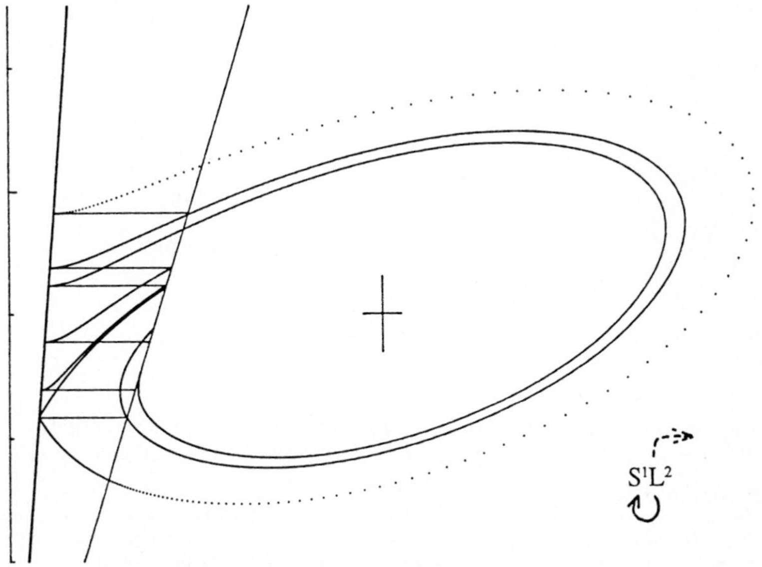

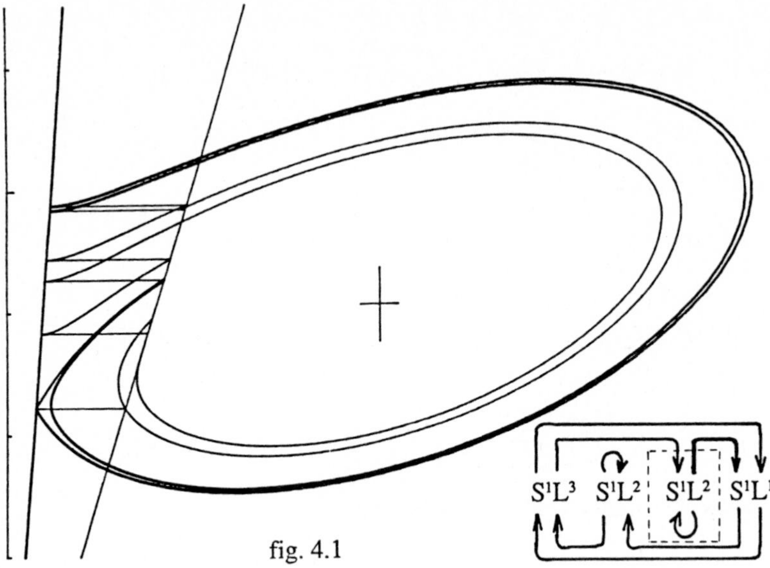

fig. 4.1

Besides the collision with the pseudo-singular point, a confinor can be destabilized when one of its boundaries loses the confining property. Among such "bifurcation" of confinors, we found a series of "bifurcations", which we named a **rainbow bifurcation**. As an example, consider a confinor of type (k, m, j, T) = (1, 1, 2, 3) in the preceeding section. In this case, the reduced vector field has an anti-confinor of an "annulus" type with exits on both of two boundaries.

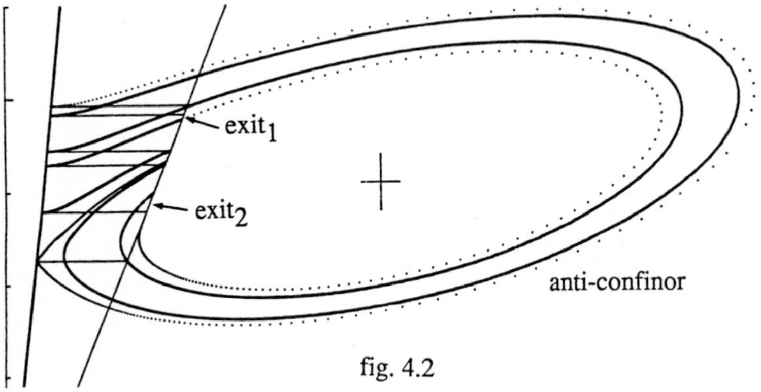

fig. 4.2

These exits of the anti-confinor are, at the same time, entrances to the confinor. If one of these exits, say $exit_1$ for example, is closed to be transformed to an entrance into this region, then the region is not an anti-confinor any more.

The orbit which "leaks" from the former confinor will run around along the former anti-confinor. If the semi-flow in the former anti-confinor has no periodic orbits in the interior of the region, so that it permits the trajectory traverse the "river", then the trajectory can enter again into the former confinor region passing by the $exit_2$ of the opposite side of the "river".

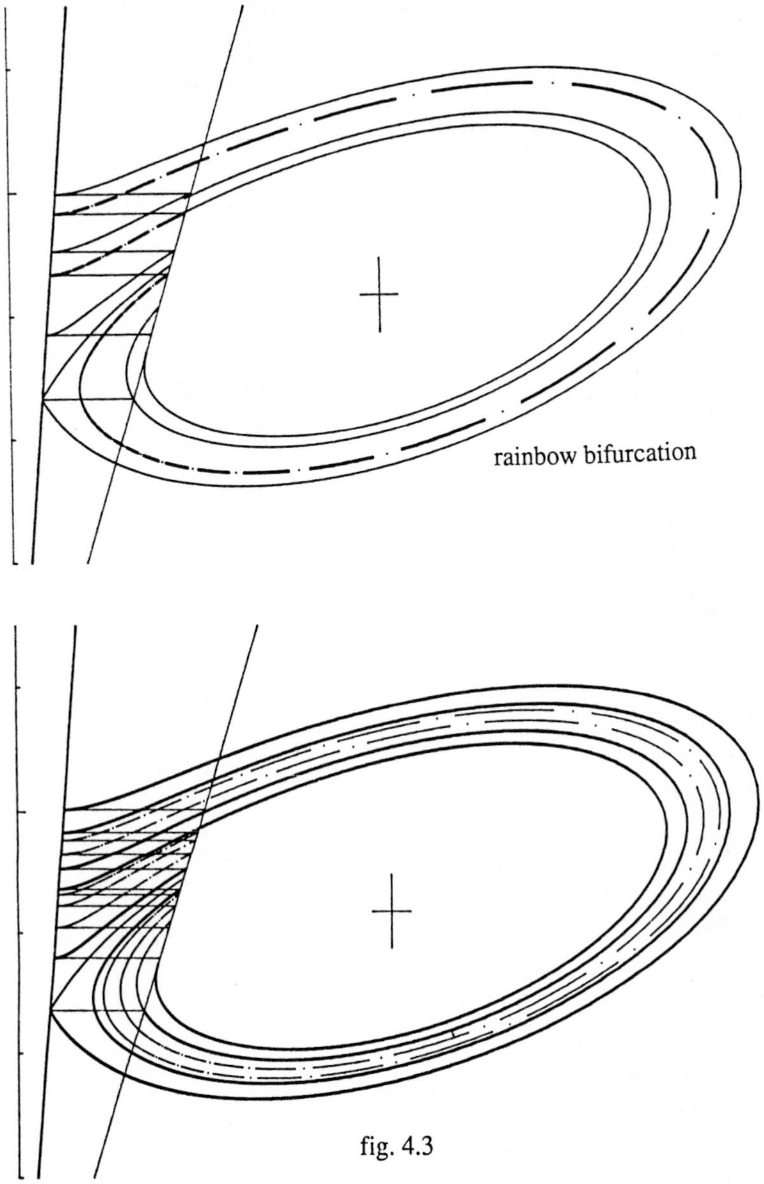

fig. 4.3

The number of turns for a trajectory to traverse the "river" is very large if the "entrance" into the "river" region is very small. This bifurcation induces a "bifurcation of schemes" indicated in fig. 4.4.

We note that similar bifurcation can be observed when the other exit of the anti-confinor is transformed into an entrance. In this case, the scheme after the rainbow bifurcation has occured is given in figure 4.5.

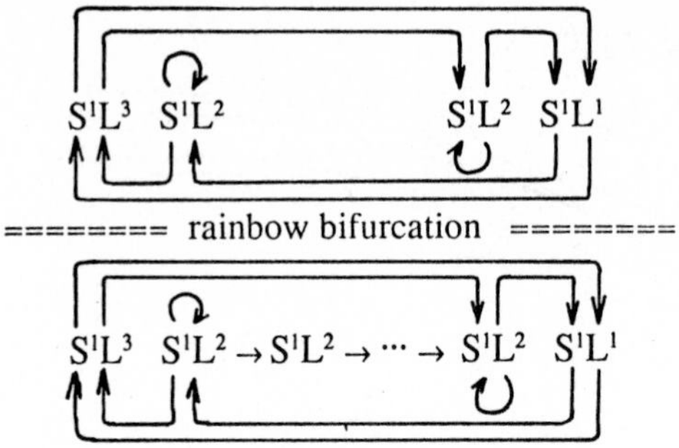

fig. 4.4

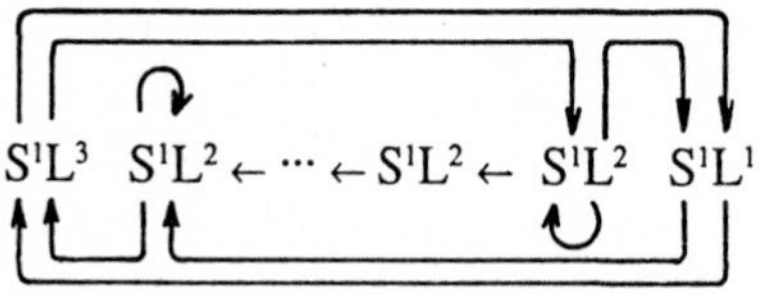

fig. 4.5

We remark that the rainbow bifurcation may be interrupted by a collision of the "river" region with the pseudo-singular point. Confinors with more or less complicated patterns and schemes can also be observed as shown in figure 4.6.

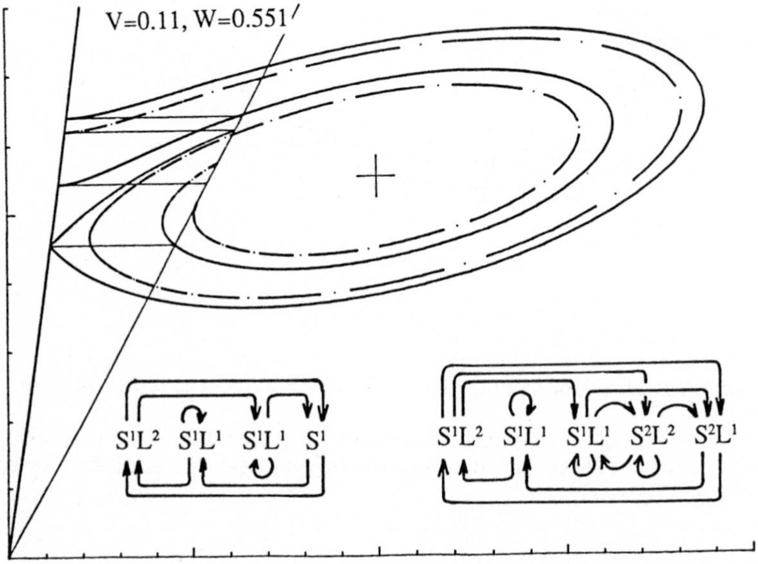

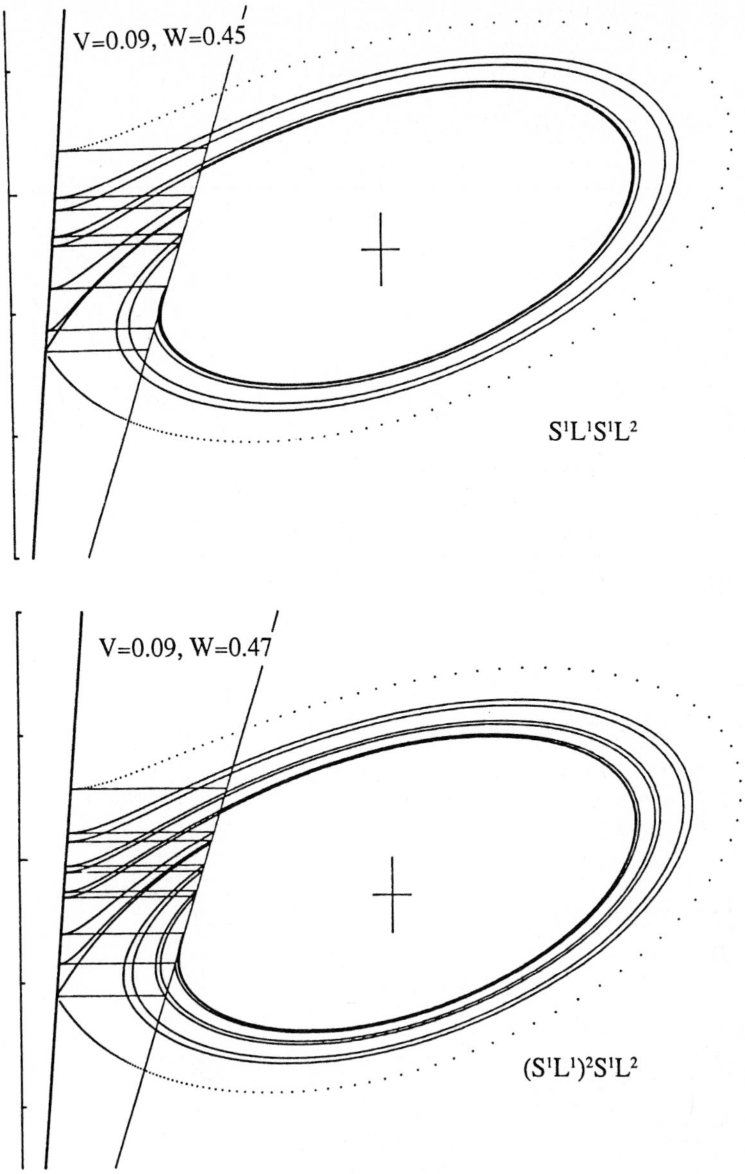

fig. 4.6

§5 Bifurcation Diagram of Confinors

Regions of parameters for various combinations of numbers (k, m, j, T=j+n) are plotted in figure 5.1. This figure is obtained numerically. In these numerical experiments, the real parts of the eigenvalues (when they are complex conjugate) of the singular point F are fixed to 0.02. The figure represents the region

$$0.1 \le W \le 0.9 \text{ and } 0.03 \le V \le 0.88.$$

Parameter D is given by $D = 2(V\text{-}0.02)/(4\text{-}W)$.

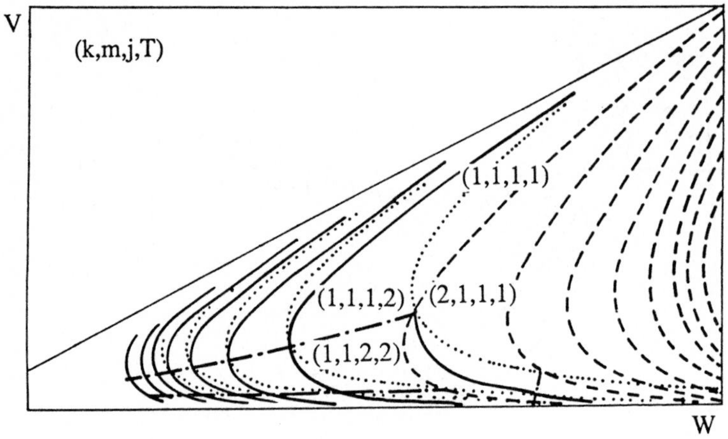

fig. 5.1

Figure 5.2 is an enlargement of figure 5.1. The region is given by :

$$0.3 \le W \le 0.9 \quad \text{and} \quad 0.03 \le V \le 0.23.$$

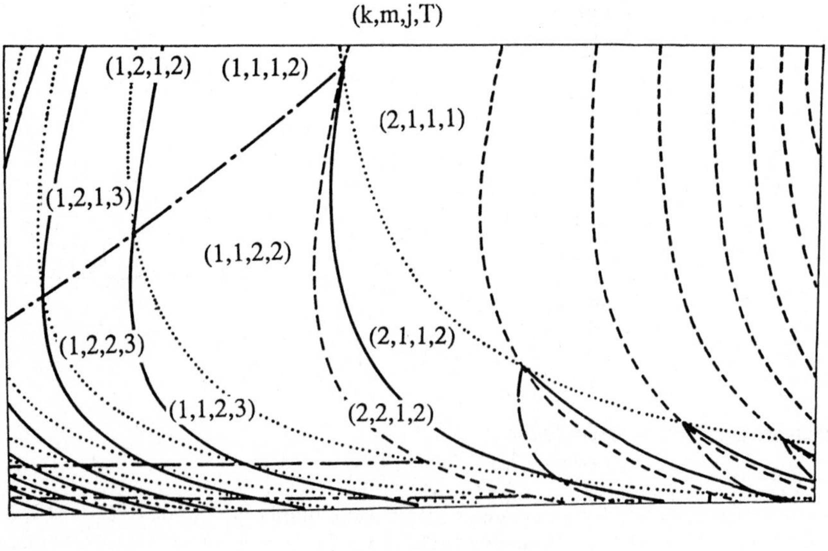

fig. 5.2

The region for (k, m, j, T) = (1, 1, 2, 3) is enlarged in figure 5.3, in which the hatched region indicates the values of parameters with a confinor of fig. 3.5 type. In this figure, curves corresponding to rainbow bifurcations are also plotted.

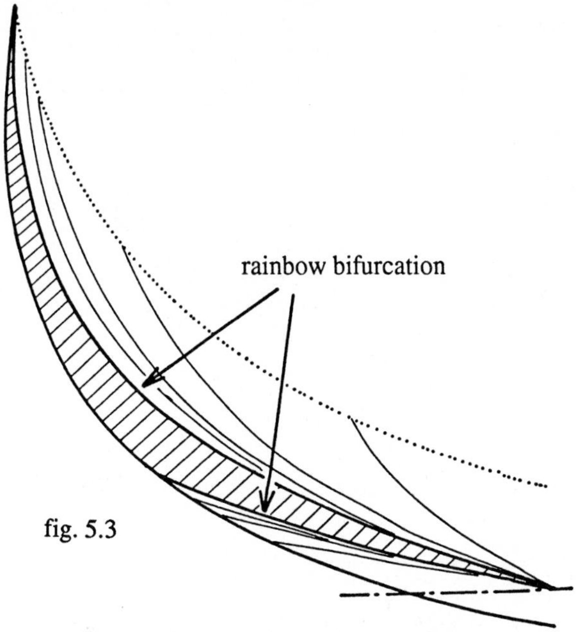

fig. 5.3

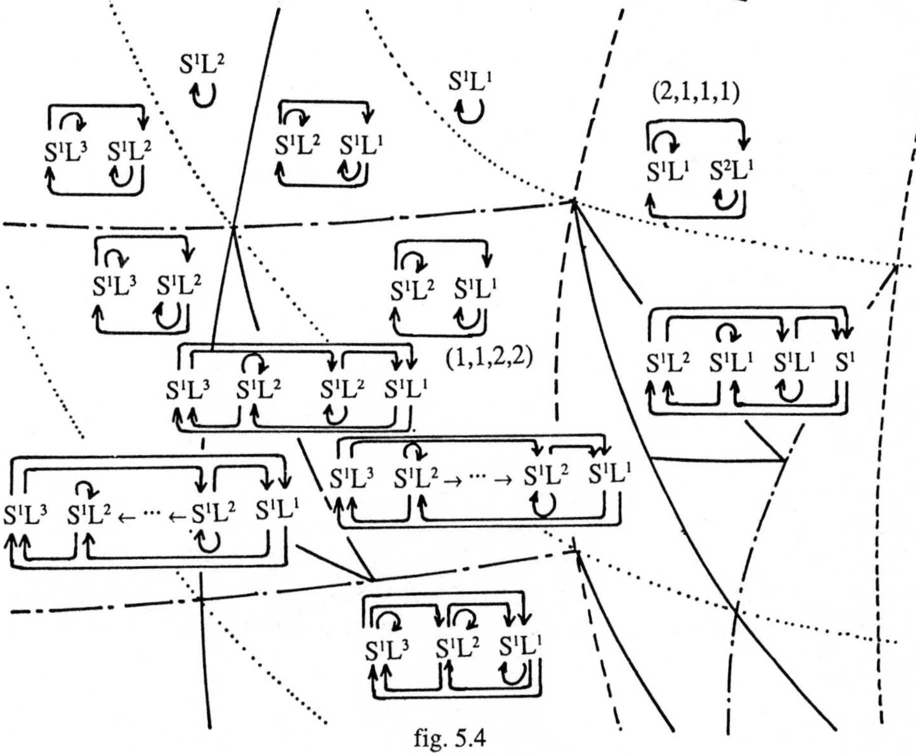

fig. 5.4

Figure 5.4 gives the schemes corresponding to these parameter regions.
Figures 5.5 to 5.7 represent bifurcation diagrams for different types of confinors.

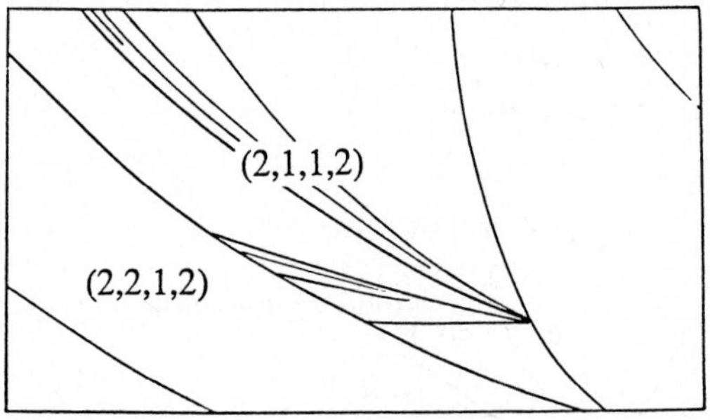

fig. 5.5

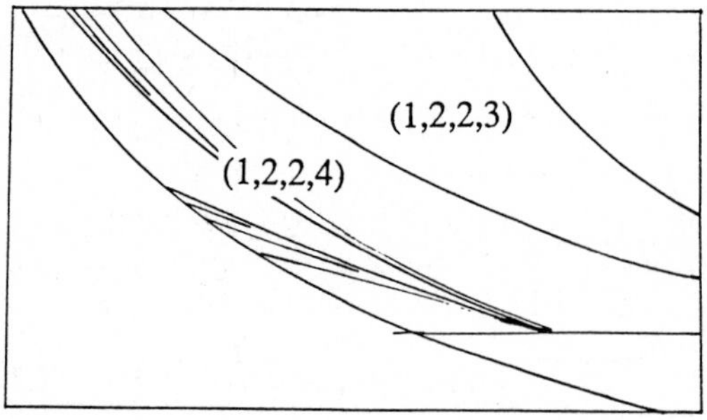

fig. 5.6

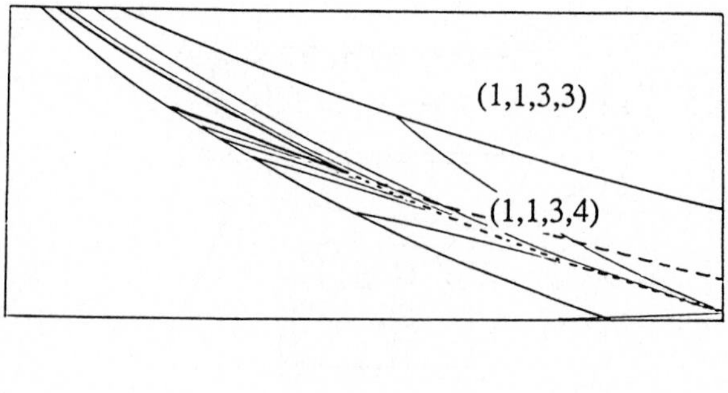

fig. 5.7

References

[1] J.Argemi : Approche qualitative d'un problème de perturbations singulières dans $\mathbf{R}^4$ Equadiff. 78, Convegno Int. su Equazioni Differenziali, Firenze, Italia, (eds. R. Conti, G. Sestini, G.Villari), 1978, 330-340.

[2] V.I.Arnold : Lectures on bifurcation in versal families, Russian Math. Surveys, 27(1972), 54-123.

[3] E.Benoit : Thesis, Univ. Nice, 1984.

[4] J.Birman & R.Williams : Knotted periodic orbits : Lorenz knots, Topology 22(1983), 47-82.

[5] J.Guckenheimer : A strange, strange attractor, Lecture Notes in Applied Math. Sci., 19(1976), 368-391.

[6] J.Guckenheimer & R.Williams : Structural stability of Lorenz attractors, Publ. Math. IHES, 50, 1979, 59-72.

[7] E.N.Lorenz : Deterministic nonperiodic flow, J.Atom. Sci., 20(1963), 130-141.

[8] R.Lozi : Thesis, Univ. Nice, 1983.

[9] R.Lozi : Sur un modèle mathématique de suite de bifurcation de motifs dans la réaction de Belousov-Zhabotinsky, C.R.A.S. Paris, 294(1982), Sér. I, 21-26.

[10] R.Lozi & S.Ushiki : Organized confinors, anti-confinors and their bifurcations in chaotic dissipative dynamical systems (in preparation).

[11] J.Masel & H.L.Swinney : A complex transition sequence in the Belousov-Zhabotinskii Reaction, Physica Scripta, T9(1985), 35-39.

[12] H.Oka & H.Kokubu : Constrained Lorenz-like Attractors, Japan. J. Appl. Math., 2(1985), 495-500.

[13] H.Oka & H.Kokubu : An approach to constrained equations and strange attractors, Pattern and Wave - Qualitative Analysis of Nonlinear Differential Equations (eds. T.Nishida, M.Mimura & H.Fujii), North Holland (Studies in Mathematics and its Applications), 1986.

[14] O.E.Rössler : Continuous chaos, New York Acad. Sci., 316, 1979, 376-392.

[15] F.Takens : Singularities of vector fields, Publ. Math. IHES, 43, 1974, 47-100.

[16] F.Takens : Constrained equations; a study of implicit differential equations and their discontinuous solutions, Lecture Notes in Math. 525, Springer, 1976, 143-234.

[17] F.Takens : Implicit differencial equation, some open problems, Lecture Notes in Math., 535, Springer, 1976, 237-253.

[18] F.Takens : Transitions from periodic to strange attractors in constrained equations, preprint, Rijksuniversiteit Groningen, 1986.

[19] S.Ushiki : Normal forms for singularities of vector fields, Japan. J. Appl. Math., 1(1984), 1-37.

[20] S.Ushiki, H.Oka & H.Kokubu : Existence d'attracteurs étranges dans le déploiement d'un champ de vecteurs invariants par translation, C.R.A.S. Paris, 298(1984), Sér. I, 39-42.

INVARIANT MANIFOLDS IN INFINITE DIMENSIONS

A. Vanderbauwhede
Institute for Theoretical Mechanics
State University of Gent
B-9000 Gent, Belgium

1. INTRODUCTION

Let X be a Banach space and $F \in C^k(X)$ $(k\geq 1)$ a mapping with $F(0)=0$. We are interested in the existence and smoothness of locally invariant manifolds passing through the fixed point $x=0$ of F. We call a manifold $M\subset X$ with $0\in M$ *locally invariant* for F if we can find a neighborhood Ω of $x=0$ such that

$$x\in M\cap\Omega\cap F^{-1}(\Omega) \Rightarrow F(x)\in M \; . \tag{1.1}$$

If M is *invariant* under a mapping $\tilde{F} : X \to X$ which coincides with F on a neighborhood of the origin, then M is locally invariant for F. This simple observation is used in most constructions of locally invariant manifolds. In this paper we will make the following assumption about $A := DF(0)$:

(S) X has a splitting $X = X_1 \oplus X_2$, with X_1 and X_2 closed subspaces of X which are invariant under A and such that

$$a := \sup_{\lambda\in\sigma(A_1)} |\lambda| < b := \inf_{\lambda\in\sigma(A_2)} |\lambda| \leq 1, \; A_i := A|_{X_i} \in \mathcal{L}(X_i) \; , \; (i=1,2)$$

We then want to find locally invariant manifolds of the form

$$W_\phi := \{x_2+\phi(x_2) \,|\, x_2\in X_2\} \; , \tag{1.2}$$

with $\phi\in C^0(X_2,X_1)$ and $\phi(0)=0$.
Comprehensive treatments of such invariant manifolds have been given by Iooss [2] and Lanford [3], among others. They show that if $F\in C^{k,1}(X)$ and if $a < b^{k+1}$, then F has a locally invariant manifold W_ϕ with $\phi \in C^{k,1}(X_2,X_1)$, $\phi(0)=0$ and $D_\phi(0)=0$. The idea of the proof is to use the contraction mapping principle to solve a fixed point equation for ϕ which is equivalent to the invariance of W_ϕ under an appropriate modification of F.

NATO ASI Series, Vol. F37
Dynamics of Infinite Dimensional Systems
Edited by S.-N. Chow, and J. K. Hale

In this contribution we want to outline a different approach, inspired by recent work on invariant manifolds for flows (see [1,4,5,6]). Our starting point is an invariant set which contains all negative orbits with a prescribed asymptotic behaviour (see further for the definitions). Then we show that this invariant set has in fact the form (1.2), with $\phi \in C^k(X_2,X_1)$ if $F \in C^k(X)$ and if $a<b^k$. In section 2 we obtain some basic global results. In section 3 we show that F can be modified in such a way that the global results apply. Finally we give in section 4 a simple example which shows that the condition $a<b^k$ cannot be relaxed if one wants an invariant manifold of class C^k.

We start with some definitions and preliminary results. Let $F \in C^0(X)$; we call a sequence $\{x_p | p\in \mathbb{N}\} \subset X$ a *positive F-orbit* if $x_{p+1}=F(x_p)$ for all $p \in \mathbb{N}$, and a *negative F-orbit* if $x_{p+1} \in F^{-1}(x_p)$ for all $p\in \mathbb{N}$. For each $\eta>0$ we denote by Y_η the Banach space of all sequences $y = \{y_p | p\in \mathbb{N}\} \subset X$ such that

$$\|y\|_\eta := \sup_{p\in\mathbb{N}} \eta^p \|y_p\| < \infty \ . \tag{1.3}$$

Remark that $Y_\eta \subset Y_\zeta$ if $0<\zeta\leqq\eta$, and that $\|y\|_\zeta \leqq \|y\|_\eta$ for all $y \in Y_\eta$, so that Y_η is continuously imbedded in Y_ζ.

For each $k\in \mathbb{N}$ we denote by $C_b^k(X)$ the space of all $f \in C^k(X)$ such that

$$|f|_m := \sup_{x\in X} \|D^m f(x)\| < \infty \qquad , \qquad 0\leqq m\leqq k \ . \tag{1.4}$$

Similarly $C_b^{0,1}(X)$ denotes the space of all $f \in C_b^0(X)$ such that

$$|f|_{Lip} = \sup_{\substack{x,y\in X\\ x\neq y}} \|x-y\|^{-1}\|f(x)-f(y)\| < \infty \ . \tag{1.5}$$

Under the assumption (S) we let $\pi_1 \in \mathcal{L}(X)$ and $\pi_2 := I-\pi_1$ be the projections in X such that

$$R(\pi_1) = N(\pi_2) = X_1 \quad , \quad R(\pi_2) = N(\pi_1) = X_2 \ . \tag{1.6}$$

In X we can use the norm $\|x\| := \max(\|\pi_1 x\|, \|\pi_2 x\|)$. Moreover we have the following.

<u>Lemma 1</u>. Assume (S). Then there exists for each $\varepsilon>0$ some $M(\varepsilon)>0$ such that

$$\|A_1^p\| \leqq M(\varepsilon)(a+\varepsilon)^p \quad , \quad \|A_2^{-p}\| \leqq M(\varepsilon)(b-\varepsilon)^{-p} \quad , \quad \forall p \in \mathbb{N} \ . \tag{1.7}$$

P r o o f. We know from spectral theory that

$$a = \lim_{p\to\infty} \|A_1^p\|^{1/p} \quad , \qquad b = \lim_{p\to\infty} \|A_2^{-p}\|^{1/p} .$$

From this (1.7) follows easily.

2. GLOBAL INVARIANT MANIFOLDS

In this section we obtain some invariant manifolds for mappings of the form $F = A+g$, with $A \in \mathcal{L}(X)$ satisfying (S) and with $g \in C_b^{0,1}(X)$.

Lemma 2. Let $F = A+g$, with $A \in \mathcal{L}(X)$ satisfying (S) and with $g \in C_b^0(X)$. Let $y = \{y_p | p\in\mathbb{N}\}$ be a negative F-orbit. Then the following statements are equivalent :

(i) $y \in Y_\eta$ for some $\eta \in (a,b)$;

(ii) $y \in Y_\eta$ for all $\eta \in (a,b)$;

(iii) $$y_p = A_2^{-p}\pi_2 y_0 - \sum_{\ell=1}^{p}{}^{(+)} A_2^{\ell-1-p}\pi_2 g(y_\ell) + \sum_{\ell=1}^{\infty} A_1^{\ell-1}\pi_1 g(y_{p+\ell}) \quad , \quad \forall p \in \mathbb{N}. \qquad (2.1)$$

(In (2.1) the notation $\Sigma^{(+)}$ indicates that the corresponding sum is only present if $p\geq 1$).

P r o o f. First we remark that the infinite sum at the right hand side of (2.1) is well-defined, by lemma 1 and the fact that $\|g(x)\|\leq |g|_0$ for all $x\in X$. The same results also give easily the implication (iii)$\Rightarrow$(ii). The implication (ii)$\Rightarrow$(i) is trivial. To prove (i)$\Rightarrow$(iii) we observe that, as a negative F-orbit, y satisfies the "variation-of-constants" formula

$$y_q = A^p y_{q+p} + \sum_{\ell=1}^{p}{}^{(+)} A^{\ell-1} g(y_{q+\ell}) \quad , \qquad \forall p,q \in \mathbb{N} . \qquad (2.2)$$

First we project (2.2) on X_2, put q=0 and apply A_2^{-p}. Next we project (2.2) on X_1, take the limit for $p\to\infty$ (using lemma 1), and then replace q by p. Adding the two results obtained in this way we find (2.1).

Lemma 3. Under the conditions of lemma 2 let $\eta \in (a,b)$ and $y \in Y_\eta$. Then y is a negative F-orbit if and only if there exists some $x_2 \in X_2$ such that

$$y_p = A_2^{-p}x_2 - \sum_{\ell=1}^{p}{}^{(+)} A_2^{\ell-1-p}\pi_2 g(y_\ell) + \sum_{\ell=1}^{\infty} A_1^{\ell-1}\pi_1 g(y_{p+\ell}) \quad , \quad \forall p \in \mathbb{N}. \tag{2.3}$$

P r o o f. The necessity of (2.3) follows from lemma 2, by taking $x_2=\pi_2 y_0$. Conversely, if (2.3) holds for some $x_2 \in X_2$, then one can directly verify that $F(y_{p+1})=Ay_{p+1}+g(y_{p+1})=y_p$ for all $p \in \mathbb{N}$.

The foregoing lemma's motivate us to consider for each $\eta \in (a,b)$ the set

$$M_\eta := \{y_0 \mid y \in Y_\eta \text{ and } y \text{ is a negative F-orbit}\} . \tag{2.4}$$

This set is obviously F-invariant, and by lemma 2 this invariant set does not change when we vary η in the interval (a,b). To determine M_η we have to find for each $x_2 \in X_2$ all $y \in Y_\eta$ satisfying (2.3). We rewrite this equation in the more compact form

$$y = Sx_2 + KG(y) , \tag{2.5}$$

where for each $\eta \in (a,b)$ the operators $S \in \mathcal{L}(X_2,Y_\eta)$ and $K \in \mathcal{L}(Y_\eta)$ and the mapping $G \in C^0(Y_\eta)$ are defined by

$$(Sx_2)_p := A_2^{-p}x_2 \quad , \quad \forall x_2 \in X_2 , \forall p \in \mathbb{N} ; \tag{2.6}$$

$$(Ky)_p := - \sum_{\ell=1}^{p}{}^{(+)} A_2^{\ell-1-p}\pi_2 y_\ell + \sum_{\ell=1}^{\infty} A_1^{\ell-1}\pi_1 y_{p+\ell} \quad , \quad \forall y \in Y_\eta, \forall p \in \mathbb{N} ; \tag{2.7}$$

and

$$G_p(y) := g(y_p) \quad , \quad \forall y \in Y_\eta , \forall p \in \mathbb{N} . \tag{2.8}$$

The fact that S and K are bounded linear operators as indicated follows from lemma 1; K has, as a linear operator on Y_η, a norm

$$\|K\|_\eta \leq \gamma(\eta) := \max\Big(\sum_{\ell=1}^{\infty} \|A_1^{\ell-1}\|\eta^{-\ell}, \sum_{\ell=0}^{\infty} \|A_2^{-\ell-1}\|\eta^{\ell}\Big) . \tag{2.9}$$

To see that $G : Y_\eta \to Y_\eta$ is continuous for $\eta \in (0,1)$ one has to use the estimate

$$\|G(y)-G(\tilde{y})\|_\eta \leq \max\Big(\max_{0\leq p\leq q} \eta^p\|g(y_p)-g(\tilde{y}_p)\|, 2|g|_0\eta^q\Big) , \tag{2.10}$$

valid for all $y,\tilde{y}\in Y$ and for all $q\in\mathbb{N}$.

Theorem 4. Let $F = A+g$, with $A\in\mathcal{L}(X)$ satisfying (S), and $g\in C_b^{0,1}(X)$. Suppose that there exists some $\eta\in(a,b)$ such that

$$\gamma(\eta)|g|_{Lip} < 1 . \tag{2.11}$$

Then there exists some $\phi\in C_b^{0,1}(X_2,X_1)$ such that $M_\eta = W_\phi$; moreover we have

$$|\phi|_0 \leqq |\pi_1 g|_0 \sum_{\ell=0}^{\infty} \|A_1^\ell\| . \tag{2.12}$$

P r o o f. If $g\in C_b^{0,1}(X)$ then

$$\|G(y)-G(\tilde{y})\|_\eta \leqq |g|_{Lip}\|y-\tilde{y}\|_\eta \quad , \quad \forall y,\tilde{y}\in Y_\eta . \tag{2.13}$$

Hence the mapping $K\circ G : Y_\eta \to Y_\eta$ is Lipschitzian, with Lipschitz constant less than or equal to $\|K\|_\eta|g|_{Lip} \leqq \gamma(\eta)|g|_{Lip}$. It follows that if (2.11) holds then $I-K\circ G$ is a homeomorphism on Y_η, with inverse $\Phi=(I-K\circ G)^{-1}$; moreover, Φ satisfies a Lipschitz condition. Consequently (2.5) has for each $x_2\in X_2$ a unique solution $y=\Phi(Sx_2)\in Y_\eta$, with $\pi_2\Phi_0(Sx_2)=x_2$. From this and (2.4) it follows that $M_\eta=W_\phi$, with $\phi\in C_b^{0,1}(X_2,X_1)$ defined by

$$\phi(x_2) := \pi_1\Phi_0(Sx_2) \quad , \quad \forall x_2\in X_2 . \tag{2.14}$$

This definition together with (2.5) also gives easily (2.12).

Theorem 5. Let $F=A+g$, with $A\in\mathcal{L}(X)$ satisfying (S), and $g\in C_b^k(X)$ for some $k\geqq 1$. Suppose that there exists some $\eta\in(a,b)$ with $a<\eta^k$ such that

$$|g|_1 \sup_{\zeta\in[\eta^k,\eta]} \gamma(\zeta) < 1 . \tag{2.15}$$

Then there exists some $\phi\in C_b^k(X_2,X_1)$ such that $M_\eta=W_\phi$.

P r o o f. We just give the main idea of the proof; for more details we refer to [5] where a full proof has been given for an analogous result for flows. By (2.15) and $|g|_{Lip}=|g|_1$ we can apply theorem 4 to obtain the mapping $\Phi=(I-K\circ G)^{-1} : Y_\eta \to Y_\eta$. If $0<\zeta<\eta$ then we can use the imbedding $Y_\eta \hookrightarrow Y_\zeta$ to consider Φ as a mapping from Y_η into Y_ζ. We claim that if $g\in C_b^k(X)$ and if (2.15)

holds, then $\Phi : Y_\eta \to Y_\zeta$ is of class C^k if $a<\zeta<\eta^k$; moreover, all derivatives of Φ are bounded on Y_η. The theorem then follows immediately from the definition (2.14) of ϕ.
To prove the claim we first observe that for each m with $1\leq m\leq k$ we can define a mapping $G^{(m)} : Y_\eta \to \mathcal{L}^{(m)}(Y_\eta, Y_{\eta^m})$ by

$$(G^{(m)}(y).(\tilde{y}_1,\ldots,\tilde{y}_m))_p := D^m g(y_p).(\tilde{y}_{1,p},\ldots,\tilde{y}_{m,p}) ,$$
$$\forall y,\tilde{y}_1,\ldots,\tilde{y}_m \in Y_\eta , \ \forall p \in \mathbb{N} . \qquad (2.16)$$

An estimate similar to the one used in (2.10) shows that $G^{(m)}$ is continuous as a mapping from Y_η into $\mathcal{L}^{(m)}(Y_\eta, Y_\zeta)$ if $0<\zeta<\eta^m$. This can then be used to show that the mapping $G : Y_\eta \to Y_\zeta$ is of class C^k if $0<\zeta<\eta^k$, and that $D^m G(y)=G^{(m)}(y)$ for all $y\in Y_\eta$ and for $1\leq m\leq k$.
Now $\Phi(y)$ satisfies the identity :

$$\Phi(y) = y + KG(\Phi(y)) \qquad , \qquad \forall y \in Y_\eta . \qquad (2.17)$$

Formal differentiation of this identity gives us for each m with $1\leq m\leq k$ a relation of the form

$$\Phi^{(m)}(y) = KG^{(1)}(\Phi(y)).\Phi^{(m)}(y)+H_m(y) , \qquad (2.18)$$

where $H_m(y)$ only depends on $\Phi(y),\Phi^{(1)}(y),\ldots,\Phi^{(m-1)}(y)$. The condition (2.15) implies that (2.18) determines for each $y\in Y_\eta$ a unique multilinear operator $\Phi^{(m)}(y) \in \mathcal{L}^{(m)}(Y_\eta, Y_{\eta^m})$. The mapping $\Phi^{(m)} : Y_\eta \to \mathcal{L}^{(m)}(Y_\eta, Y_\zeta)$ is globally bounded if $a<\zeta\leq\eta^m$ and continuous if $a<\zeta<\eta^m$. Finally one proves that $\Phi : Y_\eta \to Y_\zeta$ is of class C^k if $a<\zeta<\eta^k$, with $D^{(m)}\Phi(y)=\Phi^{(m)}(y)$ for $1\leq m\leq k$ and for all $y\in Y_\eta$. This part of the proof uses the fact that by (2.15) and the continuity of $\gamma : (a,b) \to \mathbb{R}_+$ we have $\gamma(\zeta)|g|_1 < 1$ for all ζ in an open interval containing $[\eta^k,\eta]$.

<u>Remark 1</u>. The existence of some $\eta \in (a,b)$ with $a<\eta^k$ is only possible if $a<b^k$.

<u>Remark 2</u>. If $g(0)=0$ then $0 \in M_\eta=W_\phi$ and hence $\phi(0)=0$. If moreover $g \in C_b^1(X)$ and $Dg(0)=0$, then the foregoing proof shows that also $D\phi(0)=0$.

<u>Remark 3</u>. A careful examination of the proof shows that the result of theorem 5 remains valid if we weaken the condition $g \in C_b^k(X)$ as follows. Assume

that $g \in C_b^{0,1}(X)$ and that

$$|g|_{Lip} \sup_{\zeta \in [\eta^k,\eta]} \gamma(\zeta) < 1 \tag{2.19}$$

for some $\eta \in (a,b)$ with $a<\eta^k$. Suppose also that there exists some $\beta > |\pi_1 g|_0 \Sigma_{\ell=0}^{\infty} \|A_1^{\ell}\|$ such that g is of class C^k in $\Omega := \{x\in X \,|\, \|\pi_1 x\|<\beta\}$, and that

$$|g|_{\Omega,m} := \sup_{x\in\Omega} \|D^m g(x)\| < \infty \quad , \qquad 1\leq m\leq k \ . \tag{2.20}$$

Then, using (2.12) and the fact that $|g|_{\Omega,1}\leq|g|_{Lip}$ one can prove that the conclusion of theorem 5 still holds, i.e. $M_\eta=W_\phi$ for some $\phi \in C_b^k(X_2,X_1)$. This remark is important for the application of theorem 5 to locally invariant manifolds, as we will see in section 3.

<u>Theorem 6</u>. Let $A \in \mathcal{L}(X)$ satisfy (S). Then there exists for each $\eta \in (a,b)$ a number $\delta=\delta(\eta) > 0$ such that the following holds : if $F=A+g$, with $g \in C_b^{0,1}(X)$ and $|g|_{Lip}<\delta(\eta)$, then there exists a continuous mapping $H : X \to M_\eta$ such that for each $x \in X$ we have $M_{1/\eta}(x)\cap M_\eta=\{H(x)\}$, where

$$M_{1/\eta}(x) := \{\tilde{x}\in X \,|\, \sup_{p\in\mathbb{N}} \eta^{-p}\|F^p(\tilde{x})-F^p(x)\|<\infty\} \ . \tag{2.21}$$

P r o o f. Fix some $\eta \in (a,b)$ and some $x \in X$. In order to determine $M_{1/\eta}(x)$ we have to find all $z \in Y_{1/\eta}$ such that $\{F^p(x)+z_p \,|\, p\in\mathbb{N}\}$ forms a positive F-orbit. A proof similar to the one of lemma's 2 and 3 shows that this is the case if and only if there exists some $x_1 \in X_1$ such that

$$\begin{aligned} z_p = A_1^p x_1 &+ \sum_{\ell=0}^{p-1}{}^{(+)} A_1^{p-1-\ell}\pi_1[g(F^\ell(x)+z_\ell)-g(F^\ell(x))] \\ &- \sum_{\ell=0}^{\infty} A_2^{-1-\ell}\pi_2[g(F^{p+\ell}(x)+z_{p+\ell})-g(F^{p+\ell}(x))] \ , \quad \forall p \in \mathbb{N} \ . \end{aligned} \tag{2.22}$$

We rewrite (2.22) in the more compact form

$$z = \tilde{S}x_1 + \tilde{K}\tilde{G}(x,z) \ , \tag{2.23}$$

where the definitions of $\tilde{S} \in \mathcal{L}(X_1,Y_{1/\eta})$, $\tilde{K} \in \mathcal{L}(Y_{1/\eta})$ and $\tilde{G} : X\times Y_{1/\eta} \to Y_{1/\eta}$

are similar to those used in (2.5). One verifies easily that $\|\tilde{K}\|_{1/\eta} \leq \gamma(\eta)$ and that

$$\|\tilde{G}(x,z)-\tilde{G}(x,\tilde{z})\|_{1/\eta} \leq |g|_{Lip}\|z-\tilde{z}\|_{1/\eta} \quad , \qquad \forall x \in X,\ \forall z,\tilde{z} \in Y_{1/\eta} . \qquad (2.24)$$

Consequently, if (2.11) holds then (2.23) has for each $(x,x_1) \in X\times X_1$ a unique solution $z=\tilde{z}(x,x_1) \in Y_{1/\eta}$. We have

$$\|\tilde{z}(x,x_1)-\tilde{z}(x,\tilde{x}_1)\|_{1/\eta} \leq \ell\|x_1-\tilde{x}_1\| \quad , \qquad \forall x \in X,\ \forall x_1,\tilde{x}_1 \in X_1 , \qquad (2.25)$$

where ℓ depends on η and $|g|_{Lip}$ but stays bounded as $|g|_{Lip} \to 0$. To see the continuous dependance of $\tilde{z}(x,x_1)$ on $x \in X$ we remark that (2.11) and the continuity of $\gamma(\eta)$ imply that $\gamma(\chi)|g|_{Lip} < 1$ for some $\zeta \in (a,\eta)$. Replacing η by ζ in the argument above we conclude that $\tilde{z}(x,x_1)$ belongs in fact to the space $Y_{1/\zeta}$ which is continuously imbedded in $Y_{1/\eta}$. One then proves that for fixed $z \in Y_{1/\zeta}$ the mapping $x \mapsto \tilde{G}(x,z)$ is continuous from X into $Y_{1/\eta}$; from this and $\tilde{z}(x,x_1) \in Y_{1/\zeta}$ it follows that $\tilde{z} : X\times X_1 \to Y_{1/\eta}$ is continuous.
We conclude from the foregoing that under the condition (2.11) we have for each $x \in X$ that

$$\begin{aligned} M_{1/\eta}(x) &= \{x+\tilde{z}_0(x,x_1) \mid x_1 \in X_1\}=\{x+\tilde{z}_0(x,x_1-\pi_1 x) \mid x_1 \in X_1\} \\ &= \{x_1+J(x,x_1) \mid x_1 \in X_1\} , \end{aligned} \qquad (2.26)$$

where we have used $\pi_1\tilde{z}_0(x,x_1)=x_1$ and where $J : X\times X_1 \to X_2$ is a continuous map defined by

$$J(x,x_1) := \pi_2(x+\tilde{z}_0(x,x_1-\pi_1 x)) . \qquad (2.27)$$

We also have that $M_\eta=W_\phi$ (see theorem 4), and in order to determine $M_{1/\eta}(x)\cap M_\eta$ we have to find all solutions $x_1 \in X_1$ of the equation

$$x_1 = \phi(J(x,x_1)) . \qquad (2.28)$$

Now $\phi \in C_b^{0,1}(X_2,X_1)$, with $|\phi|_{Lip} \to 0$ as $|g|_{Lip} \to 0$; also $J(x,x_1)$ is Lipschitzian in $x_1 \in X_1$, with a Lipschitz constant which goes to zero as $|g|_{Lip} \to 0$. It follows that if $|g|_{Lip}$ is sufficiently small then (2.28) has for each $x \in X$ a unique solution $x_1=\tilde{x}_1(x) \in X_1$; the mapping $\tilde{x}_1 : X \to X_1$ is continuous. This proves the theorem, by taking $H(x) := \tilde{x}_1(x)+J(x,\tilde{x}_1(x))$.

The theorem implies that for each $x \in X$ there is a unique element $H(x) \in M_\eta = W_\phi$ and a constant $C>0$ such that

$$\|F^p(H(x))-F^p(x)\| \leqq C\eta^p \quad , \quad \forall p \in \mathbb{N} . \tag{2.29}$$

Since $F^p(H(x)) \in M_\eta$ for all $p \in \mathbb{N}$ we have in particular that

$$\lim_{p\to\infty} \mathrm{dist}(F^p(x),M_\eta) = 0 . \tag{2.30}$$

3. LOCALLY INVARIANT MANIFOLDS

In this section we return to the situation described in the introduction : we consider a mapping $F \in C^k(X)$ $(k \geqq 1)$ with $F(0)=0$ and $A := Df(0)$ satisfying (S), and we are interested in the existence of locally invariant manifolds of the form W_ϕ, with $\phi \in C(X_2,X_1)$ and $\phi(0)=0$. We will apply the results of section 2, taking for g an appropriate modification of $f := F-A$; then $A+g$ coincides with F in a neighborhood of $x=0$, and invariant manifolds for $A+g$ give us locally invariant manifolds for F.

<u>Theorem 7</u>. Let $F \in C^k(X)$ for some $k \geqq 1$, and assume that $F(0)=0$ and that $A := Df(0)$ satisfies (S). Assume moreover that

(i) $a < b^k$;

(ii) there exists a mapping $\zeta \in C_b^k(X_2,\mathbb{R})$ such that $\zeta(x_2)=0$ if $\|x_2\| \geqq 1$ and $\zeta(x_2)=1$ for all x_2 in a neighborhood of $x_2=0$.

Then F has a locally invariant manifold of the form W_ϕ, with $\phi \in C^k(X_2,X_1)$, $\phi(0)=0$ and $D\phi(0)=0$. This manifold is also locally attracting : there exists a neighborhood Ω of $x=0$ such that if $F^p(x) \in \Omega$ for all $p \in \mathbb{N}$ then

$$\lim_{p\to\infty} \mathrm{dist}(F^p(x),W_\phi) = 0 . \tag{3.1}$$

P r o o f. Let $f := F-A$; then $f(0)=0$, $Df(0)=0$, and f together with its derivatives up to order k are bounded in a neighborhood of $x=0$, say a ball with radius $\rho_0>0$ around $x=0$. Let

$$\alpha(\rho) := \sup_{\|x\| \leqq \rho} \|Df(x)\| \quad , \quad 0 \leqq \rho \leqq \rho_0 . \tag{3.2}$$

Then $\alpha(\rho) \to 0$ as $\rho \to 0$, and

$$\sup_{\|x\|\leq\rho} \|f(x)\| \leqq \rho\alpha(\rho) \quad , \quad 0\leqq\rho\leqq\rho_0 . \tag{3.3}$$

Suppose first that there exists a mapping $\chi \in C_b^k(X,\mathbb{R})$ such that $\chi(x)=0$ if $\|x\|\geqq 1$ and $\chi(x)=1$ for all x in a neighborhood of $x=0$. (This is for example the case if X is a Hilbert space). Then we define for each $\rho \in (0,\rho_0]$ a mapping $g_\rho \in C_b^k(X)$ by

$$g_\rho(x) := f(x)\chi(\rho^{-1}x) \quad , \quad \forall x \in X . \tag{3.4}$$

An easy calculation shows that

$$|g_\rho|_1 \leqq \alpha(\rho)(|\chi|_0+|\chi|_1) . \tag{3.5}$$

Now fix some $\eta \in (a,b)$ with $a<\eta^k$ and let

$$\varepsilon := \min(\min_{\zeta\in[\eta^k,\eta]} \gamma(\zeta)^{-1},\delta(\eta)) , \tag{3.6}$$

with $\gamma(\eta)$ and $\delta(\eta)$ as in (2.9) and theorem 6, respectively. By (3.5) we can choose $\rho \in (0,\rho_0]$ sufficiently small such that $|g_\rho|_1<\varepsilon$. For such ρ we can apply the theorems of section 2 to $\tilde{F} := A+g_\rho$. Since $\tilde{F}(x)=F(x)$ for all x in a neighborhood of $x=0$ the theorem follows immediately.
If no such mapping χ exists and only the assumption (ii) holds then we have to modify $f(x)$ more carefully. Let $\psi : \mathbb{R} \to \mathbb{R}$ be a smooth function such that $\psi(s)=1$ if $|s|\leqq 1/2$ and $\psi(s)=0$ if $|s|\geqq 1$. Then we define $\chi : X \to \mathbb{R}$ by

$$\chi(x) = \psi(\|\pi_1 x\|)\zeta(\pi_2 x) \quad , \quad \forall x \in X , \tag{3.7}$$

where $\zeta \in C_b^k(X_2,\mathbb{R})$ is the mapping given by (ii). The mapping χ belongs to the space $C_b^{0,1}(X)$.
Next we fix some $\eta \in (a,b)$ such that $a<\eta^k$, and define $\varepsilon>0$ by (3.6). We choose some $\rho \in (0,\rho_0/2)$ such that

$$\alpha(2\rho)|\chi|_0+\alpha(\rho)|\chi|_{Lip} < \varepsilon \tag{3.8}$$

and

$$2\alpha(\rho)|\chi|_0 \sum_{\ell=0}^{\infty} \|A_1^\ell\| < 1 . \tag{3.9}$$

Then define $g_\rho \in C_b^{0,1}(X)$ by (3.4), and let $F := A+g_\rho$. One can easily verify that

$$|g_\rho|_0 \leq \rho\alpha(\rho)|X|_0 \quad \text{and} \quad |g_\rho|_{Lip} \leq \alpha(2\rho)|X|_0+\alpha(\rho)|X|_{Lip} \; . \tag{3.10}$$

By (3.8) and (3.6) we can then apply the theorems 4 and 6 to the mapping $\tilde{F}$; this gives us an attracting invariant manifold W_ϕ; by remark 2 we have $\phi(0)=0$ and $D\phi(0)=0$. In order to prove that $\phi \in C^k(X_2,X_1)$ we use remark 3. By (3.9) and (3.10) we have

$$|\pi_1 g_\rho|_0 \sum_{\ell=0}^{\infty} \|A_1^\ell\| \leq \rho\alpha(\rho)|X|_0 \sum_{\ell=0}^{\infty} \|A_1^\ell\| < \rho/2 \; ; \tag{3.11}$$

moreover, if $\|\pi_1 x\| < \rho/2$ then $\psi(\rho^{-1}\|\pi_1 x\|)=1$ and $g_\rho(x)=f(x)\zeta(\rho^{-1}x)$. Therefore $g_\rho(x)$ is of class C^k with all its derivatives up to order k bounded on the domain $\{x\in X \,|\, \|\pi_1 x\|<\rho/2\}$. By remark 3 we have then that $\phi \in C^k(X_2,X_1)$.

4. CONCLUDING REMARKS

The foregoing results remain valid if in the hypothesis (S) we assume that $a<1<b$ and if we replace b by 1 in all the statements. Similar techniques can be used to obtain similar results for the case $a\geq 1$. One can also obtain invariant manifolds of the form $\tilde{W}_\psi := \{x_1+\psi(x_1) \,|\, x_1\in X_1\}$, with $\psi \in C^0(X_1,X_2)$. The condition $a<b^k$ in theorem 7 gives a restriction on the diferentiability of the invariant manifold W_ϕ. The following example shows that this condition cannot be relaxed. We take $X=\mathbb{R}^2$ and define $F:\mathbb{R}^2 \to \mathbb{R}^2$ by

$$F(x_1,x_2) = (b^2x_1+cx_2^2, bx_2) \quad , \quad 0<b<1 \; , \; c\neq 0 \; . \tag{4.1}$$

We have $X_1 = \mathbb{R}\times\{0\}$, $X_2 = \{0\}\times\mathbb{R}$ and $a=b^2<b<1$. Now suppose that there exists some $\phi \in C^2(\mathbb{R})$ with $\phi(0)=0$, $D\phi(0)=0$ and such that $W_\phi=\{(\phi(x_2),x_2) \,|\, x_2\in\mathbb{R}\}$ is locally invariant for F. Then we have for all sufficiently small x_2 that

$$b^2\phi(x_2)+cx_2^2 = \phi(bx_2) \; . \tag{4.2}$$

Since ϕ is C^2 we have $\phi(x_2)=x_2^2\psi(x_2)$ for some continuous $\psi:\mathbb{R}\to\mathbb{R}$. Bringing this in (4.2) we obtain

$$b^2\psi(x_2)+c = b^2\psi(bx_2) ;$$

for $x_2=0$ this gives a contradiction if $c\neq0$. Hence there exists no C^2-invariant manifold of the required form.

ACKNOWLEDGEMENT

I want to thank Prof. S.-N. Chow for several discussions and for the invitation to attend the Workshop.

REFERENCES

1. G. Fischer. Zentrumsmannigfaltigkeiten bei elliptischen Differentialgleichungen. Math. Nachr. 115 (1984), 137-157.
2. G. Iooss. Bifurcations of maps and applications. Math. Studies, 36 (1979), Elsevier-North-Holland, Amsterdam.
3. O.E. Lanford III. Bifurcation of periodic solutions into invariant tori. Lect. Notes in Math. 322 (1973), 159-192. Springer-Verlag, N.Y.
4. K.R. Schneider. Hopf bifurcation and center manifolds. Coll. Math. Soc. I. Bolyai, 30, 1979, 953-970.
5. A. Vanderbauwhede. Center manifolds, normal forms and elementary bifurcations. Preprint University of Ghent, 1986.
6. A. Vanderbauwhede and S.A. Van Gils. Center manifolds and contractions on a scale of Banach spaces. J. Funct. Anal., to appear.

LINEARIZING COMPLETELY INTEGRABLE SYSTEMS ON COMPLEX ALGEBRAIC TORI

by

P. van Moerbeke
Department of Mathematics
University of Louvain
1348 Louvain-la-Neuve (Belgium)
and
Brandeis University
Waltham, Mass 02254

How does one recognize whether a Hamiltonian system is completely integrable ? Granted a system is integrable, how does one effectively integrate the problem ? The answer to these questions is unknown, up to this day! The proof of the Liouville theorem concerning integrals in involution and invariant tori is non-constructive : neither does it enable you to decide about its integrability, nor does it provide means for integrating the problem. Historically, the solutions to the classical systems that we know have required the most ingenious tricks by the best mathematical minds and they are only distinguished by their variety. The resolution of the Korteweg-de Vries equation by inverse spectral methods, some twenty years ago, has led to a number of Lie theoretical and algebraic geometrical methods for producing ordinary and partial differential equations, some of which are interesting

* The support of a National Science Foundation grant # DMS-8403136 is gratefully acknowledged.

NATO ASI Series, Vol. F37
Dynamics of Infinite Dimensional Systems
Edited by S.-N. Chow, and J.K. Hale

from the point of view of mechanics and physics. They all give rise to so-called *algebraically completely integrable* systems; this stringent, but yet quite typical notion of integrability will be defined in § 2, whereas § 1 deals with the simple but fundamental example of Euler's rigid body rotating about its center of mass. Also § 2 contains a list of important features shared by such systems and it shows how modern algebraic geometry can be brought to bear on the resolution of differential equations. Section 3 deals with the geodesic flow on the group SO(N) for a left-invariant metric. It is an important class of examples as many physical systems can ultimately be reduced to this problem. In § 4 and 5, I apply the theoretical methods developped earlier to geodesic flow on SO(4) for two different classes of metrics. An extensive account on algebraic methods in mechanics will appear in a book published by Academic Press [6]; for lectures on various facets of this work, see [16, 17, 18].

I shall not discuss here the weaker notion of analytic integrability; the perturbation techniques developped in that context are of a totally different nature. See Kozlov [15], for an expository account on this subject.

§ 1. The Euler rigid body motion.

The Euler rigid body motion is governed by the equations

$$(1.1)\qquad \dot{x} = x \wedge \lambda x \qquad x \in \mathbb{R}^3,\ \lambda x = (\lambda_1 x_1, \lambda_2 x_2, \lambda_3 x_3) \in \mathbb{R}^3,$$

where the x_i are the angular momenta about the principal axes of inertia through the center of mass and where the λ_i^{-1} are the principal moments of inertia. The system has two conserved quantities; the total angular momentum and the energy; i.e.,

$$(1.2)\quad Q_1 \equiv x_1^2 + x_2^2 + x_3^2 = c_1 \text{ and } Q_2 \equiv \tfrac{1}{2}(\lambda_1 x_1^2 + \lambda_2 x_2^2 + \lambda_3 x_3^2) = c_2.$$

It shows that the system evolves on the intersection of an ellipsoïd and a sphere. In $\mathbb{R}^3$, their intersection will be isomorphic to two circles, within an appropriate range of c_1 and c_2. Observing that the first equation of (1.1) reads

$$(1.3)\qquad \frac{dx_1}{x_2 x_3} = (\lambda_3 - \lambda_2)\,dt,$$

solving the system (1.2) linearly with regard to x_2^2 and x_3^2 and putting the result into (1.3) leads to an elliptic integral

$$(1.4)\qquad \int_{x_1(o)}^{x_1(t)} \frac{dx}{\sqrt{(x^2+\beta)(x^2+\gamma)}} = \alpha t,$$

with respect to the elliptic curve

$$y^2 = (z^2+\beta)(z^2+\gamma);$$

here α, β and γ are quantities depending on the λ and c. Therefore, the solutions $x_i(t)$ to the differential equations (1.1) substituted into the integral (1.4) are linear functions of time. Then the functions $x_i(t)$ can be expressed in terms of theta-functions of t, according to the classical inversion of Abelian integrals. Moreover, the two circles defined by the intersection $\{Q_1 = c_1\} \cap \{Q_2 = c_2\}$ form the real part of the 1-dimensional complex torus, defined by the elliptic curve above.

To mention another feature, the system (1.1) possesses a two-dimensional family of Laurent solutions

$$x_i(t) = t^{-1}(x_i^o + x_i^1 t + x_i^2 t^2 + \ldots),$$

where x_i^o and x_i^1 are specific quantities depending on λ_i and where the vector (x_1^2, x_2^2, x_3^2) depends linearly on two free parameters; the latter account for the quantities c_1 and c_2.

Another observation is that not only the variables $x_o = 1$, x_1, x_2 and x_3, but also the variables $(x_o/x_\alpha, x_1/x_\alpha, x_2/x_\alpha, x_3/x_\alpha)$ form a closed system of quadratic differential equations for each choice of $\alpha = 1$, 2 or 3. For instance, using (1.1) the derivative

$$\left(\frac{x_2}{x_1}\right)^{\cdot} = \frac{x_3}{x_1}((\lambda_1-\lambda_3)x_1^2 - (\lambda_3-\lambda_2)x_2^2)$$

$$= (2c_2 - \lambda_3 c_1)\frac{1}{x_1}\frac{x_3}{x_1}, \qquad \text{using (1.2)},$$

is indeed quadratic in $1/x_1$ and x_3/x_1.

In view of the isomorphism between $(\mathbb{R}^3, \times)$ and $(so(3), [\ ,\ ])$, the system of differential equations (1.1) can be written as a so-called *Lax pair* and leads to the *Euler-Arnold equations*

$$\text{(1.5)} \qquad \dot{A} = [A, \lambda A] \qquad A, \lambda A \in so(3)$$

with

$$A = \begin{pmatrix} 0 & -x_3 & x_2 \\ x_3 & 0 & -x_1 \\ -x_2 & x_1 & 0 \end{pmatrix} \in so(3).$$

The solution to (1.5) has the form $A(t) = O(t)A(o)O^T(t)$, where $O(t)$ is a one parameter sub-group of $SO(3)$. This is to say the spectrum of $A(t)$ and thus its characteristic polynomial

$$\det(A-zI) = -z(z^2+\Sigma x_i^2)$$

are time independent. Unfortunately, the spectrum of a 3×3 skew-symmetric matrix provides only one piece of information; the conservation of energy does not appear as part of the spectral information. Therefore one is led to considering another formulation.

Write the Lax pair (1.5) as follows

$$(1.6)\quad (A+\alpha y)^{\cdot} = [A+\alpha y,\ \lambda A+\beta y], \quad y \text{ dummy variable}$$

with $\alpha = \text{diag}(\alpha_1,\alpha_2,\alpha_3)$ and $\beta = \text{diag}(\beta_1,\beta_2,\beta_3)$.
This equation decomposes into the various powers of y :

$$y^0 : \dot{A} = [A,\lambda A] \Longleftrightarrow \text{equation (1.5)}$$

$$(1.7)\quad y^1 : 0 = [A,B] + [\alpha,\ \lambda A] \Longleftrightarrow \lambda_i = \frac{\beta_j-\beta_k}{\alpha_j-\alpha_k},\ (i,j,k) \text{ permutation of } (1,2,3)$$

$$y^2 : [\alpha,\beta] = 0 \text{ trivially satisfied.}$$

The relation (1.7) can always be satisfied by picking $\beta_j = \alpha_j^2$ and α_j such that $\lambda_i = \alpha_j + \alpha_k$. Thus the original Lax pair satisfies the extended Lax pair (1.6) and therefore, whatever be h, the spectrum of $A + \alpha h$ is preserved; therefore its characteristic polynomial

$$(1.8)\quad P(y,z) \equiv \det(A + \alpha y - zI) = \prod_1^3 (\alpha_i y - z) + \sum_1^3 x_i^2(\alpha_i y - z)$$

is preserved as well. Except for some constants, its coefficients are generated by Σx_i^2 and $\Sigma \alpha_i x_i^2$, which yield the energy :

$$(\Sigma \alpha_i)(\Sigma x_i^2) - \Sigma \alpha_i x_i^2 = \Sigma \lambda_i x_i^2 = 2Q_2.$$

The spectrum of the matrix $A + \alpha y$ as a function of $y \in \mathbb{C}$ is given by the zeroes of the polynomial $P(y,z) = 0$, thus defining an algebraic curve, called the *spectral curve*. Letting $w = y/z$, the curve (1.8) defines an elliptic curve

$$(w,z) : z^2 = - \frac{\sum_i x_i^2 (\alpha_i w - 1)}{\prod_1^3 (\alpha_i w - 1)},$$

which is shown to be isomorphic to the original elliptic curve. For more information about these methods, see Adler-van Moerbeke [1].

§ 2. Algebraically completely integrable systems and their properties.

Consider a Hamiltonian system

$$\dot{x} = f(x) = J \frac{\partial H}{\partial x}, \quad x \in \mathbb{R}^n \tag{2.1}$$

where $J = J(x)$ is a skew-symmetric matrix, such that the (Poisson) bracket

$$\{H,H'\} = \langle J \frac{\partial H}{\partial x}, \frac{\partial H'}{\partial x} \rangle$$

satisfies the Jacobi identity. Let $H_1, \ldots, H_k$ be the Casimir functions; their gradients $\partial H_i / \partial x$ are the null-vectors of J. Then the symplectic leaves

$$\bigcap_1^k \{H_i = c_i, \ x \in \mathbb{R}^n\}$$

must be even-dimensional; let $2m = n-k$. The system is called *algebraically completely integrable*, when

I. the system possesses m polynomial invariants $H_{k+1}, \ldots, H_{k+m}$ in involution (i.e., satisfying $\{H_i, H_j\} = 0$) such that the invariant manifolds

$$\bigcap_{1}^{k+m} \{H_i = c_i,\ x \in \mathbb{R}^n\}$$

are compact, connected manifolds, for generic c_i. By the Arnold-Liouville theorem [8] they are m-dimensional tori $T^m_{\mathbb{R}} = \mathbb{R}^m/\text{lattice}$. The solutions of the system (2.1) are then straight-line motions on $T^m_{\mathbb{R}}$.

II. The invariant manifolds, thought of as affine varieties in $\mathbb{C}^n$ (*non-compact*), can be completed into *complex algebraic tori*, i.e.,

$$\bigcap_{1}^{k+m} \{H_i = c_i,\ x \in \mathbb{C}^n\} = T^m_c \setminus \{\text{one or several codimension 1 subvarieties}\}^*$$

where the tori

$$T^m_c = \frac{\mathbb{C}^m}{\text{period lattice}} = \text{complex algebraic torus (Abelian variety)}$$

depend on the c's. "Algebraic" means that T^m_c can be defined as an intersection

$$\bigcap_{1}^{M} \{F_i(x_o,\ldots,x_N) = 0\}$$

involving a (large) number of homogeneous polynomials F_i. The coordinates $z_i(t)$ are required to be meromorphic on the tori T^m_c.

Condition II means, in particular, there is an algebraic map

$$(x_1(t),\ldots,x_n(t)) \rightsquigarrow (\mu_1(t),\ldots,\mu_m(t)),$$

making the following sums linear in t :

$$\sum_{i=1}^{m} \int_{\mu_i(o)}^{\mu_i(t)} \omega_k = a_k t \qquad k = 1,\ldots,m,$$

---

* called a divisor.

where $\omega_1,\ldots,\omega_m$ denote holomorphic differentials on some algebraic curve. This definition was given by Adler-van Moerbeke [2], whereas details on this section can be found in [4] and [6].

Algebraic completely integrable systems $\dot{x} = J\ \partial H/\partial x$ enjoy a number of properties, which we have observed already in the context of Euler's rigid body motion.

(i) *An a.c.i. system admits a* $n-1$*-dimensional family of solutions, which are Laurent series in* t. Indeed the variables x_i are meromorphic on the tori T^m and thus they blow up along a divisor D. The trajectories of the flow can be parametrized by the points where they hit the $m-1$-dimensional variety D and they also depend on the $k+m$ constants c_i defining T^m. Therefore each connected component where some x's blow up leads to a $k+2m = n-1$-dimensional family of Laurent solutions, called the *principal balances*,

$$(2.2)\quad x_i(t) = t^{-k_i}(x_i^{(o)}+x_i^{(1)}t+\ldots),\ k_i \in \mathbb{Z},\ \text{some } k_i > 0,$$

where the coefficients

$$(2.3)\qquad x_i^{(j)} = x_i^{(j)}(\alpha_1,\ldots,\alpha_{n-1})$$

are rational functions of the $n-1$ parameters α. The exponents j for which a free parameter α_k occurs for the first time is called a *Kowalewski exponent*. This method of finding Laurent solutions was used by S. Kowalewskaja [13,14] to detect the a.c.i. cases among the rigid body motions.

(ii) *The system of coordinates* $x_1,\ldots,x_n$ *can be enlarged to a new set* $x_o = 1, x_1,\ldots,x_N$ *having the property that for fixed but arbitrary* $0 \leqslant \alpha \leqslant N$ *we have*

$$\left(\frac{x_i}{x_\alpha}\right)^{\cdot} = \text{quadratic polynomial of } \left(\frac{1}{x_\alpha},\frac{x_1}{x_\alpha},\ldots,\frac{x_N}{x_\alpha}\right),\quad 0 \leqslant i \leqslant N.$$

The idea is the following : let D be the divisor on T^m where some x's blow up. For the sake of this exposition, let the x's have a simple pole along D. For positive integers k, define the spaces

$$L(kD) = \{\text{meromorphic functions on } T^m \text{ having a k-fold pole at worst along } D\};$$

clearly $1, x_1, \ldots, x_n \in L(D)$. A divisor D is called *projectively normal*, whenever $L(kD) = L(D)^{\otimes k}$, i.e., L(kD) is generated by homogeneous polynomials of degree k in some basis elements $x_1, \ldots, x_N$ of L(D). For $x_\alpha \neq$ constant, the ratios x_i/x_α are finite along most of D and thus their derivative $(x_i/x_\alpha)^{\cdot} = (\dot{x}_i x_\alpha - x_i \dot{x}_\alpha)/x_\alpha^2$ is finite as well. Therefore, since x_α^2 has a double pole along D, the expressions $\dot{x}_i x_\alpha - x_i \dot{x}_\alpha$ must also have a double pole (at worst), i.e., $\dot{x}_i x_\alpha - x_i \dot{x}_\alpha \in L(2D)$. Since $L(2D) = L(D)^{\otimes 2}$, the functions $\dot{x}_i x_\alpha - x_i \dot{x}_\alpha$ are expressible in terms of quadratic polynomials in the x_k :

$$\dot{x}_i x_\alpha - x_i \dot{x}_\alpha = \sum_{k,\ell} c^i_{k\ell} \, x_k x_\ell$$

and thus

$$\left(\frac{x_i}{x_\alpha}\right)^{\cdot} = \sum_{k,\ell=1}^{N} c^i_{k\ell} \frac{x_k}{x_\alpha} \cdot \frac{x_\ell}{x_\alpha},$$

which establishes the statement above. Now algebraic geometry has given us means to decide whether a divisor is projectively normal. Even if D itself is not projectively normal, 2D or 3D will have that property. The question arises now how to effectively enlarge the original coordinates $x_1, \ldots, x_n$ to a set $x_o, x_1, \ldots, x_N$ spanning L(D) : the differential equations provide the Laurent series for the coordinates $x_1, \ldots, x_n$, which were assumed to behave as 1/t in this context, and the remaining functions can be found by picking homogeneous polynomials of the Laurent series $x_1(t), \ldots, x_n(t)$ behaving like 1/t.

(iii) *How to construct a divisor* D *on the torus* T^2 *?* Consider the n-1-dimensional family of Laurent solutions (2.2) to the differential equations (2.1) depending rationally on the n-1 parameters $\alpha_1,\ldots,\alpha_{n-1}$; they lead to an explicit description of the various connected components of D by substituting the Laurent solutions $x_i(t)$ for x_i in the constants of motion H_j and by observing that the expressions thus obtained are independent of t and rational in $\alpha_1,\ldots,\alpha_{n-1}$:

$$(2.4) \qquad H_j(x(t)) \equiv F_j(\alpha_1,\ldots,\alpha_{n-1}) = c_j \qquad 1 \leqslant j \leqslant k+m.$$

These k+m equations in n-1 = k+2m-1 unknowns $\alpha_1,\ldots,\alpha_{n-1}$ provide a description of the smooth version of the divisor D associated with the family of Laurent solutions. A divisor D wrapped around an algebraic variety yields a lot of information about this variety and thus about the tori.

Consider the situation, where the invariant tori are 2-dimensional, i.e., m = 2. A torus T^2 is defined by $T^2 = \mathbb{C}^2/\Lambda$, where the lattice Λ is generated by four vectors in $\mathbb{C}^2$, defining a 2 by 4 matrix Π, called the period matrix. Algebraic geometry tells us that given a curve D wrapped around T^2, it is possible to find a basis of $\mathbb{C}^2$ and an integer basis of Λ having the property that Λ is generated by the four columns of the following matrix having a canonical form :

$$\Pi = \begin{pmatrix} \delta_1 & 0 & \alpha & \beta \\ 0 & \delta_2 & \beta & \gamma \end{pmatrix}, \quad 1 \leqslant \delta_1 \leqslant \delta_2, \text{ integers such that } \delta_1 \text{ divides } \delta_2,$$

with the integers δ_i satisfying

$$\delta_1\delta_2 = \text{genus }(D) - 1.$$

(iv) *The degree of the constants of motion and Kowalewski's exponents.* Assume the system of differential equations $\dot{x} = f(x)$ to be weight-homogeneous with a weight g_i going with each variable x_i :

$$f_i(\alpha^{g_1}x_1,\ldots,\alpha^{g_n}x_n) = \alpha^{g_i+1} f_i(x_1,\ldots,x_n), \ \forall\alpha \in \mathbb{C}.$$

For the simplicity of this exposition, take all $g_i = g$. The system $\dot{x} = f(x)$ admits formal Laurent solutions

$$x_i(t) = t^{-g}(x_i^o + x_i^1 t + \dots)$$

whose coefficients satisfy at the 0th step

$$(2.5) \qquad f_i(x_1^o,\dots,x_n^o) + g x_i^o = 0 \qquad 1 \leqslant i \leqslant n$$

and at the kth step

$$(2.6) \qquad (L-kI)x^k = \text{a polynomial in } (x^1,\dots,x^{k-1}) \qquad k \geqslant 1$$

where

$$L \equiv \frac{\partial f}{\partial x} + gI \Big|_{x=x^o}$$

is the Jacobian matrix of the equations (2.5). If the system $\dot{x} = f(x)$ is to have a n-1-dimensional family of Laurent solutions, or what is the same, if n-1 free parameters are to appear in the formal Laurent expansion, they must either come from the non-linear equations (2.5) or from the eigenvalue problem (2.6), i.e., L must have at least n-1 integer eigenvalues! One then shows these formal series are automatically convergent Laurent series (see Adler-van Moerbeke [6]).

Let me proceed to show that the *degree* α *of a homogeneous constant of motion* $H(x)$ *is an eigenvalue of* L *unless* $\partial H/\partial x$ *vanishes for every solution of* (2.5); this is due to Yoshida [21]. To see this, observe that the vector $x = x^{(o)}t^{-g}$ with $x^{(o)}$ satisfying (2.5) is a special solution of $\dot{x} = f(x)$. If ζ^o is an eigenvector of L with eigenvalue ℓ, then $\zeta = \zeta^{(o)}t^{\ell-g}$ is a solution of the variational equation

$$(2.7) \qquad \zeta = \langle \frac{\partial f}{\partial x}, \zeta \rangle$$

around the special solution $x = x^{(o)}t^{-g}$. Moreover, if $H(x)$ is a constant of the motion of degree α for $\dot{x} = f(x)$, then $\langle \partial H/\partial x, \zeta \rangle$ is a constant for any solution of the variational equation, implying that $\ell = \alpha$ must hold, unless this expression vanishes for every solution of (2.5). In the same style, one shows that if the system has two homogeneous constants of motion $H_1(x)$ and $H_2(x)$ of degree α, then α is a double eigenvalue of L, if the gradients $\partial H_1/\partial x$ and $\partial H_2/\partial x$ are independent along that solution $x^{(o)}$.

(v) *Laurent solutions of a.c.i. systems with less than* n-1 *degrees of freedom (lower balances) and the points of tangency of the flow* X_1 *to* D. Such Laurent solutions can be constructed in a systematic fashion from the principal balances. Again for simplicity assume the coordinates $x_1, \ldots, x_n$ have simple poles along D and assume $m = 2$, i.e., the tori are 2-dimensional and the divisor D represents one or several curves on T^2. The trajectories for the vector field $\dot{x} = f(x)$ are parallel straight lines on T^2 and therefore they must be tangent to the divisor D at $2g - 2$ places, where g is the genus of the divisor D. These points of tangency can explicitly be obtained as follows : Let t_1 and t_2 be the time variables respectively of the vector field $X_1 : \dot{x} = f(x) = J\,\partial H/\partial x \equiv \partial x/\partial t_1$ and a vector field $X_2 : \dot{x} = J\,\partial H'/\partial x \equiv \partial x/\partial t_2$ commuting with the first one. These variables t_1 and t_2 form a local system of coordinates on T^2 and consider their differentials dt_1 and dt_2 defined by $dt_i(X_j) = \delta_{ij}$. Then the restrictions $\omega_i = dt_i\big|_D$ of dt_i to D are holomorphic differentials on D and the zeroes of ω_2 (respectively ω_1) yield the $2g - 2$ points of tangency of X_1 (respectively X_2) to D. The differentials ω_1 and ω_2 can effectively be computed by taking the differentials of two appropriately chosen coordinates x_i and x_j/x_α with respect to t_1 and t_2 and by restricting the result to the curve D; this method is due to Haine [11]. Let $p_o \in D$ be a point of tangency of the flow X_1 to D (i.e., $\omega(p_o) = 0$) and consider the coordinate $x_\alpha = t^{-1}(x_\alpha^o + \ldots)$ such that the meromorphic function $x_\alpha^o(p)$ on D has a pole at $p_o \in D$ of maximal order (among $\alpha = 1, \ldots, n$). Then letting $p \to p_o$ in

$$\frac{1}{x_\alpha} = \frac{t}{x_\alpha^{(o)}} - \frac{x_\alpha^1}{(x_\alpha^o)^2} t^2 + \ldots,$$

one shows that at $p = p_o$, the variable $1/x_\alpha$ has a Taylor series

$$\frac{1}{x_\alpha} = t^{k+1}(y_\alpha^{(o)} + y_\alpha^{(1)} t + \ldots), \ k = \text{order of zero of } \omega_2 \text{ at } p$$

with $y_\alpha^{(o)} \not\equiv 0$ on D and $1/y_\alpha^{(o)} \not\equiv 0$ near p. Therefore inverting the latter again, we obtain

$$x_\alpha = t^{-k-1}\left(\frac{1}{y_\alpha^{(o)}} + \ldots\right) \text{ and } x_i = \text{Taylor series } \left(\frac{x_i}{x_\alpha}\right)\Big|_{p_o} \cdot \text{Laurent series } x_\alpha$$

which gives a n-2-dimensional family of Laurent solutions, versus the n-1-dimensional families discussed before! We conclude that each point of tangency of the vector field $\dot{x} = J\, \partial H/\partial x$ to D produces a n-2-dimensional family of Laurent solutions. The singular points of D produce such n-2-dimensional families as well; the arguments used here are beyond the scope of this lecture. Finally lower-dimensional families can also be constructed by considering the values c_i of the constants of motion where the complex tori T_c^2 degenerate to non-compact surfaces.

(vi) *Hamiltonian systems satisfying an extended Lax pair.* As pointed out in § 1, the Euler rigid body motion can be written as an extended Lax pair (2.6) with a dummy variable y. Assume now a Hamiltonian system having the form

$$(2.8) \qquad \dot{A} = [A,B] \text{ with } A = \sum_{i=0}^{n} A_i y^i \text{ and } B = \sum_{i=0}^{N} B_i y^i, \ A_i \text{ and } B_i \text{ matrices.}$$

When a given system can be written this way is a difficult unsolved problem. In some instances, it is quite straightforward to exhibit a Lax pair (2.8), although in most cases it requires quite a bit of ingenuity. Some Hamiltonian flows on Kostant-Kirillov coadjoint orbits in subalgebras of infinite-dimensional Lie algebras (Kac-Moody Lie algebras) yield large classes of extended Lax pairs. A general statement leading to such situations is given by the Adler-Kostant-Symes theorem [1]. An exposition of such methods and their applications can be found in the authors International Congress lecture (1983) [18].

A flow of the type above $\dot{A} = [A,B]$ preserves the spectrum of A for every $y \in \mathbb{C}$ and therefore its characteristic polynomial $P(y,z) \equiv \det(A-zI)$. The latter defines an algebraic curve $C : P(y,z) = 0$; for almost all $(y,z) \in C$, the matrix $A-zI$ has a one-dimensional null-space, defining a (holomorphic) line bundle on C. Whenever the entries of the A_i are moving in time, the curve C does not move, whereas the null-vector and thus the line bundle do move, inducing a motion on the set of line bundles. The set of holomorphic line bundles L on an algebraic curve formsa group for the operation of tensoring $L \otimes L'$ and the full set with a given topological type is parametrized by the points of a complex algebraic torus T^g, where g is the genus of the curve. This torus is called the Jacobi (or Picard) variety of the curve,in short Jac(C). Only when C is an elliptic curve, is Jac(C) isomorphic to C. Since the flow (2.8) induces deformations of line bundles, their topological type remains unchanged and therefore it induces a motion on the Jacobi variety; under some easily checkable condition on A and B, due to Griffiths [9], this motion evolves according to a straight line on Jac(C). Therefore the linearizing equations are given by

$$(2.9) \qquad \sum_{i=1}^{g} \int_{\mu_i(o)}^{\mu_i(t)} \omega_k = a_k t \qquad k = 1,\dots,g$$

where $\omega_1,\dots,\omega_g$ span the g-dimensional space of holomorphic differentials on the curve C of genus g.

§ 3. N-dimensional generalization of Euler's rigid body.

The equations (1.1) or (1.5) call for the obvious generalization

(3.1) $\dot{A} = [A,\lambda A]$ with $A = (X_{ij})$, $\lambda A = (\lambda_{ij}X_{ij})_{1\leqslant i,j\leqslant N} \in so(N)$,

with the extended Lax pair

(3.2) $$(A + \alpha y)^{\cdot} = [A + \alpha y, \lambda A + \beta y].$$

The equations (3.1) describe geodesic motion on the group SO(N) for a left-invariant metric, defined by the quadratic form

(3.3) $$H = \frac{1}{2} \sum_{1\leqslant i<j\leqslant N} \lambda_{ij}X_{ij}^2;$$

e.g., for $N = 4$, they describe the motion of solids in fluids or the motion of solids with cavities filled with fluid; some of these examples have been studied last century by Clebsch, Lyapunov and Steklov; see [19]. In order to cast (3.1) in the form (3.2) with diagonal matrices α and β, the λ_{ij} are subjected to a condition for $N \geqslant 4$ and they define the so-called *Manakov metric*; in particular for $N = 4$, the condition reads

(3.4) $$\lambda_{23}\lambda_{14}(\lambda_{13}+\lambda_{24}-\lambda_{12}-\lambda_{34}) + \lambda_{13}\lambda_{24}(\lambda_{12}+\lambda_{34}-\lambda_{23}-\lambda_{14}) + \lambda_{12}\lambda_{34}(\lambda_{23}+\lambda_{14}-\lambda_{13}-\lambda_{24}) = 0.$$

For N larger it imposes more and more relations of this nature. From the considerations above, a geodesic flow (3.1) which can be put in the form (3.2) is algebraically completely integrable and it linearizes on the Jacobi variety of the algebraic curve defined by $\det(A+\alpha y-zI) = 0$. For $N = 4$, the geodesic motion (3.1) has two Casimir functions

$$-\frac{1}{2}\operatorname{Tr} A^2 = \sum_{1\leqslant i<j\leqslant 4} X_{ij}^2 = c_1, \quad \sqrt{\det A} = X_{14}X_{23} + X_{13}X_{24} + X_{12}X_{34} = c_2,$$

the symplectic leaves being 4-dimensional and, under the condition (3.4), it has two quadratic invariants : the metric itself (3.3) and another invariant of a similar type

$$\sum_{1\leq i<j\leq 4} \mu_{ij} x_{ij}^2 .$$

Therefore the invariant manifolds and hence the tori will be 2-dimensional. The curve

$$C : P(y,z) = \det(A + \alpha y - zI) = 0$$

has genus 3 and it has a natural involution

$$\tau : (y,z) \rightsquigarrow (-y,-z)$$

due to the skew-symmetry of the matrix A. Therefore the Jacobi variety Jac(C) of C is a 3-dimensional algebraic torus, the involution τ lifts up to Jac(C) and therefore Jac(C) splits up into an even part (an elliptic curve) and an odd part, both being algebraic subtori of Jac(C). The odd part is called the Prym variety, Prym(C,τ), and is two-dimensional. Therefore the geodesic flow (3.2) for N = 4 is algebraically completely integrable; for arbitrary N, the only left-invariant metrics of the form (3.3) for which geodesic flow is a.c.i. are those which allow the formulation (3.2) (for N = 4, see Adler-van Moerbeke [2] and for N $\geq$ 4, see Haine [12]).

Since not every left-invariant metric derives from a diagonal quadratic form (3.3), the question of classifying the metrics for which geodesic flow is a.c.i. is difficult. The problem has been resolved for N = 4 (see Adler-van Moerbeke [4, 7]). The most general left-invariant metric on SO(4) is defined by a quadratic form in 6 variables $x = (x',x'')$ where $x' = (x_1,x_2,x_3)$ and $x'' = (x_4,x_5,x_6)$. In view of the decomposition so(4) = so(3) $\oplus$ so(3) of its Lie algebra, it can be reduced to a quadratic form

$$H = \sum_1^3 (\lambda_i x_i^2 + \lambda_{i+3} x_{i+3}^2) + 2 \sum_{1\leq i,j\leq 3} \lambda_{i,j+3} x_i x_{j+3}$$

depending on 15 quantities. For the sake of simplicity, consider a 9-dimensional family of metrics defined by the quadratic form

$$(3.5) \qquad H = \sum_1^3 (\lambda_i x_i^2 + \lambda_{i+3} x_{i+3}^2 + 2\, \lambda_{i,i+3} x_i x_{i+3}).$$

The geodesic flow on SO(4) for this metric, reduced to its Lie algebra, takes on the form

$$(3.6) \qquad \dot{x}' = x' \times \frac{\partial H}{\partial x'} \qquad \dot{x}'' = x'' \times \frac{\partial H}{\partial x''} \quad \text{with } x', x'' \in \mathbb{R}^3,$$

with invariant $Q_3 \equiv H$ and trivial invariants $Q_1 = \|x'\|$ and $Q_2 = \|x''\|$. The method to detect the left-invariant metrics (3.5) with a.c.i. geodesic flow (3.6) is based on the requirement that the differential equations (3.6) admit a 5-dimensional family of Laurent solutions, if (3.6) is to be a.c.i. Those solutions are shown to have a simple pole

$$(3.7) \qquad x_i(t) = t^{-1}(x_i^o + x_i^1 t + \ldots),$$

with x^j satisfying (2.5) and (2.6) with $g = 1$ (see § 2, iv).

Using Yoshida's theorem, the existence of three quadratic invariants yields a triple eigenvalue for L at $k = 2$. Moreover, the quadratic nature of the differential equations $\dot{x} = f(x)$ yields -1 as an eigenvalue of L, although it does not contribute to a degree of freedom for the Laurent solutions. Therefore, for the differential equations (3.6) to have a 5-parameter family of Laurent solutions (3.7), L must have eigenvalues 0 and 1, besides -1, 2, 2, 2, making use of $\operatorname{Tr} L = 6$. Since L is the Jacobian matrix of the equation (2.5), we conclude that the solution x^o to the system $x^o + f(x^o) = 0$ must be a curve, rather than a discrete set of points, as one would expect a priori. Therefore we are led to searching for conditions on the metric (3.5), such that the 6 non-homogeneous quadratic equations $x^o + f(x^o)$ define a curve. This occurs in exactly three cases :

<u>Metric I</u>. (Manakov metric). When the metric can be written in the diagonal form (3.3) with regard to the custom-

ary so(4) coordinates and with λ_{ij} satisfying condition (3.4) as discussed earlier in this section.

Metric II. The quadratic form (3.5) satisfies the conditions (set $A_{ij} \equiv \lambda_i - \lambda_j$)

$$(\lambda_{14}^2, \lambda_{25}^2, \lambda_{36}^2)(A_{12}A_{46} - A_{13}A_{45})^2$$

$$= A_{21}A_{54}A_{32}A_{65}A_{13}A_{46}\left(\frac{(A_{65}-A_{32})^2}{A_{65}A_{32}}, \frac{(A_{46}-A_{13})^2}{A_{46}A_{13}}, \frac{(A_{54}-A_{21})^2}{A_{54}A_{21}}\right)$$

with the following sign specification for $\lambda_{14}\lambda_{25}\lambda_{36}$:

$$\lambda_{14}\lambda_{25}\lambda_{36}(A_{12}A_{46} - A_{13}A_{45})^3$$

$$= A_{21}A_{54}A_{32}A_{65}A_{13}A_{46}(A_{54} - A_{21})(A_{65} - A_{32})(A_{46} - A_{13}).$$

Metric III. This metric takes on the form

$$(3.8) \quad H = \frac{1}{24}\sum_{i=1}^{3} (3(3c_i + d_i)x_i^2 + (c_i + 3d_i)x_{i+3}^2 + 6(d_i - c_i)x_i x_{i+3})$$

with coefficients

$$c_i = \frac{b_i}{a_i} \quad \text{and} \quad d_i = \frac{b_j - b_k}{a_j - a_k}, \quad (i,j,k) \text{ permutations of } (1,2,3),$$

parametrized by the quantities a and b subjected to

$$a_1 + a_2 + a_3 = 0 \quad \text{and} \quad b_1 + b_2 + b_3 = 0.$$

The geodesic flow (3.6) for this metric has three quadratic invariants, namely the Casimir functions $\|x'\|^2$ and $\|x''\|^2$ and the metric (3.8), and one quartic invariant to be given in § 5. Upon putting

$$3a \equiv -\frac{a_1 - a_2}{a_1 + a_2},$$

into (3.8) and upon substracting from it an appropriate linear combination of $\|x'\|^2$ and $\|x''\|^2$, the metrics defined by (3.8) turn into a one parameter family of metrics

$$(3.9) \qquad H = x_1^2 + x_4^2 a^2 + x_1 x_4 \frac{(1-a)(1+3a)}{2} - x_2 x_5 \frac{(1+a)(1-3a)}{2} - x_3 \frac{(1+a)(1-3a)^3}{16a} - x_6^2 \frac{(1+a)^3(1-3a)}{16a} - x_3 x_6 \frac{(1-a^2)(1-9a^2)}{8a},$$

with $a \in \mathbb{C}$ $(\neq 0, \pm 1, \pm 1/3)$.

Discovered and integrated by Adler and van Moerbeke in 1982 (see [4,5,18]),the geodesic flow for this metric is also associated with a Lax pair, as recently found by Reiman and Semenov-Tian-Shansky [20]. This also will be explained in § 5.

§ 4. The geodesic flow on SO(4) for metric I.

Although this geodesic flow has been integrated in § 3, using the spectral curve, we now sketch the integration of the problem using the Laurent solutions, as carried out in full detail in [6]. After some rescaling and some linear change of coordinates, the geodesic flow X_1 and a flow X_2, commuting with the first one, take on the form

$$X_1 : \dot z_1 = az_5z_6 \qquad X_2 : \dot z_1 = az_2z_3$$
$$\dot z_2 = bz_6z_4 \qquad \dot z_2 = cz_4z_6 + az_1z_3$$
$$\dot z_3 = cz_4z_5 \qquad \dot z_3 = -bz_4z_5 + az_1z_2$$
$$\dot z_4 = -bz_2z_6 + cz_3z_5 \qquad \dot z_4 = -cz_2z_6 - bz_3z_5$$
$$\dot z_5 = -cz_3z_4 + az_1z_6 \qquad \dot z_5 = bz_3z_4$$
$$\dot z_6 = -az_1z_5 + bz_2z_4 \qquad \dot z_6 = cz_2z_4$$

with $a^2 + b^2 + c^2 = 0$, $abc = 1$,

with constants of motion

(4.1)
$$Q_1 = z_2^2 - z_3^2 + z_4^2 = A_1$$
$$Q_2 = z_3^2 - z_1^2 + z_5^2 = A_2$$
$$Q_3 = z_1^2 - z_2^2 + z_6^2 = A_3$$
$$Q_4 = az_1z_4 + bz_2z_5 + cz_3z_6 = A_4/2.$$

The system of differential equations X_1 has one single 5-parameter family of Laurent solution

(4.2)
$$z_1 = -\frac{YZ}{t}[1 + \frac{UX}{YZ}t - \frac{t^2}{6}(A_1(4X^2+b^2-c^2) + A_2(4X^2-a^2) + A_3(4X^2+a^2) + \frac{3A_4X}{2YZ}(Y^2+Z^2)) + O(t^3)]$$

$$z_4 = \frac{aX}{t}[1+\frac{t^2}{6}(2A_1(Z^2-c^2) + A_2(2Z^2-c^2) + A_3(2Y^2+b^2) + \frac{3A_4YZ}{2X}) + O(t^3)]$$

the others being obtained by cyclic permutation

$z_1 \to z_2 \to z_3 \to z_1$, $A_1 \to A_2 \to A_3 \to A_1$, $A_4 \to A_4$, $X \to Y \to Z \to X$

$z_4 \to z_5 \to z_6 \to z_4$, $a \to b \to c \to a$, $U \to U$.

The equations $z^o + f(z^o) = 0$ governing the leading term z^o define an elliptic curve

$$\mathcal{E} : X^2 = Z^2 - b^2, \quad Y^2 = Z^2 + a^2$$

and in order for the Laurent series $z(t)$ to satisfy $Q_i(z(t)) = A_i$, the quantities X, Y, Z and U must be related by

$$D : U^2 + A_1 Y^2 Z^2 + A_2 Z^2 X^2 + A_3 X^2 Y^2 + A_4 XYZ = 0.$$

This defines a curve D of genus 9, a double cover of $\mathcal{E}$ ramified over 16 points. According to § 2 (iii), we must have $\delta_1 \delta_2 = \text{genus}(D) - 1 = 8$ and thus $(\delta_1, \delta_2) = (2,4)$ or $(1,8)$, the latter being excluded on different grounds. Therefore the period matrix reads

$$\begin{pmatrix} 2 & 0 & \alpha & \beta \\ 0 & 4 & \beta & \gamma \end{pmatrix}.$$

The functions of L(D) are spanned by $z_o (=1), z_1, \ldots, z_6$ and one additional function $z_7 \equiv bz_1 z_4 - az_2 z_5$. They themselves do not form a closed system of quadratic differential equations, but the 32 functions spanning L(2D) do, where

$$L(2D) = L(D) \oplus L(D)^{\otimes 2} \oplus \{\dot{z}_1 z_4 - z_1 \dot{z}_4, \; \dot{z}_2 z_5 - z_2 \dot{z}_5\}.$$

Taking the differentials of $1/z_6$ and z_2/z_6 (viewed as functions of t_1 and t_2), using the flows X_1 and X_2 and the Laurent series (4.2) and solving linearly for dt_1 and dt_2, we obtain

$$(4.3) \qquad \omega_1 = dt_1 \Big|_D = \frac{X^3 dX}{UXYZ} \text{ and } \omega_2 = dt_2 \Big|_D = \frac{XdX}{UXYZ}.$$

The holomorphic differentials ω_1 and ω_2 on D are odd with respect to the flip $(X,Y,Z,-U) \sim (X,Y,Z,U)$ interchanging the two sheets of D covering $\mathcal{E}$; thus the problem linearizes on a

Prym variety. Using the differentials (4.3), the solution reads

$$\int_{\mu_1(o)}^{\mu_1(t)} \omega_i + \int_{\mu_2(o)}^{\mu_2(t)} \omega_i = \alpha_i t \qquad i=1,2,\ \alpha_i \in \mathbb{C}$$

for some appropriate coordinates μ_i.

As pointed out in § 2 (v), there are as many four-dimensional families of Laurent solutions as points of tangency of the flow X_1 to D, at least for a smooth divisor D. These points, given by the zeroes of ω_2 are all double and are given by the 8 points covering $Z = \infty$. Therefore, by letting $Z \nearrow \infty$ in the Taylor series

$$\frac{1}{z_1} = -\frac{t}{YZ}\left(1 - \frac{UX}{YZ}t + \frac{1}{3}(A_1 + A_2 + A_3 + O(\tfrac{1}{Z}))Z^2t^2 + \ldots\right)$$

obtained from (4.2), we find the series $1/z_1$ = constant $(\neq 0)t^3 + \ldots$ and thus

$$z_1 = \frac{-3(A_1 + A_2 + A_3)^{-1}}{t^3} + O(\frac{1}{t^2}),$$

z_2 and z_3 behaving similarly, whereas z_4, z_5 and z_6 have double poles, near $t = 0$, at $Z = \infty$. This produces 8 distinct 4-dimensional families of Laurent solutions, depending on the 4 parameters $A_1,\ldots,A_4$.

§ 5. Geodesic flow on SO(4) for metric III.

The linear change of variables

$$\begin{pmatrix} x_1 \\ \\ x_4 \end{pmatrix} = i \begin{pmatrix} a-1 & -1 \\ \\ 3a+1 & 1 \end{pmatrix} \begin{pmatrix} (a-1)z_1 \\ \\ (3a-1)(a+1)z_4 \end{pmatrix}$$

$$\begin{pmatrix} x_2 \\ x_5 \end{pmatrix} = -i \begin{pmatrix} a+1 & -1 \\ 3a-1 & 1 \end{pmatrix} \begin{pmatrix} (a+1)z_2 \\ (3a+1)(a-1)z_5 \end{pmatrix}$$

$$\begin{pmatrix} x_3 \\ x_6 \end{pmatrix} = - \begin{pmatrix} a-1 & a+1 \\ 3a+1 & 3a-1 \end{pmatrix} \begin{pmatrix} (a-1)z_3 \\ (a+1)z_6 \end{pmatrix}$$

maps the geodesic flow (3.6) for metric (3.9) into the system of differential equations

$$(5.1) \quad \dot z_1 = z_3 z_5 \qquad \dot z_4 = \frac{1}{3a-1}(2a z_5 z_6 + (a-1) z_2 z_3)$$

$$\dot z_2 = z_4 z_6 \qquad \dot z_5 = \frac{1}{3a+1}(2a z_3 z_4 + (a+1) z_1 z_6)$$

$$\dot z_3 = \frac{1-a}{2} z_4 z_5 + z_1 z_5 + \frac{1+a}{2} z_1 z_2 \qquad \dot z_6 = \frac{1+a}{2} z_4 z_5 + z_2 z_4 + \frac{1-a}{2} z_1 z_2,$$

with constants of motion

$$Q_1 = aG_2 + \frac{1-a}{1+3a} G_7 \qquad Q_2 = -aG_1 + \frac{1+a}{1-3a} G_8$$

$$Q_3 = \frac{2G_6}{(1-3a)(1+3a)} + \frac{G_1}{1+3a} + \frac{G_2}{1-3a}$$

$$Q_4 = \frac{1}{(3a-1)(3a+1)} [(1-a)(1-3a)(G_1^2+G_4^2) + (1+a)(1+3a)(G_2^2+G_5^2)$$

$$+ 3(1-a^2)(2G_1G_2-G_3^2) + 4(1+a)G_2(G_6+G_8) + 4(1-a)G_1(G_6+G_7)],$$

where

$$G_1 = z_4^2 - z_2 z_5 \qquad G_2 = z_5^2 - z_1 z_4$$

$$G_4 = \frac{-2}{1-3a}(z_2 z_3 - z_5 z_6) \qquad G_5 = \frac{-2}{1+3a}(z_1 z_6 - z_3 z_4)$$

$$G_3 = z_1 z_2 - z_4 z_5$$

$$G_6 = z_1 z_4 + z_2 z_5 - z_3 z_6$$

$$G_7 = z_1^2 - z_3^2 + z_1 z_4 \qquad G_8 = z_2^2 - z_6^2 + z_2 z_5 .$$

The equations (5.1) admit the following 5-parameter family of Laurent solutions

$$(5.2)\quad z(t) = t^{-1}\zeta(\mathbb{1} + Uy^1\, t + (\gamma Z^2+\delta)^{-1}(U^2 y_o^2 + \sum_1^3 A_i y_i^2) t^2 + O(t^3))$$

where

$$\zeta = \operatorname{diag}\left(\frac{Y^2}{Z}, \frac{Z^2}{Y}, -\frac{Y}{Z}, Z, Y, -\frac{Z}{Y}\right) \text{ with } Y,Z \in \mathbb{C} \text{ such that } Y^2 + Z^2 = 1$$

(5.3)

$$\mathbb{1} = (1,1,\ldots,1)^T$$

y^1, y_o^2, y_i^2 are appropriate vectors depending on Y,Z and a only,

$\gamma \equiv 4a$ and $\delta \equiv (a-1)(3a+1)$.

The 5-dimensional family of Laurent solutions depend on the parameters Z,U,A_1,A_2 and A_3. The vectors y_i^2 can be chosen such that $Q_i(z(t)) = A_i$ for $i = 1,2,3$. As explained in § 2, (iii), the fact that $z(t)$ above must satisfy $Q_4(z(t)) = A_4$ yields a relation between the free parameters, defining a curve

$$D : (U,V,Y,Z) \text{ such that } Z^2 = V,\ Y^2 = 1 - V \text{ and}$$

$$P(U,V) \equiv (U^2(1-V)V(\alpha V+\beta))^2 - 2U^2(1-V)V\, P_3(V) + P_4(V) = 0,$$

where

$$\alpha V + \beta = 16a^3 V + (a-1)^3(3a+1),$$

and where P_3 and P_4 are respectively cubic and quartic polynomials in V, with coefficients depending on $A_1,\ldots,A_4$ and a, having the property

$$(5.4)\qquad P_3^2(V) - (\alpha V+\beta)^2\, P_4(V) = V(1-V)R_3(V),$$

$R_3(V)$ being a cubic polynomial. The curve D is an unramified

4-1 cover of the curve

$$D_o : P(U,V) = 0;$$

D_o itself is, in view of (5.4), a double cover of the hyper-elliptic curve

$$\mathcal{H} : W^2 = V(1-V)R_3(V)$$

of genus 2, ramified over 4 points where $P_4(V) = 0$. Therefore D_o has genus 5 and D has genus 17. The curve D, wrapped around T_A^2 intersects itself transversally in 8 points, adding 8 to the genus 17. Therefore the torus T^2, on which this flow linearizes, is defined by a period lattice Λ generated bu the 4 columns of the period matrix Π defined in § 2, iii, where $\delta_1\delta_2$ = genus(D) - 1 = 24; thus $(\delta_1,\delta_2) = (2,12)$ or (1,24), the latter being excluded on different grounds, and the geodesic flow (run with complex time) evolves on tori having period matrix

$$\begin{pmatrix} 2 & 0 & \alpha & \beta \\ 0 & 12 & \beta & \gamma \end{pmatrix}.$$

Differentiating $1/z_1$ and z_2/z_1 with respect to t_1 (corresponding to the flow (5.1)) and t_2 (corresponding to a quartic flow generated by Q_4), yields two differentials ω_1 and ω_2 defined on the curves D_o above and having the form

$$\omega_1 = \frac{\varphi(V)\,dV}{U\sqrt{V(1-V)R_3(V)}} \quad \text{and} \quad \omega_2 = \frac{dV}{U\sqrt{V(1-V)R_3(V)}},$$

where (for notations, see (5.3))

$$\varphi(V) \equiv \frac{\gamma V+\delta}{V(1-V)}\left((\alpha V+\beta)U^2V(1-V) - P_2(V)\right)$$

and

$$P_2(V) \equiv (-A_3(3a^2+1)+A_1+A_2)(1-V)V+A_1V+A_2(1-V).$$

Consequently, the quadratures (2.9) linearizing the flow can be expressed in terms of the differentials ω_1 and ω_2 above. Put in a more geometrical language, the linearization takes place on a two-dimensional subtorus of the three-dimensional Prym variety, Prym$(D_o/\mathcal{H})$; for details, see [5].

Finally we show that the geodesic flow (3.6) for the metric (3.8) corresponds to a Lax pair, constructed as follows : consider the group SO(4,3) of transformations of $\mathbb{R}^7$ conserving the quadratic form

$$\sum_{i=1}^{4} x_i^2 - \sum_{i=5}^{7} x_i^2.$$

The 14-dimensional simple Lie algebra G_2 can be viewed as a subalgebra of so(4,3), having the following parametrization

$$X(x',x",y,z,a) \equiv$$

$$\begin{pmatrix}
0 & -\frac{x_3+x_6/3}{2} & \frac{x_2+x_5/3}{2} & -\frac{x_1-x_4/3}{2} & -y_2 & y_1 & a_1 \\
\frac{x_3+x_6/3}{2} & 0 & -\frac{x_1+x_4/3}{2} & -\frac{x_2-x_5/3}{2} & y_3 & a_2 & z_1 \\
-\frac{x_2+x_5/3}{2} & \frac{x_1+x_4/3}{2} & 0 & -\frac{x_3-x_6/3}{2} & a_3 & z_3 & -z_2 \\
\frac{x_1-x_4/3}{2} & \frac{x_2-x_5/3}{2} & \frac{x_3-x_6/3}{2} & 0 & y_3-z_3 & y_2-z_2 & y_1-z_1 \\
-y_2 & y_3 & a_3 & y_3-z_3 & 0 & x_4/3 & -x_5/3 \\
y_1 & a_2 & z_3 & y_2-z_2 & -x_4/3 & 0 & x_6/3 \\
a_1 & z_1 & -z_2 & y_1-z_1 & x_5/3 & -x_6/3 & 0
\end{pmatrix}$$

with $a_1+a_2+a_3 = 0$. The fixed subalgebra of $G_2 \subset so(4,3)$ for the involution $X \rightsquigarrow -X^T$ is given by the matrices of the form $X(x',x'',0,0,0)$ and is isomorphic to the Lie algebra so(4). Thus we have $so(4) \subset G_2 \subset so(4,3)$. Letting

$A=X(x',x'',0,0,0)$, $B=X(u',u'',0,0,0)$, $\alpha=X(0,0,0,0,a)$, $\beta=X(0,0,0,0,b)$,

consider

$$(5.5) \qquad (A+\alpha y)^{\cdot} = [A+\alpha y, B+\beta y], \quad y \text{ dummy variable,}$$

where u' and u" are defined such that

$$[A,\beta] + [\alpha,B] = 0.$$

This relation is equivalent to

$$u_i + u_{i+3} = d_i(x_i + x_{i+3}) \text{ and } u_i - u_{i+3}/3 = c_i(x_i - x_{i+3}/3) \qquad 1 \leqslant i \leqslant 3$$

for

$$c_i = \frac{b_i}{a_i} \quad \text{and} \quad d_i = \frac{b_j-b_k}{a_j-a_k}, \qquad (i,j,k) \text{ permutation of } (1,2,3).$$

The flow (5.5) is precisely the geodesic flow (3.6) for the metric (3.8). Then the linearization takes place on the Jacobian of the spectral curve

$$C : P(y,z) = \det(A+\alpha y - zI) = 0.$$

In this instance, as in most other examples, the quadratures, obtained from the spectral curves, are different from the ones, obtained from the Laurent solutions method. The tori obtained are different as well, but one can be obtained from the other by multiplying some periods by an integer and leaving others unchanged. Two such tori are called *isogenous*. The relationship between the curves D and C is quite intricate.

References

[1] Adler, M. & van Moerbeke, P.; 1) Completely integrale systems, Euclidean Lie algebras and curves, 2) Linearization of Hamiltonian systems, Jacobi varieties and Representation theory; Adv. in Math., 38, pp. 267-317, 318-379, (1980).

[2] Adler, M. & van Moerbeke, P.; The algebraic integrability of geodesic flow on SO(4), Invent. Math., 67, pp. 297-326, (1982) with an appendix by D. Mumford.

[3] Adler, M. & van Moerbeke, P.; Kowalewski's asymptotic method, Kac-Moody Lie algebras and regularization, Comm. Math. Phys., 83, pp. 83-106, (1982).

[4] Adler, M. & van Moerbeke, P.; Geodesic flow on SO(4) and the intersection of quadrics, Proc. Natl. Acad. Sci. USA, 81, pp. 4613-4616, (1984).

[5] Adler, M. & van Moerbeke, P.; A new integrable geodesic flow on SO(4). Probability, Statistical mechanics and number theory, Adv. in Math. Suppl. Studies, vol. 9, (1986).

[6] Adler, M. & van Moerbeke, P.; A systematic Approach towards solving Integrable Systems, Perspectives in Mathematics, Academic Press (to appear in 1987).

[7] Adler, M. & van Moerbeke, P.; A full classification of algebraically completely integrable geodesic flows on SO(4), (to appear in 1986).

[8] Arnold, V.I.; Mathematical methods of classical mechanics, Springer-Verlag, New York-Heidelberg-Berlin, (1978).

[9] Griffiths, P.A.; Linearizing flows and a cohomological interpretation of Lax equations, Amer. J. of Math., 107, pp. 1445-1483, (1985).

[10] Griffiths, P. & Harris, J.; Principles of algebraic geometry. New York : Wiley-Interscience, (1978).

[11] Haine, L.; Geodesic flow on SO(4) and Abelian surfaces, Math. Ann., 263, pp. 435-472, (1983).

[12] Haine, L.; The algebraic complete integrability of geodesic flow on SO(N), Comm. Math. Phys., 94, pp. 271-287, (1984).

[13] Kowalewski, S.; Sur le problème de la rotation d'un corps solide autour d'un point fixe, Acta Math., 12, pp. 177-232, (1889).

[14] Kowalewski, S.; Sur une propriété du système d'équations différentielles qui définit la rotation d'un corps solide autour d'un point fixe, Acta Math., 14, pp. 81-93, (1889).

[15] Kozlov, V.V.; Integrability and non-integrability in Hamiltonian mechanics, Uspekhi Mat. Nauk, 38 : 1, pp. 3-67, (1983), (Transl.:Russian Math. Surveys, 38 : 1, pp. 1-76, (1983)).

[16] van Moerbeke, P.; The complete integrability of Hamiltonian system, Proceedings of the EQUADIFF conference, Würzburg (August 1982), Springer-Verlag Lecture Notes, 1017, pp. 462-475.

[17] van Moerbeke, P.; Algebraic geometrical methods in Hamiltonian mechanics, Phil. Trans. Royal Society London,

A315, pp. 379-390, (1985), (Royal Society Meeting, Nov. 1984).

[18] van Moerbeke, P.; Algebraic complete integrability of Hamiltonian systems and Kac-Moody Lie algebras, Proc. Int. Congr. of Math., Warszawa, August 1983.

[19] Perelomov, A.M.; Some remarks on the integrability of the equations of motion of a rigid body in an ideal fluid, Funct. Anal. Appl., 15, pp. 83-85, (1981), transl. 144-146.

[20] Reiman A., Semenov-Tian-Shansky, M.; A new integrable case of the motion of the 4-dimensional rigid body, Comm. Math. Phys., 105, pp. 461-472, (1986).

[21] Yoshida, H.; Necessary conditions for the existence of algebraic first integrals. I. Kowalewski's Exponents. II. Conditions for algebraic integrability, Celestial Mech., 31, pp. 363-379, 381-399, (1983).

ON SOME DYNAMICAL ASPECTS OF PARABOLIC EQUATIONS WITH VARIABLE DOMAIN

José M. Vegas
Departamento de Ecuaciones Funcionales
Facultad de Ciencias Matemáticas
Universidad Complutense
28040-Madrid. Spain

1. INTRODUCTION

We consider the scalar parabolic equation

$$(1)_\varepsilon \quad \begin{cases} u_t = \Delta u - ku + b(x,u) & \text{in } D_\varepsilon \\ \dfrac{\partial u}{\partial n} = 0 & \text{on } \partial D_\varepsilon \end{cases}$$

where $k \geq 0$ is fixed, $b: R^n \times R \to R$ is smooth, for some constant M we have $|b(x,u)| < M$, $|b_u(x,u)| < M$ for all $(x,u) \in R^n \times R$, and $\{D_\varepsilon\}$ is a family of open smooth domains in R^n such that for $0 \leq \varepsilon \leq \varepsilon' \leq 1$, D_ε is contained in $D_{\varepsilon'}$, and $|D_\varepsilon - D_{\varepsilon'}| \to 0$ as $\varepsilon \to \varepsilon'^+$, where $|.|$ denotes the Lebesgue measure in R^n. We assume also that each D_ε is connected and there is a ball B in R^n which contains every D_ε.

For each ε in $[0,1]$, $(1)_\varepsilon$ defines a dissipative gradient semiflow on $H^1(D_\varepsilon)$. If $\{\partial D_\varepsilon\}$ vary continuously with ε, $(1)_\varepsilon$ can be studied by regular perturbation methods (see, e.g., Garabedian and Schiffer [4]). For some special families of domains which are continuous in ε in a weaker sense, this problem has been studied by Hale and Vegas [8], Matano and Mimura [10], Vegas [12,13,14] and others. The perturbation now becomes singular, but it still can be analyzed by the methods of bifurcation theory, because the first eigenvalues and eigenfunctions of the Laplacian on D_ε with Neumann boundary conditions are continuous in ε, so the Lyapunov-Schmidt reduction can be applied.

When the only condition we impose on D_ε is their continuity in measure, the problem becomes much harder, since there is no control on the eigenvalues and eigenfunctions of Δ on D_ε (remember that we are dealing with the Neumann problem). In 14 we have applied a variational technique in order to analyze the continuation of equilibrium points of $(1)_0$ to $(1)_\varepsilon$ for ε small, concluding that, if u_0 is either a stable equilibrium or a hyperbolic saddle point of $(1)_0$, then, if ε is sufficiently small, $(1)_\varepsilon$ has an equilibrium point (not necessarily unique) which is close to u_0.

NATO ASI Series, Vol. F37
Dynamics of Infinite Dimensional Systems
Edited by S.-N. Chow, and J. K. Hale

In this paper we study the semigroup associated to $(1)_\varepsilon$, which is defined on $L^2(D_\varepsilon)$, by "embedding" it in a semigroup defined on $L^2(B)$ which depends continuously on ε, and whose dynamical properties mimic those of equation $(1)_\varepsilon$. However, this continuity in ε cannot be strengthened to C^1, and this is why we cannot apply standard differential techniques to the continuation problem. Fortunately, this dependence in ε turns out to be sufficient to guarantee some collective compactness of these semigroups, which implies the upper semicontinuity of the attractors associated to them.

2. CONSTRUCTION OF THE SEMIGROUP. ITS CONTINUITY IN ε.

For each ε we denote $R_\varepsilon = D_\varepsilon - D_0$, and $E_\varepsilon = B - D_\varepsilon$, so B can be written as the union of D_0, R_ε and E_ε, and $|R_\varepsilon| \to 0$ as $\varepsilon \to 0$.

Let $D(A_\varepsilon) = \{u \in L^2(B)$: $u|D_\varepsilon \in H^2(D_\varepsilon)$, $u|E_\varepsilon \in H^2(E_\varepsilon)$,

$$\frac{\partial u}{\partial n} = 0 \text{ on } \partial D_\varepsilon = \partial E_\varepsilon \text{ and on } \partial B\}.$$

Let $A_\varepsilon: D(A_\varepsilon) \to L^2(B)$ be defined as $\Delta u - ku$ both on D_ε and on E_ε.

<u>Remark</u>: If for an arbitrary smooth domain D we define

$H^2_N(D) = \{u \in L^2(D)$: $u \in H^2(D)$, $\frac{\partial u}{\partial n} = 0$ on $\partial D\}$, and

$A(D)$: $H^2_N(D) \to L^2(D)$ is given by $A(D)u = \Delta u - ku$, then we may write

$L^2(B) = L^2(D_\varepsilon) \oplus L^2(E_\varepsilon)$; $D(A_\varepsilon) = H^2_N(D_\varepsilon) \oplus H^2_N(E_\varepsilon)$, and

$A = A(D_\varepsilon) \oplus A(E_\varepsilon)$.

That is, there is no condition on the interface ∂D_ε for the fonctions of the domain of A_ε. Thus, it is clear that A_ε <u>generates a semigroup</u> $T_\varepsilon(t)$ on $L^2(B)$: given $u_0 \in L^2(B)$, $T_\varepsilon(t)u$ is just the solution of $u_t = \Delta u - ku$ on D_ε and on E_ε <u>independently</u>, with initial value u_0 (and Neumann boundary conditions on each subdomain).

<u>THEOREM 1</u>. $T_\varepsilon(t)u$ <u>is jointly continuous in</u> (ε,t,u).

<u>Proof</u>: It is a consequence of the Trotter-Kato convergence theorem. Let $f \in L^2(B)$ be given, and let $\varepsilon_n \to \varepsilon_0$ (which we take to be 0, for simplicity). We denote A_{ε_n}, R_{ε_n}, E_{ε_n} by A_n, R_n and E_n, respectively. All we have to show, then, is that $A_n^{-1}f \to A_0^{-1}f$ strongly in $L^2(B)$.

Call $u_n = A_n^{-1}f|D_n$, $v_n = A_n^{-1}f|E_n$. Then, for all $g \in H^1(B)$ we have

$$\int_{D_n} (\nabla u_n . \nabla g + k u_n g) dx = \int_{D_n} f g dx \tag{2.1}$$

$$\int_{E_n} (\nabla v_n . \nabla g + k v_n g) dx = \int_{E_n} f g dx \tag{2.2}$$

Also, u_n minimizes $J_n^{(1)}$ on $H^1(D_n)$, and v_n minimizes $J_n^{(2)}$ on $H^1(E_n)$, where

$$J_n^{(1)}(u) = \int_{D_n} (|\nabla u|^2 + k u^2 - 2fu) dx , \quad u \in H^1(D_n) \tag{2.3}$$

$$J_n^{(2)}(v) = \int_{E_n} (|\nabla v|^2 + k v^2 - 2fv) dx \quad v \in H^1(E_n) \tag{2.4}$$

There exist u_0 in $H^1(D_0)$ and a subsequence of u_n (still labeled as u_n) which converges weakly to u_0 in $H^1(D_0)$ and strongly in $L^2(D_0)$. By letting $n \to \infty$ in (2.1), it is easy to see that u_0 must be precisely $A_0^{-1} f | D_0$, and then the whole sequence converges to u_0.

On the other hand, if U_0 is any extension in H^1 of u_0, we have

$$\int_{D_0} (|\nabla u_n|^2 + k u_n^2 - 2 f u_n) dx + \int_{R_n} (|\nabla u_n|^2 + k u_n^2 - 2 f u_n) dx$$

$$\leq \int_{D_0} (|\nabla u_0|^2 + k u_0^2 - 2 f u_0) dx + \int_{R_n} (|\nabla U_0|^2 + k U_0^2 - 2 f U_0) dx$$

$$\leq \int_{D_0} (|\nabla u_n|^2 + k u_n^2 - 2 f u_n) dx + \int_{R_n} (|\nabla U_0|^2 + k U_0^2 - 2 f U_0) dx,$$

and these inequalities imply that u_n converges to u_0 strongly in $H^1(D_0)$ and $\int_{R_n} u_n^2 dx \to 0$ as $n \to \infty$.

Let us now consider E_1. There exists a subsequence of v_n which converges weakly in H^1 and strongly in L^2 to some $z_1 \in H^1(E_1)$. From this we can select a further subsequence which converges (in the same senses) to some $z_2 \in H^1(E_2)$, which must coincide with z_1 on E_1. Proceeding in this way and applying a diagonal selection, we find a subsequence of v_n and a function z on E_0 to which it converges weakly in $H^1(E_j)$ and strongly in $L^2(E_j)$ for each j fixed. Since $\|v_n\|_{H^1(E_j)}$ is bounded independently of n and j, we see that z is in $H^1(E_0)$ and $\|v_n - z\|_{L^2(E_j)} \to 0$ as $n \to \infty$ for every fixed j. This, together with the fact that $\|u_n\|_{L^2(R_n)} \to 0$ imply that $A_n^{-1} f\, E_0 \to z$ strongly in L^2. Again, it is easily seen that z must be $A_0^{-1} f | E_0$, and then the whole sequence converges to z. This finishes the proof.

Let us now turn to the nonlinear equation $(1)_\varepsilon$.

THEOREM 2. *Let $S_\varepsilon(t)$ denote the nonlinear semigroup associated to $(1)_\varepsilon$ in the following way: For $u \in L^2(B)$, $S_\varepsilon(t)u|D_\varepsilon$ is the solution of $(1)_\varepsilon$ with initial value $u|D_\varepsilon$, while $S_\varepsilon(t)u|E_\varepsilon$ is just $T_\varepsilon(t)u|E_\varepsilon$. Then, $S_\varepsilon(t)$ is globally Lipschitz with constant at most $e^{(M-k)t}$, and $S_\varepsilon(t)u$ is jointly continuous in the three variables* (ε, t, u).

Proof: The first part follows immediately from the variation of constants formula. The second part can also be proven by this method, but it is more direct to apply the nonlinear version of the Trotter-Kato Theorem for operators of the form "monotone - kI" (see Brézis-Pazy [2]).

3. UPPER SEMICONTINUITY OF THE ATTRACTOR

The nonlinear semigroup $S_\varepsilon(t)$ has the following properties:

1) There exists a bounded set in $L^2(B)$ which is independent of ε and attracts compact sets of $L^2(B)$ for each ε.

This can be shown by regarding $S_\varepsilon(t)$ as $S_\varepsilon(t)|L^2(D_\varepsilon) \oplus T_\varepsilon(t)|L^2(E_\varepsilon)$; standard a priori estimates imply that $\|S_\varepsilon(t)u\|_{L^2(D_\varepsilon)}$ decreases in t if $\|u\|_{L^2(D_\varepsilon)} < \frac{M}{k}$, while $\|T_\varepsilon(t)u\|_{L^2(D_\varepsilon)}$ always decreases. Thus, the M/k-ball centered at zero in $L^2(B)$ satisfies the required properties.

2) For each ε, $S_\varepsilon(t)$ is a compact operator for $t > 0$.

This is a standard result from the theory of parabolic equations on bounded domains.

These propeties imply (see, e.g., Billotti-LaSalle [1], Hale [5,6,7] or Massatt [9]) the existence of a compact set $\hat{K}_\varepsilon$ (the "attractor"), which consists of the union of all bounded solutions defined on $(-\infty, \infty)$, is uniformly asymptotically stable and attracts all compact sets of $L^2(B)$. $\hat{K}_\varepsilon$ carries all the asymptotic information concerning $(1)_\varepsilon$, and it would be interesting to know whether there is some "continuity" of $\hat{K}_\varepsilon$ with respect to ε. Now, even if $(1)_0$ is Morse-Smale, the type of continuity of $S_\varepsilon(t)$ in ε involves a topology in the space of compact maps of $L^2(B)$ in itself for which such systems are not structurally stable (the C^0 topology); hence, the theory of Morse-Smale systems cannot be directly applied to our situation. However, the family $\{S_\varepsilon(t)\}$ enjoy an additional property which will give us a partial result.

There is a result of Cooperman [3] (see also Hale [5]) which states that, if a semigroup which depends continuously on a parameter ε has a bounded set independent of ε which attracts all compact sets, and the semigroup is "collectively contracting" (in the sense of Theorem 3 below), then the attractor is upper semicontinuous in ε. In our case, the first property is satisfied. Let us see that the second holds too. First of all, we need a lemma:

LEMMA: Let $u_n \in L^\infty(B)$ be a sequence of functions such that there exist a constant M and a sequence $\varepsilon_n \to 0$ which satisfy:

1) $u_n|D_{\varepsilon_n} \in H^1(D_{\varepsilon_n})$, $u_n|E_{\varepsilon_n} \in H^1(E_{\varepsilon_n})$ and

$$\|u_n\|_{H^1(D_{\varepsilon_n})} \leq M, \quad \|u_n\|_{H^1(E_{\varepsilon_n})} \leq M \quad \text{for all } n.$$

2) $\|u_n\|_{L^\infty(B)} \leq M$ for all n.

Then u_n has a subsequence which converges strongly in $L^2(B)$.

Proof: It follows exactly the same steps as the proof of the second part of Theorem 1, with the only remark that the crucial fact $\int_{R_n} u_n^2 dx \to 0$ is now guaranteed by the uniform L^∞ bound.

COROLLARY: Let $b_\varepsilon : L^2(B) \to L^2(B)$ denote the map $u \to b(x,u)\chi_{D_\varepsilon}$. Then the family of operators

$$U_\varepsilon(t)u := \int_0^t T_\varepsilon(t-s)b_\varepsilon(S_\varepsilon(s)u)ds = \int_0^t T_\varepsilon(s)b_\varepsilon(S_\varepsilon(t-s)u)ds$$

is collectively completely continuous; that is, for each bounded set V in $L^2(B)$, the set $\bigcup\{U_\varepsilon(t)V: 0 \leq \varepsilon \leq 1\}$ is precompact.

Proof: It is well known that

$$\|T_\varepsilon(s)b_\varepsilon(S_\varepsilon(t-s)u)\|_{H^1(D_\varepsilon)} \le s^{-1/2}e^{-ks}\|b_\varepsilon(S_\varepsilon(t-s)u\|_{L^2(D_\varepsilon)}$$
$$\le s^{-1/2}e^{-ks}M|D_\varepsilon|, \text{ and}$$
$$\|T_\varepsilon(s)b_\varepsilon(S_\varepsilon(t-s)u)\|_{L^\infty(B)} \le M.$$

Hence, there is a uniform upper bound (independent of ε) for $\|U_\varepsilon(t)u\|_{H^1(D_\varepsilon)}$, $\|U_\varepsilon(t)u\|_{H^1(E_\varepsilon)}$ and $\|U_\varepsilon(t)u\|_{L^\infty(B)}$, and the Lemma gives us the result.

THEOREM 3. The family of semigroups $S_\varepsilon(t)$ is collectively α-contracting; that is, for each V bounded in $L^2(B)$,

$$\alpha(\{S_\varepsilon(t)v:\ 0 \le \varepsilon \le 1,\ v\in V\}) \le e^{-kt}\,\alpha(V),$$

where α is the Kuratowski measure of noncompactness.

Proof: $S_\varepsilon(t) = T_\varepsilon(t) + U_\varepsilon(t)$, $T_\varepsilon(t)$ is a linear contraction of norm e^{-kt} and $U_\varepsilon(t)$ is collectively completely continuous.

Now we can apply Cooperman's result, obtaining

THEOREM 4. The family of attractors $\hat{K}_\varepsilon$ is upper semicontinuous; that is, for every neighborhood of $\hat{K}_0$ there exists ε_0 such that $\hat{K}_\varepsilon$ is contained in that neighborhood for $0 < \varepsilon < \varepsilon_0$.

Finally, it is clear that K_ε, the attractor of $(1)_\varepsilon$ in $L^2(D_\varepsilon)$, satisfies: $\hat{K}_\varepsilon = K_\varepsilon \oplus 0$; in other words, $\hat{K}_\varepsilon$ is just the extension by zero of K_ε outside D_ε. Therefore, Theorem 4 is also a theorem on upper semicontinuity of K_ε in a certain sense (the spaces $L^2(D_\varepsilon)$ or $H^1(D_\varepsilon)$ change with ε).

4. REMARK: APPLICABILITY OF THE HOMOTOPY INDEX

In this last section we will only mention that Theorems 2 and 4 enable us to apply the Conley index as generalized by Rybakowski [11] to noncompact phase spaces, giving a new proof of the continuation of hyperbolic equilibrium solutions of $(1)_0$ to nearby invariant sets (and, hence, equilibria) for $(1)_\varepsilon$.

REFERENCES

1. BILLOTTI, J. and LASALLE, J.P.: Periodic dissipative systems. Bull. Am. Math. Soc., 6 (1971) 1082-1089.

2. BREZIS, H. and PAZY, A.: Convergence and approximation of semigroups of nonlinear operators in Banach spaces. J. Funct. Anal., 9 (1972) 63-74.

3. COOPERMAN, G.D.: α-condensing maps and dissipative systems. Ph. D. Thesis, Brown University, June 1978.

4. GARABEDIAN, P.R. and SCHIFFER, M.: Convexity of domain functionals. J. d'Analyse Mathématique, 2 (1952-53) 281-368.

5. HALE, J.K.: Some recent results on dissipative processes. Lecture Notes in Math., 799, Springer-Verlag, New York, 1979. pp 152-172.

6. HALE, J.K.: Asymptotic behavior and dynamics in infinite dimensions, in Nonlinear Differential Equations, Research Notes in Math., 132, Pitman, 1985, pp 1-42.

7. HALE, J.K., MAGALHAES, L. and OLIVA, W.: An introduction to infinite dimensional systems: geometric theory. Springer-Verlag, 1984.

8. HALE, J.K. and VEGAS, J.: A nonlinear parabolic equation with varying domain. Arch. Rat. Mech. Anal., 86 (1984) 99-123.

9. MASSATT, P.: Attractivity properties of α-contractions. J. Diff. Eq., 48 (1983) 326-333.

10. MATANO, H. and MIMURA, H.: Pattern formation in competition-diffusion systems in nonconvex domains. Publ. RIMS, Kyoto Univ., 19 (1983), 1049-1079.

11. RYBAKOWSKI, K.: On the homotopy index for infinite-dimensional semiflows. Trans. Am. Math. Soc., 269 (1982), 351-383.

12. VEGAS, J.: Bifurcations caused by perturbing the domain in an elliptic equation. J. Diff. Eq., 48(2) (1983) 189-226.

13. VEGAS, J.: A Neumann elliptic problem with variable domain. Contributions to Nonlinear PDE, Research Notes in Math., 89, Pitman, 1983. pp 264-273.

14. VEGAS, J.: Irregular perturbations of the domain in elliptic problems with Neumann boundary conditions. To appear in Contributions to Nonlinear PDE, vol II. Pitman-Longman.

Acknowledgement: This work has been financed by a grant of the Comisión Asesora para la Investigación Científica y Técnica of Spain, CAICYT 3308683.

BIFURCATION FROM HOMOCLINIC TO PERIODIC SOLUTIONS BY AN INCLINATION LEMMA WITH POINTWISE ESTIMATE

Hans-Otto Walther

Mathematisches Institut
Universität München

D 8000 München 2, F.R.G.

Bifurcation from homoclinic to periodic orbits in two dimensions has been known for a long time [1,4]. L.P. Šil'nikov [8] obtained the first result for arbitrary finite dimension. His idea was to consider a point on the homoclinic trajectory as fixed point of a suitably constructed map so that continuation by the implicit function theorem yields fixed points which define periodic solutions. The difficulty involved is to show smoothness of Šil'nikov's map. This requires a careful investigation of trajectories close to a hyperbolic equilibrium. The underlying vectorfields have to be at least C^2-smooth [7].

In [9] we proved a result in infinite dimension: For the functional differential equation

$$\dot{x}(t) = af(x(t-1))$$

with periodic nonlinearity $f:\mathbb{R} \to \mathbb{R}$ in a certain class of functions, there exists a critical parameter $a = a_0$ with a heteroclinic solution, and for $a > a_0$, periodic solutions of the second kind bifurcate off. This was done without recourse to Šil'nikov's idea, and required only C^1-smoothness of f, but questions of uniqueness and stability for the bifurcating solutions remained unanswered.

Assuming more smoothness, M. Blazquez [2], S.N. Chow and B. Deng [3] and the author [11] recently obtained results for parabolic [2,3] and functional [3,11] differential equations which include uniqueness and stability. All these proofs employ modifications of Šil'nikov's map, but the crucial parts - how to derive smoothness - are different.

In [11] we tried to give a conceptually relatively simple proof of smoothness. The key is a sharpened inclination lemma,

NATO ASI Series, Vol. F37
Dynamics of Infinite Dimensional Systems
Edited by S.-N. Chow, and J. K. Hale

or λ-lemma, for C^2-maps in Banach spaces. Before being more explicit, let us describe the general idea how to get bifurcation from homoclinic to periodic solutions using a map of Šil'nikov type.

Let $X:\Omega \to B$ be a parameterized local semiflow in a Banach space B, $\Omega \subset [0,\infty) \times B \times A$ with open parameter interval $A \ni 0$. Assume that X has continuous partial derivatives with respect to B and A, that (1) $X(t,0,a) = 0$ for all $(t,a) \in [0,\infty) \times A$, and that (2) at the critical parameter $a = 0$ we have a homoclinic trajectory $x^0:\mathbb{R} \to B$, i.e. $x^0(t) = X(t-s,x^0(s),0)$ for all real $t \geqq s$, $x^0(0) \neq 0$, $x^0(t) \to 0$ as $|t| \to +\infty$.

The linearized semiflow $T:[0,\infty) \times B \times A \to B$ given by $T(t,x,a) = D_2X(t,0,a)x$ is a family of strongly continuous semigroups of continuous linear operators $T(t,\cdot,a)$. Suppose (3) that the spectra σ_a of the generators contain precisely one positive, simple eigenvalue u_a, and that there are constants $\lambda_1 < \lambda < 0 < \mu$ with $\lambda < -\mu$ and with $\mathrm{Re}\, z < \lambda_1 < 0 < u_a < \mu$ for all $z \in \sigma_a \setminus \{u_a\}$, $a \in A$. Let Φ_a denote a continuously differentiable family of unit eigenvectors of the eigenvalues u_a. By (3),

$$B = P_a \oplus Q_a$$

with T-invariant stable and unstable subspaces Q_a and $P_a = \mathbb{R}\Phi_a$. Write p_a and q_a for the associated projections onto P_a and Q_a.

In order to avoid a coordinate transformation here, we simply assume that X is locally normalized, i.e. that (4) there is an open neighborhood $\mathcal{B}$ of $0 \in B$ such that for each $a \in A$, the local stable manifold of the nonlinear semiflow $X(\cdot,\cdot,a)$ at $0 \in B$ coincides with $\mathcal{B} \cap Q_a$, and that the local unstable manifold at $0 \in B$ is $\mathcal{B} \cap P_a$.

Construction of a return map: Suppose $x^0(0) = r\Phi_0 \in \mathcal{B} \cap P_0$ for some $r > 0$. Consider the transversals $H_a := r\Phi_a + Q_a$ to P_a, and small parallelograms

$$E_a := E(\eta,\delta,a) := \{x \in B: |p_a x| < \eta, |q_a x| < \delta\} \subset \mathcal{B},$$

such that $E_a \cap H_a = \emptyset$, for all $a \in A$. Trajectories which start in open upper halves

$$E_a^+ := \{x \in B: p_a x \in (0,\eta)\cdot\Phi_a, |q_a x| < \delta\} \subset E_a$$

reach the transversal H_a at some first $t = \sigma(x,a) > 0$ (provided r,η,δ,A are sufficiently small). Set $\Sigma(x,a) := X(\sigma(x,a),x,a) \in$

H_a, for all $x \in E_a^+$, $a \in A$. There exists $\theta > 0$ such that the homoclinic trajectory (which starts at $r\Phi_0 \in H_0$) satisfies $x^0(t) \in E_0 \cap Q_0$ for all $t \geq \theta$. By continuity, there are open neighborhoods $\tilde{\mathcal{B}}$ of $r\Phi_0$ and $\tilde{A}$ of $a = 0$ with $X(\theta,x,a) \in E_a$ on $\tilde{\mathcal{B}} \times \tilde{A}$.

Observe that trajectories in $E_a \cap Q_a$ converge to 0 and never reach H_a (if η,δ are small enough), and that trajectories which start in $E_a \setminus E_a^+$ may or may not intersect with H_a: Following trajectories which start in $\tilde{\mathcal{B}}$ does not lead to a map which has $r\Phi_0$ as a fixed point. - According to Šil'nikov, we continue the maps $\Sigma(\cdot,a):E_a^+ \to H_a$ to maps $\check{\Sigma}(\cdot,a)$ on all of E_a by $\check{\Sigma}(x,a) := r\Phi_a$ for $p_a x \in (-\eta,0]\cdot\Phi_a$. Now the composite map

$$\check{S}:(x,a) \to \check{\Sigma}(X(\theta,x,a),a)$$

is defined on $\tilde{\mathcal{B}} \times \tilde{A}$, $\check{S}(x,a) \in H_a$ on $\tilde{\mathcal{B}} \times \tilde{A}$, and $x^0(0) = r\Phi_0$ is a fixed point of $\check{S}(\cdot,0)$.

Furthermore we see that if $\check{\Sigma}$ were C^1-smooth, then $D_1\check{\Sigma}(x,a) = 0$ for $p_a x \in (-\eta,0]\cdot\Phi_a$, hence $D_1\check{S}(r\Phi_0,0) = 0$, and the implicit function theorem would imply existence of a C^1-curve $a \to x_a^*$ of stable and attractive fixed points of $\check{S}(\cdot,a)$ passing through $r\Phi_0$ at $a = 0$. If in addition (+) $X(\theta,x_a^*,a) \in E_a^+$ then $\check{S}(x_a^*,a)$ is given by translation along the trajectory, and we have a periodic trajectory with period $\theta + \sigma(X(\theta,x_a^*,a),a)$.

A sufficient condition that (+) holds for $a > 0$ small is that for such a (5) the trajectories $X(\cdot,r\Phi_a,a)$ in the unstable manifolds are at time $t = \theta$ above the stable manifold, i.e. in E_a^+.

In the following sections A - G we give a detailed exposition of our approach from [11] to smoothness of the map $\check{\Sigma}$, in the simplest nontrivial situation, without parameters. The last section H states a result with parameters, closing with a sketch of its application to the bifurcation problem.

We begin with a version of the sharpened inclination lemma.

<u>A. Inclinations of tangent vectors</u>. Lemma 2.1 and subsequent remarks in [10] imply the following result with pointwise estimate.

<u>Lemma</u> Let $\tilde{g}:\tilde{U} \to B$ be a C^2-map on an open set U of a Banach space B, with $\tilde{g}(0) = 0 \in \tilde{U}$. Let $L := D\tilde{g}(0)$. Assume (A.1) $B = P \oplus Q$ with L-invariant subspaces, and that there exist positive

reals $\alpha < 1$, $\gamma \geqq \beta > 1$ with (A.2) $|Lx| \leqq \alpha|x|$ on Q, $\beta|x| \leqq |Lx| \leqq \gamma|x|$ on P, and with (A.3) $(\alpha\gamma)/\beta < 1$. Let (A.4) $\tilde{g}(\tilde{U} \cap P) \subset P$, $\tilde{g}(\tilde{U} \cap Q) \subset Q$.

Then there exist an open neighborhood $U \subset \tilde{U}$ of $0 \in B$ and $\bar{c} > 0$, $\tilde{\beta} \in (1,\beta)$ such that for all $p_2 \geqq p_1 > 0$ there is a constant $\hat{c} > 0$ with the following property:

If a set $H \subset U$ satisfies

$$\left.\begin{array}{l} p_1 \leqq |px| \leqq p_2 \text{ on } H,\ q\chi \neq 0 \text{ and } \Lambda(\chi) := \dfrac{|p\chi|}{|q\chi|} \leqq \bar{c} \\ \text{on } T_xH \setminus \{0\} \text{ for all } x \in H \end{array}\right\} \qquad (A.5)$$

with the projections $p:B \to P$, $q:B \to Q$ given by (A.1), then

$$|px| \leqq p_2\tilde{\beta}^{-k}, \qquad (A.6)$$

$$q\chi \neq 0 \text{ and } \Lambda(\chi) \leqq \hat{c}|px| \qquad (A.7)$$

for all $k \in \mathbb{N}_0$, $x \in H_k := (\tilde{g}|U)^{-k}(H)$, $\chi \in T_xH_k \setminus \{0\}$.

<u>Remarks</u> <u>For</u> $\dim P = 1$, $|Lx| = |L||x|$ on P, and we may assume $\beta = |L| = \gamma$ so that (A.3) <u>is automatically satisfied</u>. - For an arbitrary set $Z \subset B$, the set T_xZ of tangent vectors at $x \in Z$ is defined as usual, by derivatives of differentiable curves which pass through x and have trace in Z. In general, T_xZ is not a vector space - but always, $0 \in T_xZ$. - (A.6) and (A.7) imply that inclinations $\Lambda(\chi)$, $\chi \in T_xH_k \setminus \{0\}$, tend to 0 uniformly with respect to $x \in H_k$ as $k \to +\infty$.

For other inclination lemmas in infinite-dimensional spaces, see [5,6].

<u>B</u>. <u>Preparations</u>. In order to avoid technicalities we present the core of our approach in the simplest nontrivial situation, without parameters. Consider a local C^2-flow $X:\Omega \to B$, $\Omega \subset \mathbb{R} \times B$, on a finite-dimensional space B, with stationary point $0 = X(t,0)$ for all $t \in \mathbb{R}$. We assume that the generator of the linearization $T:\mathbb{R} \times B \ni (t,x) \to T_tx \in B$ at $0 \in B$ has a simple positive eigenvalue u, and that there are constants

$$\left.\begin{array}{l} \lambda < -\mu < 0 \text{ with } \operatorname{Re} z < \lambda < 0 < u < \mu \text{ for all} \\ \text{eigenvalues } z \neq u. \end{array}\right\} \qquad (B.1)$$

Then (B.2) $B = P \oplus Q$ with T_t-invariant spaces $P = \mathbb{R}\Phi$ and Q, where Φ is a unit eigenvector of the eigenvalue u, and there is a constant $c_1 > 0$ such that for all $t \geqq 0$,

$$T_t x = e^{ut} x \text{ on } P, \quad |T_t x| \leq c_1 e^{\lambda t} |x| \text{ on } Q \tag{B.3}$$

We assume in addition that (B.4) P and Q are invariant under the nonlinear flow X, i.e. $(t,x) \in \Omega$ and $x \in P$ imply $X(t,x) \in P$, and analogously for Q. Write $X(t,x) = T_t x + R(t,x)$, with a remainder $R:\Omega \to B$ which is C^2-smooth. Note that (B.5) R leaves P and Q invariant. We have (B.6) $R(t,0) = 0$, $D_2R(t,0) = 0$, $D_1D_2R(t,0) = 0$ on $\mathbb{R}$.

We prepare both the construction of a shift Σ along trajectories close to $0 \in B$, and the application of the preceding lemma to a restriction of a time-N-map of X.

It is not hard to find a positive integer N, positive reals $\alpha < 1$ and $\gamma = \beta > 1$, and an open set $\tilde{U}$ such that the C^2-map

$$\tilde{g}:\tilde{U} \ni x \to X(N,x) \in B, \quad L := D\tilde{g}(0) = T_N,$$

satisfies conditions (A.1) - (A.4), and furthermore

$$\alpha < e^{\lambda N}, \quad \beta < e^{\mu N}, \quad \alpha < 1/\beta \tag{B.7}$$

In particular, (B.8) $[0,N] \times \tilde{U} \subset \Omega$.

As in section A, write $x = px + qx$ with $px \in P$, $qx \in Q$, for all $x \in B$. We choose $c > 0$ and a convex open neighborhood $U \subset \tilde{U}$ of $0 \in B$ with $pU \subset U$, $qU \subset U$ so small that the lemma applies with constants $\bar{c}$ and $\tilde{\beta}$, and such that we have

$$|D_2pR(t,x)| + |D_2qR(t,x)| < c \text{ on } [0,N] \times U, \tag{B.9}$$

$$|D_2(D_1qR)(0,x)| + |D_2(D_1pR)(0,x)| < c \text{ on } U, \tag{B.10}$$

and (B.11) $\alpha + c < e^{\lambda N}$, (B.12) $c < u$, (B.13) $c < \beta$. - It follows that there is some $c^* > 0$ with (B.14) $|D_2X(t,x)| < c^*$ for t in $[0,N]$, $x \in U$. Set $g := \tilde{g}|U$. We have

$$\left.\begin{array}{l}(\beta-c)|px| \leq |pg(x)| \leq (\beta+c)|px| \text{ and} \\ |qg(x)| \leq (\alpha+c)|qx| \text{ for all } x \in U.\end{array}\right\} \tag{B.15}$$

Proof of the first estimate: Set $r := g - L$. Note $pr(qx) = 0$. Apply the mean value theorem to $pr(x) = pr(x) - pr(qx)$, use (B.9) and (A.2).

Similarly, (B.9) implies $|pR(t,x)| \leq c|px|$, $|qR(t,x)| \leq c|qx|$ on U. Using (B.3) and (B.15) one shows - <u>without the variation-of-constants formula</u> - that there exist $\rho > 0$, $c_2 > 0$, $c_3 > 0$ such that for all $x \in U$ with $X(s,x) \in U$ on $[0,t]$,

$$c_2 e^{\rho t}|px| \leq |pX(t,x)| \leq c_3 e^{\mu t}|px|, \tag{B.16}$$

$$|qX(t,x)| \leq c_3 e^{\lambda t}|qx|. \tag{B.17}$$

Finally, there are $c_4 > 0$, $c_5 > 0$ such that for all $x \in U$,

$$c_4|px| \leqq |pD_1X(0,x)| \leqq c_5|px| \text{ and } \quad (B.18)$$

$$|qD_1X(0,x)| \leqq c_5|qx|. \quad (B.19)$$

Proof of the first estimate in (B.18): (B.5) yields $pR(t,qx) = 0$ on $[0,N] \times U$, hence $D_1pR(0,qx) = 0$ on U. (B.10) and the mean value theorem for $D_1pR(0,x) - D_1pR(0,qx)$ give $|pD_1R(0,x)| \leqq c|px|$ on U. $pT_tx = T_tpx = e^{ut}px$ on $\mathbb{R} \times B$ implies $D_1pT(0,x)(1) = u\cdot px$. Use $X(t,x) = T_tx + R(t,x)$ on Ω, and (B.12).

<u>C</u>. <u>The</u> <u>map</u> Σ. Observe first that (B.16) yields

$$\left.\begin{array}{l} pX(t,x) \in (0,\infty)\cdot\Phi \quad \text{for all } x \in U \text{ with } px \in (0,\infty)\cdot\Phi \\ \text{and } X(s,x) \in U \text{ on } [0,t]. \end{array}\right\} \quad (C.1)$$

Fix $r > 0$ with $r\Phi \in U \cap P$. Set $H := r\Phi + Q$. Choose $\eta_1 > 0$ and $\delta_1 > 0$ so small that the open box

$$E^+ := E^+(\eta,\delta) := \{x \in B\colon px \in (0,\eta)\cdot\Phi, |qx| < \delta\}$$

with $\eta = \eta_1$, $\delta = \delta_1$ satisfies $E^+ \subset U$, $E^+ \cap H = \emptyset$, and that for every $x \in E^+$ there exists $\sigma = \sigma(x) > 0$ with $X(t,x) \in U$ on $[0,\sigma]$, $pX(t,x) \in (0,r)\cdot\Phi$ on $[0,\sigma)$, $pX(\sigma,x) = r\Phi$ (or $X(\sigma,x) \in H$), $D_1X(\sigma,x)(1) \notin Q$. Furthermore, we can achieve that the map $\sigma\colon E^+ \to (0,\infty)$ is continuously differentiable. - Let $x \in E^+$. By (B.16), $r = |pX(\sigma,x)| \leqq c_3e^{\mu\sigma}|px|$, or

$$\frac{1}{\mu} \log \frac{1}{c_3|px|} \leqq \sigma(x). \quad (C.2)$$

With (B.17), we obtain

$$|qX(\sigma(x),x)| \leqq c_3|qx|\left(\frac{r}{c_3}\right)^{\lambda/\mu}|px|^{-\lambda/\mu}. \quad (C.3)$$

Consider the C^1-map $\Sigma\colon E^+ \ni x \to X(\sigma(x),x) \in H \subset B$ and its "Šil'nikov continuation" $\check{\Sigma}$ to the set

$$E := E(\eta_1,\delta_1) := \{x \in B\colon |px| < \eta_1, |qx| < \delta_1\}$$

defined by $\check{\Sigma}(x) := r\Phi$ on $E \setminus E^+$. $\check{\Sigma}$ is C^1-smooth on $E \setminus Q$. Proof of differentiability at points $x \in E \cap Q$, with $D\check{\Sigma}(x) = 0$: Suppose $\lim_{n\to\infty} x_n = x$, $x_n \in E \setminus \{x\}$ for all $n \in \mathbb{N}$, and $\bar{x}_k = x_{n_k} \in E^+$ for a subsequence $(n_k)_{k\in\mathbb{N}}$. Clearly $\check{\Sigma}(x_n) - \check{\Sigma}(x) = 0$ if $x_n \notin E^+$. For all $k \in \mathbb{N}$, $\check{\Sigma}(\bar{x}_k) - \check{\Sigma}(x) = X(\sigma(\bar{x}_k),\bar{x}_k) - r\Phi = qX(\sigma(\bar{x}_k),\bar{x}_k)$, and $|\bar{x}_k - x| \geqq |p\bar{x}_k - 0| - |q||\bar{x}_k - x|$, or $(1 + |q|)|\bar{x}_k - x| \geqq |p\bar{x}_k| > 0$. (C.3) and (B.1) show that difference quotients for $\check{\Sigma}$ tend to 0 as $k \to +\infty$.

In the next sections we shall show that there exists δ_3 in $(0,\delta_1)$ with

$$\sup_{x \in E^+(\eta,\delta_3)} |D\Sigma(x)| \to 0 \quad \text{as } \eta \to 0. \tag{C.4}$$

This implies continuity of $D\check{\Sigma}$ at points $x \in Q$ with $|qx| < \delta_3$.

<u>D</u>. <u>Discretization</u>. The preimages $H_k^* := \sigma^{-1}(kN) \subset E^+$, $k \in \mathbb{N}$, are nonempty for k sufficiently large and satisfy $H_k^* \subset g^{-k}(H) =: H_k$ for all $k \in \mathbb{N}$, with

$$g^k(x) = X(kN,x) = \Sigma(x) \quad \text{on each } H_k^*. \tag{D.1}$$

Let $x \in E^+$. The tangent vector $w_x := D_1X(0,x)(1)$ to the trajectory $X(\cdot,x)$ at $t = 0$ satisfies

$$D\sigma(x)w_x = -1, \tag{D.2}$$

since $\sigma(X(t,x)) - \sigma(x) = -t$ for small $t > 0$. Therefore σ and H_k^* are transversal whenever $H_k^* \neq \emptyset$, and H_k^* is a submanifold of codimension 1. (D.2) and $T_{kN}\{kN\} = \{0\}$ imply $w_x \notin T_xH_k^*$ for all $x \in H_k^*$, $k \in \mathbb{N}$, so that

$$B = \mathbb{R}w_x \oplus T_xH_k^*. \tag{D.3}$$

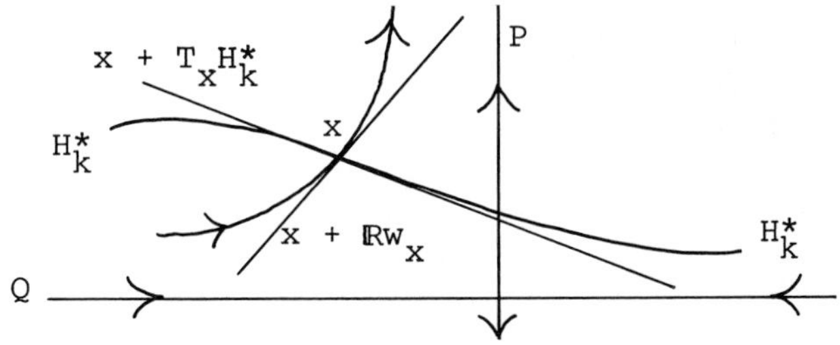

Σ is constant along trajectories. Therefore (D.4) $D\Sigma(x)w_x = 0$ on each $H_k^* \neq \emptyset$. (D.1) gives (D.5) $D\Sigma(x)\chi = Dg^k(x)\chi$ for all $\chi \in T_xH_k^*$, $x \in H_k^*$, $k \in \mathbb{N}$. <u>The basic idea for the proof of</u> (C.4) is to use (D.3) and (D.4) for an estimate of $|D\Sigma(x)|$, $x \in H_k^*$, in terms of $|(Dg^k(x)|T_xH_k^*)|$ and of the angle between the decomposing spaces in (D.3), and to apply (A.6) and the pointwise estimate (A.7) for the inclination of T_xH_k to the majorizing terms.

E. Estimate of $Dg^k(x)$ on T_xH_k. Set $p_1 := p_2 := r$ and recall $\Lambda(\chi) = 0 < \bar{c}$ if $0 \neq \chi \in T_xH = Q$, $x \in H$. We have (A.6) and some $\hat{c} > 0$ so that (A.7) holds. Let us sketch how to find $c_6 > 0$ with

$$|Dg^k(x)\chi| \leq c_6 e^{\lambda kN}|\chi| \text{ for all } k \in \mathbb{N}, x \in H_k, \chi \in T_xH_k. \qquad (E.1)$$

The first inequality in (B.11), (A.6) and (A.7) permit to choose $j \in \mathbb{N}$ such that for all integers $k \geq j$, $x \in H_k$, $\bar{x} \in H_{k-1}$, $\chi \in T_xH_k \setminus \{0\}$, $\bar{\chi} \in T_{\bar{x}}H_{k-1} \setminus \{0\}$ we have

$$(\alpha + c + c\cdot\Lambda(\chi))\,\frac{1 + \Lambda(\bar{\chi})}{1 - \Lambda(\chi)} < e^{\lambda}. \qquad (E.2)$$

Consider $\chi \in T_xH_k \setminus \{0\}$, $x \in H_k$, $k \geq j$. Set $\bar{x} := g(x)$ and $\bar{\chi} := Dg(x)\chi \in T_{\bar{x}}H_{k-1} \setminus \{0\}$. We show (E.3) $|\bar{\chi}| \leq e^{\lambda N}|\chi|$: With $q\bar{\chi} \neq 0 \neq q\chi$,

$$\frac{|\bar{\chi}|}{|\chi|} = \frac{|q\bar{\chi}|}{|q\chi|}\,\frac{\left|\frac{1}{|q\bar{\chi}|}\,p\bar{\chi} + \frac{1}{|q\bar{\chi}|}\,q\bar{\chi}\right|}{\left|\frac{1}{|q\chi|}\,p\chi + \frac{1}{|q\chi|}\,q\chi\right|}.$$

(A.2) and (B.9) give $|q\bar{\chi}| = |qL\chi + qDr(x)\chi| \leq \alpha|q\chi| + c|\chi| \leq (\alpha+c)|q\chi| + c|p\chi|$. Using this and (E.2), we get (E.3). - Finally, iteration and an appropriate choice of c_6 yield (E.1).

For points $x \in H_k^*$, $k \in \mathbb{N}$, and vectors $\chi \in T_xH_k^*$, we combine (D.5), $T_xH_k^* \subset T_xH_k$, (E.1), $\sigma(x) = kN$ and (C.2) and infer

$$|D\Sigma(x)\chi| \leq c_7|px|^{-\lambda/\mu}|\chi| \quad \text{with } c_7 := c_6(c_3/r)^{-\lambda/\mu}. \qquad (E.4)$$

F. Estimate of $D\Sigma(x)$ at $x \in H_k^*$. We choose $\delta_2 \in (0,\delta_1)$ with (F.1) $c_5\delta_2 < c_4/(2\hat{c})$, $\varepsilon \in (0,1)$ with (F.2) $\varepsilon|q|/(1-\varepsilon) < 1/2$, and $j \in \mathbb{N}$ with (F.3) $|p\chi| \leq \varepsilon|\chi|$ for all integers $k \geq j$ and all $x \in H_k^*$, $\chi \in T_xH_k^*$. - The latter is possible because of (A.6) and (A.7).

Let $x \in H_k^*$, $k \geq j$, $|qx| < \delta_2$. (F.3) and $|p\Phi| = |\Phi| = 1 > \varepsilon$ imply $\Phi \notin T_xH_k^*$ so that we have another decomposition

$$B = P \oplus T_xH_k^* \qquad (F.4)$$

The associated projections p_x onto P and q_x onto $T_xH_k^*$ satisfy

$$\left.\begin{array}{l} p_x\chi = p\chi \text{ if } \chi \in P,\ p_x\chi = p\chi - (|q\chi|/|qq_x\chi|)\cdot pq_x\chi \\ \text{if } \chi \in B \setminus P. \end{array}\right\} \qquad (F.5)$$

(The equation for $\chi \in B \setminus P$ follows from $p_x\chi = pp_x\chi = p(\chi -$

$- q_x\chi)$, $0 \neq q\chi = qq_x\chi + qp_x\chi = qq_x\chi$.) Proof of

$$|p_x - p| < 1/2, \; |q_x - q| = |id - p_x - (id - p)| < 1/2 \quad (F.6)$$

- consider $\chi = y\Phi + \bar{\chi}$ with $|\chi| = 1$, $y \in \mathbb{R}$, $\bar{\chi} \in T_x H_k^*$. Then $|(p_x - p)\chi| = |y\Phi - p(y\Phi + \bar{\chi})| = |p\bar{\chi}|$. By (F.3), $|p\bar{\chi}| \leq \varepsilon|\bar{\chi}| \leq \varepsilon(|p\bar{\chi}| + |q\bar{\chi}|)$. Note $|q\bar{\chi}| = |q(y\Phi + \bar{\chi})| = |q\chi| \leq |q|$. Hence $|p\bar{\chi}| \leq \varepsilon(|p\bar{\chi}| + |q|)$, $|p\bar{\chi}| \leq \varepsilon|q|/(1-\varepsilon) < 1/2$, with (F.2).

It follows that

$$|\tilde{L}| \leq c_8 \cdot \sup_{-1 \leq y \leq 1, \chi \in T_x H_k^*, |\chi|=1} |\tilde{L}(y\Phi + \chi)| \quad (F.7)$$

for all continuous linear maps $\tilde{L}: B \to B$, where $c_8 := 1 + |p| + |q|$. - (D.2) gives $p_x w_x \neq 0$. Next, we show

$$|D\Sigma(x)| \leq c_9 |px|^{-\lambda/\mu}\left(1 + \frac{1}{|p_x w_x|}\right) \quad (F.8)$$

with $c_9 := c_8 c_7(1 + (1 + |q|)\cdot c_5 \cdot (\eta_1 + \delta_1))$:
(F.7) implies $|D\Sigma(x)| \leq c_8(|D\Sigma(x)\Phi| + \sup_{\chi \in T_x H_k^*, |\chi|=1} |D\Sigma(x)\chi|)$.
From (D.4), $|D\Sigma(x)\Phi| = |D\Sigma(x)p_x w_x|/|p_x w_x| = |-D\Sigma(x)q_x w_x|/|p_x w_x|$. Using (E.4), we arrive at $|D\Sigma(x)| \leq c_8(c_7|px|^{-\lambda/\mu}|q_x w_x| \cdot |p_x w_x|^{-1} + c_7|px|^{-\lambda/\mu})$. (F.6), (B.18) and (B.19) yield $|q_x w_x| \leq (1 + |q|) \cdot (|pw_x| + |qw_x|) \leq (1 + |q|)c_5(|px| + |qx|)$, and (F.8) becomes obvious.

Now the pointwise estimate (A.7) becomes crucial. We derive

$$|p_x w_x|^{-1} \leq (2/c_4)|px|^{-1} \quad (F.9)$$

- in case $w_x \in P$, this is a trivial consequence of $p_x w_x = pw_x$ and of the lower estimate in (B.18). For $w_x \notin P$, (F.5) gives $|p_x w_x| \geq |pw_x| - |qw_x|\Lambda(q_x w_x)$. Using (B.18) as before, (B.19) for $|qw_x|$, $|qx| < \delta_2$, (F.1) and finally (A.7), we get $|p_x w_x| \geq c_4|px| - (c_4/2\hat{c})\cdot\hat{c}\cdot|px|$.

Altogether, we have shown in this section that for all integers $k \geq j$ and all $x \in H_k^*$ with $|qx| < \delta_2$,

$$|D\Sigma(x)| \leq c_{10}(|px|^{-\lambda/\mu} + |px|^{(-\lambda/\mu) - 1}) \quad (F.10)$$

where $c_{10} := c_9(1 + (2/c_4))$.

G. <u>Estimate at arbitrary points</u> $x \in E^+(\eta_2,\delta_3)$. Choose positive reals $\delta_3 < \delta_2$ and $\eta_2 < \eta_1$ so small that

$$\left.\begin{array}{l} \text{for all } x \in E^+(\eta_2,\delta_3),\ \sigma(x) > jN \text{ and} \\ X(t,x) \in E^+(\eta_1,\delta_2) \text{ on } [0,N]. \end{array}\right\} \tag{G.1}$$

Let $x \in E^+(\eta_2,\delta_3)$. The largest integer k with $kN \leq \sigma(x) < kN + N$ satisfies $k \geq j$. Set $\bar{x} := X(\sigma(x) - kN,x)$. Then $\bar{x} \in E^+(\eta_1,\delta_2)$ and $\sigma(\bar{x}) = kN$, or $\bar{x} \in H_k^*$. We have $\Sigma(x) = \Sigma(\bar{x}) = \Sigma(X(\sigma(x) - kN,x))$, and there is a neighborhood $W \subset E^+(\eta_2,\delta_3)$ of x such that for all $y \in W$, $\Sigma(y) = \Sigma(X(\sigma(x) - kN,y))$. Hence $|D\Sigma(x)| \leq |D\Sigma(\bar{x})|\cdot|D_2X(\sigma(x) - kN,x)|$. (B.14) for $t = \sigma(x) - kN < N$, x in $E^+(\eta_2,\delta_3) \subset U$, and (F.10) yield $|D\Sigma(x)| \leq c_{10}(|p\bar{x}|^{-\lambda/\mu} + |p\bar{x}|^{(-\lambda/\mu)-1})\cdot c^*$. With (G.1) and (B.16) - and with (B.1) - we obtain

$$|D\Sigma(x)| \leq c_{11}(|px|^{-\lambda/\mu} + |px|^{(-\lambda/\mu)-1}) \tag{G.2}$$

where $c_{11} := c_{10}c^*((c_3e^{\mu N})^{-\lambda/\mu} + (c_3e^{\mu N})^{(-\lambda/\mu)-1})$. Finally, (G.2) and the hypothesis (B.1) on the spectrum imply (C.4).

H. <u>Bifurcation</u>. The simplest nontrivial situation <u>with parameters</u> occurs for a local <u>flow</u> $(t,x,a) \to X(t,x,a)$ of class C^2 in a finite-dimensional space B, with parameters in an open interval $A \ni 0$. Suppose (1) $0 \in B$ is a stationary point, the spectral hypothesis (3) is satisfied, and X is locally normalized (4). Then one can make the previous considerations locally uniform with respect to the parameter. The result is that there exist $\eta_2 > 0$ and $\delta_3 > 0$ and an open interval $A_1 \ni 0$ with

$$\sup_{x \in E^+(\eta,\delta_3,a),\, a \in A_1} |D_1\Sigma(x,a)| \to 0 \quad \text{as } \eta \to 0 \tag{H.1}$$

where $\Sigma(x,a) = X(\sigma(x,a),x,a) \in H_a \subset B$ for $x \in E^+(\eta_2,\delta_3,a)$, a in A_1.

As a consequence one obtains that $D_1\check{\Sigma}$ exists on the whole domain of $\check{\Sigma}$, and is continuous, now with respect to (x,a). As in section C, $D_1\check{\Sigma}(r\Phi_0,0) = 0$.

We show that existence of the partial derivative $D_2\check{\Sigma}$ follows from the analogue of (C.3), asserting that there is $c_3' > 0$ with

$$|q_a X(\sigma(x,a),x,a)| \leq c_3' |q_a x| (r/c_3')^{\lambda/\mu} |p_a x|^{-\lambda/\mu} \qquad (C.3')$$

for all $x \in E^+(\eta_2,\delta_3,a)$ and all $a \in A_1$: Consider $x \in E(\eta_2,\delta_3,a)$ with $p_a x = 0$, $a \in A_1$, so that $\Sigma(x,a) = r\Phi_a$, and sequences of points $a_n \in A_1 \setminus \{a\}$ with $\lim_{n\to\infty} a_n = a$. Differentiability of $a^* \to \Phi_{a^*}$ shows that in case $p_{a_n} x \in (-\eta_2,0]\cdot\Phi_{a_n}$ for all n, difference quotients $D_n := (a_n - a)^{-1}(\check{\Sigma}(x,a_n) - \check{\Sigma}(x,a))$ tend to $r\cdot\lim_{h\to 0} h^{-1}(\Phi_{a+h} - \Phi_a)$. - If $p_{a_n} x \in (0,\eta_2)\cdot\Phi_{a_n}$ for all n, write $D_n = r(a_n - a)^{-1}(\Phi_{a_n} - \Phi_a) + d_n$ with $d_n := (a_n - a)^{-1}(\check{\Sigma}(x,a_n) - r\Phi_{a_n}) = (a_n - a)^{-1} q_{a_n}\Sigma(x,a_n)$. By C^1-smoothness of $a^* \to p_{a^*}$, there is $c_p > 0$ with $|p_{a_n} x| = |(p_{a_n} - p_a)x| \leq |p_{a_n} - p_a||x| \leq c_p(\eta_2 + \delta_3)|a_n - a|$. Using this and (C.3') and $\lambda < -\mu$, we obtain $\lim_{n\to\infty} d_n = 0$. - Now it is easy to complete the argument.

We return to bifurcation. Suppose in addition that (2) there is a homoclinic trajectory x^0 for $a = 0$. Then the map

$$\check{S}: \tilde{\mathcal{B}} \times \tilde{A} \ni (x,a) \to \check{\Sigma}(X(\theta,x,a),a) \in B$$

is defined, with some fixed $\theta > 0$ so that $X(t,r\Phi_0,0)$ is in $E(\eta_2,\delta_3,0)$ for all $t \geq \theta$. $\check{S}$ is as smooth as $\check{\Sigma}$. We have $\check{S}(r\Phi_0,0) = r\Phi_0 = x^0(0)$, and (H.2) $D_1\check{S}(r\Phi_0,0) = 0$. Altogether, we can use a version of the implicit function theorem which guarantees existence of a locally unique differentiable curve $a \to \psi_a$ of solutions to an equation $F(\psi,a) = 0$ through a given solution ψ^* at $a = 0$, provided both derivatives D_1F and D_2F exist, D_1F is continuous, and $D_1F(\psi^*,0)$ is an isomorphism.

We obtain a differentiable curve of fixed points x_a^* of $\check{S}(\cdot,a)$ through $x^0(0)$, which are all stable and attractive, due to (H.2).

If finally (5) $X(\theta,r\Phi_a,a) \in E^+(\eta_2,\delta_3,a)$ for $a > 0$, then it can be shown that for $a > 0$ these fixed points define periodic trajectories which are unique in a neighborhood of the homoclinic orbit, and stable and attractive with asymptotic phase.

Precise statements and complete proofs, in a more difficult situation with semiflows in an infinite-dimensional space, are contained in [11].

References

[1] A.A. Andronov, C.E. Chaikin. Theory of oscillations. Princeton Univ. Press, Princeton,N.J., 1949
[2] M. Blazquez. Bifurcation from a homoclinic orbit in parabolic differential equations. Preprint, Brown Univ., Providence, R.I., 1985
[3] S.N. Chow, B. Deng. Homoclinic and heteroclinic bifurcation in Banach spaces. Preprint, Mich. State Univ., East Lansing, Mi., 1986
[4] S.N. Chow, J.K. Hale. Methods of bifurcation theory. Springer, New York et al., 1982
[5] J.K. Hale, X.B. Lin. Symbolic dynamics and nonlinear semiflows. Preprint LCDS 84-8, Brown Univ., Providence, R.I., 1984
[6] D. Henry. Invariant manifolds. Notes, Math. Inst., Univ. of Sao Paulo, 1983
[7] E. Reyzl. Diplomarbeit. Math. Inst., Univ. München, 1986
[8] L.P. Šil'nikov. On the generation of periodic motion from trajectories doubly asymptotic to an equilibrium state of saddle type. Mat. Sb. 77 (119), 1968; Engl. transl. in: Math. USSR - Sbornik 6 (1968), 427 - 438
[9] H.O. Walther. Bifurcation from a heteroclinic solution in differential delay equations. Trans. AMS 290 (1985), 213 - 233
[10] H.O. Walther. Inclination lemmas with dominated convergence. Research report 85-03, Sem. f. Angew. Math., ETH Zürich, 1985 (submitted)
[11] H.O. Walther. Bifurcation from a saddle connection in functional differential equations: An approach with inclination lemmas. Preprint, Math. Inst., Univ. München, 1986 (submitted)

APPROXIMATE METHODS FOR SET VALUED DIFFERENTIAL EQUATIONS WITH DELAYS

F. WILLIAMSON
16, Avenue de la Commune de Paris, 94400 VITRY SUR SEINE (FRANCE)

Introduction

In this paper a finite difference method is used to study the existence of solutions for a kind of set valued differential equations involving delays.
More precisely we consider differential problems of the form :

$$(P)\quad \begin{cases} \frac{dx}{dt} \in S(x(t),\dot{x}(t-r)) & \text{a.e. in a bounded interval} \\ x(\theta) = y(\theta) & \text{for any } \theta \in [-r,o] \\ x(t) \in Q \end{cases}$$

where S is a continuous set valued mapping with non necessarily convex values in $\mathbb{R}^{\ell}$ having for arguments the position x(t) at time t and the speed $\dot{x}(t-r)$ at the retarded time t-r (we set for convenience $\dot{x} = \frac{dx}{dt}$), Q is a compact set of $\mathbb{R}^{\ell}$ and y a continuously differentiable mapping defined on $[-r,o]$ with values in Q.

Such problems can be met when dealing with control problems involving a state equation containing delays of the form :

$$\frac{dx}{dt} = f(x(t), \dot{x}(t-r),u),\ u \in U$$

where U denotes the set of controls.

We are interested in absolutely continuous solutions x(t) such that $||\dot{x}(t)|| \leq M$ for almost every t, where M denotes a positive constant such that $M \geq \mathrm{Sup}||\dot{y}(\theta)||$, $\theta \in [-r,o]$.

We will have to assume that a tangential condition of the following form is satisfied by the set valued mapping S :

$(\mathcal{T}_c)$ $S(x,v) \subset T_Q(x)$ for any $x \in Q$ and $||v|| \leq M+1$, where $T_Q(x)$ denotes the Clarke tangent cone to the set Q at the point $x \in Q$.

We will associate with (P) the following discrete set valued problem in $\mathbb{R}^{\ell}$:

$$(P_m)\quad \begin{cases} \dfrac{x_m^{j+1} - x_m^j}{h_m} \in S(x_m^j, v_m^{j-Lm}) + 2^{-m}\mathbf{B} & j = 0,1,2\ldots \\ x_m^{j-Lm} = y(jh_m - r) & j = 0,1,\ldots L_m \end{cases}$$

with $v_m^j = \dfrac{x_m^{j+1} - x_m^j}{h_m}$ and $h_m = \dfrac{r}{L_m}$, L_m integer, and use it to construct a sequence of approximate solutions x_m linearly interpolating the points x_m^j and defined through :

$$(\Sigma_m)\quad \begin{cases} x_m(\tau) = x_m^j + (\tau - jh_m)\, v_m^j & \text{for } \tau \in [jh_m, (j+1)h_m],\ j = 0,1,2\ldots \\ x_m(\theta) = y(\theta) & \text{for } \theta \in [-r,0], \end{cases}$$

NATO ASI Series, Vol. F37
Dynamics of Infinite Dimensional Systems
Edited by S.-N. Chow, and J. K. Hale

a subsequence of which will be shown to converge on any given interval $[0,T]$ with $T = Nr$ to an exact solution of (P).

This convergence will be proved by means of a compactness argument due to FILIPPOV (1971) and involving nested regular nets.

1. Construction of the approximate solutions of Problem (P)

As precised above we assume that the set valued mapping $S: R^{\ell} x R^{\ell} \to 2^{R^{\ell}}$ satisfies the tangential condition $(\mathcal{T}_c)$ where the Clarke tangent cone $T_Q(.)$ to the set Q is defined by :

(1) $T_Q(x) = \{v \in R^{\ell} \mid D_c d_Q(x)(v) < o\}$.
with $d_Q(x) = \operatorname{Inf}_{q \in Q} ||x-q||$ and D_c denotes the Clarke derivative defined for a given function f as :

$$(2) \quad D_c f(x)(v) = \limsup_{\substack{y \to x \\ \theta \to o^+}} \frac{f(y+\theta v)-f(y)}{\theta}$$

(where we set for an arbitrary real valued function φ:

$$\limsup_{\substack{y \to x \\ \theta \to o^+}} \varphi(y,\theta) = \lim_{\theta \to o^+} \sup_{y \in B(x,\theta)} \varphi(y,\theta))$$

We will first derive a consequence of the preceding tangential condition to be used for constructing approximate solutions of problem (P). So as to take into account the condition $||\dot{x}(t)|| \leqslant M$ for any t, we will replace the set valued mapping S(.) by its intersection $S(.) \cap B(o,M)$ with the compact ball B(o,M), where M satisfies : $\sup_{\theta \in [-r,o]} ||\dot{y}(\theta)|| \leqslant M$.

The announced property can be stated as follows :

Proposition 1.

We assume as above that the subset $Q \subset R^{\ell}$ is compact and that the mapping $S : R^{\ell} \times R^{\ell} \to 2^{R^{\ell}}$ is uppersemicontinuous with compact values and satisfies the tangential condition $(\mathcal{T}_c)$. Then with any $\varepsilon > o$ is associated some $\eta(\varepsilon) > o$ such that for any $w \in S^c(x,v)$ with $(x,v) \in Q \times B(o,M+1)$ and $h \leqslant \eta(\varepsilon)$ there exists some $q \in Q$ such that :

$$(3) \quad \left|\left|\frac{q-x}{h} - w\right|\right| \leqslant \varepsilon .$$

Proof : We may introduce the following set :

$$(4) \quad \mathcal{E} = \{(u_1, u_2, w) \in Q \times B(o, M+1) \times R^{\ell} \mid w \in S(u_1,u_2)\}$$
$$= \text{Graph } S \cap [Q \times B(o,M+1) \times S(Q \times B(o,M+1))]$$

which is compact, as the image under the uppersemicontinuous set valued map S with compact values of the compact set Q x B(o,M+1) is itself compact. Further we infer from $(\mathcal{T}_c)$ that the vector w involved in $(\mathcal{E})$ belongs to $T_Q(u_1)$ and hence satisfies : $D_c d_Q(u_1)(w) \leqslant o$. It next follows from the definition of the Clarke derivative (see F.H. CLARKE, J.P. AUBIN (1977)) that for any $\varepsilon > o$ there exists neighborhoods $\mathcal{U}(u_1)$ and $\mathcal{V}(w)$ together with $\eta(u_1,w) > o$ such that for $x \in \mathcal{U}(u_1) \cap Q$, $p \in \mathcal{V}(w)$ and $h < \eta(u_1,w)$ the following inequality holds :

$$(5) \quad d_Q(x+hp) = d_Q(x+hp) - d_Q(x) \leqslant h[D_c d_Q(u_1)(w) + \varepsilon] \leqslant \varepsilon h.$$

Denoting by $\mathcal{N}(u_2)$ an open neighborhood of the vector u_2 we may consider the open covering $\Omega(u_1,u_2,\mathbf{w}) = \mathcal{U}(u_1) \times \mathcal{N}(u_2) \times \mathcal{V}(w)$ of the compact set $\mathcal{E}$. It admits a finite subcovering $\Omega_i(u_{1,i},u_{2,i},\mathbf{w}_i)$ $i = 1,2,\dots p$. Then any triple $(x,v,\mathbf{w}) \varepsilon \mathcal{E}$ belongs to some neighborhood $\Omega_j(u_{1,j},u_{2,j},w_j)$ of this finite subcovering with which is associated a positive constant $\eta_j(u_{1,j},\mathbf{w}_j)$. We thus have $x \,\varepsilon\, \mathcal{U}(u_{1,j}) \cap Q$ together with $w \,\varepsilon\, \mathcal{V}(w_j)$ and by choosing $h < \eta(\varepsilon) = \min_{1 \leq i \leq p} \eta_i(u_{1,i},\mathbf{w}_i)$ it follows from (5) that : $d_Q(x+hw) \leq \varepsilon h$. As by the compactness of Q there exists $q_x \varepsilon Q$ for which $d_Q(x+hw) = \|x+hw-q_x\|$, we get finally :

$\left\|\frac{q_x-x}{h} - \mathbf{w}\right\| \leq \boldsymbol{\varepsilon}$, where $\mathbf{w} \,\varepsilon\, S(x,v)$, $x \varepsilon Q$, $v \varepsilon B(o,M+1)$ are arbitrary. Q.E.D.

In connection with the preceding result we may associate with problem (P) the discrete problem (P_m) appearing in the introduction. So as to prepare for the proof of convergence of its solutions to a solution of problem (P) by means of the FILIPPOV compactness result, we will choose the steplength h_m in such a way that $\frac{h_{m-1}}{h_m}$ is <u>integer</u> for all m(and not equal to 1).

We will have to introduce the following coefficients :

(6) $a_m(k) = \mathrm{Min}\ \{j \varepsilon N \mid \frac{kh_m}{h_{j+1}} \varepsilon N\}$

(7) $b_m(k) = \frac{\ell h a_m(k)}{h_m}$ with $\ell = \mathrm{Max}\ \{j \varepsilon N \mid jh_{a_{m(k)}} < kh_m\}$

(we set by convention : $a_m(o) = b_m(o) = o$).

These coefficients satisfy the following properties.

<u>Proposition 2</u>

<u>The coefficients $a_m(k)$ and $b_m(k)$ defined by (6) and (7) satisfy for any integers k and m :</u>

(8) <u>$a_m(k) < m$</u>

(9) <u>$b_m(k) < k$</u>

(10) <u>$[k-b_m(k)]\ h_m \leq h_{a_{m(k)}}$</u>

<u>where $h_{a_{m(k)}}$ denotes the steplength associated with $a_m(k)$.</u> In addition <u>for any integer $k > L_m$ the following relations hold</u> :

(11) <u>$a_m(k) = a_m(k-L_m)$</u>

(12) <u>$b_m(k) = b_m(k-L_m)$</u> $+ L_m$

Proof. Properties (8),(9) and (10) follow easily from the definition of the coefficients $a_m(k)$ and $b_m(k)$.

To prove assertion (11) let us observe that as $a_m(k-L_m)$ is the smallest integer j such that $\frac{(k-L_m)h_m}{h_{j+1}}$ is integer and as moreover $r=L_m h_m = L_{j+1}\, h_{j+1}$, where L_m and L_{j+1} are positive integers, we have : $\frac{(k-L_m)h_m}{h_{j+1}} = \frac{kh_m - L_{j+1}\, h_{j+1}}{h_{j+1}} = \frac{kh_m}{h_{j+1}} - L_{j+1}$. Thus $\frac{kh_m}{h_{j+1}}$ is integer if and only if the same holds for $\frac{(k-L_m)h_m}{h_{j+1}}$, which proves (11). Concerning assertion (12) we may observe that : $b_m(k-L_m) = \frac{\ell' h_{a_m(k-Lm)}}{h_m} = \frac{\ell' h_{a_m(k)}}{h_m}$, where

$\ell' = \text{Max}\,\{j \in N \mid jh_{a_m(k)} < (k-L_m)h_m\} = \text{Max}\{ j\in N \mid (j+L_{a_m(k)})h_{a_m(k)} < kh_m\}$,

from which we deduce by (7) that $\ell = \ell' + L_{a_m(k)}$.

As a consequence we have as announced :

$$b_m(k) = \frac{\ell h_{a_m}(k)}{h_m} = \frac{\ell'+L_{a_m(k)}}{h_m}\, h_{a_m(k)} = b_m(k-L_m) + L_m . \qquad \text{Q.E.D.}$$

We are now able to prove the following result for the existence of solutions of the discrete problem (P_m).

Proposition 3

Let us assume that the set valued map S having compact images contained in the ball B(o,M) satisfies the following condition :

(13) $\begin{cases} \text{for any } x,y \in Q \text{ and } v,w \in B(o,M+1) \\ \text{Max}(\|x-y\|,\|v-w\|) \leqslant \delta \text{ implies } S(x,v) \subset S(y,w) + \frac{1}{2} A\delta B \end{cases}$ where B denotes the unit ball of IR^{ℓ} and A a positive constant. Let us assume further that for any m the steplength h_m is chosen in such a way that:

(14) $h_m \leq \text{Min}\left(\frac{2^{-m}}{M+1}, \eta(2^{-m})\right)$

where $\eta(\varepsilon)$ denotes the constant of Proposition 1 and that in addition for any $\theta,\theta' \in [-r,o]$ the following holds :

(15) $|\theta - \theta'| \leqslant 2h_m$ implies $\|\dot{y}(\theta) - \dot{y}(\theta')\| \leqslant 2.\ 2^{-m}$.

If we take as initial values :

(16) $x_m^{j-L_m} = y(jh_m - r)$ for $o \leqslant j \leqslant L_m$,

then the following system is satisfied by the x^j_m and $v^j_m = \frac{x_m^{j+1} - x_m^j}{h_m}$ for $j = o,1,\ldots, \alpha L_m - 1$, where α denotes a positive integer and $h_m = \frac{r}{L_m}$ with L_m integer:

(17) $||v^j_m - v_m^{b_m(j)}|| \leqslant \pi_\alpha(A).\, 2^{-a_m(j)}$

(18) $v^j_m \in S(x^j_m, v_m^{j-L_m}) + 2^{-m}B$ if $j > 1$, $v^o_m \in S(y(o), \dot{y}(-r)) + 2^{-m}B$ if $j=0$

(19) $x^j_m + h_m v^j_m \in Q$.

The coeffients $\pi_\alpha(A)$ defined as :

(20) $\pi_\alpha(A) = \frac{1}{1-\frac{A}{2}}\left[1 + (\frac{A}{2})^\alpha(1-A)\right]$ $\qquad \alpha = 1,2,3,\dots$

remain bounded for $1 \leqslant A < 2$ and tend to $+\infty$ for $A \geqslant 2$. when $\alpha \to +\infty$. They are non decreasing for fixed $A \geqslant 1$.

Proof

We will construct inductively the sequences x^j_m and $v^j_m = \frac{x^{j+1}_m - x^j_m}{h_m}$ satisfying the relations (17) through (19). For this purpose we assume that these relations are satisfied for any $j \leqslant k-1$ and set :

$x^k_m = x^{k-1}_m + h_m v^{k-1}_m \in Q$. We will show that it is possible to choose v^k_m in such a way that it satisfies (17) and (18). We will set for abbreviation : $\gamma = a_m(k)$, $\lambda = b_m(k)$ and $G = M+1$. Taking into account the fact that for $j \leqslant k-1$:

$$v^j_m = \frac{x^{j+1}_m - x^j_m}{h_m} \in S(x^j_m, v_m^{j-L_m}) + 2^{-m}B \subset B(o,G)$$

it follows by (10) that :

(21) $||x^k_m - x^\lambda_m|| \leqslant ||x^k_m - x^{k-1}_m|| + ||x^{k-1}_m - x^{k-2}_m|| + \dots + ||x^{\lambda+1}_m - x^\lambda_m||$

$$\leqslant G(k-\lambda)h_m \leqslant Gh_\gamma .$$

We will use this estimate repeatedly to prove (17). To this end we will as a preliminary step derive an estimate for $||v_m^{k-L_m} - v_m^{\lambda-L_m}||$ in the cases $o \leqslant k < L_m$ and $\alpha L_m \leqslant k < (\alpha+1)L_m$ with $\alpha \geqslant 1$ arbitrary.

We begin with the case $k = o$ which has to be considered separately. We set $x^o_m = y(o)$ and choose some $w^o \in S(y(o), \dot{y}(-r))$ arbitrarily. Since $y(o) \in Q$, $||\dot{y}(-r)|| \leqslant M$ and $h_m < \eta(2^{-m})$, it follows from Proposition 1 that there exists $x^1_m \in Q$ satisfying : $||v^o_m - w^o|| \leqslant 2^{-m}$ with $v^o_m = \frac{x^1_m - x^o_m}{h_m}$. We thus have $v^o_m \in S(y(o), \dot{y}(-r)) + 2^{-m}B \subset B(o,M+1)$.

As by construction $x_m^o + h_m v_m^o = x_m^1 \varepsilon Q$, the couple (x_m^o, v_m^o) satisfies the relations (17) (with $b_m(o) = o$), (18) and (19).

Case $o < k < L_m$ By definition of the v_m^j together with (16) we have :

$$(22)\quad v_m^{k-L_m} = \frac{1}{h_m}\left[y([k+1]h_m - r) - y(kh_m - r)\right] = \dot{y}(kh_m - r + \theta_k h_m)$$

$$(23)\text{ resp. } v_m^{\lambda-L_m} = \dot{y}(\lambda h_m - r + \theta_\lambda h_m) \text{ with } \theta_k, \theta_\lambda \varepsilon]0,1[.$$

As by (8) $h_m < h_{a_m(k)} = h_\gamma$, it follows by (10) that :

$$|(kh_m - r + \theta_k h_m) - (\lambda h_m - r + \theta_\lambda h_m)| \leqslant |(k-\lambda)h_m + (\theta_k - \theta_\lambda)h_m| \leqslant 2h_\gamma.$$

Therefore we infer from (15) together with (22) and (23) that :

$$(24)\quad \| v_m^{k-L_m} - v_m^{\lambda-L_m}\| = \|\dot{y}(kh_m - r + \theta_k h_m) - \dot{y}(\lambda h_m - r + \theta_\lambda h_m)\| \leqslant 2.2^{-\gamma}$$

which is the required preliminary estimate for $0 \leqslant k < L_m$.
It further follows from (21), (24) and (14) that :

$$(25)\quad \mathrm{Max}(||x_m^k - x_m^\lambda||, ||v_m^{k-L_m} - v_m^{\lambda-L_m}||) \leqslant 2.2^{-\gamma}$$

and hence by (13) :

$$(26)\quad S(x_m^\lambda, v_m^{\lambda-L_m}) \subset S(x_m^k, v_m^{k-L_m}) + (2^{-\gamma}A)B .$$

We infer from this inclusion that: $v_m^\lambda \varepsilon S(x_m^\lambda, v_m^{\lambda-L_m}) + 2^{-m}B \subset S(x_m^k, v_m^{k-L_m}) + (2^{-m} + 2^{-\gamma}A)B$ and thus there exists $w_m^k \varepsilon S(x_m^k, v_m^{k-L_m}) \subset T_Q(x_m^k)$ such that :

$$(27)\quad ||v_m^\lambda - w_m^k|| \leqslant 2^{-m} + 2^{-\gamma}A.$$

As above there exists by Proposition 1 an $x_m^{k+1} \varepsilon Q$ satisfying:

$$(28)\quad ||v_m^k - w_m^k|| \leqslant 2^{-m} \quad \text{with } v_m^k = \frac{x_m^{k+1} - x_m^k}{h_m} .$$

The same inequality joined to the definition of w_m^k yields:

$$(29)\quad v_m^k \;\varepsilon\; S(x_m^k, v_m^{k-L_m}) + 2^{-m}B \subset B(0,G)$$

which proves assertion (18) for $j = k$. Further as $\gamma = a_m(k) < m$, relations (27) and (28) yield together :

$$||v_m^k - v_m^\lambda|| \leqslant ||v_m^k - w_m^k|| + ||w_m^k - v_m^\lambda|| \leqslant 2^{-\gamma}[A + 2^{\gamma-m+1}] \leqslant (A+1)2^{-\gamma}$$

and we have obviously by definition of x_m^{k+1} : $x_m^k + h_m v_m^k \varepsilon Q$. Therefore the pair (x_m^k, v_m^k) satisfies the relations (17) through (19) with $j = k$, $0 \leqslant k < L_m$ and $\pi_1(A) = A + 1$, where we may assume $A \geqslant 1$.

Case $\alpha L_m \leqslant k < (\alpha+1)L_m$ with $\alpha \geqslant 1$ integer.

We now assume that the relations (17) through (19) are satisfied with $\pi_\alpha(A)\geq 2$ for all j such that: $(\alpha-1)L_m\leq j<\alpha L_m$. We will prove that these relations remain true for $\alpha L_m\leq k<(\alpha+1)L_m$ and $\pi_\alpha(A)$ replaced with $\pi_{\alpha+1}(A) = 1+\frac{1}{2}A\,\pi_\alpha(A)$ if in addition the same relations hold for all $j<k$.

We first notice that as a consequence of (11) and (12) :

$$||v_m^{k-L_m} - v_m^{b_m(k)-L_m}|| = ||v_m^{k-L_m} - v_m^{b_m(k-L_m)}|| \leq \pi_\alpha(A)2^{-a_m(k-L_m)} = \pi_\alpha(A)2^{-a_m(k)}$$

and this together with (21) implies :

(30) $\mathrm{Max}(||x_m^\lambda - x_m^k||, ||v_m^{\lambda-L_m} - v_m^{k-L_m}||) \leq \mathrm{Max}(Gh_\gamma, \pi_\alpha(A)2^{-\gamma}) \leq 2^{-\gamma}\pi_\alpha(A)$

from where it follows by (13) :

$$v_m^\lambda \in S(x_m^\lambda, v_m^{\lambda-L_m}) + 2^{-m}B \subset S(x_m^k, v_m^{k-L_m}) + [\tfrac{1}{2}A.\pi_\alpha(A)2^{-\gamma} + 2^{-m}]\,B.$$

Hence there exists as above $w_m^k \in S(x_m^k, v_m^{k-L_m}) \subset T_Q(x_m^k)$ satisfying : $||v_m^\lambda - w_m^k|| \leq \frac{1}{2}A\pi_\alpha(A)2^{-\gamma} + 2^{-m}$ and by Proposition 1 there exists further $x_m^{k+1} \in Q$ for which $||v_m^k - w_m^k|| \leq 2^{-m}$ with

$$v_m^k = \frac{x_m^{k+1} - x_m^k}{h_m}.$$

This inequality may be rewritten as :

(31) $v_m^k \in S(x_m^k, v_m^{k-L_m}) + 2^{-m}B \subset B(0,G)$

and hence (18) is satisfied, while by definition of x_m^{k+1} the vector $x_m^k + h_m v_m^k$ belongs to Q. Moreover we have :

$$||v_m^k - v_m^\lambda|| \leq ||v_m^k - w_m^k|| + ||w_m^k - v_m^\lambda|| \leq 2.2^{-m} + \frac{1}{2}A\,\pi_\alpha(A)2^{-\gamma}$$
$$\leq 2^{-\gamma}\left[1+\frac{1}{2}A\pi_\alpha(A)\right] = 2^{-\gamma}\pi_{\alpha+1}(A)$$

where we have set as announced :

(32) $\pi_{\alpha+1}(A) = 1 + \frac{1}{2}A\pi_\alpha(A)$.

Therefore the relations (17) through (19) are satisfied for any integers $\alpha\geq 1$ and k such that $\alpha L_m\leq k<(\alpha+1)L_m$. The behaviour of the coefficients $\pi_\alpha(A)$ for $\alpha\to+\infty$ may be deduced from their explicit expression (20) obtained from (32), together with its consequence :

(33) $\pi_{\alpha+1}(A) - \pi_\alpha(A) = (A-1)(\frac{A}{2})^\alpha$. Q.E.D.

Remark. It is to be noticed that mere continuity of the set valued mapping S is not sufficient to derive the estimate (17) needed for the proof of convergence of the approximate solutions, since the respective growths of x and v have to be matched to obtain this estimate. Thus the more precise property (13) has to be assumed for S.

2. Convergence of the approximate solutions to a solution of problem (P)

The proof of convergence will be based on a compactness result due to FILIPPOV [1971] that we recall below.

Proposition (FILIPPOV)

Let $B(o,\infty;U)$ denote the vector space consisting of all bounded functions defined on $[o,+\infty]$ with values in the Banach space U supplied with the topology of uniform convergence on compact sets of R. Let (v_m) denote a sequence of step functions with values in U defined on a regular net with step length h_m, where we assume that h_{m-1}/h_m is integer for all m. If the functions v_m satisfy for any positive integers k and m the following properties, where C denotes a fixed positive constant :

α) $||v_m(kh_m) - v_m(b_m(k)h_m)|| \leq C.2^{-a_m(k)}$

β) for any $t \geq o$ $\{v_m(t)\}_m$ is a precompact set of U,

then the sequence (v_m) is precompact in the space $B(0,\infty;U)$.

We will establish the following result.

Proposition 4.

Let us assume that the set Q is compact, that the set valued map S defined on $\mathbb{R}^\ell \times \mathbb{R}^\ell$ and having compact images contained in the ball $B(0,M)$ of $\mathbb{R}^\ell$ satisfies (13) together with the tangential condition $(\mathcal{T}_c)$ and finally that the initial data y of problem (P) is continuously differentiable on $[-r,o]$ and such that $M \geq \sup_{\theta\in[-r,o]} ||\dot{y}(\theta)||$. Then if $T = Nr$ is an integer multiple of r, there exists a function $x(.) \in C([-r,T]; \mathbb{R}^\ell)$ which solves problem (P) and has for derivative a regulated function i.e. some limit of a uniformly converging sequence of step functions.

Proof

By the definition (Σ_m) of the piecewise linear functions x_m interpolating the x_m^j their weak derivative $\dot{x}_m$ satisfies : $\dot{x}_m(t) = v_m^j$ for any $t\in[jh_m,(j+1)h_m[$ and $j = 0,1,2...$ It thus follows from (29) and (31) that : $||v_m^j|| \leq G$ for any j and m ; since (17) is satisfied with $\alpha = N$ for any integer j such that $jh_m < Nr$, we infer from FILIPPOV'S result that the derivatives $\dot{x}_m$ are contained in a compact subset of $B(0,T; \mathbb{R}^\ell)$. In addition for any $t \in [o,T]$ the vectors $x_m(t)$ belong to the compact set Co Q. We have further for any $t_1,t_2 \in [o,T]$:

$$||x_m(t_2)-x_m(t_1)|| = ||\int_{t_1}^{t_2} \dot{x}_m (S)dS|| \leq G(t_2-t_1)$$

which implies that the x_m are uniformly equicontinuous in $[o,T]$. Therefore by ASCOLI'S theorem the sequence x_m is relatively compact in $C([o,T]: \mathbb{R}^\ell)$ and hence there exists a subsequence (x_μ) of (x_m) converging uniformly to a limit x in $C([o,T]; \mathbb{R}^\ell)$ and such that the derivatives $\dot{x}_\mu$ converge uniformly in $[o,T]$ to a limit v which necessarily coïncides with the weak derivative $\dot{x}$ of x. We now consider any fixed $t_o \in [o,T]$ and denote by (h_μ) a subsequence of (h_m) such that : $t_o \in [k_\mu h_\mu,(k_\mu+1)h_\mu[$ for any μ; then for the same subsequence we also have $t_o-r \in [(k_\mu-L_\mu)h_\mu,(k_\mu+1-L_\mu)h_\mu[$. The uniform convergence of the derivatives $\dot{x}_\mu$ to $\dot{x}$ on $[-r,T]$ implies that for any given $\varepsilon > o$ there exists some integer μ_o such that $\mu \geq \mu_o$ implies :

$$(34) \qquad \sup_{t\in[-r,T]} ||\dot{x}_\mu (t) - \dot{x}(t)|| \leq \varepsilon.$$

Further as $\dot{x}_\mu$ coïncides with $\dot{y}$ on $[-r,o]$ and takes on $[k_\mu h_\mu,(k_\mu+1)h_\mu[$ resp. on $[(k_\mu-L_\mu)h_\mu,(k_\mu+1-L_\mu)h_\mu[$ the value $v_\mu^{k_\mu}$ resp. $v_\mu^{k_\mu-L_\mu}$ if $k_\mu h_\mu > r$, we have by (34) for $\mu \geq \mu_0$: $||\dot{x}_\mu(k_\mu h_\mu)-\dot{x}(t_0)|| \leq \varepsilon$ resp. $||\dot{x}_\mu([k_\mu-L_\mu]h_\mu)-\dot{x}(t_0-r)|| \leq \varepsilon$, and hence the sequences $\dot{x}_\mu([k_\mu-L_\mu]h_\mu)$ and $\dot{x}_\mu(k_\mu h_\mu)$ converge to $\dot{x}(t_0-r)$ and $\dot{x}(t_0)$ respectively. Moreover by the uniform equicontinuity of the (x_μ) and their uniform convergence to x on $[-r,T]$ the sequence $x_\mu(k_\mu h_\mu)$ converges to $x(t_0)$. Rewriting now relation (18) for $m=\mu$ and $j=k_\mu$ as :

(35) $\dot{x}_\mu(\tau_\mu) \in S(x_\mu(\tau_\mu), \dot{x}_\mu(\tau_\mu-r))+2^{-\mu}B$ with $\tau_\mu = k_\mu h_\mu$, it follows by (13) that $\dot{x}_\mu(\tau_\mu) \in S(x(t_0),\dot{x}(t_0-r))+ \varepsilon_\mu B$. with $\varepsilon_\mu = 2^{-\mu}+\frac{1}{2}$ A Max$(||x_\mu(\tau_\mu)-x(t_0)||, ||\dot{x}_\mu(\tau_\mu-r)-\dot{x}(t_0-r)||)$. As $\dot{x}_\mu(\tau_\mu)$ converges to $\dot{x}(t_0)$ for $\mu \to +\infty$ and as S has closed images we deduce from (35) that: $\dot{x}(t_0) \in S(x(t_0),\dot{x}(t_0-r))$ where $t_0 \in [0,T[$ is arbitrary. Finally as $x_\mu^k \in Q$ for any μ and $kh_\mu \leq Nr$, it follows that $x(t) \in Q$ for any $t \in [o,T]$. Q.E.D.

Remark : As $0 \in [o,h_\mu[$ we have $\dot{x}_\mu(o) = v_\mu^o$ for any integer μ; further we have $||v_\mu^o - w^o|| \leq 2^{-\mu}$ where $w^o \in S(y(o),\dot{y}(-r))$ has been chosen arbitrarily. This property joined to (34) implies that $\dot{x}(o) = w^o$. In conclusion for any given $w^o \in S(y(o), \dot{y}(-r))$ there exists at least one solution of Problem (P) satisfying :

$$\frac{dx}{dt}(o) = w^o$$

References

F.H.CLARKE, J.P. AUBIN [1977] Monotone invariant solutions to differential inclusions.
J. of the London Math. Society (2) 16 p. 357-366.

A.F. FILIPPOV [1971] On the existence of solutions of multivalued differential equations.
Math. Zametki. 10 p. 307-313.

J. HALE [1971] Theory of Functional Differential Equations.
Springer-Verlag.

N. OGUZTÖRELI [1966] Time lag Control Systems.
Academic Press.

Bounds for the Chaotic Behavior of Newton's Method

Dr. Helena S. Wisniewski

Dir. Appl. & Comp. Mathematics

Defense Advanced Research Projects Agency

Introduction: Given a real-valued function of the real line f(x) an elementary algorithm for finding zeroes of f is Newton's Method.

$$x_n = Tx_{n-1} \text{ where } Tx = x-f'(x)/f(x).$$

It is well-known that for suitable f this algorithm converges locally to zeroes of f. [Thomas pg. 466]. In particular this result is true for polynomials.

This paper discusses the case where f is a polynomial with all real coefficients and all real roots. The main result is that starting with almost all real x Newton's Method converges to a root of the polynomial, and moreover, the size of the set of points which do not lead to an orbit reaching a prescribed neighborhood of the roots through n iterations is bounded exponentially in n.

Theorem. Let p(x) be a polynomial with all real coefficients and all real, distinct roots. Let θ be a neighborhood of the roots of p. Let N_n be the set of x_0 such that $x_0, x_1, \ldots, x_n \notin \theta$ and $x_{n+1} \in \theta$. Then the measure of N_n is bounded by Ke^{-cn} where K,c are constants and the measure is calculated by mapping the real line onto the closed interval $[-\pi/2,\pi/2]$ by $s = \arctan(bx)$.

The constant c in the exponential in fact can be any number smaller than the the minimum of absolute value of T' evaluated on a finite set of exceptional points. (Section iv). The fact that Newton's Method converges for almost all x originally appeared in Barna [1] and again in Smale [4]. Other papers relevant

NATO ASI Series, Vol. F37
Dynamics of Infinite Dimensional Systems
Edited by S.-N. Chow, and J.K. Hale

to this topic are Barna [2] and Saari [3].

The proof of this theorem is in section iv. Section ii discusses the polynomial x^3-x in detail to illustrate the method of analysis used in obtaining this result. Section iii analyzes a fourth degree polynomial showing how "chaotic" behavior arises for this elementary algorithm. Section v contains a sketch of a proof using filtration arguments.

This result is an extension of the author's previous results concerning Diffeomorphisms on Compact Riemannian Manifolds [6], [7].

ii. Let p(x) = x3-x. Then Newton's Method for p is iteration via the rational function

$$Tx = 2x^3/(3x-1).$$

Figure 1 is a graph of T. It has asymptotes at the roots of p'(x), $\pm 1/\sqrt{3}$. Newton's Method produces a sequence of points on this graph. Throughout this paper Newton's Method will be considered variously to produce a sequence on (a) the real line, (b) the graph of T, or (c) the vertical axis. All are equivalent views.

The roots of p are -1, 0, and 1. The points (-1,-1), (0,0), and (1,1) are precisely the points where the line y=x intersects the graph of Tx. Newton's Method consists of starting at a point (x,Tx), tracing along the horizontal line x=Tx until it intersects the line y=x, and moving vertically to the graph of Tx. The new point produced is (Tx,T(Tx)) (figure 2).

The horizontal lines $x= \pm 1/\sqrt{3}$ partition the graph. Any number x for which (x,Tx) is below the line $x=-1/\sqrt{3}$ produces an orbit which converges to (-1,-1). Any number whose point on the graph of T falls above $x=1/\sqrt{3}$ produces an orbit which converges to (1,1) (figure 3a). The interesting behavior occurs in $(-1/\sqrt{3},1/\sqrt{3})$. There is, of course, an interval of convergence around zero. The interval of convergence for zero is $(-1/\sqrt{5},1/\sqrt{5})$ with $\{\pm 1/\sqrt{5}\}$ forming a period 2 orbit (figure 3b). In the intervals $(-1/\sqrt{3},-1/\sqrt{5})$ and $(1/\sqrt{5},1/\sqrt{3})$ almost all points lead to convergence to either -1 or 1. The set of starting points for which Newton's Method does not converge consists of $\pm 1/\sqrt{3}$ and preimages for all orders of $\pm 1/\sqrt{3}$. Tracing the preimages reveals that the two intervals ($(-1/\sqrt{3},-1/\sqrt{5})$ and $(1/\sqrt{5},1/\sqrt{3})$) consist of alternating subintervals -- that is alternating in the sense that the preimages of $\pm 1/\sqrt{3}$ alternate on either side of zero and the subintervals determined by these preimages converge respectively to -1 and 1. The alternating subintervals can be labelled in terms of the number of iterations under T required for orbits originating in them to leave the inner component of the graph (figure 4). Note that the fundamental interval on the left maps onto the portion of the inner component below the line $x = -1/\sqrt{3}$, and that on the right maps onto the portion above the line $x = 1/\sqrt{3}$. In fact the intervals formed by $\pm 1/\sqrt{3}$ and the first preiterates of $\pm 1/\sqrt{3}$ (labelled L and R respectively) form a fundamental domain for T at the period two points in the sense that any orbit origniating near the period two points passes through one of these two intervals (for one iteration) before leaving the inner component of the graph of T.

Knowing in detail the ultimate fate of every point under T, then estimating the measure of the set N_n is straightforward. Given $\epsilon > 0$ we are interested for any n in the set of points whose orbits under Newton's Method remain outside of ϵ neighborhoods of the three roots through n iterations. We iterate the three ϵ neighborhoods backwards to determine the components of N_n (figure 4). These are all intervals. For large n the size of the subintervals of N_n is a function of the value of T' at $\pm 1/\sqrt{5}$ and infinity. This is because orbits which do not reach the neighborhood of some root must remain (or originate) near the period two orbit or far out near $\pm\infty$.

Since

$$T'x = \frac{6x^4-6x^2}{9x^4-6x^2+1}$$

T' is less than one (2/3) at infinity. However, mapping the real line to $[-\pi/2,\pi/2]$ via arctan reciprocates this to 3/2. More precisely, instead of using Tx we consider θs defined by

$$\theta s = \arctan[T(\tan(s)], \text{ and}$$

$$\theta'(s) = T'(x)\ [(1+x^2)/(1+T^2x)].$$

As x goes to $\pm\infty$ s goes to $\pi/2$ and θ' goes to 3/2. More generally for any polynomial θ' approaches n/n-1 for x approaching $\pm\infty$. For the value of the derivative at the period two points we have

$$T'(\pm 1/\sqrt{5}) = -6.$$

However, since these are period two points for estimating the effect of iteration on length, we should use the square root of the derivative of T composed with itself. Thus, the formula for θ' above gives

$$d/ds\ \theta(\theta s)|_{1/\sqrt{5}} = T'(1/\sqrt{5})\ T'(-1/\sqrt{5}) = 36.$$

The square root of this value, 6, is the value which indicates the expansion around the period two orbit.

The estimate is

$$\mu(N_n) < Ke^{-cn}$$

with c any number less than 3/2 and $K = K(c)$ accounting for the bounded number of iteration needed for the derivatives over all the subintervals making up N_n to be greater than e^c. Even if the derivatives were uniformly bounded below by 3/2, c must be less 3/2, since the number of intervals in N_n grows algebraically in n.

This argument proves the theorem for all cubic polynomials with all real roots and real coefficients. The only point in doubt is whether the period two orbit is always an expanding source. A simple argument that this is so results from considering the rectangle formed by horizontal and vertical lines determined by the period two points (figure 5). The slopes of the lines from the root to the period two points are reciprocals. Since they are also lower bounds (in absolute value) for the derivative of T at the period two points, then the product of the derivatives at the period two points is greater than one. The derivative of T composed with itself at either period two point is the product of the derivatives of T at the period two points so the period two orbit is an expanding source. For polynomials of degree greater than three we will need the stronger result that each of the derivatives at the period two points is greater than one in absolute value.

iii. $p(x)=(x^2-1)(x^2-4)$. Interestingly enough this simple example will show just how complicated Newton's Method can get. The graph of T is shown in figure 6. Much of the analysis of section ii is the same for this example. For one thing there is a period two orbit which bounds the local interval of convergence of each interior root of p(x). However, since the graph of T for this quartic polynomial has two interior sections, the accounting of components of the sets N_n becomes much more complicated. To see this clearly note the regions of the graph labelled I, II, III, IV. Region I is that portion of the graph between the horizontal lines through the first asymptote and the first period two point.etc. Points below region I lead to convergence to (-2,-2); points above region IV lead to convergence to (2,2); points between I and II and III and IV converge to (-1,-1) and (1,1) respectively. The points which do not lead to convergence are in one of the four bands I, II, III, IV. Moreover, their orbits must remain in the four bands for all iterations.

Consider the folling relations:

$$T(I) \supseteq II \cup III \cup IV;$$

$$T(II) \supseteq I;$$

$$T(III) \supseteq IV; \text{ and}$$

$$T(IV) \supseteq III \cup II \cup I.$$

Using these relations and the fact that T is expanding in the bands I, II, III, and IV establishes the existence of periodic points of all orders. In particular, there is an orbit of period three with points in I, III, and IV (figure 7). The argument for the existence of this period three orbit goes as follow.

The mapping relations

I --> III,

III --> IV, and

IV --> I

mean that I maps onto III, III maps on IV, and IV maps onto I. Taking inverse images starting with I yields

$$I1 = I \cap T^{-1}(III \cap T^{-1}(IV \cap T^{-1}(I))).$$

Here the set I_1 is a proper subset of I since $T'<-1$ on I, III, and IV. In fact, it is a nested subinterval of an interval in I. This relationship produces a sequence $\{I_n\}$ of nested (closed) intervals whose lengths decrease to zero. By Cantor's Theorem they have a unique point of intersection -- say x_1. Then

$$Tx_1 = x_2, \; Tx_2 = x_3, \text{ and } Tx_3 = x_1 \qquad \text{with}$$

$$x_1 \in I, \quad x_2 \in III, \text{ and } x_3 \in IV.$$

Any unique repeating sequence constructed with the mapping rules between I, II, III and IV defines a unique periodic orbit. Additionally, any sequence constructed with the rules, even if not repeating defines an orbit which remains for all n in the bands I, II, III and IV and hence does not converge to any root. The non-convergent set is uncountable (It is a Cantor set!). The fact that it has measure zero is a consequence of the main theorem of this paper. It was first proved by Barna [1].

The sets N_n for this example consist of three parts.

1) Those points in the local convergence regions --- four of them --- whose orbits do not reach ϵ neighborhoods of the roots in n iterations;

2) Those points in the bands I, II, III, IV whose orbits do not leave the bands in n iterations (this includes the non-convergent set); and

3) Those points in the bands I, II, III, IV whose orbits leave the bands but do not reach the e neighborhoods in n iterations.

Because of the mixing of the bands I, II, III, and IV, accounting for the components of N_n becomes complicated. The estimate of the size of N_n will be taken from the main theorem, proven in the next section. Note that the period three example above reveals that $T'x<-1$ is required for all period two points which bound the local intervals of convergence.

iv. <u>Theorem</u>. Let p(x) be a dth degree polynomial with all real coefficients and d distinct real roots -- $r_1, r_2, \ldots, r_d$. Let Tx be the rational function

$$Tx = x-p(x)/p'(x).$$

Let O be a neighborhood of the roots of p(x) and

$$N_n = \{x: T^i(x) \notin O,\ i=1,2,\ldots,n,\ Tn+1x \in O\}.$$

Let $\theta s = \arctan(bT(\tan s)/b)$. Then the measure of the sets $\theta(N_n)$ is bounded by $K \exp(-cn)$ where $c = \min\{d/(d-1)$ and T'over all period two points$\}$.

Proof: The proof consists of a series of lemmas and observations.

Observation: Since

$$Tx = x - p(x)/p'(x)$$

and

$$T'x = [p(x)p''(x)]/[p'(x)]^2,$$

The graph of T consists of n connected components separated by asymptotes at the roots of $p'(x)$ -- $s_1,\ldots,s_d-1$. Each of the inner d-2 components also contains one of the roots of $p''(x)$ -- $t_1, t_2, \ldots, t_{d-2\theta}$ $T'>0$ to the left of r_1, to the right of $r^d\surd$ and between r_i and t^i-1 for $i=2,\ldots,d-1$.

Lemma: (Barna) For the inner d-2 connected components of the graph of T the slope equals every negative number exactly twice, once to the left of the contained root of $p(x)$ and once to the right.

The previous observation leads to the conclusion that $T'<0$ for exactly 2d-2 intervals, with two in each of the n-2 inner components of the graph of T. Moreover, in each interval for which $T'<0$ T' takes on all negative values. The fact that it takes on each negative value exacly once per interval results from the following equation.

$$[p(x)p''(x)]/[p'(x)]^2 = -a^2, \text{ or}$$

$$p(x)p''(x) + a2[p'(x)]^2 = 0.$$

This is a polynomial equation of degree 2d-2. It is known to have 2d-2 roots from above. Those are all its roots.

Lemma. All points for which $T'>0$ produce a sequence under iteration by T which converges to the root of p(x) contained in the same conponent of the graph.

Proof. T' is positive

1) To the left of r_1,
2) To the right of r_d, and
3) Between r_i and t_{i-1} for $i=2,...,d-1$.

The graph of T is above the line y=x to the left of r_1, below the line y=x to the right of r_d, and either below or above the line between r_i and t_{t-i} depending on whether $r_i<t_{i-1}$. Assume without loss of generality that the graph of T lies below the line y=x. Given any point s_0 in this region of increasing T, $r<s_0$ which implies that $r<Ts_0 = s_1$. Since the graph is below the line y=x, then the horizontal line through (s_0,Ts_0) intersects y=x to the left of s_0. Thus

$$r < s_1 = Ts_0 < s_0.$$

The infimum for this sequence s_n is clearly r. (figure 8)

Lemma. Each interior root $r_2, \ldots, r_{d-1}$ has a local interval of convergence entirely within the same component of the graph of Tx bounded by a period two orbit. All orbits begining in the leftmost(rightmost) component of the graph of Tx -- $(-\infty, r_1)$ resp. $(r_d, +\infty)$ -- converge to $r_1(r_d)$ without leaving the component.

Proof. From elementary calculus we have the existence of an interval of convergence around each root in which iteration by Tx converges to the contained root. In fact the entire interval (r_i, t_{i-1}) or (t_{i-1}, r_i) is in the interval of convergence by the previous lemma.

Now let x_1 be a point outside of (r_i, t_{i-1}) which converges to r_i -- say $r_i - \delta$ for small δ. Construct a sequence by choosing preimages of x_1 which are contained in the same component of the graph of T. Since T and hence T^{-1} is alternating and monitone about the root for this sequence, the two subsequences of odd and even preimages are monotone and bounded (figure 9). Hence they converge to x_l* and x_r* with $x_l^* < x_r^*$. Also

$$T(x_{2n+1}) = x_{2n} \text{ so that in the limit}$$

$$T(x_l) = x_r. \text{ Similiarly } T(x_r) = x_l.$$

Hence the pair x_l and x_r forms a period two orbit and the interval (x_l, x_r) is the maximal local interval of convergence.

Observation: For the leftmost component of the graph $(-\infty, r_1)$ is in the local interval of convergence for r_1 because $T'>0$ over the entire interval. Any point in (r_1,s_1) maps in one iteration to a point in $(-\infty,r_1)$. Thus the entire leftmost component is the "local" interval of convergence for r_1. The same argument works for r_d.

Observation: Any point whose orbit does not converge to a root is in the intervals between roots of $p'(x)$ and the local intervals of convergence of the inner d-2 roots. Moreover, all preimages of these points also must be in the intervals (bands I, II, III and IV in section iii).

Lemma. The period two points which bound the local interval of convergence for an inner root form a source for iteration by T.

Proof: This was proved by the geometric argument at the end of section ii.

A stronger and needed result is

Lemma. (Barna [2]) The value of T' evaluated at the period two points which bound the local interval of convergence for an inner root is less than -1.

This lemma is needed for the estimation because of the mixing between the bands of nonconvergence.

Observation: There are periodic orbits of all periods for the dynamical system defined by iteration under T. This follows from the fact that the images of the nonconvergent bands contain other nonconvergent bands and that T is expanding on the nonconvergent bands.

Definition: Let $\Theta s = \arctan(bT(\tan s)/b)$.

Throughout the remainder of this proof all calculations will be made using

θ (on the interval $[-\pi/2,\pi/2]$). For economy of notation r_i, s_i, and t_i will represent both the zeroes of p, p' and p'' and their images under arctan(bx).

Observation: Including the constant b>0 in the definition of θ allows one to make the values of θ' at the period 2 points (for θ) as close as desired to the values of T' at the period 2 points. This is because of the formula for θ' in terms of T'.

$$\theta' s = T'x\ (1+[bx]^2)/(1+[bTx]^2).$$

The constant b does not alter the value of θ' at $\pm\infty$. That remains d/(d-1).

Observation.

N_n is a finite union of intervals, since it consists of preiterates of the d^2-d+2 intervals in N_1 which is $\theta^{-1}[M_1]$ - int M_1. The ϵ intervals around r_1 and r_d lead to two each in $(-\pi/2,r_1)$ and $(r_d,\pi/2)$ and d-2 each in (r_1,r_d). The ϵ intervals around the d-2 inner roots account for (d-1) each or (d-2)(d-1) total (figure 10).

Discussion: (refer to figure 11a)

On the inner d-2 components of the graph of T (θ) and outside the period 2 orbit (say $\{s_1,s_r\}$)

$$|\theta' s| > |\theta' s_1| \quad \text{or} \quad |\theta' s| > |\theta' s_r|.$$

Inside the period 2 orbit this is not true because θ' is monotone where it is negative. However, given any $\delta > 0$, there exist intervals I_1 and I_r around s_1 and s_r respectively such that

$$|\theta' s| > |\theta' s_1| - \delta \quad \text{and} \quad |\theta' s| > |\theta' s_r| - \delta$$

for all s in I_1 and I_r.

Given any interval J in either I_1 or I_r for which a preimage K lies entirely in I_1 or I_r --- i.e. $\theta(K) = J$ with both K,J contained in I_1 I_r ---, then the length of K is bounded by the length of J divided by the minimum of the derivative of θ over J.

Let $|K|$ = length of K = $|b-a|$ and $|J| = |\theta b - \theta a|$. By the mean-value theorem

$$|\theta b - \theta a| = |b-a|\ \theta' c \quad \text{for some } c \in K.$$

Thus

$$|J| \geq |b-a|\ m = |K|\ m \qquad \text{where}$$

$$m = \min \{|\theta' s| : s \in (a,b). \quad \text{So}$$

$$|K| \leq |J|/m.$$

A similar situation is present on the two outer components of the graph (figure 11b). There is an interval I_1 to the left of r_1 such that

$$s \in I_1 \qquad \theta' s > (d/d-1)\ \delta.$$

Therefore, there exist 2d-2 intervals in which any point satisfies

$$|\theta' s| > \min\{\theta' t\} - \delta$$

where the minimum is taken over all the period 2 points and $\pm\pi/2$ (∞).

<u>Observation</u>: Any connected component of N_n between a fundamental period 2 point on an inner lobe of the graph and the associated asymptote, is a multiple preiterate of one of the fundamental intervals for the component of the graph in which it lies. The length estimate from above applies to every such iteration. Therefore, any component of N_n has the length estimate apply to it for at least n-k-1 iterations. This can be summarized as the following lemma.

<u>Lemma</u>: N_n is the union of a finite number of intervals which for $n>k$ are all in the regions of $[-\pi/2,\pi/2]$ in which $|\theta'|$ is bounded above 1. Forward iterates of the components of N_n are also in the regions for at least n-k-1 iterations. The number of components of N_n grows algebraically (actually polynomially) in n.

<u>Proof</u>: This has been mostly proved already. The fact that the number of components grows algebraically in n follows from the fact that T (and hence θ) is multi-valued. The k+1 for which an iterate of N_n is not in the regions where $|\theta'|>1$ is necessary to cover the last k iterations until the iterates reach 0 and the one iteration during which iterates mapping onto (r1,s1) or (sd,rd) and hence to I_1 and I_r are not in the regions.

<u>Observation</u>: To complete the proof simply requires that one sum up the estimates of the sizes of the components of N_n. Each component, J, satisfies

$$\mu(J) \leq Ke\text{-}cn \quad \text{with } K=K(\epsilon) \text{ accounting for the } k+1$$

iterations for which an iterate is not in the regions of $|\theta'|>1$, and c any number smaller than the minimum of $|\theta'|$ at the period two points and d/(d-1). Since the number of components in N_n grows only algebraically, the same estimate holds for N_n with K and c redefined. That is,

$$\mu(N_n) \leq K_1 \exp(-c_1 n) \quad \text{with } c_1 < c.$$

However, since the only restriction on c is that it be smaller than the minimum of $|\theta'|$ at the exceptional points, c_1 faces only that same restriction. Also, by appropriate choice of θ the derivative of θ at the exceptional points can be made arbitrarily close to the derivative of T at the exceptional points. Therefore, the estimate

$$\mu(N_n) \leq Ke\text{-}cn \quad \text{holds for } K = K(\epsilon) \text{ and}$$

$$c < \min\{|T'x| \text{ for x a period two point}, d/(d-1)\}.$$

This completes the proof.

v. The proof is more succinct when made using the notion of a filtration.

Definition: Given a endomorphism f on a compact manifold M, a filtration for f is a sequence of compact submanifolds with boundary such that:

$$M = Mk \supseteq M_{k-1} \quad \ldots . \supseteq M_0 = 0,$$

$$\text{Dim } M_i = \text{Dim } M \text{ for all } i,$$

$$f(M_k) \subseteq \text{int } M_k.$$

Observation: θ is an endomorphism on the interval $[-\pi/2, \pi/2]$ and constructing a filtration for θ is the next order of business.

Lemma. Given $\epsilon > 0$ let M_1 be the union of the closures of the ϵ neighborhoods of the roots, let M_2 be the union of the closures of all the local intervals of convergence (including the intervals $(-\pi/2, r_1)$ and $(r_d, \pi/2)$), and let M_3 be the entire real line. Then M_0, M_1, M_2, and M_3 form a filtration for the dynamical system defined by iteration by θ (figure 12).

Proof. The only thing to check is that $\theta(M_i)$ is contained in int M_i. But this is obvious.

Lemma: Any interval J for which $\theta^i(J)$, i=1,...,m, are contained in M_j - int M_{j-1} for m > k, j = 2,3 satisfies

$$|J| \leq |\theta m\text{-}_k(J)| \exp[-c(m-k)].$$

For j=2 c is any number less than d/(d-1). For j=3 c is any number less than the minimum of $|\theta'|$ taken over the period two points.

Observation: The intervals I_1, I_r, etc. defined above are in either M_2 - int M_1 or in M_3 - int M_2. Any connected component of N_n which remains in either M_2 - int M_1 or M_3 - int M_2 for a bounded number of iterations -- say k iterations -- will be inside one of the intervals.

Lemma: Any interval J for which

$$J \quad M_3 - \text{int } M_2, \text{ and}$$

$$\theta(J) \quad M_2 - \text{int } M_1, \text{ satisfies}$$

$$\theta^2(J) \quad (-\pi/2, r_1) \text{ or } (r_d, \pi/2).$$

Observation: N_n consists of three parts:

C1) Those points in M_3 - int M_2 whose orbits remain in M_3 - int M_2 for all n iterations;

C2) Those points in M_2 - int M_1 whose orbits remain in M_2 - int M_1 for all n iterations; and

C3) Those points in M_3 - int M_2 whose orbits reach M_2 but not int M_1 through n iterations.

<u>Lemma</u>: The components of N_n satisfy

$$\mu(C_1) \geq K_1 \exp[-c_1 n],$$

$$\mu(C_3) \leq K_3 \exp[-c_3 n], \text{ and}$$

$$\mu(C_2) \leq K_2 \exp[-c_2 n].$$

c_1 is the minimum of $|\theta'|$ over the period two points,

c_3 is $d/(d-1)$, and

c_2 is $\min\{c_1, c_2\}$.

FIGURE 1

GRAPH OF $Tx = (2x^3)/(3x^3 - 1)$

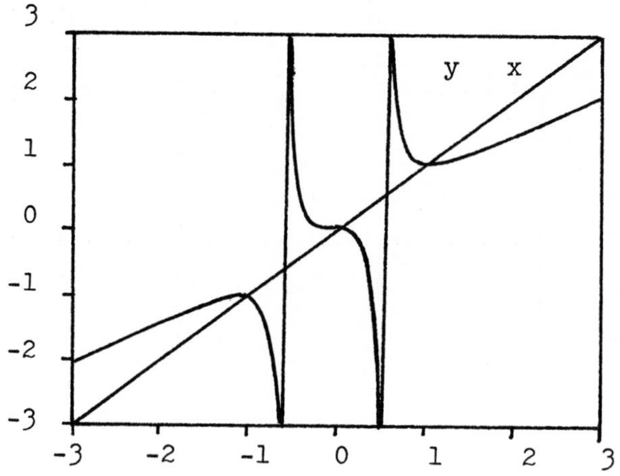

FIGURE 2

NEWTON'S METHOD

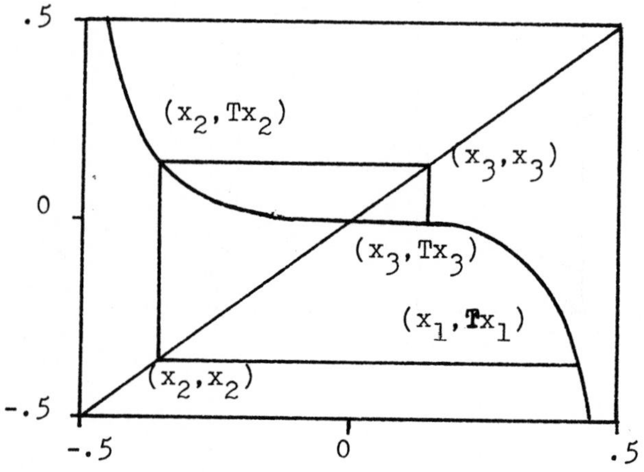

FIGURE 3a

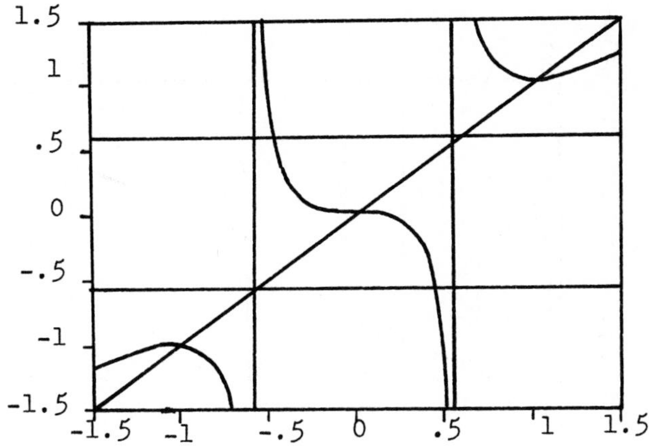

FIGURE 3b
PERIOD 2 ORBIT

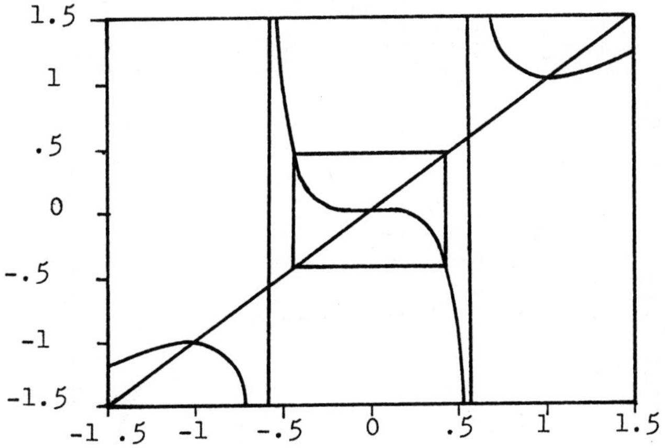

FIGURE 4

COMPONENTS OF N_n FOR n 1,2

L,R FUNDAMENTAL DOMAINS

* PERIOD 2 ORBIT

GRAPH OF $Tx = (2x^3)/(3x^2 - 1)$

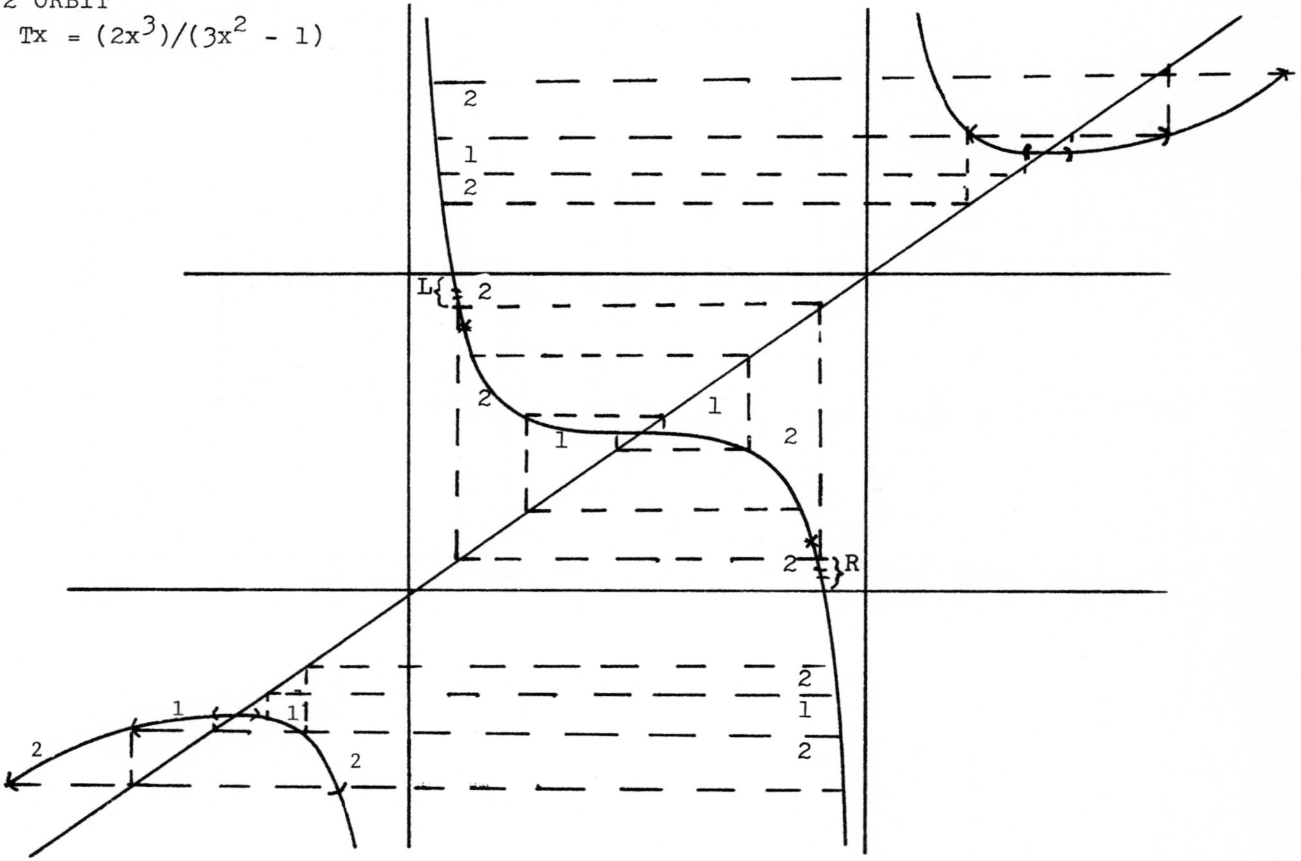

FIGURE 5

SLOPE $\ell_1 = -b/a$ SLOPE $\ell_2 = -a/b$

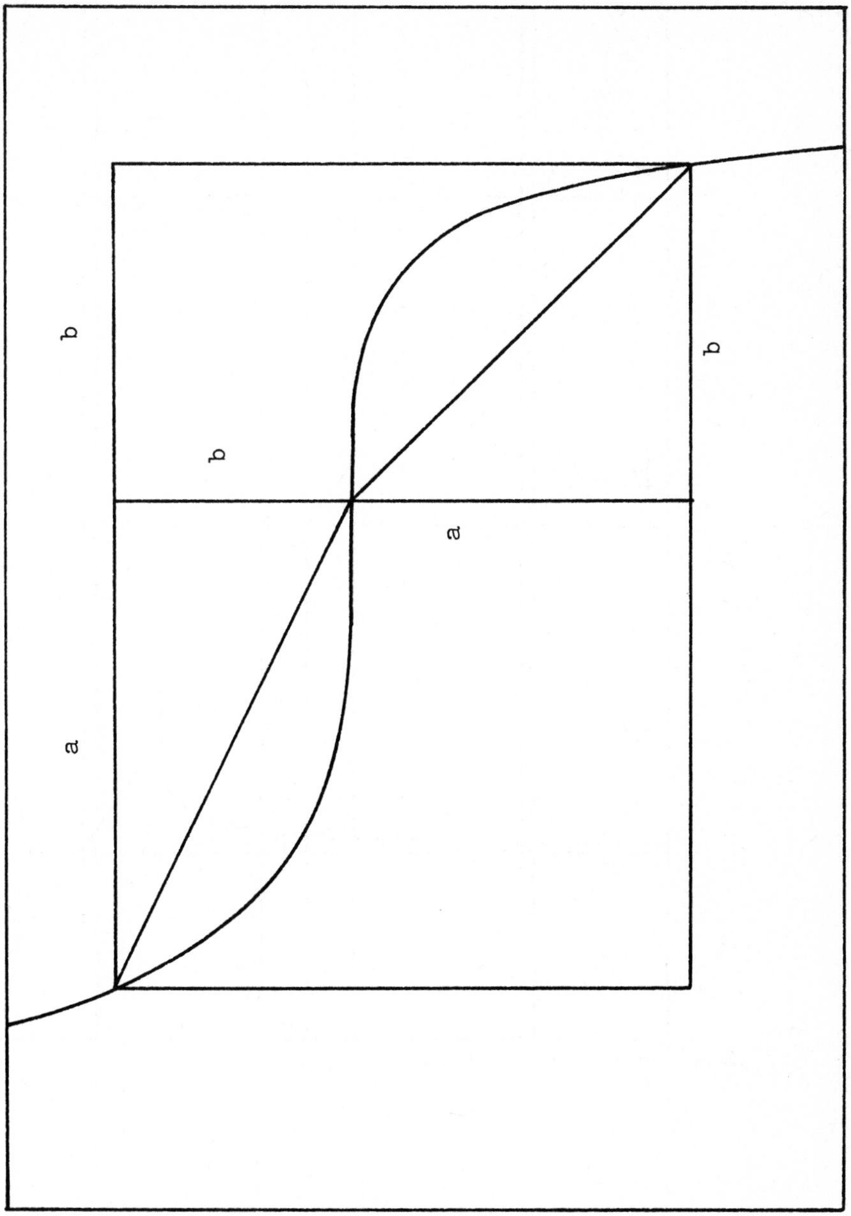

FIGURE 6

GRAPH OF Tx FOR $P(x) = (x^2 - 1)(x^2 - 4)$

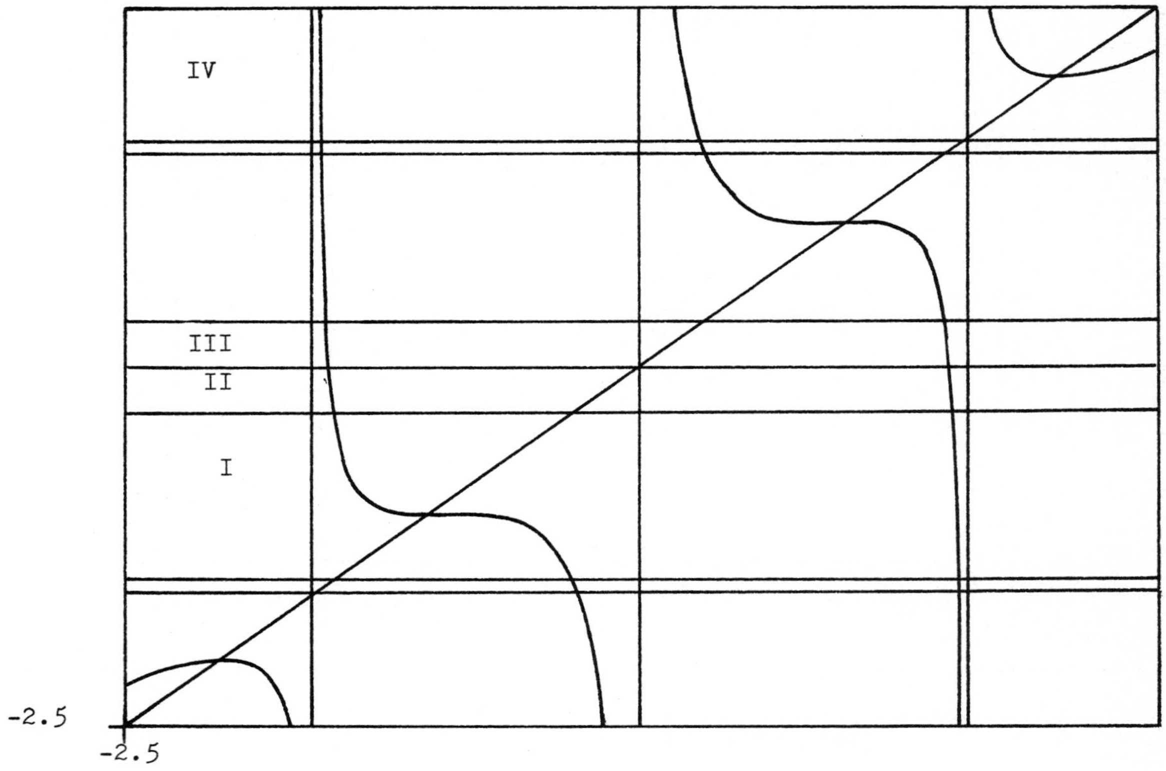

FIGURE 7
PERIOD 3 ORBIT

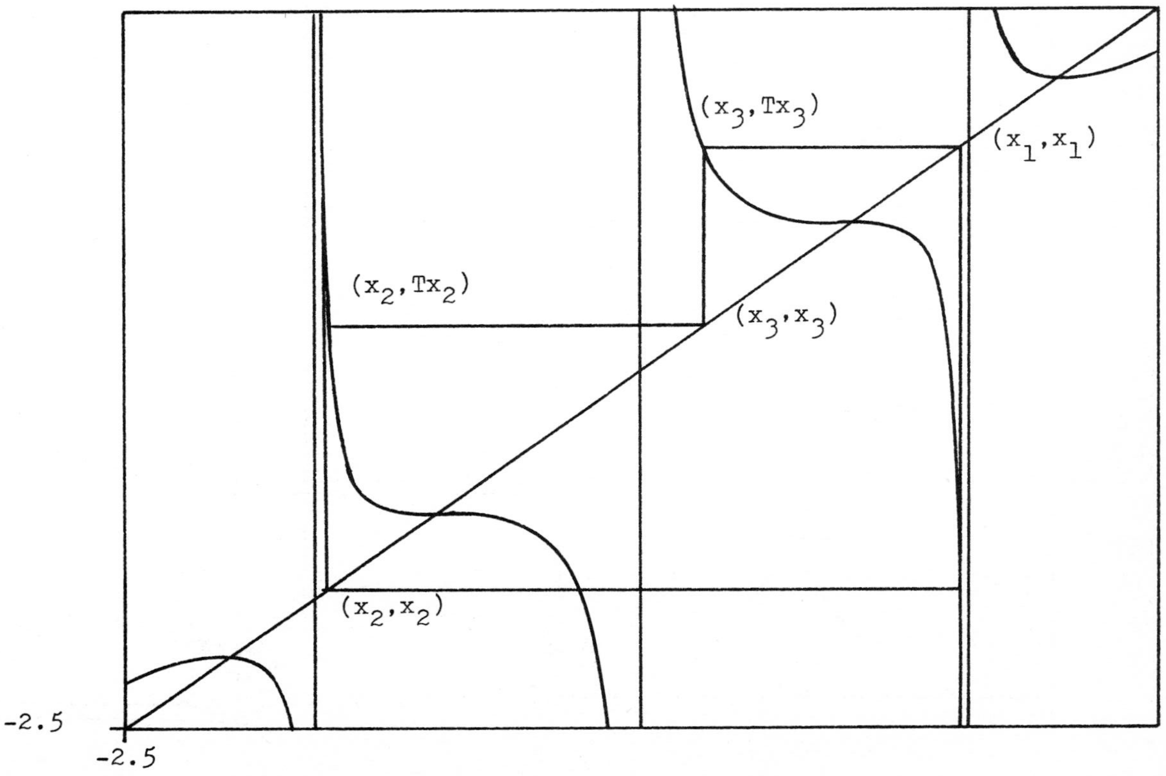

FIGURE 8

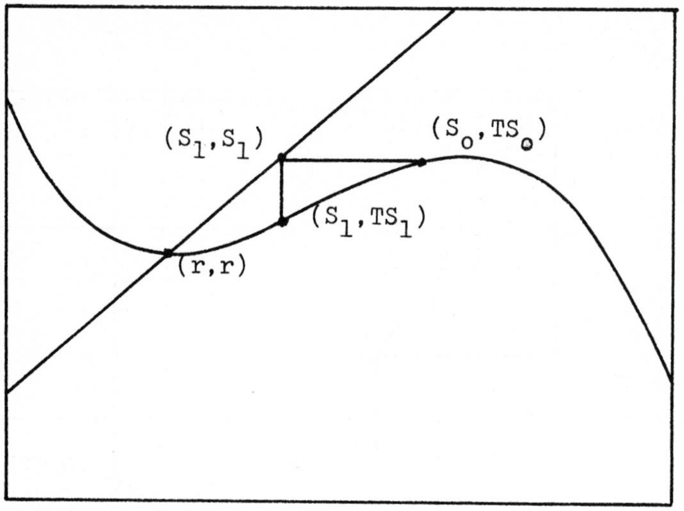

FIGURE 9

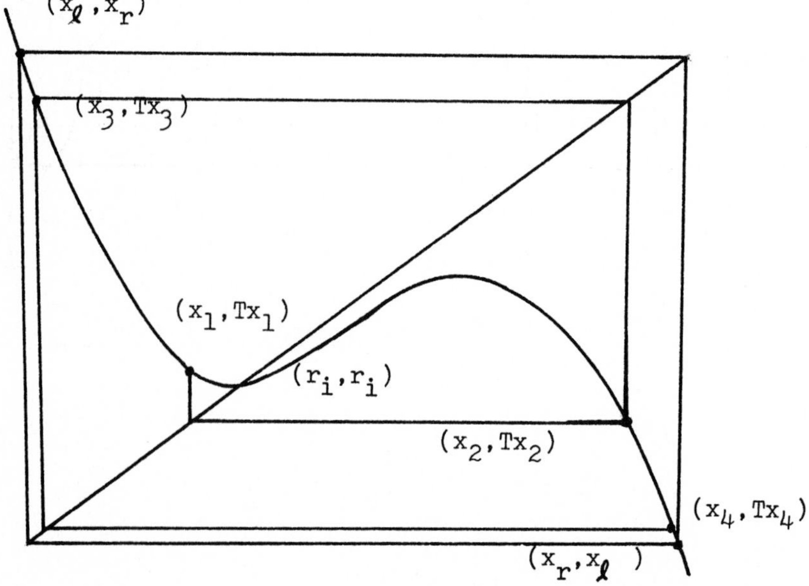

FIGURE 10

COMPONENTS OF N_1 FOR THE QUARTIC CASE. E_1, E_2, E_3, E_4 ARE THE ε NBHDS OF THE ROOTS.

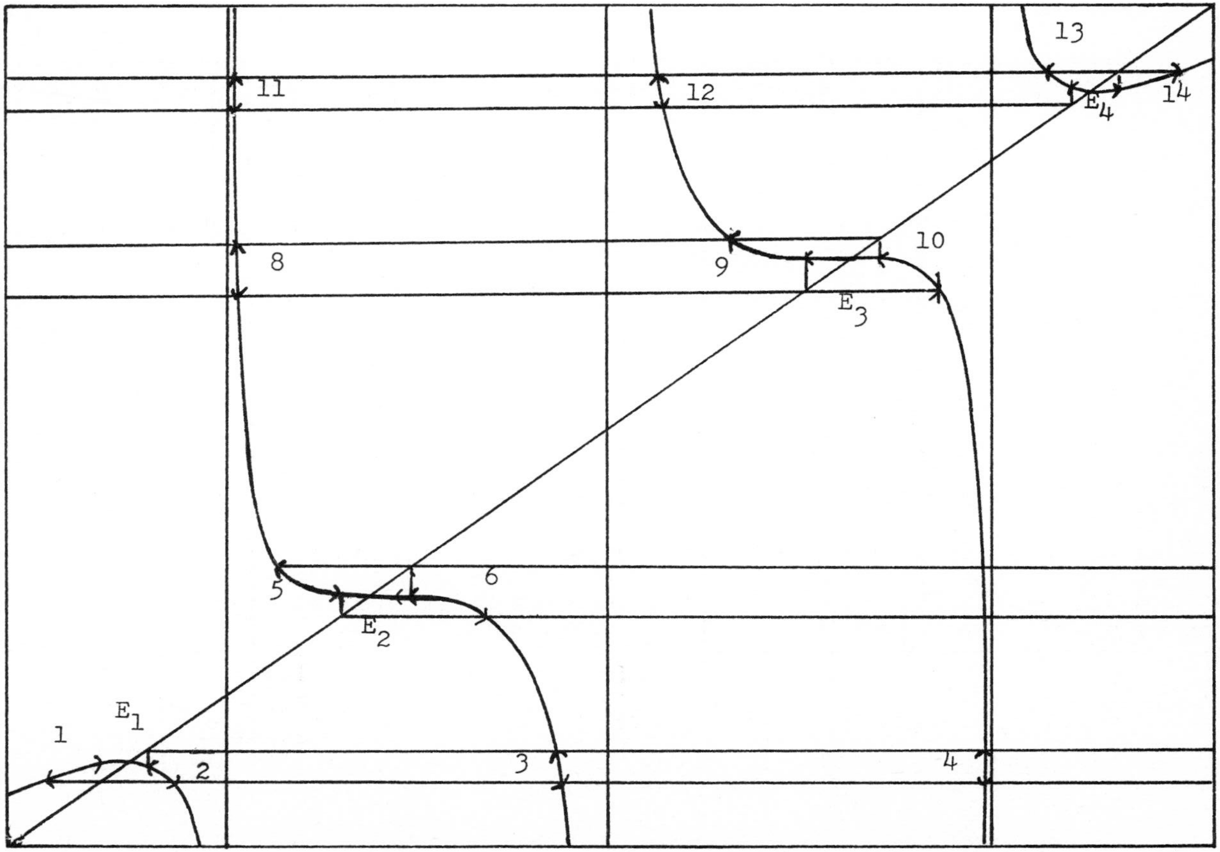

FIGURE 11a

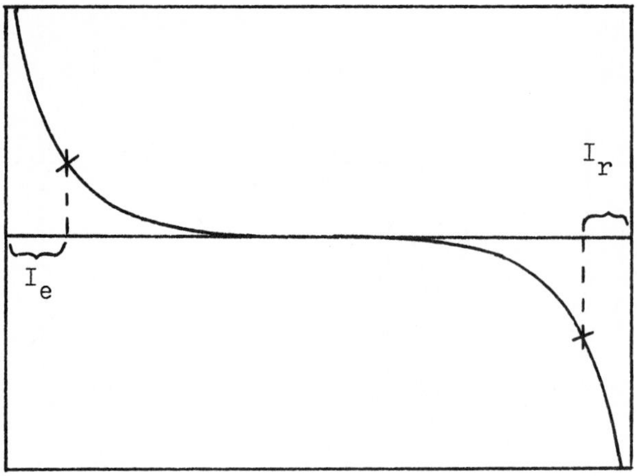

FIGURE 11b

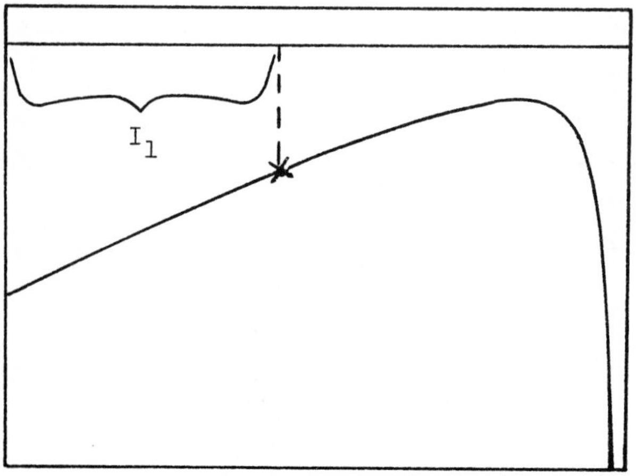

FIGURE 12

$\phi \subset M_1 \subset M_2 \subset M_3 = R$

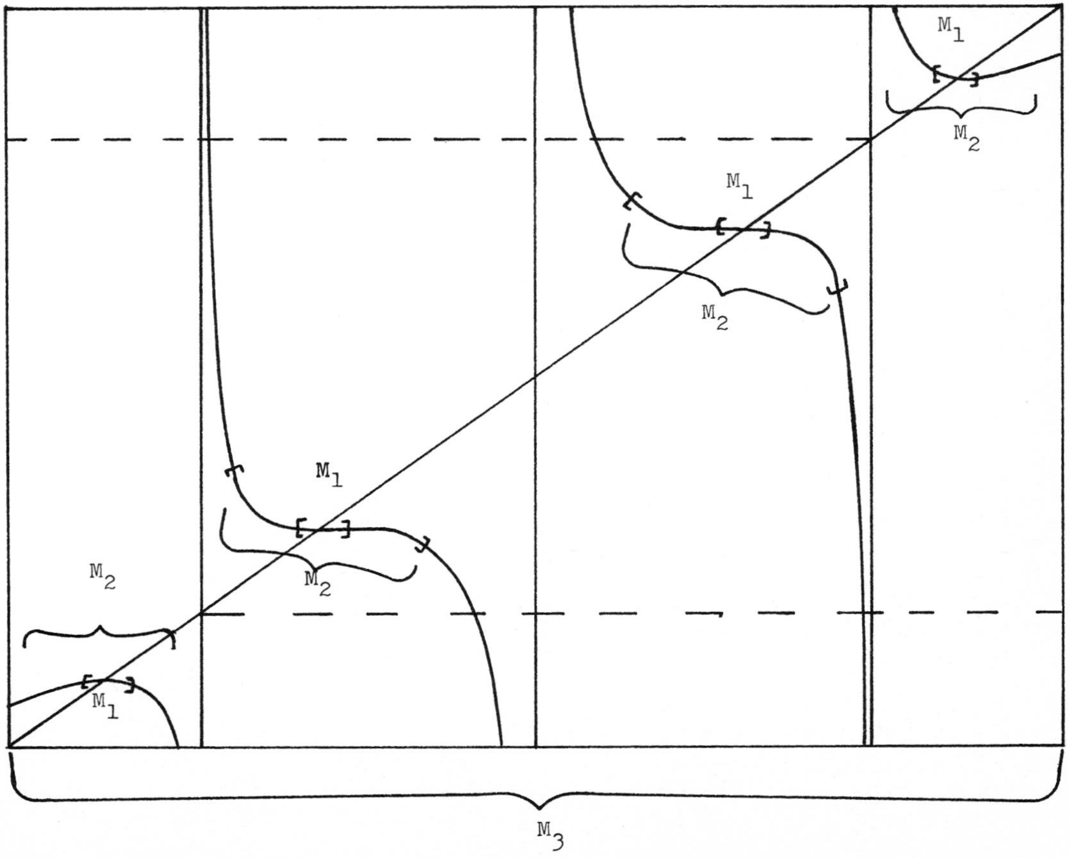

References

1. B. Barna, Uber die diverfenzpunkte des Newtonschen Verfahrens zur Bestimmung von Wurzelen algebraischer Gleichungen II, Publicationes Mathematicae, Debrecen, 4 (1956) 384-397.

2. B. Barna, Uber die diverfenzpunkte des Newtonschen Verfahrens zur Bestimmung von Wurzelen algebraischer Gleichungen III, Publicationes Mathematicae, Debrecen, 8(1961) 193-207.

3. D. Saari and J. Urenko, Newton's Method, Circle Maps, and Chaotic Motion, Amer. Math. Monthly, 91 (1984) 3-17.

4. S. Smale, The fundamental theorem of algebra and complexity theory, Bull. Amer. Math. Soc., 4 (1981) 1-37.

5. G. Thomas, Calculus and Analytic Geormetry, Alternate Edition Addison-Wesley, 1972.

6. H. Wisniewski, Rate of Approach to Minima and Sinks - The C^2 Axiom A No Cycle Case, Geometric Dymamics, Proc. Int. Symp. Dynam. Sys., Rio de Janiero, Brazil, 1981, Springer-Verlag Lecure Notes in Math., vol. 1007.

7. H. Wisniewski, Rate of Approach to Minima and Sinks - The Morse Smale Case, Trans. Amer. Math. Soc., vol 284, number 2, 1984.

LIST OF PARTICIPANTS

Amann, H., Mathematisches Institut, Universitaet Zürich, Raemistrasse 74, CH-8001, Zürich, Switzerland

Angenent, S.B., Department of Mathematics, University of Leiden, Niels Bohr Weg 1, Leiden, The Netherlands

Aronson, D.G., School of Mathematics, University of Minnesota, Minneapolis, MN 55455, U.S.A.

Azenha, A., Inst. Sup. Técnico, Universidade Tecnica de Lisboa, Av. Rovisco Pais, 1096 Lisboa Codex, Portugal

Ball, J.M., Department of Mathematics, Heriot-Watt University, Riccarton, Edingburgh, EH14 4AS Scotland, U.K.

Baser, U., Mathematics Department, Marmara University, Findilkzade, Istambul, Turkey

Bates, P., Department of Mathematics, Brigham Young University, Provo, UT 84602, U.S.A.

Burton, T.A., Department of Mathematics, Southern Illinois University, Carbondale, IL 62901, U.S.A.

Calsina, A., Department de Matematiques, Universitat Autonoma de Barcelona, Bellaterra, Barcelona, Spain

Cañada, A., Departamento Análisis Matemático, Universidad de Granada, 18071, Granada, Spain

Carr, J., Department of Mathematics, Heriot-Watt University, Riccarton, Edingburgh EH14 4AS, Scotland, U.K.

Carvalho, C., Departmento de Matematica, Faculdade de Ciencias de Lisboa, Lisboa, Portugal

Casal, A., Universidad Complutense de Madrid, 28040, Madrid, Spain

Chossat, P., Department of Mathematics, University of Nice, Parc Valrose, 06034 Nice Cedex, France

Chow, S.-N., Department of Mathematics, Michigan State University, Wells Hall, East Lansing, MI 48824, U.S.A.

Dafermos, C., Division of Applied Mathematics, Brown University, Providence, RI 02912, U.S.A.

Diekmann, O., Centre for Mathematics and Computer Science, P.O. Box 4079, 1009 AB Amsterdam, The Netherlands

Fachada, J.L., Faculdade de Ciencias de Lisboa, Lisboa, Portugal

Fiedler, B., Inst. of Applied Mathematics, Im Neuenheimer Feld 326, Universitaet Heidelberg, D-6900 Heidelberg, FRG

Ferreira, J.M., Centro de Fisica da matéria, Condensada, Av. Gama Pinto, 1699 Lisboa Codex, Portugal

Fitzgibbon, W.E., Department of Mathematics, University of Houston, University Park, Houston, TX 77004, U.S.A.

Freitas, P.S., Departmento de Matematica, Instituto Superior Tecnico, 1096 Lisboa Codex, Portugal

Fujii, H., Institute of Computer Science, Kyoto Sangyo University, Kyoto 603, Japan

Fusco, G., Dipartimento di Metodi e Modelli Matematici, Università di Roma "LaSapienza", Via A. Scarpa, 10, 00161 Roma, Italy

Grossinho, M.R., Centro de Fisica da Materia Condensada, Av. Gama Pinto, 2, 1699 Lisboa Codex, Portugal

Hale, J., Division of Applied Mathematics, Brown Univerity, Providence, R.I. 02912, U.S.A.

Henry, D.B., Instituto de Matematica e Estatistica, Universidade de Sao Paulo, CX Postal, 20570 Sao Paulo, Brazil

Iooss, G., Department of Mathematics, University of Nice, Parc Valrose, 06034 Nice Cedex, France

Jaeger, W., SFB 123, Universitaet Heidelberg, D-6900 Heidelberg, West Germany,

Keilhofer, H., Institut fuer Mathematik, Universitaet Augsburg, Memminger Str. 6, D-8900, Augsburg, West Germany

Kirchgassner, K.W., Mathematisches Institut A, Universitaet Stuttgart, Pfaffenwaldring 57, 7000 Stuttgart 80, West Germany

Knops, R.J., Department of Mathematics, Heriot-Watt University, Edinburgh, United Kingdom

Langford, W.F., Department of Mathematics & Statistics, University of Guelph, Guelph, Ontario Canada NIG2WI

Lauterbach, R., Institut fuer Mathematik, Universitaet Augsburg, Memminger Str. 6, D-8900 Augsburg, West Germany

Lunel, S.M.V., Centre for Mathematics and Computer Science, P.O. Box 4079, 1009 AB Amsterdam, The Netherlands

Magalhães, L., Departamento de Matemática, Inst. Superior Técnico, Universidade Técnica de Lisboa, 1096 Lisboa Codex, Portugal

Mallet-Paret, J., Division of Applied Mathematics, Brown University, Providence, RI 02912, U.S.A.

Mascarenhas, M.L., Departmento de Matematica, Faculdade de Ciencias de Lisboa, Lisboa, Portugal

Marcati, P., Dept. of Pure and Appl. Mathematics, University of L'Aquila, 67100 L'Aquila, Italy

Mawhin, J., Université de Louvain, Institut Mathématique, B-1348 Louvain-La-Neuve, Belgium

Mischaikow, K., Lefschetz Center for Dynamical Systems, Division of Applied Mathematics, Brown University, Providence, RI 02912, U.S.A.

Mielke, A., Math. Institut A, Pfaffenwaldring 57, D-7000 Stuttgart 80, West Germany

Mora, X., Departament de Matemátiques, Universitat Autonoma de Barcelona, Bellaterra, Barcelona, Spain

Nachman, A., AFOSR/NM, Boling Air Force Base, Washington, D.C. 20332, U.S.A.

Nishiura, Y., Institute of Computer Science, Kyoto Sangyo University, Kyoto 603, Japan

Norbury. J., Mathematics Institute, University of Oxford, 24/9 St. Giles, Oxford, United Kingdom

Nussbaum, R.D., Mathematics Department, Rutgers University, New Brunswick, NJ 08903, U.S.A.

Oliva, W.M., Departamento de Matemática Aplicada, Universidade de Sao Paulo, Caixa Postal 20570, Sao Paulo, Brazil

Oliveira, J.T., Faculdade de Ciencias Sociais e Humanas, University Nova de Lisboa, Portugal

Papanicolaou, G., Courant Institute, 251 Mercer Street, N.Y, N.Y., U.S.A.

Peitgen, H., Department of Mathematics, University California, Santa Cruz, CA 95060, U.S.A.

Perello, C., Departament Matematiques, Universitat Autonoma de Barcelona, Bellaterra, Barcelona, Spain
Ribeiro, J.A., Departamento de Matematica, Instituto Superior Técnico, 1096 Lisboa Codex, Portugal
Ricou, M., Departamento de Matematica, Instituto Superior Técnico, Universidade Técnica de Lisboa, Av. Rovisco Pais, 1096 Lisboa Codex, Portugal
Rocha, C., Departamento de Matematica, Instituto Superior Tecnico,Universidade Tecnica de Lisboa, Av. Rovisco Pais, 1096 Lisboa Codex, Portugal
Rodrigues, J.F., Departmento de Matematica, Faculdade de Ciencias, University de Lisboa, Portugal
Sanchez, L., C.M.A.F., Av. Gama Pinto 2, 1699 Lisboa Codex, Portugal
Sanders, J.A., Department of Mathematics and Computer Science, Free University, PO Box 7161, 1007 Mc Amsterdam, The Netherlands
Sattinger, D.H., School of Mathematics, University of Minnesota, Minneapolis, MN 55455, U.S.A.
Sell, G.R., Institute for Mathematics, University of Minnesota, Minneapolis, MN 55455, U.S.A.
Smoller, J.A., Mathematics Department, University of Michigan, Ann Arbor, MI, U.S.A.
Sola-Morales, J., Department de Matematiques, Universitat Aut. de Barcelona, Bellaterra, Barcelona, Spain
Sougandinis, P.E., Lefschetz Center for Dynamical Systems, Division of Applied Mathematics, Brown University, Providence, RI 02912, U.S.A.
Staffans, O.J., Institute of Mathematics, Helsinky University Technology, SF-02150, Espoo 15, Finland
Stavrakakis, N., Division of Applied Mathematics, Brown University, Providence, RI 02912, U.S.A.
Stech, H., Department of Mathematics, Virginia Polytechnic Institute, Blacksburg, VA 24060, U.S.A.
Tartar, L., Centre d'Etudes de Limeil-Valenton, B.P. 27, 94190 Villeneuve Saint Georges, France
Terman, D., Department of Mathematics, Michigan State University, East Lansing, MI 48824, U.S.A.
Thiel, U., Division de Matematicas, Universidad Aut. de Madrid, 28049 Madrid, Spain
Ulrich, K., Institut fuer Angewandte Mathematik, Universitaet Hannover, Welfengarten 1, D-3000 Hannover 1, West Germany
Ushiki, S., Institute of Mathematics, Yoshida College, Kyoto University, Kyoto 606, Japan
Vanderbauwhede, A., Institute for Theoretical Mechanics, State University of Gent, Krijgslaan 281, D-9000 Gent, Belgium
Van Moerbeke, P.J., Department of Mathematics, University of Louvain, 1348 Louvain-la-Neuve, Belgium
Vazquez, J., Division Matematicas, University Aut. Madrid, 28049 Madrid, Spain
Vegas, J.M., Departamento de Ecuaciones Funcionales, Facultad de Ciencias Matemáticas, Universidad Complutense de Madrid, 28040 Madrid, Spain
Viegas, V., Department Matematica, Instituto Superior Tecnico, 1096 Lisboa Codex, Portugal
Walther, H.O., Mathematisches Institut, Universität Muenchen,

Theresienstr. 39, D-8000 Muenchen 2, West Germany
Williamson, F., 16 Av. de La Commune de Paris, 94400 Vitry sur Seine, France
Wisniewski, H.S., Dir. Appl. and Comp. Mathematics, DARPA, 1400 Wilson Blvd., Arlington, VA 22209, U.S.A.

NATO ASI Series F

Vol. 1: Issues in Acoustic Signal – Image Processing and Recognition. Edited by C. H. Chen. VIII, 333 pages. 1983.

Vol. 2: Image Sequence Processing and Dynamic Scene Analysis. Edited by T. S. Huang. IX, 749 pages. 1983.

Vol. 3: Electronic Systems Effectiveness and Life Cycle Costing. Edited by J. K. Skwirzynski. XVII, 732 pages. 1983.

Vol. 4: Pictorial Data Analysis. Edited by R. M. Haralick. VIII, 468 pages. 1983.

Vol. 5: International Calibration Study of Traffic Conflict Techniques. Edited by E. Asmussen. VII, 229 pages. 1984.

Vol. 6: Information Technology and the Computer Network. Edited by K. G. Beauchamp. VIII, 271 pages. 1984.

Vol. 7: High-Speed Computation. Edited by J. S. Kowalik. IX, 441 pages. 1984.

Vol. 8: Program Transformation and Programming Environments. Report on an Workshop directed by F. L. Bauer and H. Remus. Edited by P. Pepper. XIV, 378 pages. 1984.

Vol. 9: Computer Aided Analysis and Optimization of Mechanical System Dynamics. Edited by E. J. Haug. XXII, 700 pages. 1984.

Vol. 10: Simulation and Model-Based Methodologies: An Integrative View. Edited by T. I. Ören, B. P. Zeigler, M. S. Elzas. XIII, 651 pages. 1984.

Vol. 11: Robotics and Artificial Intelligence. Edited by M. Brady, L. A. Gerhardt, H. F. Davidson. XVII, 693 pages. 1984.

Vol. 12: Combinatorial Algorithms on Words. Edited by A. Apostolico, Z. Galil. VIII, 361 pages. 1985.

Vol. 13: Logics and Models of Concurrent Systems. Edited by K. R. Apt. VIII, 498 pages. 1985.

Vol. 14: Control Flow and Data Flow: Concepts of Distributed Programming. Edited by M. Broy. VIII, 525 pages. 1985.

Vol. 15: Computational Mathematical Programming. Edited by K. Schittkowski. VIII, 451 pages. 1985.

Vol. 16: New Systems and Architectures for Automatic Speech Recognition and Synthesis. Edited by R. De Mori, C.Y. Suen. XIII, 630 pages. 1985.

Vol. 17: Fundamental Algorithms for Computer Graphics. Edited by R.A. Earnshaw. XVI, 1042 pages. 1985.

Vol. 18: Computer Architectures for Spatially Distributed Data. Edited by H. Freeman and G.G. Pieroni. VIII, 391 pages. 1985.

Vol. 19: Pictorial Information Systems in Medicine. Edited by K. H. Höhne. XII, 525 pages. 1986.

Vol. 20: Disordered Systems and Biological Organization. Edited by E. Bienenstock, F. Fogelman Soulié, G. Weisbuch. XXI, 405 pages.1986.

Vol. 21: Intelligent Decision Support in Process Environments. Edited by E. Hollnagel, G. Mancini, D.D. Woods. XV, 524 pages. 1986.

Vol. 22: Software System Design Methods. The Challenge of Advanced Computing Technology. Edited by J.K. Skwirzynski. XIII, 747 pages. 1986.

NATO ASI Series F

Vol. 23: Designing Computer-Based Learning Materials. Edited by H. Weinstock and A. Bork. IX, 285 pages. 1986.

Vol. 24: Database Machines. Modern Trends and Applications. Edited by A.K. Sood and A.H. Qureshi. VIII, 570 pages. 1986.

Vol. 25: Pyramidal Systems for Computer Vision. Edited by V. Cantoni and S. Levialdi. VIII, 392 pages. 1986.

Vol. 26: Modelling and Analysis in Arms Control. Edited by R. Avenhaus, R.K. Huber and J.D. Kettelle. VIII, 488 pages. 1986.

Vol. 27: Computer Aided Optimal Design: Structural and Mechanical Systems. Edited by C.A. Mota Soares. XIII, 1029 pages. 1987.

Vol. 28: Distributed Operating Systems. Theory und Practice. Edited by Y. Paker, J.-P. Banatre and M. Bozyiğit. X, 379 pages. 1987.

Vol. 29: Languages for Sensor-Based Control in Robotics. Edited by U. Rembold and K. Hörmann. IX, 625 pages. 1987.

Vol. 30: Pattern Recognition Theory and Applications. Edited by P.A. Devijver and J. Kittler. XI, 543 pages. 1987.

Vol. 31: Decision Support Systems: Theory and Application. Edited by C.W. Holsapple and A.B. Whinston. X, 500 pages. 1987.

Vol. 32: Information Systems: Failure Analysis. Edited by J.A. Wise and A. Debons. XV, 338 pages. 1987.

Vol. 33: Machine Intelligence and Knowledge Engineering for Robotic Applications. Edited by A.K.C. Wong and A. Pugh. XIV, 486 pages. 1987.

Vol. 34: Modelling, Robustness and Sensitivity Reduction in Control Systems. Edited by R.F. Curtain. IX, 492 pages. 1987.

Vol. 35: Expert Judgment and Expert Systems. Edited by J.L. Mumpower, L.D. Phillips, O. Renn and V.R.R. Uppuluri. VIII, 361 pages. 1987.

Vol. 36: Logic of Programming and Calculi of Discrete Design. Edited by M. Broy. VII, 415 pages. 1987.

Vol. 37: Dynamics of Infinite Dimensional Systems. Edited by S.-N. Chow and J.K. Hale. IX, 514 pages. 1987.